Intelligent Control of Robotic Systems

Intelligent Control of Robotic Systems

Laxmidhar Behera
Swagat Kumar
Prem Kumar Patchaikani
Ranjith Ravindranathan Nair
Samrat Dutta

CRC Press
Taylor & Francis Group
Boca Raton London New York

CRC Press is an imprint of the
Taylor & Francis Group, an **informa** business

MATLAB® is a trademark of The MathWorks, Inc. and is used with permission. The MathWorks does not warrant the accuracy of the text or exercises in this book. This book's use or discussion of MATLAB® software or related products does not constitute endorsement or sponsorship by The MathWorks of a particular pedagogical approach or particular use of the MATLAB® software.

CRC Press
Taylor & Francis Group
6000 Broken Sound Parkway NW, Suite 300
Boca Raton, FL 33487-2742

© 2020 by Taylor & Francis Group, LLC
CRC Press is an imprint of Taylor & Francis Group, an Informa business

No claim to original U.S. Government works

Printed on acid-free paper

International Standard Book Number-13: 978-1-138-59771-6 (Hardback)

This book contains information obtained from authentic and highly regarded sources. Reasonable efforts have been made to publish reliable data and information, but the author and publisher cannot assume responsibility for the validity of all materials or the consequences of their use. The authors and publishers have attempted to trace the copyright holders of all material reproduced in this publication and apologize to copyright holders if permission to publish in this form has not been obtained. If any copyright material has not been acknowledged please write and let us know so we may rectify in any future reprint.

Except as permitted under U.S. Copyright Law, no part of this book may be reprinted, reproduced, transmitted, or utilized in any form by any electronic, mechanical, or other means, now known or hereafter invented, including photocopying, microfilming, and recording, or in any information storage or retrieval system, without written permission from the publishers.

For permission to photocopy or use material electronically from this work, please access www.copyright.com (http://www.copyright.com/) or contact the Copyright Clearance Center, Inc. (CCC), 222 Rosewood Drive, Danvers, MA 01923, 978-750-8400. CCC is a not-for-profit organization that provides licenses and registration for a variety of users. For organizations that have been granted a photocopy license by the CCC, a separate system of payment has been arranged.

Trademark Notice: Product or corporate names may be trademarks or registered trademarks, and are used only for identification and explanation without intent to infringe.

Library of Congress Cataloging-in-Publication Data

Names: Behera, Laxmidhar, author.
Title: Intelligent control of robotic systems / Laxmidhar Behera, Swagat
Kumar, Prem Kumar Patchaikani, Ranjith Ravindranathan Nair and
Samrat Dutta.
Description: First edition. | Boca Raton, FL : CRC Press/Taylor & Francis
Group, 2019. | Includes bibliographical references.
Identifiers: LCCN 2019015139| ISBN 9781138597716 (hardback : acid-free
paper) | ISBN 9780429486784 (ebook)
Subjects: LCSH: Robots--Control systems. | Intelligent control systems.
Classification: LCC TJ211.35 .B46 2019 | DDC 629.8/92--dc23
LC record available at https://lccn.loc.gov/2019015139

Visit the Taylor & Francis Web site at
http://www.taylorandfrancis.com

and the CRC Press Web site at
http://www.crcpress.com

Contents

Preface		xvii
Acknowledgment		xxi
Authors		xxiii
1	**Introduction**	**1**
1.1	Vision-Based Control	3
1.2	Kinematic Control of a Redundant Manipulator	6
	1.2.1 Redundancy Resolution using Null Space of the Pseudo-inverse	8
	1.2.2 Extended Jacobian Method	8
	1.2.3 Optimization Based Redundancy Resolution	9
	1.2.4 Redundancy Resolution with Global Optimization	9
	1.2.5 Neural Network Based Methods	10
1.3	Visual Servoing	11
	1.3.1 Image Based Visual Servoing (IBVS)	12
	1.3.2 Position Based Visual Servoing (PBVS)	12
	1.3.3 2-1/2-D Visual Servoing	13
1.4	Visual Control of a Redundant Manipulator: Research Issues	13
1.5	Learning by Demonstration	16
	1.5.1 DS-Based Motion Learning	19
1.6	Stability of Nonlinear Systems	21
1.7	Optimization Techniques	22
	1.7.1 Genetic Algorithm	24
	1.7.2 Expectation Maximization for Gaussian Mixture Model	25
1.8	Composition of the Book	27
I	**Manipulators**	**29**
2	**Kinematic and Dynamic Models of Robot Manipulators**	**31**
2.1	PowerCube Manipulator	31
2.2	Kinematic Configuration of the Manipulator	32
2.3	Estimating the Vision Space Motion with Camera Model	35
	2.3.1 Transformation from Cartesian Space to Vision Space	36

		2.3.2	The Camera Model	38
		2.3.3	Computation of Image Feature Velocity in the Vision Space	39
	2.4	Learning-Based Controller Architecture		40
	2.5	Universal Robot (UR 10)		41
		2.5.1	Mechatronic Design	41
			2.5.1.1 Platform	41
			2.5.1.2 End-Effector	43
			2.5.1.3 Perception Apparatus	43
		2.5.2	Kinematic Model	43
	2.6	Barrett Wam Manipulator		45
		2.6.1	Overview of the System	45
		2.6.2	Experimental Setup	46
		2.6.3	Dynamic Modeling	47
		2.6.4	System Description and Modeling	49
		2.6.5	State Space Representation	53
	2.7	Summary		54

3 Hand-eye Coordination of a Robotic Arm using KSOM Network   55

	3.1	Kohonen Self Organizing Map		56
		3.1.1	Competitive Process	57
		3.1.2	Cooperative Process	57
		3.1.3	Adaptive Process	58
	3.2	System Identification using KSOM		60
	3.3	Introduction to Learning-Based Inverse Kinematic Control		66
		3.3.1	The Network	68
		3.3.2	The Learning Problem	69
		3.3.3	The Approach	69
		3.3.4	The Formulation of Cost Function	69
		3.3.5	Weight Update Laws	70
	3.4	Visual Motor Control of a Redundant Manipulator using KSOM Network		89
		3.4.1	The Problem	92
	3.5	KSOM with Sub-Clustering in Joint Angle Space		94
		3.5.1	Network Architecture	95
		3.5.2	Training Algorithm	96
		3.5.3	Testing Phase	97
		3.5.4	Redundancy Resolution	98
		3.5.5	Tracking a Continuous Trajectory	99
	3.6	Simulation and Results		100
		3.6.1	Network Architecture and Workspace Dimensions	100
		3.6.2	Training	101
		3.6.3	Testing	101

		3.6.3.1	Reaching Isolated Target Positions in the Workspace	103

		3.6.3.2	Tracking a Straight Line Trajectory	105
		3.6.3.3	Tracking an Elliptical Trajectory	107
	3.6.4	Real-Time Experiment		108
		3.6.4.1	Redundant Solutions	109
		3.6.4.2	Tracking a Circular and a Straight Line Trajectory	110
		3.6.4.3	Multi-Step Movement	111
3.7	Summary			111

4 Model-based Visual Servoing of a 7 DOF Manipulator 113
4.1 Introduction 113
4.2 Kinematic Control of a Manipulator 113
    4.2.1 Kinematic Control of Redundant Manipulator 114
4.3 Visual Servoing 115
    4.3.1 Estimating the Vision Space Motion with Camera Model 116
    4.3.2 Transformation from Cartesian Space to Vision Space 117
    4.3.3 The Camera Model 119
    4.3.4 Computation of Image Feature Velocity in the Vision Space 120
4.4 Kinematic Control of a Manipulator Directly from Vision Space 121
4.5 Image Moments 122
4.6 Image Moment Velocity 126
4.7 A Pinhole Camera Projection 128
4.8 Image Moment Interaction Matrix 132
4.9 Experimental Results using a 7 DOF Manipulator 139
4.10 Summary 141

5 Learning-Based Visual Servoing 145
5.1 Introduction 145
5.2 Kinematic Control using KSOM 148
    5.2.1 KSOM Architecture 149
    5.2.2 KSOM: Weight Update 149
    5.2.3 Comments on Existing KSOM Based Kinematic Control Schemes 150
5.3 Problem Definition 151
5.4 Analysis of Solution Learned Using KSOM 151
    5.4.1 KSOM: An Estimate of Inverse Jacobian 152
    5.4.2 Empirical Verification 152
        5.4.2.1 Inverse Jacobian Evolution in Learning Phase 153

		5.4.2.2	Testing Phase: Inverse Jacobian Estimation at each Operating Zone	153
		5.4.2.3	Inference	154
5.5	KSOM in Closed Loop Visual Servoing			156
	5.5.1	Stability Analysis		157
5.6	Redundancy Resolution			159
5.7	Results			160
	5.7.1	Learning Inverse Kinematic Relationship using KSOM		160
	5.7.2	Visual Servoing		161
	5.7.3	Redundancy Resolution		164
		5.7.3.1	Tracking a Straight Line	165
		5.7.3.2	Tracking an Elliptical Trajectory	168
5.8	Summary			172
5.9	Reinforcement Learning-Based Optimal Redundancy Resolution Directly from the Vision Space			172
5.10	Introduction			172
5.11	Redundancy Resolution Problem from the Vision Space			174
5.12	SNAC Based Optimal Redundancy Resolution from Vision Space			175
	5.12.1	Selection of Cost Function		176
	5.12.2	Control Challenges		177
5.13	T-S Fuzzy Model-Based Critic Neural Network for Redundancy Resolution from Vision Space			179
	5.13.1	Fuzzy Critic Model		179
	5.13.2	Weight Update Law		181
	5.13.3	Selection of Fuzzy Zones		182
	5.13.4	Initialization of the Fuzzy Network Control		183
		5.13.4.1	Remark	184
5.14	KSOM Based Critic Network for Redundancy Resolution from Vision Space			185
	5.14.1	KSOM Critic Model		185
	5.14.2	KSOM: Weight Update		188
	5.14.3	Initialization of KSOM Network Control		188
5.15	Simulation Results			190
	5.15.1	T-S Fuzzy Model		190
	5.15.2	Kohonen's Self-organizing Map		191
5.16	Real-Time Experiment			195
	5.16.1	Tracking Elliptical Trajectory		196
		5.16.1.1	T-S Fuzzy Model	196
		5.16.1.2	KSOM	199
	5.16.2	Grasping a Ball with Hand-manipulator Setup		201
5.17	Summary			202

6 Visual Servoing using an Adaptive Distributed Takagi-Sugeno (T-S) Fuzzy Model — 205
- 6.1 T-S Fuzzy Model — 206
- 6.2 Adaptive Distributed T-S Fuzzy PD Controller — 208
  - 6.2.1 Offline Learning Algorithm — 209
  - 6.2.2 Online Adaptation Algorithm — 212
  - 6.2.3 Stability Analysis — 214
- 6.3 Experimental Results — 216
  - 6.3.1 Visual Servoing for a Static Target — 220
  - 6.3.2 Compensation of Model Uncertainties — 222
  - 6.3.3 Visual Servoing for a Moving Target — 223
- 6.4 Computational Complexity — 225
- 6.5 Summary — 225

7 Kinematic Control using Single Network Adaptive Critic — 229
- 7.1 Introduction — 229
  - 7.1.1 Discrete-Time Optimal Control Problem — 230
  - 7.1.2 Adaptive Critic Based Control — 231
    - 7.1.2.1 Training of Action and Critic Network — 232
  - 7.1.3 Single Network Adaptive Critic (DT-SNAC) — 234
  - 7.1.4 Choice of Critic Network Model — 235
    - 7.1.4.1 Costate Vector Modeling with MLN Critic Network — 235
    - 7.1.4.2 Costate Vector Modeling with T-S Fuzzy Model-Based Critic Network — 236
- 7.2 Adaptive Critic Based Optimal Controller Design for Continuous-time Systems — 241
  - 7.2.1 Continuous-time Single Network Adaptive Critic (CT-SNAC) — 242
  - 7.2.2 Critic Network: Weight Update Law — 243
  - 7.2.3 Choice of Critic Network — 245
    - 7.2.3.1 Critic Network using MLN — 245
    - 7.2.3.2 T-S Fuzzy Model-Based Critic Network with Cluster of Local Quadratic Cost Functions — 246
  - 7.2.4 CT-SNAC — 248
- 7.3 Discrete-Time Input Affine System Representation of Forward Kinematics — 257
- 7.4 Modeling the Primary and Additional Tasks as an Integral Cost Function — 259
  - 7.4.1 Quadratic Cost Minimization (Global Minimum Norm Motion) — 260
  - 7.4.2 Joint Limit Avoidance — 260
- 7.5 Single Network Adaptive Critic Based Optimal Redundancy Resolution — 261

		7.5.1	T-S Fuzzy Model-Based Critic Network for Closed Loop Positioning Task	262
		7.5.2	Training Algorithm	263
	7.6	Computational Complexity		264
	7.7	Simulation Results		265
		7.7.1	Global Minimum Norm Motion	266
		7.7.2	Joint Limit Avoidance	272
	7.8	Experimental Results		276
		7.8.1	Global Minimum Norm Motion	276
		7.8.2	Joint Limit Avoidance	278
	7.9	Conclusion		280
8	**Dynamic Control using Single Network Adaptive Critic**			**283**
	8.1	Introduction		283
	8.2	Optimal Control Problem of Continuous Time Nonlinear System		284
		8.2.1	Linear Quadratic Regulator	285
		8.2.2	Hamilton-Jacobi-Bellman Equation	287
		8.2.3	Optimal Control Law for Input Affine System	288
		8.2.4	Adaptive Critic Concept	289
	8.3	Policy Iteration and SNAC for Unknown Continuous Time Nonlinear Systems		291
		8.3.1	Policy Iteration Scheme	291
		8.3.2	Optimal Control Problem of an Unknown Dynamic . .	292
		8.3.3	Model Representation and Learning Scheme	295
			8.3.3.1 TSK Fuzzy Representation of Nonlinear Dynamics	295
			8.3.3.2 Learning Scheme for the TSK Fuzzy Model .	295
		8.3.4	Critic Design and Policy Update	296
			8.3.4.1 Construction of Initial Critic Network using Lyapunov Based LMI	296
			8.3.4.2 Lyapunov Function	297
			8.3.4.3 Conditions for Stabilization	298
			8.3.4.4 Design of Fitness Function	301
		8.3.5	Learning Near-Optimal Controller	301
			8.3.5.1 Update of Critic Network	304
			8.3.5.2 Fitness Function for PI Based Training . . .	305
		8.3.6	Examples	307
			8.3.6.1 Simulated Model	307
			8.3.6.2 Example using Real Robot	310
	8.4	Summary		317

| Contents | xi |

9 Imitation Learning — 319
- 9.1 Introduction — 319
- 9.2 Dynamic Movement Primitives — 320
  - 9.2.1 Mathematical Formulations — 321
    - 9.2.1.1 Choice of Mean and Variance — 322
    - 9.2.1.2 Spatial and Temporal Scaling — 322
  - 9.2.2 Example — 323
- 9.3 Motion Encoding using Gaussian Mixture Regression — 324
  - 9.3.1 SED: Stable Estimator of Dynamical Systems — 326
    - 9.3.1.1 Learning Model Parameters — 326
    - 9.3.1.2 Log-likelihood Cost — 327
- 9.4 FuzzStaMP: Fuzzy Controller Regulated Stable Movement Primitives — 327
  - 9.4.1 Motion Modeling with C-FuzzStaMP — 328
    - 9.4.1.1 Fuzzy Lyapunov Function — 329
    - 9.4.1.2 Learning Fuzzy Controller Gains — 331
    - 9.4.1.3 Design of Fitness Function — 333
    - 9.4.1.4 Example — 333
  - 9.4.2 Motion Modeling with R-FuzzStaMP — 335
    - 9.4.2.1 Stability Analysis of the Motion System — 339
    - 9.4.2.2 Design of the Fuzzy Controller — 342
  - 9.4.3 Global Validity and Spatial Scaling — 346
    - 9.4.3.1 Examples — 348
- 9.5 Learning Skills from Heterogeneous Demonstrations — 354
  - 9.5.1 Stability Analysis — 357
    - 9.5.1.1 Asymptotic Stability in the Demonstrated Region — 361
    - 9.5.1.2 Ensuring Asymptotic Stability outside Demonstrated Region — 363
  - 9.5.2 Learning Model Parameters from Demonstrations — 364
    - 9.5.2.1 Motion Modeling using GMR — 364
    - 9.5.2.2 Motion Modeling using LWPR — 367
    - 9.5.2.3 Motion Modeling using ϵ-SVR — 368
    - 9.5.2.4 Complete Pipeline — 370
  - 9.5.3 Spatial Error Calculation — 371
  - 9.5.4 Examples — 371
    - 9.5.4.1 Example of Monotonic and Non-monotonic State Energy — 372
    - 9.5.4.2 Example of Multitasking with Single and Multiple Task-equilibrium — 375
  - 9.5.5 Summary — 382

10 Visual Perception — 385

- 10.1 Introduction — 385
- 10.2 Deep Neural Networks and Artificial Neural Networks — 386
  - 10.2.1 Neural Networks — 387
    - 10.2.1.1 Multi-layer Perceptron — 389
    - 10.2.1.2 MLP Implementation using Tensorflow — 392
  - 10.2.2 Deep Learning Techniques: An Overview — 395
    - 10.2.2.1 Convolutional Neural Network (Flow and Training with Back-propagation) — 395
  - 10.2.3 Different Architectures of Convolutional Neural Networks (CNNs) — 399
- 10.3 Examples of Vision-Based Object Detection Techniques — 404
  - 10.3.1 Automatic Annotation of Object ROI — 405
    - 10.3.1.1 Image Acquisition — 407
    - 10.3.1.2 Manual Annotation — 407
    - 10.3.1.3 Augmentation and Clutter Generation — 407
    - 10.3.1.4 Two-class Classification Model using Deep Networks — 409
    - 10.3.1.5 Experimental Results and Discussions — 411
  - 10.3.2 Automatic Segmentation of Objects for Warehouse Automation — 412
    - 10.3.2.1 Network Architecture — 413
    - 10.3.2.2 Base Network — 416
    - 10.3.2.3 Single Shot Detection — 416
  - 10.3.3 Automatic Generation of Artificial Clutter — 417
  - 10.3.4 Multi-Class Segmentation using Proposed Network — 417
- 10.4 Experimental Results — 417
  - 10.4.1 System Description — 417
    - 10.4.1.1 Server — 418
  - 10.4.2 Ground Truth Generation — 418
  - 10.4.3 Image Segmentation — 419
- 10.5 Summary — 421

11 Vision-Based Grasping — 423

- 11.1 Introduction — 423
- 11.2 Model-Based Grasping — 425
  - 11.2.1 Problem Statement — 425
  - 11.2.2 Hardware Setup — 426
  - 11.2.3 Dataset — 427
  - 11.2.4 Data Augmentation — 427
  - 11.2.5 Network Architecture and Training — 428
  - 11.2.6 Axis Assignment — 428
  - 11.2.7 Grasp Decide Index (GDI) — 428
  - 11.2.8 Final Pose Selection — 431
  - 11.2.9 Overall Pipeline and Result — 431

Contents   xiii

    11.3 Grasping without Object Models 433
        11.3.1 Problem Definition 433
        11.3.2 Proposed Method. 434
              11.3.2.1 Creating Continuous Surfaces in 3D Point
                        Cloud 434
        11.3.3 Finding Graspable Affordances 438
        11.3.4 Experimental Results 443
              11.3.4.1 Performance Measure 443
        11.3.5 Grasping of Individual Objects 445
        11.3.6 Grasping Objects in a Clutter 446
        11.3.7 Computation Time 451
    11.4 Summary 452

12 Warehouse Automation: An Example   453
    12.1 Introduction 453
    12.2 Problem Definition 456
    12.3 System Architecture 457
    12.4 The Methods 459
        12.4.1 System Calibration 459
        12.4.2 Rack Detection 460
        12.4.3 Object Recognition 462
        12.4.4 Grasping 465
        12.4.5 Motion Planning 466
        12.4.6 End-Effector Design 469
              12.4.6.1 Suction-based End-effector 469
              12.4.6.2 Combining Gripping with Suction 470
        12.4.7 Robot Manipulator Model 471
              12.4.7.1 Null Space Optimization 473
              12.4.7.2 Inverse Kinematics as a Control Problem . . 474
              12.4.7.3 Damped Least Square Method 475
    12.5 Experimental Results 476
        12.5.1 Response Time 477
        12.5.2 Grasping and Suction 478
        12.5.3 Object Recognition 478
        12.5.4 Direction for Future Research 480
    12.6 Summary 482

II   Mobile Robotics   483

13 Introduction to Mobile Robotics and Control   485
    13.1 Introduction 485
    13.2 System Model: Nonholonomic Mobile Robots 486
    13.3 Robot Attitude 487
        13.3.1 Rotation about Roll Axis 487
        13.3.2 Rotation about Pitch Axis 488
        13.3.3 Rotation About Yaw Axis 489

13.4 Composite Rotation	490
13.5 Coordinate System	491
13.5.1 Earth-Centered Earth-Fixed (ECEF) Co-ordinate System	491
13.6 Control Approaches	492
13.6.1 Feedback Linearization	493
13.6.2 Backstepping	495
13.6.3 Sliding Mode Control	496
13.6.4 Conventional SMC	498
13.6.5 Terminal SMC	499
13.6.6 Nonsingular TSMC (NTSMC)	500
13.6.7 Fast Nonsingular TSMC (FNTSMC)	501
13.6.8 Fractional Order SMC (FOSMC)	502
13.6.9 Higher Order SMC (HOSMC)	503
13.7 Summary	505

14 Multi-robot Formation   507

14.1 Introduction	507
14.2 Path Planning Schemes	509
14.3 Multi-Agent Formation Control	518
14.3.1 Fast Adaptive Gain NTSMC	519
14.3.2 Fast Adaptive Fuzzy NTSMC (FAFNTSMC)	524
14.3.3 Fault Detection, Isolation and Collision Avoidance Scheme	527
14.4 Experiments	530
14.5 Summary	535

15 Event Triggered Multi-Robot Consensus   537

15.1 Introduction to Event Triggered Control	537
15.2 Event Triggered Consensus	539
15.2.1 Preliminaries	541
15.2.2 Sliding Mode-Based Finite Time Consensus	544
15.3 Event Triggered Sliding Mode-based Consensus Algorithm	544
15.3.1 Consensus-based Tracking Control of Nonholonomic Multi-robot Systems	549
15.4 Experiments	552
15.5 Summary	554

16 Vision-Based Tracking for a Human Following Mobile Robot   555

16.1 Visual Tracking: Introduction	555
16.1.1 Difficulties in Visual Tracking	555
16.1.2 Required Features of Visual Tracking	555
16.1.3 Feature Descriptors for Visual Tracking	556

Contents   xv

16.2 Human Tracking Algorithm using SURF Based Dynamic Object Model 558
   16.2.1 Problem Definition 559
   16.2.2 Object Model Description 560
      16.2.2.1 Maintaining a Template Pool of Descriptors 561
   16.2.3 The Tracking Algorithm 562
      16.2.3.1 Step 1: Target Initialization 563
      16.2.3.2 Step 2: Object Recognition and Template Pool Update 563
      16.2.3.3 Step 3: Occlusion Detection, Target Window Prediction 564
   16.2.4 SURF-Based Mean-Shift Algorithm 564
   16.2.5 Modified Object Model Description 565
   16.2.6 Modified Tracking Algorithm 566
16.3 Human Tracking Algorithm with the Detection of Pose Change due to Out-of-plane Rotations 567
   16.3.1 Problem Definition 567
   16.3.2 Tracking Algorithm 568
   16.3.3 Template Initialization 569
   16.3.4 Tracking 570
      16.3.4.1 Scaling and Re-positioning the Tracking Window 571
   16.3.5 Template Update Module 571
   16.3.6 Error Recovery Module 572
      16.3.6.1 KD-tree Classifier 572
      16.3.6.2 Construction of KD-Tree 573
      16.3.6.3 Dealing with Pose Change 573
      16.3.6.4 Tracker Recovery from Full Occlusions 574
16.4 Human Tracking Algorithm Based on Optical Flow 576
   16.4.1 The Template Pool and its Online Update 577
      16.4.1.1 Selection of New Templates 578
   16.4.2 Re-Initialization of Optical Flow Tracker 580
   16.4.3 Detection of Partial and Full Occlusion 580
16.5 Visual Servo Controller 581
   16.5.1 Kinematic Model of the Mobile Robot 582
   16.5.2 Pinhole Camera Model 582
   16.5.3 Problem Formulation 582
   16.5.4 Visual Servo Control Design 583
   16.5.5 Simulation Results 584
      16.5.5.1 Example: Tracking an Object which Moves in a Circular Trajectory 584
16.6 Experimental Results 585
   16.6.1 Experimental Results for the Human Tracking Algorithm Based on SURF-based Dynamic Object Model 585

	16.6.2	Tracking Results	586
	16.6.3	Human Following Robot	589
	16.6.4	Discussion on Performance Comparison	590
	16.6.5	Experimental Evaluation of Human Tracking Algorithm Based on Optical Flow	591
16.7	Summary		593

Exercises   **595**

Bibliography   **603**

Index   **645**

Preface

Robots are artificial agents that exhibit some aspect of sentient behavior. They come in all forms, shapes and sizes. Some may even be form-less, e.g., software-bots that filter out spam in your email boxes or a chat-bot that answers to queries on websites. In this book, we will primarily look into physical robots that cannot only perceive their environment but also alter it by manipulating objects around it. These robots are no more confined to the cages on a factory shop floors and are moving to other places of human habitation such as home, offices and hospitals where they work alongside humans, sharing each other's workspace. These robots are becoming smarter with each passing day and someday, they will replace humans in all kinds of dull, dirty and dangerous jobs, relieving them for more creative pursuits. Industry 4.0 paints an optimistic future of smart and flexible factories where the production pipeline can change in real-time in response to variations arising from factors like weather, socio-economic and political changes. This would be made possible by having robots that can independently take local decisions based on global cues provided by a centralized ERP[1]/ WMS[2] decision maker. Such robots can work in unstructured and dynamic environments and can learn to cooperate and collaborate with humans and other robots, while learning new skills from and sharing knowledge with them. Such intelligent robots could then be rented cheaply by small and medium enterprises (SMEs) who cannot afford to buy these robots, thereby allowing them to reap the benefits of robot-based automation at an affordable cost. Such changes will give rise to new business opportunities in the form of "Robotics-as-a-Service" where the end-users will pay for the services offered and not for robots which will be managed and maintained by the service provider. It is also envisaged that such changes will make it possible to achieve mass personalization in contrast to mass standardization that is being offered by today's industries. These aspirations of Industry 4.0 can be realized through advancement in multiple fields such as additive manufacturing, machine learning, artificial intelligence, signal processing, computer vision, cloud computing, embedded systems, and mechatronics etc.

Historically, Robotics and AI have grown in parallel ways complementing and enriching each other. This book is the outcome of our sincere endeavor to show the synergy between the two disciplines while asking the hard questions on naturalization of human intelligence that make robots mimic humans in

---

[1] Enterprise Resource Planner
[2] Warehouse Management System

complex tasks such as warehouse automation, surveillance, imitation learning, and multi-robot systems.

Traditionally, methods for robot control relied heavily on physics-based models which benefited from strong mathematical foundations available in the control literature. However, these methods had limited ability to deal with uncertainty associated with non-deterministic factors like parameter variation, sensor noise, extraneous disturbances, model nonlinearity, and model approximation. In contrast, machine learning-based methods relied on data generated by the system to understand the underlying model and then use it to develop necessary control strategies for systems. The latter approach is more commonly known as a "data-driven" approache which was pioneered by computer science researchers who had neither any background nor any interest in physics-based models. On the other hand, researchers and engineers from Electrical and Mechanical Engineering were more comfortable with traditional methods based on physics-based models and took casual interest in learning-based approaches. Over the years, the authors of this book have tried to marry these two schools by augmenting physics-based models with learning-based control approaches. This approach has the following benefits:

- Imprecise machines guided by learning-based algorithms can achieve high level precision which has the advantage of reducing the cost of robotic systems.

- Elimination of manual programming of the robots: In the proposed scheme, robots autonomously localize themselves, they learn to adaptively interact with the environment, and these programs are mostly independent of robotic platforms. Hence there is significant reduction of customized programming of each robot in different contexts.

- In transfer learning, most of the learning models are developed in simulation environments. The fine-tuning of these models in physical robots requires very little data. Thus the process reduces the robot cycle time significantly as huge number of robots can learn their own kinematics and dynamics in simulation environment only.

- Usually researchers work only on specifics-kinematic control or visual perception or dynamic control or reinforcement learning. In this book, we take an approach which will guide readers to build a complete integrated robotic system that combines kinematics, dynamics, visual perception, and manipulation.

In this book, we will focus on five major aspects of robotics. First being the perception where we will describe various computer vision techniques for object detection, recognition, and tracking, etc. In the process, we will provide an overview of deep learning-based methods and demonstrate their impact on the performance of these algorithms. The second aspect is related to manipulation and motion planning which aims at solving the inverse kinematics of a manipulator in the Cartesian as well as image plane. In the process,

Preface

we will describe several methods for solving the hand-eye coordination and visual servoing problem. The third aspect is related to mobile robots which will demonstrate vision-based algorithms for mobile robot navigation. The fourth aspect will focus on vision-based techniques for grasping where we will describe a model-free approach for computing graspable handles directly from 3D point cloud. The fifth and final aspect will be multi-robot coordination which becomes essential when one has to deal with multiple robots. In this context, we will describe several methods to achieve formation control in a group of robots which is resource optimal and fault-tolerant.

The precise and lucid presentation of tools, techniques, and associated engineering science as provided in this book will help scientists, researchers and practicing engineers to get an in-depth understanding of techniques required for developing integrated robotic systems for various applications, such as robots for automating pick and place tasks, automated mobile robots for movement of goods in warehouses, a robot for assisting patients, drones for infrastructure monitoring, and surveillance, etc. In the process, the researchers will get exposed to some of the niche areas such as deep learning, programming by demonstration (PbD), visual servoing, and multi-robot control. The accompanied source codes and examples will help the readers in getting a good grasp of the concept.

Scientists, researchers, and graduate students alike will benefit from both machine learning and control theoretic frameworks as presented in this book. We have tried to make every chapter self-contained by including introductory primers and examples. We provide an introductory chapter that includes background material on robotics covering topics such as kinematics, dynamics, and control. Simultaneously, the readers are expected to have a basic background in machine learning techniques such as back-propagation networks, Kohonen self-organizing map, adaptive critic networks, and deep learning. Readers can refer to the book on Intelligent Systems and Control by Laxmidhar Behera and Indrani Kar, which is published by Oxford University Press as a good primer for the subjects dealt with in this book.

A repository of colour images can be downloaded from https://www.crcpress.com/9781138597716

MATLAB® is a registered trademark of The MathWorks, Inc. For product information, please contact:

The MathWorks, Inc.
3 Apple Hill Drive
Natick, MA, 01760-2098 USA
Tel: 508-647-7000
Fax: 508-647-7001
E-mail: info@mathworks.com
Web: www.mathworks.com

Acknowledgment

Most of the works as presented in this book were done in the Intelligent Systems and Control Lab at IIT Kanpur. All authors gratefully acknowledge the contributions of funding agencies such as DST, DEITY, UKIERI, ADNOC-GRC and TCS that helped to carry out the underlying research and experimentation.

All of us gratefully acknowledge the contributions from our colleagues who all worked, in some way or another, in the lab. Special thanks are due to Dr. Indrazno Siradjuddin, Dr. Meenakshi Gupta, and Dr. Anima Mazumder for their significant contribution to Chapters 4 and 6, 16, and 10, respectively.

Laxmidhar Behera thanks all his family members, specifically his parents for their blessings, his wife Gopali Priyadarsini, who has been such a brilliant companion, and his three daughters, Yamuna, Lalita, and Visakha for bringing aesthetics into his life. Last, but not least, he acknowledges the association of Bhakti-Vedanta club at IIT Kanpur which inspires him to inculcate the ideal human values.

Swagat Kumar would like to acknowledge the contribution of Ms. Olyvia Kundu toward providing the content for the Chapter 14 on "Vision-based Grasping." In addition, he would like thank his friends and colleagues who have always been a source of motivation and encouragement for him.

Prem Kumar Patchaikani would like to thank his parents and family members for their constant support in all his endeavors. He would also like to thank Prof. Behera for giving him the opportunity to work on this book, and also the fellow authors; without their support and cooperation, this book couldn't have been completed successfully on time.

Ranjith Ravindranathan Nair would like to express his sincere gratitude to all the co-authors of this book, especially Prof. Behera, and all the colleagues of intelligent systems and control lab at IIT Kanpur for their support. He would also like to extend his heartfelt thanks to his beloved parents, for their endless support, prayers, and blessings, his soulmate Devika for her fathomless support, patience, and care, sister and family for their constant support, and last but not least, he would like to thank his son Samarth, who made him feel more fulfilled than he has ever imagined.

Samrat Dutta would like to extend his deep gratitude to Prof. Laxmidhar Behera for giving him the opportunity to be a co-author of this book. He is extremely grateful to his parents for their love and support, his wife Anima for making his life a beautiful journey and his beloved daughter Ahana for giving the feeling of completeness. He is also thankful to his friends in Intelligent Systems and control Lab (IITK), colleagues in TCS Innovation Lab for their encouragement and last but not the least, he expresses his gratitude to publication team for their effort.

Authors

Laxmidhar Behera is working as Poonam and Prabhu Goel Chair Professor at IIT Kanpur having research and teaching experience of more than 24 years. He has received his BSc (engineering) and MSc (engineering) degrees from NIT Rourkela in 1988 and 1990, respectively. He received his PhD degree from IIT Delhi in 1996. He pursued his postdoctoral studies in the German National Research Center for Information Technology, GMD, Sank Augustin, Germany, during 2000–2001. Previously, he has worked as an assistant professor at BITS Pilani during 1995–1999 and as a reader at Intelligent Systems Research Center (ISRC), University of Ulster, UK during 2007–2009. He has also worked as a visiting researcher/professor at FHG, Germany and ETH, Zurich, Switzerland. His research work lies in the convergence of machine learning, control theory, robotic vision, and heterogeneous robotic platforms. He has received more than INR 170 million research grants to support his research activities. He has established industrial collaboration with TCS, Renault Nissan, BEL, Bangalore, and ADNOC, Abu Dhabi while making significant technological development in areas such as robotics-based ware-house automation, vision and drone-guided driver assistance systems, and drone-guided pipeline inspection systems. He has published more than 250 papers in journals and conference proceedings. He has supervised 16 PhD theses to completion. He is a Fellow of INAE and Senior Member of IEEE. He is a Technical Committee member on IEEE SMC on Robotics and Intelligent Systems. His other research interests include intelligent control, semantic signal/music processing, neural networks, control of cyber-physical systems, and cognitive modeling.

Swagat Kumar obtained his Bachelor's degree in Electrical Engineering from North Orissa University in 2001 and his Master's and his Ph.D. degree in Electrical Engineering from IIT Kanpur in 2004 and 2009 respectively. He was a postdoctoral researcher at Kyushu University in Japan during 2009-10. He, then, worked as an assistant professor at IIT Jodhpur for about two years before joining TCS Research in 2012. He currently heads the robotics research group at TATA Consultancy Services in New Delhi. His research interests include Robotics, Computer Vision and Machine Learning. He is a member of IEEE Robotics and Automation Society. He has co-authored about forty articles in peer-reviewed conferences and journals and filed several patents.

Prem Kumar Patchaikani received the B.E. degree in Electrical and Electronics Engineering from Thiagarajar College of Engineering, Madurai, India, in 2003, the M.Tech. degree in Power and Control from IIT Kanpur, Kanpur, India, in 2005, and the Ph.D. degree from the Department of Electrical Engineering, IIT Kanpur, in 2012. He was a Design Engineer with Larsen & Toubro Ltd., Chennai, India, from 2005 to 2006. He was a Visiting Researcher with the Intelligent System Research Center, University of Ulster, Londonderry, U.K., from 2008 to 2009, and in 2011. He is currently a Lead Engineer with General Electric, Bengaluru, India. His current research interests include visual servoing, redundant manipulators, neural networks and fuzzy logic-based control, adaptive critic, system identification, and IC engine controls.

Ranjith Ravindranathan Nair received the Master's degree in Guidance and Navigational Control from the College of Engineering, Thiruvananthapuram, Kerala, India. He received his Ph.D in Control and Automation from the Department of Electrical Engineering, Indian Institute of Technology (IIT) Kanpur, Kanpur, India. After completing his Ph.D, he was working as a post doctoral research fellow in Intelligent Systems and Control Lab at IIT Kanpur. Currently, he is working as an Assistant Professor in the Department of Electronics & Communication Engineering at Indian Institute of Information Technology Pune, India. His primary research interests include multiagent systems, formation control, Intelligent control, nonlinear control systems, cyber-physical systems, and multi-robot/ multi-vehicle systems.

Samrat Dutta received the B.Tech. degree in Electrical and Electronics Engineering from the Siliguri Institute of Technology, Siliguri, India, in 2005, and the M.Tech. degree in Electrical Engineering with a specialization in control system from the Department of Electrical Engineering, Jadavpur University, Kolkata, India, in 2010. He has received his Ph.D. degree in Control and Automation from the Department of Electrical Engineering, IIT Kanpur, Kanpur, India in 2018. He was a Cluster Head with Sterling Communications Ltd., Chennai, India, from 2005 to 2007, where he was involved in Tata Teleservices mobile network infrastructure projects in India. He is currently working as a research scientist with TCS innovations Labs, Bangalore, India. His current research interests include Imitation learning, Reinforcement Learning, Intelligent Robots, Robotic Vision, Deep Learning, neural networks and fuzzy logic-based control, and system identification.

1
Introduction

Robot-assisted applications in warehouse, agriculture and health care are highly multi-disciplinary. Mechanical design, robot dynamics and control, kinematics and control, visual servoing, visual perception, motor skill learning, grasping, machine learning and AI and sensors are some of the key areas that contribute to the development of a wholesome robotic product. Often researchers work on a specific area – an approach that lead to the neglect of systemic integration. Amazon Picking Challenges (APC-2015 and APC-2016) and Amazon Robotics Challenge – ARC-2017 – have motivated researchers to look at the problem of of developing autonomous robots at the systemic integration level as well. Based on this approach, the first part of the book will take an integrative approach to the design of three key areas – visual control, motor skill learning and visual perception – to build autonomous manipulation systems using control theoretic framework.

Development of intelligent autonomous robots that can assist humans in their daily needs or can be used as replacement for human labor in various tasks, has always been a challenging and interesting problem in the field of robotics. These applications require robots to have human-like learning capability. Starting from turning off the alarm clock using our hand when we wake up in the morning, to switching off the light before going to bed at night, we perform uncountable number of tasks consciously or unconsciously through out the day while making use of our limbs. These tasks, such as moving an object, playing an instrument, writing a letter, and playing soccer require skill and dexterity which we have been acquiring through rigorous training since the day we are born. Initial abrupt movements of the limbs of a child, interact with the environment and get rewarded by the consequence. This works as the reinforcement signal in the learning process. The temporal error between the expected and the received rewards are encoded in the dopamine signal of the mammalian brain [1–3]. In the training process, children learn independent and controlled movement of their limbs. These movements are adaptive enough to deal with the dynamically changing environment. Generally, the training could be either *independent* or *supervisory*. The former is about correcting one's own actions continuously to optimize a performance index based on the reinforcement signal as received for an arbitrary action performed in the environment and the latter is about improving one's own actions based on the expert's instructions. In this way, we learn to control our limbs at the joint level and eventually we learn higher level motions in the task space. In fact,

FIGURE 1.1: The Mobile Manipulator System for warehouse automation at IIT Kanpur.

like humans, many life-forms on this planet have this capability of learning from experience. Thus, when we think of a robotic system co-existing with humans in society, it must have human-like adaptability in its behavior. They should also be able to learn from their experience as humans do.

Our team IITK-TCS participated in Amazon Picking Challenge 2016 and Amazon Robotics Challenge 2017. The team secured fifth position in ARC 2016. The team secured fifth, third, and 4th position in stow task, pick task, and stow-cum-pick task respectively in ARC 2017. The robotic system that was used in APC 2016 is shown in Figure (1.1). The system consisted of Barrett Arm mounted on a mobile platform that could automatically adjust its height to reach any box within the shelf. The robot arm was mounted with a Kinect camera. The suction was used to grasp the desired object. RCNN was used to learn the object categories and to identify the centroid of 2-d image of the desired object where the suction gripper is guided with help of cloud data to grasp the object.

The system consisted of UR ten robot manipulators, a server with eight GPUs, Suction gripper, RGB-D sensor, and Ensenso Camera. A customized deep network was designed that is capable of semi-supervised leveling, automatic cluttering, and semantic segmentation as shown in Figure (1.2). Among all teams, our algorithm had the highest grasp rate.

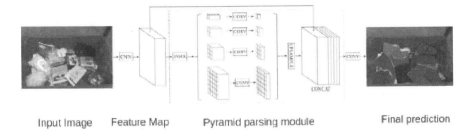

FIGURE 1.2: Visual perception using deep neural network.

1.1 Vision-Based Control

The schematic of the various components associated with the warehouse automation system can be given as in Figure 1.3. The target item is segmented using a vision system comprising of a RGB-D sensor. The goal configuration is used to generate a dynamic trajectory from a library of previously demonstrated trajectories to reach the item. The controller then finds an optimal control to execute the motion. The grasping is finally accomplished using feedback-based reinforced training and learning in a physics-based simulator.

An automated precision agriculture system – for example, a mobile manipulator employed for pruning, spraying and plucking in horticulture – will have similar functional blocks as shown for the automated warehouse system as shown above. The same is true for an automated system for manufacturing.

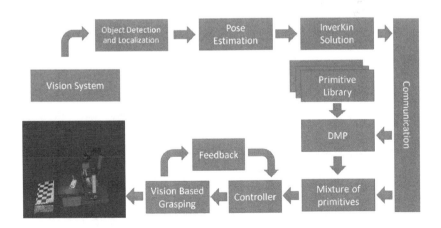

FIGURE 1.3: The proposed robot assisted ware-house automation system.

In practical applications of a robotic manipulator such as an automated warehouse, the robotic movements are controlled in two stages in the control hierarchy:

- There is an *inner loop* controller (also called dynamic controller) that converts the joint space command to the joint torque command as the manipulators are controlled in joint space. The electrical controller, that is a part of the robot's hardware, converts this joint torque to electrical signal and subsequently the robot moves. The controller here we are interested in, is not just *any* but from a specific class of controllers that help the robot to achieve the desired joint states while optimizing a certain performance index.

- There is an *outer loop* controller (also called kinematic controller) that plans and executes the robot's movement in the Cartesian space of the robot. This movement is defined by the task assigned to the robot. Design of these controllers has always been an intriguing and challenging control engineering problem.

The core of such an automated system is the dynamic and kinematic control of the manipulator while guaranteeing the closed loop stability. In this book, we will be concerned with data-driven learning frameworks that enable the robot to learn controllers in the *inner loop* which is dynamic control and *outer loop* of the control hierarchy (see Fig. 1.4) that is referred to kinematic control. The controller in the inner loop involves robot's dynamics and is

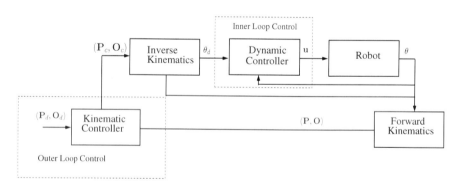

FIGURE 1.4: The *inner loop* controller converts the joint space command to the joint space torque and the *outer loop* controller plans the robot's end-effector trajectory in the task space. The robot's joint position θ is converted to Cartesian position and orientation $(\mathbf{P}, \mathbf{O})$ using forward kinematics. The *outer loop* controller provides position and orientation command $(\mathbf{P}_c, \mathbf{O}_c)$ based on the current $\mathbf{P}, \mathbf{O}$. The inverse kinematics algorithm finds corresponding joint position command θ_d based on $\mathbf{P}_c, \mathbf{O}_c$. The *inner loop* controller converts θ_d to joint torque $\mathbf{u}$.

responsible for providing joint torques which are required for the robot to achieve commanded joint positions while optimizing some performance index.

Classical techniques, for decades, have given us the ability to solve problems associated with nonlinear systems and its control. These techniques have been efficiently used in industrial robots with a high accuracy of task completion. However, they reflect strong dependence upon precise mathematical modeling of the entire process [4, 5]. With the increasing diversified applications of robots, in many real-life situations it becomes difficult to work with these techniques as the precise mathematical description of the process is not available. They have limited ability to incorporate effects of the dynamically changing environment interacting with the robot. Additionally, one needs to possess good technical knowledge and skill to work with these techniques. It is very difficult for a general user, who lacks in-depth knowledge in this domain, to make use of these techniques in order to prepare the robotic system for the assigned task.

Given the progress in AI and machine learning, researchers have been interested to emulate human learning processes in robots. We will thus present learning-based kinematic and dynamic control strategies in this book while well positioning with the state of the art techniques available in the literature. As the *outer loop* controller, we will discuss various data-driven motion planners that plan reaching motions for the manipulator in the Cartesian space. Besides neural network-based schemes, one of the approaches is to develop a motion planner in terms of movement primitives [6], which can be combined to produce more complicated robotic motions. The movement primitives are represented by a *dynamical system* (DS). A *learning by demonstration* (LbD) based approach is adopted to identify the movement primitives from the demonstrations given by a user. As we learn DS using real data from the robot, ensuring stability of the learned system is challenging. Using learning-based approaches, both the dynamic and kinematic controllers have been designed. In this sense, techniques developed, work can make a robot learn its own kinematic and dynamic controllers as it performs from simple to complex tasks.

It is presumed that the dextrous manipulation is achieved by continuous learning with the reinforcement signal obtained from the environment. It is argued in [7–9] that the temporal error between the expected and the received rewards is encoded in the dopamine signal of the mammalian brain. The successful implementation of the robotic systems to assist the human beings in real-life requires adaptability to the dynamic environment by continuous learning and robustness to the model and sensor inaccuracies. Hence, learning-based control schemes which could adapt according to the rewards are imperative in the robotic systems to operate in a dynamic cluttered environment. These observations have led to the development of learning-based visual automation of robotic arms, which enables the robotic systems to serve well in both structured and unstructured environments. To achieve the excellence similar to humans, the redundant manipulator guided through visual feedback must necessarily work in a dynamically changing environment. Vision can provide continuous information about a dynamic environment, and kinematic redundancy is introduced in the manipulator for handling complex tasks which occur in changing environments.

Vision-based control of a redundant manipulator is a challenging task which involves two sub-tasks: (i) visual servoing [10, 11] and (ii) redundant manipulator control [12, 13]. In general, visual servoing computes the end-effector velocity required to reach the desired position from the image features obtained through the visual feedback. It basically assumes that there exists a non-redundant manipulator which can generate the desired end-effector velocity with its own inverse kinematic algorithm. A non-redundant manipulator can achieve the end-effector velocity estimated from visual servoing with unique joint angle configuration. In a dynamic environment, the available unique joint angle configuration may become infeasible to position the end-effector due to the presence of obstacles and the physical constraints. This necessitates the use of redundant manipulators for vision-based control in dynamic environments, which have excess DOF than that required for the given task. Theoretically infinite choices of joint angle configuration exist for redundant manipulators to achieve the estimated end-effector velocity. The excess DOF can be effectively utilized in performing additional constraints introduced by the dynamic environment. An optimal joint angle configuration needs to be selected, while satisfying these additional constraints. This is popularly known as redundancy resolution.

1.2 Kinematic Control of a Redundant Manipulator

The position of the end-effector and the associated joint angle configuration are coupled with forward and inverse kinematic relationship of the manipulator. The forward kinematic map expresses the Cartesian space position of the end-effector $\mathbf{x}$ for the given joint angle configuration $\boldsymbol{\theta}$, as,

$$\mathbf{x} = \mathbf{f_x}(\boldsymbol{\theta}) \tag{1.1}$$

where the dimension of the task space $\mathbf{x}$ is n, and that of the joint angle space $\boldsymbol{\theta}$ is m. In case of redundant manipulators $n < m$ and the degree of redundancy is given by $m - n$. $\mathbf{f_x}(\boldsymbol{\theta})$ is highly nonlinear and is obtained from the geometry of the manipulator using Denavit-Hartenberg (D-H) parameters [14].

The inverse kinematic relation computes the joint angle space configuration $\boldsymbol{\theta}$ which is required to reach the desired position $\mathbf{x}_d$. The closed form inverse kinematic relationship exists only for simple manipulator configurations. The problem becomes much more difficult for redundant manipulator since infinite number of solutions exist to reach the given workspace position. The control of a kinematically redundant manipulator to reach the object is a highly challenging task owing to the one-to-many inverse kinematic relationships. The redundancy resolution schemes discuss about different methodologies to exploit the available redundancy for performing the additional tasks which occur in real-time.

In general inverse kinematic control is achieved with forward differential kinematic relationship, since it expresses a linear relationship between the joint angular velocity $\dot{\theta}$ and the Cartesian space velocity $\dot{x}$. The forward differential kinematic relationship between $\dot{\theta}$ and $\dot{x}$ is represented as,

$$\dot{x} = J\dot{\theta} \qquad (1.2)$$

where $J = \frac{\partial f_x}{\partial \theta}$ is the kinematic Jacobian of the manipulator. In case of redundant manipulators, J is not a square matrix and theoretically infinite joint angular velocity $\dot{\theta}$ exists to generate the given end-effector velocity. The joint angular velocity required for the given end-effector velocity is computed using inverse Jacobian. Inverse Jacobian does not exist in case of redundant manipulators since the associated Jacobian is not square and hence, the pseudo-inverse has been employed. Inverse kinematic control of the redundant manipulator using generalized pseudo-inverse was first proposed by Whitney [15]. The pseudo-inverse method computes the value of $\dot{\theta}$ as,

$$\dot{\theta} = J^+\dot{x}_d \qquad (1.3)$$

where J^+ is the pseudo-inverse of the kinematic Jacobian, and $\dot{x}_d$ is the desired end-effector velocity. Henceforth, the notation $(.)^+$ will be used to indicate the generalized pseudo-inverse of $(.)$. The open-loop solution obtained using the above equation unavoidably leads to solution drift due to numerical integration and hence, results in task space error $e = x_d - x$. To overcome this drawback in open-loop control, the closed loop kinematic control is proposed with the task space error e. In closed loop kinematic control the joint velocity is computed as,

$$\dot{\theta} = k_p J^+ e \qquad (1.4)$$

where $k_p > 0$ is proportional gain which controls the speed of the convergence to the desired position x_d. The pseudo-inverse based solution results in lazy arm movement, i.e., it minimizes the joint angular velocity in least square sense. Baillieul showed in [16] that the pseudo-inverse solution may reach a singular configuration when implemented without any modification.

Pseudo-inverse based kinematic control is widely popular since the relationship between the various joint angular velocities, which can generate the desired end-effector velocity can be established using the pseudo-inverse of Jacobian J. J^+ obeys the property that the matrix $(I - J^+J)$ projects onto the null space of J and, hence, the vector $J(I - J^+J)\phi = 0$ for all vectors ϕ. A joint angular velocity computed as $\dot{\theta} = (I - J^+J)\phi$ for any vector $\phi \in R^m$ does not generate any end-effector motion but only changes the internal joint angle configuration of the manipulator. The internal reconfiguration of the manipulator is popularly known as self-motion of the manipulator. The different joint angular velocities which can generate the given end-effector velocity are given by the relationship,

$$\dot{\boldsymbol{\theta}} = \mathbf{J}^+\dot{\mathbf{x}} + k_n(\mathbf{I} - \mathbf{J}^+\mathbf{J})\boldsymbol{\phi} \tag{1.5}$$

where $\mathbf{I}$ is the identity matrix of order m, and k_n is the gain which determines the magnitude of the self-motion.

1.2.1 Redundancy Resolution using Null Space of the Pseudo-inverse

The self-motion of the redundant manipulator is used to achieve the additional tasks required in the dynamic environment by optimizing certain performance measure J_i. In general $\boldsymbol{\phi}$ is chosen as $-\frac{\partial J_i}{\partial \boldsymbol{\theta}}$ to minimize J_i and as $\frac{\partial J_i}{\partial \boldsymbol{\theta}}$ to maximize J_i. Liegeois [17] used equation (1.5) to resolve the redundancy by using the null space of the Jacobian for avoiding the kinematic limits. The self-motion has been exploited to perform additional tasks such as satisfying kinematic constraints [18], maximizing the manipulability index [19], minimizing the infinity norm of the joint angular velocity [20] and obstacle avoidance [21].

The value of gain k_n determines the speed of convergence to the optimal joint angle configuration. It is clear from equation (1.5) that the manipulator may exhibit self-motion even after $\dot{\mathbf{x}}_d = \mathbf{0}$ because of improper choice of the gain k_n. A high value of gain k_n results in undesirable oscillation and a low value of gain k_n may end up with sub-optimal solutions. The value of gain k_n is to be properly chosen to avoid oscillations. Li et al. [22] proposed an analytical method to calculate the limiting values of k_n, while constraining the manipulator within its kinematic limits. The Jacobian null space based redundancy resolution schemes optimize an instantaneous cost function for resolving the redundancy and results in local optimum [23]. A detailed discussion about various pseudo-inverse based redundancy resolution methods and the associated challenges are available in [12, 24].

Pseudo-inverse based methods are widely used in the literature but they perform poor near singularities due to numerical instability. The damped least square method avoids problems associated with singularities by not exactly following the desired trajectory [25,26]. Damped least square method has been tested on redundant manipulators to achieve torque minimization [27] and obstacle avoidance [28]. The challenges associated with various damped-least square based control schemes are available in [29].

1.2.2 Extended Jacobian Method

The solution obtained with the null space projection methods is not periodic, i.e., the closed trajectory in the Cartesian space does not result in a periodic joint angle space trajectory. Hence, the null space projection methods are not suitable for repetitive tasks. Extended Jacobian methods [16, 30, 31] are proposed to achieve closed joint angle space trajectories. Extended Jacobian methods form a square Jacobian matrix by augmenting the kinematic Jacobian with the additional task space constraints. The inverse kinematic solution is obtained by computing the inverse of the augmented Jacobian. Though extended Jacobian methods result in cyclic trajectories, the methods suffer from algorithmic singularity, i.e., the augmented Jacobian may become singular even if the kinematic Jacobian is not rank-deficient.

English et al. [32] discussed a single framework integrating both the null space projection and the extended Jacobian based kinematic control schemes.

1.2.3 Optimization Based Redundancy Resolution

The null space projection methods and the extended Jacobian methods require the computation of inverse of the Jacobian. Alternatively optimization based methods are developed which resolve the redundancy without explicit computation of the inverse of the Jacobian. In optimization based methodologies [33, 34], the redundancy resolution scheme is formulated as a time-varying optimization task with equality and inequality constraints and the necessary condition for the optimality is derived. Various dynamic neural network (NN) architectures are proposed such that the equilibrium point of the dynamic neural network (DNN) corresponds to the necessary condition of the optimality. At each instant the DNN is presented with the desired end-effector velocity and the network evolves from the initial joint angle configuration to the optimal solution of the redundant manipulator. Quadratic program formulation has been used for kinematic limit avoidance [35], torque optimization [36], obstacle avoidance [37], and acceleration level resolution [38]. Infinity norm minimization [39, 40] is achieved by formulating the control task as a linear program and by minimizing the convex energy function with a recurrent neural network (RNN). The major drawback with DNN based optimization approaches is that the convergence speed of the network to the optimal solution is not known and, hence, the approaches are computationally intensive.

1.2.4 Redundancy Resolution with Global Optimization

All the aforementioned approaches minimize an instantaneous cost function and achieve local optimum. The solution obtained is locally optimum since it may not be optimum for the whole trajectory. Local optimization may lead to control instabilities, which result in high torque and joint angular velocity for relatively long trajectories which cover the whole workspace [41]. Instantaneous cost minimization is generally preferred since it is computationally simple in terms of the current joint angle configuration. Global optimization with an integral cost function over the entire trajectory is developed to cope up with the instabilities occurring for long trajectories. Kinematic and dynamic redundancy resolution by minimizing a global cost function is discussed using Pontryagin's maximum principle in [42]. Kazerounian and

Wang [43] analytically showed that the local minimization of the joint angular acceleration is equivalent to the global minimization of the joint angular velocity in least square sense. Suh and Hollerbach achieved global torque minimization [44] using the principles of calculus of variation. Minimum time control is discussed for redundancy resolution in [45]. The integral cost based optimization has been implemented for path planning in [46]. The redundancy resolution is formulated as an optimal control problem in [47], and the $2n$ first order partial differential equations with boundary conditions have been derived. It has been shown that it is equivalent to n second order differential equations resolving the redundancy in acceleration level.

The major drawback with the global cost optimization is that the accurate knowledge of the forward kinematics is required and the optimal solution is obtained by solving the associated two-point boundary value problem numerically for individual trajectories. Hence, all the aforementioned global optimization methods are offline processes and cannot be implemented in real-time over the whole workspace.

1.2.5 Neural Network Based Methods

All the above local and global optimization schemes require the accurate forward kinematic model to resolve the redundancy and perform poorly with inaccurate models. Model inaccuracies pose a major challenge for pseudo-inverse computation since the pseudo-inverse is sensitive to parameter variations, and the parameter variation may eventually result in controller instability. The universal function approximation property of NN is used to learn either the inverse kinematic map or the forward kinematic map to control the manipulator without the complete knowledge of the kinematic model. The major challenge associated in learning inverse kinematic map of the redundant manipulator is that it is a one-to-many relationship. Existing NN architectures learn only a single joint angle configuration for the given end-effector position by converging toward the arithmetic mean of all the joint angle configurations available in the data. Such convergence results in poor positioning accuracy and the redundancy available in the data is lost.

Ahmad and Guez [48] learned the inverse kinematic map with multi-layer perceptron (MLP) and the map is used as a seed generator for redundancy resolution with pseudo-inverse method. The method to learn an exact joint angle configuration is not discussed and, hence, the learned map is a poor approximation of the inverse kinematics. Martin and Millan [49] suggested the distal learning approach to learn the inverse map with an NN approximating the forward kinematic map. Though a self-motion network is suggested in addition to inverse network for redundancy resolution, the method requires the geometric knowledge of the manipulator configuration, and it is difficult to generalize for n-link manipulator. Two neural networks are used for resolving the redundancy in [50]. The first network learns the null space projection vector ϕ which optimizes the chosen additional task. The second network is used to compute the output of the damped least square pseudo-inverse. The method requires the computation of pseudo-inverse during the training phase.

All the aforementioned redundancy resolution strategies compute the joint angular velocity from the end-effector. Apart from neural network approaches, accurate kinematic model is needed for control and require computationally intensive pseudo-inverse. Global optimal solution has been achieved only through offline methods.

1.3 Visual Servoing

Vision is employed in robotics owing to its flexibility during manipulation. Visual feedback gives dynamic information about the environment and the object. Typically, vision-based manipulator control is executed in open loop fashion, "looking" and then "moving." [51] This results in poor positioning accuracy due to the model inaccuracies. An alternative approach is to use a visual control loop which is generally referred to as visual servoing. A detailed survey on visual servoing can be found in [10], [11], and [52]. Vision-based manipulator control use either single camera or multiple cameras to give visual feedback to the manipulator system. Visual servoing systems use one of two camera configurations: eye-in-hand or eye-to-hand. In the eye-in-hand configuration [53, 54], a camera is mounted on the end-effector while in the eye-to-hand configuration [55, 56], the cameras are fixed in the workspace. Eye-to-hand configuration is also known as stand-alone camera system [52].

Visual servoing schemes use the image features $\mathbf{u}$ to represent the position of the end-effector and the object in the vision space. The desired position $\mathbf{x}_d$ and the current position $\mathbf{x}$ of the end-effector are observed through the camera as the desired image feature vector $\mathbf{u}_d$ and the current image feature vector $\mathbf{u}$ respectively. In general, visual servoing uses the linear relationship between the change in the image feature vector $\mathbf{u}$ and the change in the Cartesian space position of the end-effector $\mathbf{x}$ for controlling the manipulator. The image Jacobian $\mathbf{L}$, represents the relationship between the end-effector motion and the image feature motion as,

$$\dot{\mathbf{u}} = \mathbf{L}\dot{\mathbf{x}} \qquad (1.6)$$

where $\mathbf{u}, \dot{\mathbf{u}} \in R^p$. $\mathbf{L}$ is a $p \times n$ matrix and is also referred to as interaction matrix in literatures. The control task is to compute the necessary Cartesian space velocity motion such that the end-effector will reach the desired position in vision space asymptotically.

The simple proportional control law which results in asymptotic stabilization is expressed as,

$$\dot{\mathbf{x}}_d = k_p.\mathbf{L}^+(\mathbf{u}_d - \mathbf{u}) \qquad (1.7)$$

where k_p is a proportional gain, $\mathbf{L}^+$ is the pseudo-inverse of $\mathbf{L}$ and $\mathbf{u}_d - \mathbf{u}$ is the error between the desired and the current image features. Here afterwards, the error vector in the visions space is expressed as $\mathbf{e}_u$ and, hence, $\mathbf{e}_u = \mathbf{u}_d - \mathbf{u} : \in R^p$. The above controller requires the exact knowledge of $\mathbf{L}$ and its pseudo-inverse though it ensures global stability. The exact knowledge of $\mathbf{L}$ requires the complete knowledge of the 3 D Cartesian space which may not be available in a dynamic environment. The depth information of the object has to be estimated for visual control of the manipulator [57–59]. Hence, the image Jacobian $\mathbf{L}$ is to be estimated at each instant. To reduce the computational complexity, the image Jacobian is estimated at the desired position and then the pseudo-inverse is evaluated for the estimated image Jacobian to implement the controller. This eliminates the continuous estimation of $\mathbf{L}$ and the computation of the pseudo-inverse $\mathbf{L}^+$ in real-time. But, this results in a locally stabilizing controller since the sufficient positivity condition of stability is valid in local region only [60]. The learning-based servoing scheme proposed in [61] for non-redundant manipulator focusses on learning the inverse Jacobian at the chosen operation point only.

Visual servoing systems are classified based on the method of using the visual information as follows:

1.3.1 Image Based Visual Servoing (IBVS)

In image based visual servoing, the 2 dimensional image features are used to estimate the motion of the manipulator directly. The error signal is specified in the 2 dimensional image plane as the difference between the actual and the desired features. IBVS is also known as 2-D visual servoing [62–65] since the control input is computed directly from 2-D image features. Since the control input is generated from the vision space directly, IBVS may result in poor Cartesian space trajectories [60, 66].

1.3.2 Position Based Visual Servoing (PBVS)

In position based visual servoing [67, 68], features extracted from the image are used to estimate the position and orientation of the object with respect to the camera (or world) coordinate system. Using these values, an error between the current and the desired pose of the robot manipulator is defined in the task space to compute the necessary end-effector motion. In this scheme, the control task is completely separated from the estimation process involved in computing the Cartesian pose from the image data. Since the control is actuated using the Cartesian pose information, it is also called 3-D visual servoing. Position based visual servoing for eye-in-hand configuration is discussed in [67, 69, 70], and the stand-alone camera configuration is discussed in [71–73]. Position based visual servoing for hybrid eye-in-hand and eye-to-hand multi-camera system is discussed in [74]. In contrast to IBVS, desirable

Cartesian space velocity is obtained in PBVS, since the joint angular velocity is computed from the estimated current pose of the robot manipulator in the Cartesian space. But the algorithm suffers with poor positioning accuracy, if the accurate camera model is not known, since there is no closed loop control over the pose estimation.

1.3.3 2-1/2-D Visual Servoing

The combination of both image and position based approaches are developed to avail the benefit of both the methods [75, 76] in the visual control process. The error is defined by combining both the vision space and the Cartesian space information, and then the end-effector velocity is computed. It is known as 2-1/2-D visual servoing since both the image and the Cartesian space features are used to control the manipulator. Such hybrid schemes [77–79] decouple the translation and rotation velocity of the end-effector with proper feature selections which results in desirable Cartesian space trajectories and accurate positioning. Hybrid visual servoing is proposed for both the eye-to-hand and eye-in-configuration using homography in [80].

1.4 Visual Control of a Redundant Manipulator: Research Issues

As discussed in previous sections, classical approaches estimate the Cartesian space velocity of the end-effector from the vision space with the visual servoing schemes, and then the redundancy is resolved for the chosen additional task while following the Cartesian space trajectory generated by the visual servoing scheme. The schematic diagram of the two stage control process for vision-based redundant manipulator control is shown in Figure 1.5. The visual servoing scheme uses the image features $\mathbf{u}_d$, $\mathbf{u}$ and the Cartesian space information $\mathbf{x}$ to compute the end-effector velocity $\dot{\mathbf{x}}$. Redundancy resolution schemes compute the joint angular velocity from the end-effector velocity $\dot{\mathbf{x}}$, using the current joint angle $\boldsymbol{\theta}$ and the environmental constraints.

Alternatively the vision space trajectories can be directly controlled from the joint angle space by combining visual servoing with redundancy resolution in a single framework. The redundancy is achieved for the trajectories specified in the vision space while satisfying the additional constraints introduced by the environment. The schematic diagram of visual controller with integrated visual servoing and redundancy resolution is shown in Figure 1.6. The controller computes the joint angle configuration directly from the visual feedback resulting in a direct and efficient control over the vision space.

14   Introduction

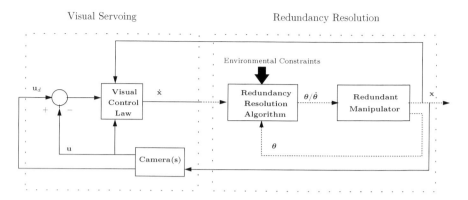

FIGURE 1.5: Classical vision-based redundant manipulator control scheme.

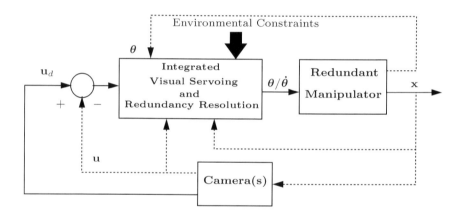

FIGURE 1.6: Visual control: Integrated visual servoing and redundancy resolution.

The relationship between image feature velocity and joint angular velocity is obtained by combining equations (1.2) and (1.6) as,

$$\begin{aligned} \dot{\mathbf{u}} &= \mathbf{LPJ}\dot{\boldsymbol{\theta}} \\ &= \mathbb{J}\dot{\boldsymbol{\theta}} \end{aligned} \qquad (1.8)$$

where $\mathbf{P}$ is the transformation matrix representing the coordinate transformation between the world coordinate frame and the camera coordinate frame, and $\mathbb{J} = \mathbf{LPJ}$ is a $p \times m$ Jacobian matrix from the joint angle space to the vision space. Here afterwards, the notation $\mathbb{J}$ will be used to represent the Jacobian from the joint angle space to the vision space.

Visual Control of a Redundant Manipulator: Research Issues

The closed loop proportional controller resulting in asymptotic stabilization is given as,

$$\dot{\boldsymbol{\theta}} = k_p \, \mathbb{J}^+ \mathbf{e}_u. \tag{1.9}$$

Since the pseudo-inverse of the Jacobian $\mathbb{J}$ is used in the control law, the controller would result in "lazy-arm movement." The null space of $\mathbb{J}$ can be used to satisfy the additional constraints required in the dynamic environment.

In this context, model-based redundancy resolution for visually controlled manipulator is proposed in [81] for trajectories defined in vision space. The trajectories are defined in vision space from a single camera in eye-in-hand configuration, and then task sequencing is used to prioritize the task for achieving kinematic limit avoidance. Mansard and Chaumette [82] achieved obstacle avoidance by task sequencing while following vision space trajectories in eye-in-hand configuration. Later the approach is extended for multiple-task considering occlusion and kinematic limit avoidance together in [83]. These approaches need accurate knowledge of the model and pseudo-inverse of the Jacobian at each instant.

The redundancy can be resolved either for the Cartesian space end-effector velocity estimated from visual servoing scheme or for the trajectories directly defined in vision space. Both of these control schemes can be implemented in model-based control framework. Development of learning-based visual control schemes is motivated due to following the drawbacks of model-based control paradigm.

1. Visual servoing requires the exact computation of image Jacobian **L**, and kinematic Jacobian **J** is needed for the redundancy resolution. The accurate knowledge about the workspace is required to compute **L**, which may not be available in the dynamic environment and, hence, image Jacobian is to be estimated at each instant in real-time. In addition, classical approaches need pseudo-inverse of both image and kinematic Jacobian at each instant, which is computationally intensive. As degree-of-freedom increases, the pseudo-inverse computation poses a major challenge in real-time implementation. Thus the efficient combined estimate of pseudo-inverse of **J** and **L** using learning methodologies is an interesting research problem within a learning-based paradigm.

2. Existing redundancy resolution schemes minimize an instantaneous cost function and achieve local optimum in real-time. Minimizing an instantaneous cost function leads to instabilities resulting in high torque and joint angular velocity, for long trajectories spanning the entire workspace. Optimization of a global cost function as an integral cost function over the entire trajectory leads to an offline optimal control problem. Such global optimization schemes are robust to instabilities but cannot be implemented in real-time as required in a visual servoing situation. Hence the development of real-time optimal redundancy resolution schemes for global cost function is one of the unsolved research problems in this area and worth investigating within the learning paradigm.

3. Just like kinematic visual control, optimality plays an important role in dynamic visual servoing. Optimization in dynamic visual servoing results in minimization of torque effort while avoiding the obstacles and kinematic limits. An optimal control strategy which is computationally efficient and robust to model inaccuracies and adaptive to environmental changes is highly desirable. A learning-based paradigm for dynamic visual servoing also poses an interesting research problem.

These challenges necessitate developing computationally efficient visual control schemes which are robust to model inaccuracies and could directly affect the error in vision space from joint space. These goals can be achieved through learning-based approaches. Learning-based control schemes can adapt their behavior according to the feedback received from the environment and are robust to model inaccuracies. An autonomous system which can improve its performance with time, using the available feedback and cope up with model inaccuracies and changes occurring in a dynamic environment is highly preferred for real-life implementation of the robotic systems. Robots are active agents and are amenable for learning-based schemes, which do not require the complete knowledge of the camera, manipulator, and the environment model.

This book is thus concerned with development of learning-based approaches for controlling the redundant manipulators kinematically and dynamically, directly from the vision space. Learning-based algorithms are developed to resolve the redundancy while controlling the manipulator directly from the vision space. Learning-based approaches are developed to resolve the redundancy either by minimizing a local or a global performance index. NN architectures which model the nonlinear controller as a cluster of local linear controllers are used for redundancy resolution. The joint angle configuration required to reach the desired position is computed as the output of the neural networks using the positioning error in vision space and the additional environmental constraints. The proposed learning-based control algorithms are tested on a 7 DOF kinematically redundant PowerCube manipulator controlled with stereo vision in eye-to-hand configuration.

1.5 Learning by Demonstration

Learning by demonstration (LbD) is also known as programming by demonstration (PbD) or imitation learning (IL). As we have already mentioned, classical approaches of robot motion planning involves hard-coded programming or pre-programming of robotic motions for predefined tasks, which in fact limits the adaptability of the robotic motion in an dynamic environment. Such manual motion planning also requires in-depth technical skill. With the increasing degrees of freedom (DOF) in new generation robotic manipulators,

Learning by Demonstration

FIGURE 1.7: Learning by demonstration in human life: a) a child observes how to kick a football; b) she kicks the football as demonstrated.

the robotic motion is expected to be more adaptive and intuitive while accomplishing a wide variety of tasks as the humans do. Given the scenario, manual coding based solution is not viable anymore. In contrast, the learning-based approaches are more intuitive. By making the robot capable of learning to perform a task from instances such as watching the expert's movement of limbs, the rigorous manual programming can be avoided.

Learning in robotics is fundamentally motivated by the biological systems. In motion learning, the robot is expected to learn new *task-oriented policies*[1]. The learning could be solely trial-and-error basis where the policy is improved based on the reward collected for performing an action. This reward based learning is *independent* and falls under the category of *reinforcement learning* (RL) [84–86] where an agent mainly learns suitable policies to achieve the goal. In contrast, in LbD or IL the agent learns the policies by observing the demonstrations given by the teacher to accomplish a specified task [87, 88]. In this case the policies are learned such that the error between the demonstrated and the executed profile is minimized. In other words, policies are learned under the *supervision* of the demonstrator.

LbD is a data-driven learning technique which humans use very frequently in their life time. As a child grows up, LbD or IL plays a vital role in his/her learning of numerous behaviors/skills such as communicating through gesture, movements of limbs during various sports, common social behaviors, playing an instrument (Figure 1.7 presents such examples). LbD being a powerful alternative to classical motion planning techniques, it is employed in both symbolic and trajectory level. In the symbolic level planning, the high-level representation or the concept is encoded, i.e., the robot learns what the demonstrator is trying to achieve, whereas in the trajectory level planning, the robot encodes the trajectory itself, provided during human demonstrations.

---

[1] A policy is a map between the state-space and the action-space.

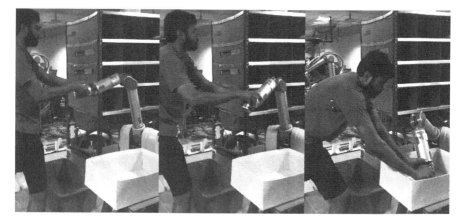

FIGURE 1.8: A robotic manipulator is given kinesthetic demonstrations for pick and place task.

It learns the nonlinear relationship between the sensory information and the motor action. LbD at trajectory level is intended to reproduce demonstrated trajectories as *similar* as possible. By the word *similar* it is meant that the pattern of the demonstrated trajectory is preserved in the reproduced trajectory. In both the cases, ultimately the intended task is accomplished. Therefore, LbD based motion planning is implemented in two stages:

- Collecting data from demonstrations
- Learning a model from the data

The robot collects the relevant data during the demonstrations. The data is generally recorded in the form of *state-action-pair*[2]. In the LbD paradigm, the demonstrations can be provided to the robot by the following ways [89]:

- **Kinesthetic**
  In kinesthetic teaching, the demonstrations are provided to the robot by physically holding the robotic arm while guiding through the intended trajectory to accomplish a task (see Figure 1.8). The main advantage of this approach is that each point in the demonstrated trajectory is associated with a joint space measurement. For instance, if the demonstrations involve end-effector trajectories, then the associated inverse kinematic solutions always exist.

---

[2]The *state-action-pair* is defined as the combination of the current state and commanded action. For example, suppose a robotic manipulator is demonstrated a picking task; the *state-action-pair* could be the position of the robot (joint/Cartesian space) and the velocity at each sampling interval during the demonstrations.

Learning by Demonstration 19

- **Teleoperation**
  Teleoperation is another technique of providing demonstrations to the robot. It is quite popular in medical/surgical robots. The demonstration is given to the robot by operating it using some joystick-like device for a particular task. However, demonstrations through teleoperation are not very accurate in the case of robots with high degrees of freedom.

- **Observation**
  Demonstrations through observation is the most used technique in human learning. Humans learn various limb movements by observing others. In robots, when a demonstrator provides demonstrations, the activities are recorded using a motion tracking system. The robot uses the data to learn the motion profiles.

We mainly focus on kinesthetic demonstrations as the motion tracking system is not required to record the data and these are more realistic than the teleoperated demonstrations.

The trajectories are encoded in nonlinear functions represented by regressive models (learned using Gaussian mixture regression, Gaussian processes, support vector regression, etc.). The parameters of these regressive models are learned from the demonstration data. These nonlinear functions are in fact referred to as *dynamical systems* (DS) when represented as differential equations, as they evolve through time to reproduce the robot's trajectory.

1.5.1 DS-Based Motion Learning

Classical approaches of motion planning for robots divide a particular task in two separate parts as *planning* and *execution* [90]. The *planning* part is responsible for exploring the robot's trajectory in the workspace and the *execution* part realizes the planned path as a trajectory tracking problem, which employs various control theories to minimize the tracking error. Using DS-based approach, these two parts are integrated in a single generalized model [91–94]. The tedious manual programming part is avoided as the DS actually encodes the demonstrations in the form of *movement primitives* (MPs). MPs are the building blocks of a motion learning system. In LbD approach of motion learning, MPs are identified by learning the parameters of the DS from demonstrations. Figure: 1.9 presents an example of such representation. A DS-based system is represented as follows:

$$\dot{\mathbf{x}} = \zeta(\mathbf{x}, \boldsymbol{\theta}) \tag{1.10}$$

$\mathbf{x}$ is the state of the system (for example, it could be position when the DS is modeled for reaching motion), $\boldsymbol{\theta}$ is the parameter and ζ learns the map between two spaces. Equation (1.10) is learned from demonstrations where the expert shows how to accomplish an assigned task. To learn the DS, optimization techniques are employed, which search for the appropriate parameters that fit the demonstrations.

FIGURE 1.9: This plot is an example of movement primitives in a demonstrations. The dots as a whole represent the demonstrated trajectory (could be in the joint or Cartesian space of a robotic a manipulator). All the trajectories end at the black '*'. The grey regions can be regarded as the movement primitives learned from the demonstrations. These movement primitives are learned in the form of dynamical systems which are locally linear and valid in their neighborhood. The weighted combination of these local linear systems represents complex motion profiles. The trajectory here is encoded using a Gaussian mixture model where the '+' in the grey regions represents mean of the Gaussian function.

DS-based systems are not guaranteed to be stable as they are learned in unconstrained optimization process from the demonstration data which includes sensor noise. An unstable system is never guaranteed to end up to the target / equilibrium state while unfolding in time. On the other hand, a globally stable DS is guaranteed to reach the the target state anywhere from the state space even if the trajectory is perturbed. Hence, like human, a globally stable DS-based motion model has the capability of providing spontaneous directional command to the robot to reach its goal anywhere in the reachable workspace. The main advantage of using a DS-based system is that it can be easily modulated as desired with resilience to perturbation and instant adaptability [95]. During the last decade, DS-based techniques have been found to be useful for many applications such as discrete motions [96–98], rhythmic motions [99–101], and hitting motions [102]. DS models are learned from multiple demonstrations to encode a specific behavior. Due to the modular nature of the DS-based approach, many formulations have been suggested in the recent past such as Vector Integration To Endpoint (VITE) model for arm reaching movements [103–105], Central Pattern Generators (CPGs) to model rhythmic behaviors [106–109], Reservoir Computing [110], and Recurrent Neural Network (RNN) [111–113].

In this context, another class of DS-based formulation is presented in [92, 114] which can be modulated as per the task requirement. The Dynamic Movement Primitive (DMP) is proposed to learn the DS from demonstration. DMP consists of a PD controller and a nonlinear term that actually captures

Stability of Nonlinear Systems   21

the features of the demonstration. DMP is quite fast as it learns from a single demonstration and is globally stable due to the PD controller. DMP has an explicit dependence on time (which is its own clock) which in fact controls the switching between the nonlinear term and the PD controller.

1.6 Stability of Nonlinear Systems

While designing a controller for a nonlinear system[3] such as a robot, the foremost priority is to analyze whether the overall system is stable or not. Any dynamical system is associated with equilibrium point/state(s). The behavior of the system in the neighborhood of the equilibrium state defines the system's stability.

Definition 1.1. *A state $\mathbf{x}^*$ is called as an equilibrium state of a dynamical system if once the state of the system $\mathbf{x}(t)$ reaches at $\mathbf{x}(t) = \mathbf{x}^*$, it remains there for all the future time.*

The equilibrium state of a dynamical system can be calculated by setting the state equation to zero at $\mathbf{x}(t) = \mathbf{x}^*$. We shall define a few terms those are defined to analyze the stability of the dynamical systems. These definitions are in frequent use in this thesis.

First, let us define a few notations: $\mathbf{B}_R$ represents the ball given by $\|\mathbf{x}\| < R$ in the state space defined by $\mathbf{S}_R$ with $\|\mathbf{x}\| = R$. The basic definition of stability is given as follows [115]:

Definition 1.2. *The equilibrium state $\mathbf{x} = 0$ is said to be* stable *in the sense of Lyapunov if, for any $R > 0$ there exists $r > 0$, such that if $\|\mathbf{x}(0)\| < r$, then $\|\mathbf{x}\|(t) < R$ for all $t \geq 0$. Otherwise the equilibrium point is* unstable.

Definition 1.2 essentially states that the equilibrium state is called stable if a state starting within the radius r, remains in the ball $\mathbf{B}_R$ with arbitrarily chosen R. However, for the kind of problem we are interested in, it requires the states of the system not only remain in a region, but also attain the desired value. This type of requirement is addressed by the concept of *asymptotic stability* which is defined as [115]:

Definition 1.3. *The equilibrium state $\mathbf{x} = 0$ is said to be* locally asymptotically stable *if it is stable, and if there exists $r > 0$, such that $\|\mathbf{x}(0)\| < r$ and eventually $\mathbf{x}(t) = 0$ as $t \to \infty$.*

Definition 1.3 states the asymptotic stability of a dynamical system locally in the state space. The global asymptotic stability of the dynamical system is given as follows:

---

[3]In this thesis we work with autonomous systems. Any system we mention here should be considered an autonomous system unless it is explicitly described otherwise.

Definition 1.4. *The equilibrium state* $\mathbf{x} = 0$ *is said to be* globally asymptotically stable *if the* asymptotic stability *holds for any initial state* $\mathbf{x}(0) = \mathbf{x}_0$. *It is also termed as* asymptotically stable in the large.

Lyapunov stability theorems are used in this thesis to determine the stability criteria of the learned DS and are given as follows [115]:

Theorem 1.1. *If there exists a scalar function* $V(\mathbf{x})$ *associated with* (1.10) *has continuous first order partial derivatives in a ball* $\mathbf{B}_R$ *and such that*

- $V(\mathbf{x})$ *is positive definite in* $\mathbf{B}_R$
- $\dot{V}(\mathbf{x})$ *is negative semi-definite in* $\mathbf{B}_R$

then the equilibrium point $\mathbf{x} = 0$ *is locally stable. If* $\dot{V}(\mathbf{x})$ *is negative definite in* $\mathbf{B}_R$, *then the equilibrium point is locally asymptotically stable.*

Theorem 1.2. *If there exists a scalar function* $V(\mathbf{x})$ *associated with* (1.10) *has continuous first order partial derivatives such that*

- $V(\mathbf{x})$ *is positive definite*
- $\dot{V}(\mathbf{x})$ *is negative definite*
- $V(\mathbf{x}) \to \infty$ *as* $\|\mathbf{x}\| \to \infty$

then the equilibrium point $\mathbf{x} = 0$ *is globally asymptotically stable.*

Theorem 1.2 essentially suggests that if the dynamical system is globally asymptotically stable, $\mathbf{B}_R$ spans over the entire state space. Let $V : S \to \Re$ is a continuously differentiable positive definite function in S where S contains the origin of the state space. We can note the following on stability of the system.

- If $\frac{\partial V}{\partial \mathbf{x}} \zeta$ is negative semi-definite in $\mathbf{B}_R$, then the equilibrium state of the system (1.10) is stable in $\mathbf{B}_R$.
- If $\frac{\partial V}{\partial \mathbf{x}} \zeta$ is negative definite in $\mathbf{B}_R$, then the equilibrium state of the system (1.10) is asymptotically stable in $\mathbf{B}_R$.
- If $\frac{\partial V}{\partial \mathbf{x}} \zeta$ is negative definite in $\mathbf{B}_R$ and $\mathbf{B}_R$ spans the entire state space, then the equilibrium state of the system (1.10) is globally asymptotically stable.

1.7 Optimization Techniques

Optimization has been used as an important tool in this thesis to learn the parameters of the data-driven models. The role of the optimization algorithm

Optimization Techniques

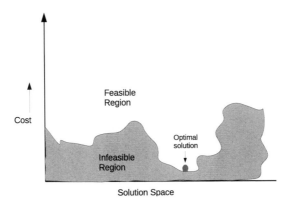

FIGURE 1.10: The optimization algorithm searches for a solution from the *white* region with minimum associated cost. The solutions from the *grey* region may promise a lower objective values but they are not useful as they do not satisfy the constraints. The black circle represents the global optimal solution in this case.

in the context is to minimize the error between the model prediction and the actual measurement by searching appropriate set of parameters in the model. As we have already discussed that an unconstrained search process does not guarantee stability of the models, the search space needs to be constrained by imposing the desired criteria. In general, an optimization problem is formulated as follows:

$$\underset{\mathbf{w}}{\text{minimize}} \quad P(\mathbf{w})$$
$$\text{subject to} \quad \mathcal{C}^c_{eqality}(\mathbf{w}) = 0 \quad c = 1, ..., n \quad (1.11)$$
$$\mathcal{C}^c_{ineqality}(\mathbf{w}) \leq 0 \quad c = 1, ..., m$$

where, $\mathbf{w}$ is called the design parameter and $P(\mathbf{w})$ is called the performance index or objective function of the system parameterized by $\mathbf{w}$. The intention is to find a $\mathbf{w}$ ($= \mathbf{w}_{optimal}$) that is associated with the minimum value of $P(\mathbf{w})$. $\mathcal{C}^c_{eqality}(\mathbf{w})$ and $\mathcal{C}^c_{ineqality}(\mathbf{w})$ represent the equality and inequality constraints respectively. The solution to this problem must satisfy these constraints. Figure 1.10 explains the constrained optimization where the solutions from the *grey* region do not satisfy the constraints in (1.11). The optimization algorithm is expected to find a solution from the *white* region of the solution space, associated with minimum value of $P(\mathbf{w})$.

In general optimization problems can be categorized as convex and non-convex problems. An optimization problem is convex when both its objective and the constraints are convex function. The characteristic of a convex optimization problem is if there exists an optimal solution to the problem, it is associated with the global optima [116]. But the same is not true about the non-convex optimization problems. A non-convex optimization problem is that, which has at least one non-convex function as the objective or as constraint. A non-convex optimization problem suffers from *local optima* problem and thereby making it difficult to solve [117].

Techniques have been developed to find the optimal solution of the non-convex problems. Interior-point methods, active-set techniques, sequential quadratic programming, etc., [118–120] are quite efficient and fast to find solutions which are locally optimal. Being local optimization techniques, these approaches end up as different solutions with different initial guesses. The chances of finding the global optima is high if the initial guess is close to the global optima. On the other hand genetic algorithm (GA) [121, 122], particle swarm optimization [123], and simulated annealing [124] reaches global optima (ideally) given a non-convex optimization problem. However, this achievement comes with expensive computation when the variable size is high. In this thesis work we mostly deal with non-convex constrained optimization problems as the work involves finding of appropriate means and covariances of nonlinear membership functions, matrices with certain sign, vectors having certain angles with some arbitrary vector, etc. GA has been frequently used in this thesis as it is good at handling such constraints while minimizing the objective value.

1.7.1 Genetic Algorithm

Genetic algorithm (GA) is a heuristic search approach applicable to a wide range of optimization problems. The algorithms have the capability of discovering the global minimum in an optimization problem if it is run for sufficient number of generations. The GA is motivated by the natural evolution of the living creatures, which makes them able to adapt the changing environment by creating complex structure over generations. Hence, evolution is the fundamental policy in GA. Mating of two different individuals and getting different offspring is the key to success of the natural evolution. GA also comprises similar steps. In this section we briefly discuss all these steps to give an overview of GA. A GA is driven by a few genetic operator such as crossover, mutation

---

Algorithm 1 Basic architecture of GA

1: Create a set of initial population
2: **for** Each generation **do**
3:    **for** Each chromosome **do**
4:       Perform crossover
5:       Perform mutation
6:       Compute fitness
7:    **end for**
8:    Select parent chromosomes for mating based on fitness
9: **end for**

Optimization Techniques

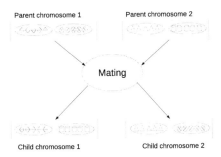

FIGURE 1.11: The creation of the offspring in a new generation is depicted here. The chromosomes in the new generation have the possibility to retain better fitness value than the previous generation as the genes are shuffled between parents.

which play the pivotal role in creating better offspring. Algorithm 1 presents the basic architecture of the GA.

Crossover: Crossover is a genetic operator that amalgamates the genes of two chromosomes from the parents in order to create new chromosome. In GA, a chromosome is represented by a bit string. In n-point crossover, n is selected randomly. The strings are split up in two segments at the n position and the segments are exchanged to create two new strings as shown in the Fig. 1.11. **Mutation:** Mutation is another genetic operator. It brings random changes in a chromosome with a hope that the change will give good fitness to the new chromosome. Mutation is performed based on a probability factor. This sometimes saves the algorithm from being stuck in the local minima. **Fitness:** The fitness of a chromosome is evaluated based on a fitness function. The fitness function defines the optimization problem. After the crossover and mutation, the new chromosome goes through the fitness test. The fitness tells how close the chromosome is to the optimal solution. **Selection:** The elite offsprings are selected based on the fitness values of the chromosomes. These elites are considered the parental population for the new generation.

Termination: The algorithm is terminated based on some termination criteria such as number of generations, minimum cost achieved, number of stall generations. The chromosome with the best fitness gives the solution to the optimal problem.

1.7.2 Expectation Maximization for Gaussian Mixture Model

Gaussian mixture model (GMM) is efficient in capturing the underlying distribution of real datasets as it uses linear superposition of Gaussian components [125]. Such superposition can be developed as probabilistic models known as *mixture distributions* [126, 127]. By tuning the means, covariances, and the priors in a GMM of sufficient number of Gaussian components, almost any continuous density can be estimated with arbitrary precision. The Gaussian mixture distribution is given by superimposing K Gaussian components and is represented as follows:

$$p(\mathbf{x}) = \sum_{k=1}^{K} \pi_k \mathcal{N}(\mathbf{x}|\boldsymbol{\mu_k}\boldsymbol{\Sigma_k}) \tag{1.12}$$

where, each $\mathcal{N}(\mathbf{x}|\boldsymbol{\mu_k}\boldsymbol{\Sigma_k})$ is the Gaussian component with their own mean $\boldsymbol{\mu_k}$ and covariance $\boldsymbol{\Sigma_k}$. The superposition of the components is parameterized by the prior (*mixing coefficient*) π_k. In order to capture the data distribution, the parameters $\boldsymbol{\mu_k}$, $\boldsymbol{\Sigma_k}$ and π_k need to be properly adjusted.

Expectation maximization (EM) is a sophisticated and powerful optimization technique to find maximum log-likelihood solutions for models with latent variables [128,129]. For a dataset $\mathbf{X} = [\mathbf{x}_1\ \mathbf{x}_2\ ...\ \mathbf{x}_N]^T$, $\mathbf{x}_i \in \Re^d$, with the corresponding latent variable $\mathbf{Z} = [\mathbf{z}_1\ \mathbf{z}_2\ ...\ \mathbf{z}_N]^T$, $\mathbf{z}_i \in \Re^K$, the log-likelihood associated with the Gaussian mixture distribution of K components is given by

$$\ln p(\mathbf{X}|\boldsymbol{\pi}, \boldsymbol{\mu}, \boldsymbol{\Sigma}) = \sum_{n=1}^{N} \ln \left[\sum_{k=1}^{K} \pi_k \mathcal{N}(\mathbf{x}_n|\boldsymbol{\mu_k}\boldsymbol{\Sigma_k}) \right] \tag{1.13}$$

where π_k is the mixing coefficient and $\boldsymbol{\mu_k}$ and $\boldsymbol{\Sigma_k}$ are the mean and covariance of the associated Gaussian. The expression for the mean and the covariance matrix can be obtained by applying the optimality condition to the log-likelihood and are given as follows:

$$\boldsymbol{\mu}_k = \frac{1}{N_k} \sum_{n=1}^{N} \gamma(z_{nk}) \mathbf{x}_n \tag{1.14}$$

$$\gamma(z_k) = \frac{\pi_k \mathcal{N}(\mathbf{x}_n|\boldsymbol{\mu_k}\boldsymbol{\Sigma_k})}{\sum_{j=1}^{K} \pi_j \mathcal{N}(\mathbf{x}_n|\boldsymbol{\mu_j}\boldsymbol{\Sigma_j})} \tag{1.15}$$

$$N_k = \sum_{n=1}^{N} \gamma(z_{nk}) \tag{1.16}$$

and

$$\boldsymbol{\Sigma}_k = \frac{1}{N_k} \sum_{n=1}^{N} \gamma(z_{nk}) (\mathbf{x}_n - \boldsymbol{\mu}_k)(\mathbf{x}_n - \boldsymbol{\mu}_k)^T \tag{1.17}$$

and finally,

$$\pi_k = \frac{N_k}{N} \tag{1.18}$$

Algorithm 2 describes the steps in the EM algorithm for finding the optimal set of parameter in GMM.

Algorithm 2 Steps in EM algorithm

1: Initialize $\boldsymbol{\mu}_k$, $\boldsymbol{\Sigma}_k$ and $\boldsymbol{\pi}_k$.
2: **while** EM has not converged **do**
3:    **E step:** Evaluate $\gamma(z_k)$ as given in (1.15).
4:    **M step:** Re-estimate the parameters $\boldsymbol{\mu}_k$, $\boldsymbol{\Sigma}_k$ and $\boldsymbol{\pi}_k$ (as given in (1.14),(1.17) and (1.16)) that maximize the log-likelihood.
5:    Compute the log-likelihood as given in (1.13)
6: **end while**

1.8 Composition of the Book

The remaining chapters are organized as follows. The kinematic and dynamic models of the robot manipulator used for experimentations are included in Chapter 2. Chapter 3 deals with the hand-eye coordination of a Robotic Arm using KSOM Network. The Model-based visual servoing of a 7 DOF manip- ulator is detailed in Chapter 4. Chapter 5 focuses on development of optimal redundancy resolution scheme for visually controlled redundant manipula- tor. Visual servoing using an adaptive distributed Takagi-Sugeno (T-S) fuzzy model has been detailed in Chapter 6. The kinematic control of the robotic manipulator using Single Network Adaptive Critic (SNAC) has been briefed in Chapter 7. Chapter 8 deals with the dynamic control of manipulators using SNAC. Imitation learning techniques and their application in robotic systems are detailed in Chapter 9. Chapter 10 deals with the deep learning-based visual perception techniques. Chapter 11 deals with visual grasping techniques. Warehouse automation, an experimental example for intelligent control of robotic systems, has been detailed in Chapter 12. Chapter 13 gives an introduction to the mobile robotics and control. The various multi-robot formation coordination and control techniques are given in Chapter 14. Chapter 15 deals with the event triggered-based multi-robot consensus. Various vision-based tracking algorithms for a human following mobile robot and their experimental demonstrations are detailed in Chapter 16.

Part I

Manipulators

2

Kinematic and Dynamic Models of Robot Manipulators

There are many available benchmark robot manipulators which are used in the laboratory for research purposes. Three such robot manipulators have been selected for experimentations. All appropriate control algorithms as presented in this book are tailormade for these three sets of manipulators. These are as follows:

- Seven degrees of freedom PowerCube manipulator from Schunk
- Six degrees of freedom Manipulator - UR10 - from Universal Robotics
- Seven degrees of freedom Manipulator - direct drive whole arm Manipulator (WAM) from Barrett

It is important that readers become familiar with these models in terms of both kinematics and dynamics before reading subsequent chapters.

The vision-based redundant manipulator control strategies as presented in Chapters 3 and 4 have been implemented on a 7 DOF PowerCube™ robot manipulator supplied by SCHUNK [130, 131], whose end-effector is visually seen through a stereo-vision setup fixed on the workspace. A brief introduction of this experimental set-up is given in this chapter for easier understanding of the simulation and the experimental results presented in the book. The kinematic model of this manipulator, along with the model and the image Jacobian of the stereo-vision setup is presented in this chapter as well.

The dynamic control using SNAC as presented in Chapter 8 uses the Barrett Arm. The complete dynamic model has been presented in this chapter as well. The vision-based picking and stowing using UR10 arm has been presented in Chapter 12. The kinematic model of this UR10 manipulator has been briefly described in this chapter as well.

2.1 PowerCube Manipulator

The 7 DOF PowerCube™ manipulator and the workspace comprising the stereo-vision set-up are shown in Figure 2.1. The end-effector of the manipulator is seen through the two Fire-i™ digital cameras [132] fixed in the

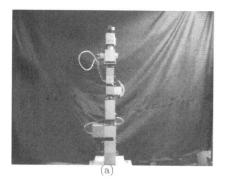 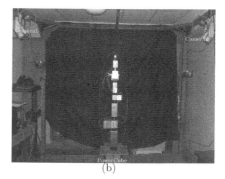

FIGURE 2.1: Experimental setup: (a) PowerCube™ Manipulator (b) Workspace with stereo vision

workspace. The cameras are located at the top corners of the Figure 2.1(b) and the two cameras are mounted such that a large workspace is available for real-time implementation. The end-effector is identified in vision space with a red tape wrapped around it, and the desired position is represented using a yellow ball during experimentation. The manipulator and the workspace are observed through the cameras with an image frame of dimension 320 × 240 pixels. The current position of the end-effector gets projected in the image plane and the regions of interest are extracted using image processing techniques such as thresholding and filtering. The centroid of the identified region is used to identify the current position of the end-effector and the desired position. The image processing and the learning-based control scheme are implemented on a PC (personal computer) with Intel Core 2 Duo E7300 CPU with 2.66 GHz clock and 4GB RAM. The computer is operated with Debian 4.02 operating system running in multi-user graphics mode with all the services enabled. The proposed learning-based schemes are to be tested in the Cartesian space visible in both the cameras. Typical workspace visible through stereo-vision is shown in Figure 2.2. A cubic volume is chosen within the workspace to learn the inverse kinematic solution, so that it will be easier to choose the desired position in real-time implementation.

2.2 Kinematic Configuration of the Manipulator

The coordinate frames of individual joint of the PowerCube manipulator is shown in Figure 2.3. The forward kinematic relationship of 7 DOF PowerCube manipulator is obtained from D-H parameters [14] given in Table 2.1, where

Kinematic Configuration of the Manipulator

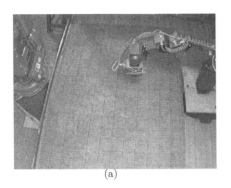
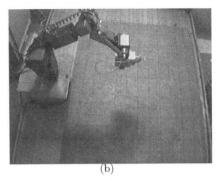

(a) (b)

FIGURE 2.2: View of the workspace from stereo-vision. The robot manipulator end-effector is identified with red tape (a) Left Camera (b) Right Camera.

TABLE 2.1: D-H Parameters of PowerCube™

link (i)	α_i	a_i	d_i	θ_i
1	-90^0	0	d_1	θ_1
2	90^0	0	0	θ_2
3	-90^0	0	d_3	θ_3
4	90^0	0	0	θ_4
5	-90^0	0	d_5	θ_5
6	-90^0	0	0	θ_6
7	180^0	0	d_7	θ_7

the dimensions of the manipulator links are: $d_1 = 0.368m$, $d_3 = 0.3815m$, $d_5 = 0.3085m$, and $d_7 = 0.2656m$.

D-H parameter computes the position of the end-effector with respect to the world coordinate frame, whose origin O_0 is located at the base of the manipulator. The end-effector position is obtained using the aforementioned D-H parameters as follows:

$$
\begin{aligned}
x = &-d_7((-(c_1c_2c_3 - s_1s_3)s_4 - c_1s_2c_4)c_6 - ((-c_1c_2s_3 - s_1c_3)s_5 \\
&+ ((c_1c_2c_3 - s_1s_3)c_4 - c_1s_2s_4)c_5)s_6) \\
&+ d_5((c_1c_2c_3 - s_1s_3)s_4 \\
&+ c_1s_2c_4) + d_3c_1s_2
\end{aligned}
$$

$$
\begin{aligned}
y = &-d_7((-(c_1s_3 + s_1c_2c_3)s_4 - s_1s_2c_4)c_6 \\
&- ((c_1c_3 - s_1c_2s_3)s_5 + ((c_1s_3 + s_1c_2c_3)c_4 - s_1s_2s_4)c_5)s_6) \\
&+ d_5((c_1s_3 + s_1c_2c_3)s_4 \\
&+ s_1s_2c_4) + d_3s_1s_2
\end{aligned}
$$

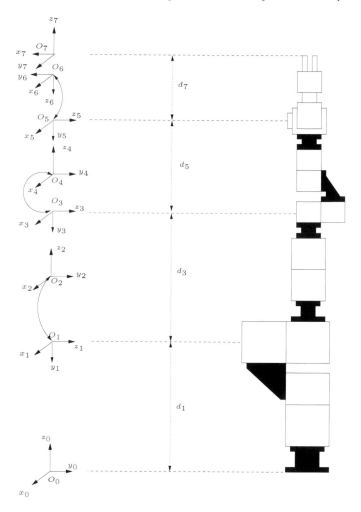

FIGURE 2.3: Assignment of DH frames for the UR10 robot associated to table above. Except for x_2 and x_3, all other x_i points inside the sheet.

$$\begin{aligned} z = & -d_7((s_2c_3s_4 - c_2c_4)c_6 - (s_2s_3s_5 + (-c_2s_4 - s_2c_3c_4)c_5)s_6) \\ & + d_5(c_2c_4 - s_2c_3s_4) \\ & + d_3c_2 + d_1 \end{aligned} \quad (2.1)$$

where $c_i = \cos\theta_i$, $s_i = \sin\theta_i$, $i = 1, 2, \ldots 6$. The end-effector position is independent of θ_7 since the seventh link generates roll motion for the manipulator

TABLE 2.2: Kinematic limits of the manipulator

Joint angle	Joint Velocity (rad/sec)
$-160° \leq \theta_1 \leq 160°$	$1.7e-5 \leq \dot{\theta}_1 \leq 2.618$
$-95° \leq \theta_2 \leq 95°$	$1.7e-5 \leq \dot{\theta}_2 \leq 2.618$
$-160° \leq \theta_3 \leq 160°$	$1.7e-5 \leq \dot{\theta}_3 \leq 2.618$
$-90° \leq \theta_4 \leq 90°$	$1.7e-5 \leq \dot{\theta}_4 \leq 2.618$
$-160° \leq \theta_5 \leq 160°$	$1.7e-5 \leq \dot{\theta}_5 \leq 2.618$
$-120° \leq \theta_6 \leq 120°$	$1.7e-5 \leq \dot{\theta}_6 \leq 4.189$
$-720° \leq \theta_7 \leq 720°$	$1.7e-5 \leq \dot{\theta}_7 \leq 6.283$

configuration. We mainly focus on the positioning the end-effector at a desired location. The orientation of the end-effector is not considered and, hence, $\mathbf{x} \in R^3$ in all the experiments is carried out in this thesis. Since θ_7 does not contribute to a change in the position, θ_7 will not be shown while discussing the experimental results and it is assumed as to be 0.

The physical kinematic limits of the manipulator are tabulated in Table 2.2. These limits constrain the implementation of kinematic control scheme and the control schemes can be physically realized only if they generate joint angle trajectories which satisfy the tabulated physical limits.

2.3 Estimating the Vision Space Motion with Camera Model

The positional coordinates of the end-effector in the Cartesian space get projected as pixel coordinates in the frame buffer of the image plane. The position $\mathbf{x} = [x \ y \ z]^T$ in the Cartesian space gets projected into the camera frame buffer as (x_f, y_f), which corresponds to the $x - y$ coordinates of the camera frame buffer respectively.

The positions of the end-effector in both the Cartesian and the vision space are used during learning phase and in simulations. The position of the end-effector in the vision space is obtained through series of transformations. These transformations are computed as a camera model, which computes the position of the point in the vision space, from the point's position in the Cartesian space. This necessitates a camera model to compute $\mathbf{u}$ from $\mathbf{x}$ in simulations.

2.3.1 Transformation from Cartesian Space to Vision Space

The transformation associated with computing a point's position in the vision from the Cartesian space is shown in Figure 2.4. The origin of the world coordinate frame and that of the camera coordinate frame are shown as O_w and O_c respectively. The origin of the camera coordinate frame is located at $[T_x\ T_y\ T_z]^T$ in the world coordinate frame and the orientation is represented using $\mathbf{R}_c$. The origin of the camera coordinate frame (x_c, y_c, z_c) coincides with front nodal point of the camera and the z_c axis coincides with the camera's optical axis. The image plane is assumed to be parallel to the (x_c, y_c) plane at a distance of f from the origin, where f is the effective focal length of the camera. The position (x_f, y_f) of a point P in the camera plane is obtained from the point's position $\mathbf{x} = [x\ y\ z]^T$ in the world coordinate frame as follows:

The position $\mathbf{x} = [x\ y\ z]^T$ is transformed from the world coordinate frame to the position $\mathbf{x}_c = [x_c\ y_c\ z_c]^T$ in the camera coordinate frame through rotation $\mathbf{R}_c$ and translation $\mathbf{T}_c$. The transformation is expressed in the form

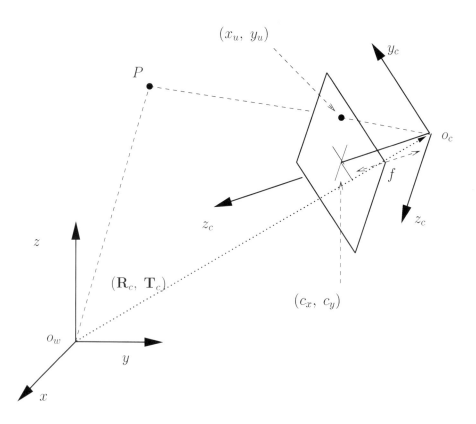

FIGURE 2.4: Transformation from Cartesian space to vision space.

of equation as,

$$\begin{bmatrix} x_c \\ y_c \\ z_c \end{bmatrix} = \mathbf{R}_c \begin{bmatrix} x \\ y \\ z \end{bmatrix} + \begin{bmatrix} T_x \\ T_y \\ T_z \end{bmatrix}, \quad \text{where } \mathbf{R}_c = \begin{bmatrix} r_1 & r_2 & r_3 \\ r_4 & r_5 & r_6 \\ r_7 & r_8 & r_9 \end{bmatrix}. \quad (2.2)$$

In the above equation, $\mathbf{R}_c$ describes the orientation of the camera in the world coordinate frame, and $\mathbf{T}_c = [T_x \ T_y \ T_z]^T$ is the translational position of the camera in the world coordinate frame. The projected position of the point P in the image plane is computed using an ideal pinhole camera model. This transformation is obtained by perspective projection as follows:

$$\begin{aligned} x_u &= f\frac{x_c}{z_c} \\ y_u &= f\frac{y_c}{z_c}. \end{aligned} \quad (2.3)$$

The position (x_u, y_u) is computed with the assumption of an ideal pinhole camera. But, there exists distortion and the position obtained with ideal pinhole model is not accurate. This is compensated by using the lens distortion coefficient κ. The true position of the point's image (x_d, y_d) in the sensor plane is computed from the ideal undistorted position as,

$$\begin{aligned} x_u &= x_d(1 + \kappa\rho^2) \\ y_u &= y_d(1 + \kappa\rho^2) \end{aligned} \quad (2.4)$$

where $\rho = \sqrt{x_d^2 + y_d^2}$. Finally the image of the point is transformed from the sensor plane to its coordinates in the camera's frame buffer (x_f, y_f) as,

$$\begin{aligned} x_f &= \frac{s_x x_d}{d_x} + c_x \\ y_f &= \frac{y_d}{d_y} + c_y \end{aligned} \quad (2.5)$$

where

c_x, c_y : Pixel coordinates of optical center;

s_x : Scale factor to account for any uncertainty due to imperfections in hardware timing for scanning and digitization;

d_x : Dimension of camera's sensor element along x coordinate direction (in mm/sel);

d_y : Dimension of camera's sensor element along y coordinate direction (in mm/sel).

The computation of a point's position in the frame buffer requires the geometric and camera parameters used in the aforementioned transformations.

The parameters which specify the position and the orientation of the camera relative to the world coordinate frame are commonly known as extrinsic or external parameters. The camera parameters which project the point from the camera coordinate frame to the frame buffer are known as intrinsic or internal parameters. The camera calibration is the process of estimation of a model for camera overlooking a workspace.

2.3.2 The Camera Model

Tsai's algorithm [133] is a popularly known camera calibration technique, and an online implementation of the Tsai calibration algorithm is proposed by R. Willson [134]. The Tsai model is based on the pinhole perspective projection discussed above and estimates eleven parameters: f, κ, c_x, c_y, s_x, T_x, T_y, T_z, R_x, R_y, and R_z. Tsai model represents the rotation angles for the transformation between the world and camera coordinates with (R_x, R_y, R_z). The elements of the rotation matrix $\mathbf{R}_c$ is computed from $[R_x, R_y, R_z]$ as follows:

$$\begin{aligned}
r_1 &= c_\beta c_\gamma \\
r_2 &= c_\gamma s_\alpha s_\beta - c_\alpha s_\gamma \\
r_3 &= s_\alpha s_\gamma + c_\alpha c_\gamma s_\beta \\
r_4 &= c_\beta s_\gamma \\
r_5 &= s_\alpha s_\beta s_\gamma + c_\alpha c_\gamma \\
r_6 &= c_\alpha s_\beta s_\gamma - c_\gamma s_\alpha \\
r_7 &= -s_\beta \\
r_8 &= c_\beta s_\alpha \\
r_9 &= c_\alpha c_\beta
\end{aligned} \quad (2.6)$$

where $c_\alpha = \cos(R_x)$, $c_\beta = \cos(R_y)$, $c_\gamma = \cos(R_z)$, $s_\alpha = \sin(R_x)$, $s_\beta = \sin(R_y)$, $s_\gamma = \sin(R_z)$.

In addition to the above eleven variable camera parameters, Tsai's model uses the following six fixed intrinsic camera constants:

d_x : Size of camera's sensor element in x coordinate direction (in mm/sel),

d_y : Size of camera's sensor element in y coordinate direction (in mm/sel),

N_{cx} : Number of sensor elements in camera's x direction (in sels),

N_{fx} : Number of pixels in frame grabber's x direction (in pixels),

d_{px} : Effective number of pixel in y coordinate direction of the frame buffer (in mm/pixel), and

d_{py} : Effective number of pixel in y coordinate direction of the frame buffer (in mm/pixel).

These six parameters can be obtained from the manufacturer's data sheet.

The chess board based calibration algorithm available in OpenCV [135] is used to obtain the data points over the workspace seen through the stereo-vision setup.

The image feature vector $\mathbf{u}$ is obtained using the estimated model for stereo-vision as $\mathbf{u} = (u_1\ u_2\ u_3\ u_4)^T$ where (u_1, u_2) and (u_3, u_4) are the $x-y$ coordinates of the first and the second camera respectively. Hence, (u_1, u_2) is the (x_f, y_f) of the first camera, and (u_3, u_4) corresponds to the (x_f, y_f) of the second camera respectively. Hence, the control vectors $\mathbf{u}_d$ and $\mathbf{u}$ belong to R^4 in the all the experiments presented in this thesis.

2.3.3 Computation of Image Feature Velocity in the Vision Space

The image Jacobian which represents the motion of the image features with respect to the motion in the Cartesian space is given by,

$$\mathbf{L} = \begin{bmatrix} \frac{k_{cx}}{z_c} & 0 & \frac{-k_{cx}(x_f - c_x)}{z_c} \\ 0 & \frac{k_{cy}}{z_c} & \frac{-k_{cy}(y_f - c_y)}{z_c} \end{bmatrix} \quad (2.7)$$

where (k_{cx}, k_{cy}) are the gains associated to transform the Cartesian space position to the x-y coordinate of the vision space, and z_c is the distance between the image plane and the object in the camera coordinate frame. The camera gains (k_{cx}, k_{cy}) are computed using the camera parameters as follows:

$$k_{cx} = \frac{f s_x}{d_x}$$

$$k_{cx} = \frac{f}{d_y} \cdot \quad (2.8)$$

The vision space velocity is computed from the Cartesian space velocity as,

$$\begin{bmatrix} \dot{u}_1 \\ \dot{u}_2 \\ \dot{u}_3 \\ \dot{u}_4 \end{bmatrix} = \begin{bmatrix} \mathbf{L}_1 & 0 \\ 0 & \mathbf{L}_2 \end{bmatrix} \begin{bmatrix} \dot{\mathbf{x}}_{c_1} \\ \dot{\mathbf{x}}_{c_2} \end{bmatrix} \quad (2.9)$$

where $\mathbf{L}_i$ is the image Jacobian of the ith camera, $\dot{\mathbf{x}}_{c_i} = \begin{bmatrix} \dot{x}_{c_i} & \dot{y}_{c_i} & \dot{z}_{c_i} \end{bmatrix}^T$ represents the velocity of the end-effector in the coordinate frame of the ith camera.

The end-effector velocity in the coordinate frame of the ith camera is computed as,

$$\begin{bmatrix} \dot{x}_{c_i} \\ \dot{y}_{c_i} \\ \dot{z}_{c_i} \end{bmatrix} = \mathbf{R}_{c_i} \begin{bmatrix} \dot{x} \\ \dot{y} \\ \dot{z} \end{bmatrix}$$

$$\dot{\mathbf{x}}_{c_i} = \mathbf{R}_{c_i}\, \dot{\mathbf{x}} \quad (2.10)$$

where $\mathbf{R}_{c_i}$ is the rotational transformation between the robot coordinate frame and the camera coordinate frame. The parameters (k_{cx}, k_{cy}), (c_x, c_y), and $\mathbf{R}_c$ are obtained from the camera model estimated with Tsai algorithm.

2.4 Learning-Based Controller Architecture

The kinematic and camera model described above are used during the training phase of the proposed learning-based control methodologies, and the learned controller is then tested in both simulations and real-time experiments. The schematic of a typical vision-based manipulator control scheme in learning paradigm is shown in Figure 2.5. It consists of a stereo-vision system and a

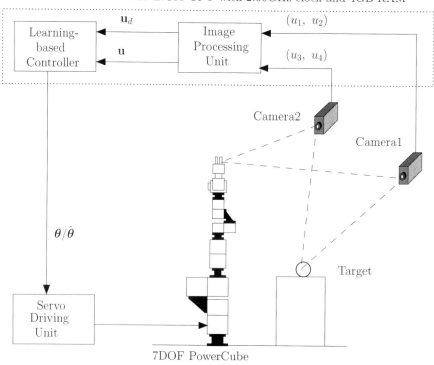

FIGURE 2.5: Schematic of visual servo control i) u_1, u_2, u_3, u_4 : Camera coordinates seen through stereo-vision system ii) $\mathbf{u}_d$: Desired position (object) iii) $\mathbf{u}$: Current position of end-effector iv) Control input $\boldsymbol{\theta}$ - joint angle, $\dot{\boldsymbol{\theta}}$ - joint angular velocity.

robot manipulator. Image processing as well as the learning-based controller are executed on a personal computer. The image processing unit is used to extract 4-dimensional image coordinate vectors to represent the current end-effector position $\mathbf{u}$, and the desired position $\mathbf{u}_d$. The learning-based controller generates either the joint angle $\boldsymbol{\theta}$ or the joint angular velocity $\dot{\boldsymbol{\theta}}$, which is given to the servo unit to drive the robot manipulator so that the end-effector reaches the desired position.

2.5 Universal Robot (UR 10)

Amazon Robotics Challenge 2017:

ARC' 17 posed a simplified version of the task that humans face in warehouses across the globe, namely, stowing items from tote into a storage system and then picking items from storage system and putting those items into Amazon packing boxes. Each team was asked to design a fully autonomous robot to perform such task. A set of forty items was provided, referred to as `known-set`. In addition, a set of novel items was also provided before forty-five minutes from the start of task. Each task involved picking or stowing of a set of items referred as `competition-set`, having equal numbers of known and novel items, and have a maximum physical volume of $95,000$ cm^3, and *ii*) Design a visual perception system which can perform object recognition in the presence of cluttered known and novel items. Sixteen teams were selected worldwide for this challenge. In the stow task, twenty items were provided, all of which had to be stowed into the designed storage system. In the pick task, thirty-two items were given and ten of them had to be picked and placed in a tote. Top eight teams, based on the combined performance in previous tasks, were selected for the final stow-pick task in which a total of thirty-two items was provided. In this task, the teams were required to first perform a stow task and then a pick task. This task was relatively challenging as the errors of the stow task could propagate to the pick task. Apart from this, an item dropping from above a specified height or any item protruding from the storage system by more than 2.5 cm was penalized. Moreover, reporting incorrect location of the items in the storage systems also contributed to penalty.

2.5.1 Mechatronic Design

2.5.1.1 Platform

Our robot platform setup shown in Figure 2.6 consists of a UR10 robot manipulator with its controller box/internal computer and a host PC/external computer. The UR10 robot manipulator is a 6 DOF robot arm designed to safely work alongside and in collaboration with a human. This arm can

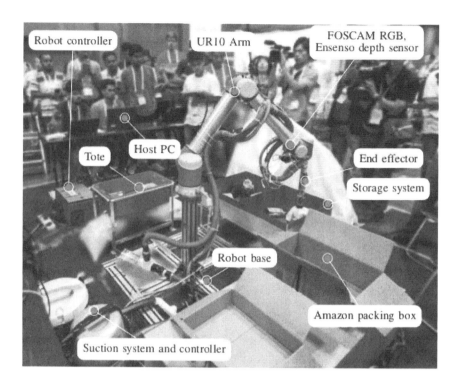

FIGURE 2.6: Our robotic system workspace for ARC'17. Image courtesy: Amazon Robotics.

follow position commands like a traditional industrial robot, as well as take velocity commands to apply a given velocity in/around a specified axis. The low level robot controller is a program running on UR10's internal computer broadcasting robot arm data, receiving and interpreting the commands and controlling the arm accordingly. There are several options for communicating with the robot low level controller to control the robot including the teach pendent or opening a TCP socket (C++/Python) on a host computer. We used open source C++ based UrDriver wrapper class integrated with ROS on a host PC (Intel i7 processor with 16 GB of system RAM) to implement our proposed velocity based kinematic control scheme. The host PC streams joint velocity commands via URScript to the robot real-time interface over Ethernet at $125Hz$. The driver was configured with necessary parameters like IP address of the robot at startup using ROS parameter server.

Universal Robot (UR 10) 43

2.5.1.2 End-Effector

Amazon provided a large variety of items most of which could be grasped using suction, few of which were deformable, transparent, book, etc. Therefore, our end-effector design is suction based grasp. It has a rectangular suction hose followed by a nozzle which can rotate between 0° to 90°. The nozzle tip contains a bellow and its angle is governed by a linear actuator. The suction hose and nozzle are connected through a flexible rubber tube. The two vacuum cleaners were employed to generate required suction to grasp an item. A custom designed bleed valve with linear actuation was used to ensure instant release of a grasped item, otherwise the item remains attached to the end-effector until the suction pressure drops entirely. Furthermore, a manifold air pressure sensor, is also inserted near the vacuum cleaners to reflect the air pressure as an analog voltage which is converted to a digital value and provides feedback for a firm grasp. To close the loop for sensing grasped items, we used flow meter reading to sense pressure difference and force-torque sensor mounted on end-effector (`wrist-3-link` for UR10) for force feedback.

2.5.1.3 Perception Apparatus

We used eye-in-hand approach, i.e., the vision hardware consisting of RGB-D Ensenso camera with a HD Foscam camera was mounted on the manipulator itself in contrast to the other teams which used the vision hardware externally. This offered us an advantage of an extremely simplified system with minimal external components while relegating the need of complex external sensor calibration procedures. However, for the proper realization of the eye-in-hand approach, the vision and manipulator system should be calibrated. We achieved this by developing a semi-autonomous procedure based on which the system can self-calibrate itself while requiring minimal human effort.

2.5.2 Kinematic Model

TABLE 2.3: DH parameters (in mm or rad), with the value of $\theta \in \mathbb{R}^6$ in the shown configuration below

Link i	α_i	a_i	d_i	θ_i
1	$-\frac{\pi}{2}$	0	$d_1 = 128$	$\theta_1 = 0$
2	0	$a_2 = -612.7$	0	$\theta_2 = \frac{\pi}{2}$
3	0	$a_3 = -571.6$	0	$\theta_3 = 0$
4	$\frac{\pi}{2}$	0	$d_4 = 163.9$	$\theta_4 = -\frac{\pi}{2}$
5	$-\frac{\pi}{2}$	0	$d_5 = 115.7$	$\theta_5 = 0$
6	0	0	$d_5 = 92.2$	$\theta_6 = 0$

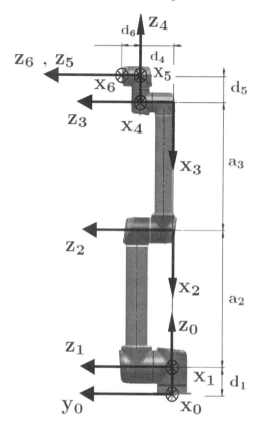

FIGURE 2.7: Assignment of DH frames for the UR10 robot associated to table above. Except for x_2 and x_3, all other x_i points inside the sheet.

The homogeneous transformation matrix (arm matrix) T_{0n} for an n DOF manipulator, which represents the final position and orientation of end-effector with respect to the base coordinate system, can be obtained by chain product of successive coordinate transformation matrices using standard Denavit–Hartenberg(D-H) parameters [14]. Let $T_{i1,i}$ for $i = 1, 2...., n$ be the transformation matrices between successive arms, the final arm matrix can be expressed as $T_{0n} = \prod_{i=0}^{n} T_{i1,i}$.

Jacobian of a Robotic Structure, which can be derived directly from the relation between joint positions and end-effector position, is the mapping between velocities in each coordinate system. It is a very useful relation especially in kinematic control. The velocity space is easier to operate in when we want to determine the inverse kinematics iteratively. In practical terms this implies what end-effector velocities will occur, relative to the base-frame, corresponding to certain joint velocities. This relationship is established through the *Jacobian* [14].

Consider forward kinematics of an n DOF robot manipulator with task workspace configuration consisting of position and orientation with respect to base coordinate system and designated as $\mathbf{X} \in \mathbb{R}^{N_w}$, where N_w represents number of independent variables in robot task space. We represent a manipulator joint configuration with variable $\theta \in \mathbb{R}^{N_c}$, where N_c represents number of independent variables in robot joint configuration space.

The kinematic relationship of a robotic structure can be represented by following sets of equations.

$$x_1 = f_1(\theta_1, \theta_2, ...\theta_{N_c})$$
$$x_1 = f_2(\theta_1, \theta_2, ...\theta_{N_c})$$
$$\vdots$$
$$x_{N_w} = f_{N_w}(\theta_1, \theta_2, ...\theta_{N_c})$$

The forward kinematic model for the manipulator for a given joint configuration $\widehat{\theta}$ can be written as,

$$\mathbf{X} = \mathbf{f}(\widehat{\theta}) \quad (2.11)$$

Here, $\widehat{\theta} \in \mathbb{R}^{N_c}$ is the vector of joint configuration involved in the forward kinematic model and $\mathbf{X} \in \mathbb{R}^{N_w}$ is the vector of position and orientation with respect to base coordinate system.

The derivative of $\mathbf{X}$ w.r.t. time t is

$$\dot{\mathbf{X}} = \frac{\partial \mathbf{f}(\widehat{\theta})}{\partial \widehat{\theta}} \dot{\widehat{\theta}} = \mathbf{J}\dot{\widehat{\theta}} \quad (2.12)$$

where, $\mathbf{J} = \frac{\partial \mathbf{f}(\widehat{\theta})}{\partial \widehat{\theta}}$ is the $N_w \times N_c$ manipulator Jacobian matrix. Formally, the Jacobian is a set of partial differential equations - a multidimensional form of a derivative. We can split the Jacobian into a linear velocity contribution $\mathbf{J_v}$ and an angular velocity contribution $\mathbf{J_v}$ [14].

$$\mathbf{J} = \begin{bmatrix} \mathbf{J_v} \\ \mathbf{J_w} \end{bmatrix} \quad (2.13)$$

2.6 Barrett Wam Manipulator

2.6.1 Overview of the System

The proposed dynamic motion generation module along with the developed control scheme to generate motor skills for novel situation was demonstrated using a basic ball hitting experiment. We employed a 4 DOF Barrett WAM

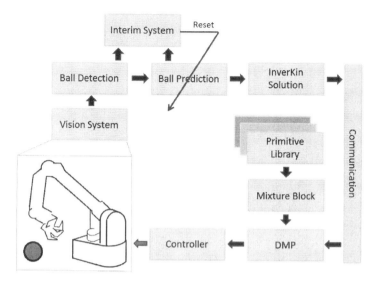

FIGURE 2.8: System overview.

robot manipulator for the same. The ball was suspended from a fixed height with a string. Vision system consisted of two off the shelf Basler's cameras. The ball is detected using standard segmentation techniques and an extended kalman filter (EKF) was employed to predict the ball interception point. Since at the ball interception point there is a discrete jump in the dynamics of the moving ball, an interim system is employed to take care of this hybrid dynamics and ensure the convergence of the EKF. Once the ball prediction touches the ball interception plane the goal parameters $(\theta_g, \dot{\theta}_g)$ corresponding to the predicted ball interception point is passed to the trajectory generation module in real time. Then a novel trajectory is generated by mixing the a priori demonstrated primitives in the library. This trajectory is then fed to the control scheme for accurate and stable execution. Once the ball is returned the follow up DMP trajectory from the ball interception point to the home position is joined with the striking trajectory and and the robot returns to the home position. Execution foe of the entire system is depicted in Figure (2.8). In the next section the background of DMPs and FSMC is discussed.

2.6.2 Experimental Setup

Experimental hardware setup consists of a 4 DOF Barrett Wam robot manipulator, a hanging ball, and two off the shelf Basler's $acA800 - 550uc$ cameras at 200 frames per second (FPS) and $800X600$ resolution. The ball is detected using standard segmentation techniques and EKF was employed to predict the

Barrett Wam Manipulator

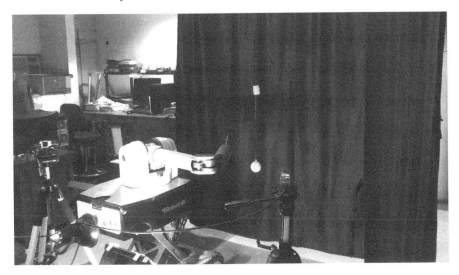

FIGURE 2.9: Hardware setup consisting of two off the shelf cameras, a hanging ball and 4 DOF Barrett Wam Robot Manipulator.

ball interception point. Since at the ball interception point there is a discrete jump in the dynamics of the moving ball, an interim system is employed to take care of this hybrid dynamics and ensure the convergence of the EKF. There are two phases of the robot motions, hitting phase and follow up phase. In the hitting phase, the robot starts from a fixed home position to the ball interception point and then the follow up phase proceeds in which robot returns from the ball interception point to the home position. The robot motions are initiated in accordance with the different phases and dynamic goal parameters (e.g. ball position and velocity at the interception point).

2.6.3 Dynamic Modeling

Precise model-based control necessitates the requirement of an accurate model of the robotic system. This section focuses on the development of a dynamic model for an $n-$ DOF operation of the Barrett Whole Arm Manipulator (WAM). We use the recursive Newton-Euler technique to achieve the same. A detailed description of the parameters required for implementing the Newton-Euler algorithm is presented next.

The Newton-Euler formulation is based on three important laws of mechanics:

- Every action has an equal and opposite reaction. Thus, if link i exerts a force f and a torque τ on link $i+1$, then link $i+1$ in turn exerts a force $-f$ and a torque $-\tau$ on link i.

- The rate of change of linear momentum equals the total force applied to the link.

- The rate of change of angular momentum equals the total torque applied to the link.

Based on these basic principles, the governing equations for this technique can be derived [14]. We present the main steps of implementation of the Newton-Euler algorithm.

Algorithm 3 Newton-Euler Algorithm

Forward Recursion : Computing ω_i, α_i and $a_{c,i}$
Initial Conditions: $\omega_0 = 0$, $\alpha_0 = 0$, $a_{c,0} = 0$ and $a_{e,0} = 0$
for each $i := 1$ to n **do**
  $\omega_i \leftarrow R_{i-1}^i \omega_{i-1} + b_i \dot{q}_i$; where $b_i = R_{i-1}^i z_0$
  $\alpha_i \leftarrow R_{i-1}^i \alpha_{i-1} + b_i \ddot{q}_i + \omega_i \times b_i \dot{q}_i$
  $a_{e,i} \leftarrow R_{i-1}^i a_{e,i-1} + \dot{\omega}_i \times r_{i,i+1} + \omega_i \times (\omega_i \times r_{i,i+1})$
  $a_{c,i} \leftarrow R_{i-1}^i a_{e,i-1} + \dot{\omega}_i \times r_{i,c_i} + \omega_i \times (\omega_i \times r_{i,c_i})$
end for
Backward Recursion : Computing f_i and τ_i
Terminal Conditions: $f_{n+1} = 0, \tau_{n+1} = 0$
for each $j := n$ to 1 **do**
  $g_i \leftarrow R_0^i g_0$
  $f_i \leftarrow R_{i+1}^i f_{i+1} + m_i a_{c,i} - m_i g_i$
  $\tau_i \leftarrow R_{i+1}^i \tau_{i+1} - f_i \times r_{i,c_i} + \left(R_{i+1}^i f_{i+1}\right) \times r_{i+1,c_i} + I_i \alpha_i + \omega_i \times (I_i \omega_i)$
end for

As can be seen from Algorithm 3, there are two key steps in the Newton-Euler technique: The Forward Recursion, which involves the computation of the angular velocity, angular acceleration and linear acceleration of each link, starting from the first link and moving outwards, and the Backward Recursion, which involves computation of the forces and torques at each link, starting from the n-th link and moving inwards. Figure 2.10 shows a random link and the forces and torques acting on it.

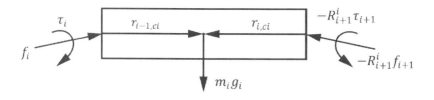

FIGURE 2.10: Forces and torques acting on a random link [14].

Barrett Wam Manipulator 49

Notation : The notation used is described as follows:

- $a_{c,i}$ — Acceleration of center of mass of link i in frame i.
- $a_{e,i}$ — Acceleration of end of link i in frame i.
- ω_i — Angular velocity of frame i w.r.t frame i.
- α_i — Angular acceleration of frame i w.r.t frame i.
- g_i — Acceleration due to gravity in frame i.
- f_i — Force exerted by link $i-1$ on link i in frame i.
- τ_i — Torque exerted by link $i-1$ on link i in frame i.
- R_i^{i+1} — Rotation matrix from frame $i+1$ to frame i.
- m_i — Mass of link i.
- I_i — Inertia matrix of link i, about a frame parallel to frame i, whose origin is at the center of mass of link i.
- r_{i,c_i} — Vector from joint i to the center of mass of link i.
- r_{i+1,c_i} — Vector from joint $i+1$ to the center of mass of link i.
- $r_{i,i+1}$ — Vector from joint i to joint $i+1$.

2.6.4 System Description and Modeling

Barrett, the leader in advanced robotic manipulators, provides information about the D-H parameters of the WAM and also about its inertial specifications in its data sheets. However, a dynamic model of the WAM is not disclosed. In our work, we have derived a dynamic model of the Barrett WAM for 4 degree-of-freedom operation. The parameters required for obtaining this model are detailed in this section. Schematics of the Barrett WAM with its seven revolute joints are shown in Figure 2.11 (copyright has been obtained from Barrett). The necessary D-H parameters of the Barrett WAM are provided in Table 2.4.

For obtaining the necessary rotation matrices, we use the generalized D-H transform matrix

$$\mathbf{T}_i^{i-1} = \begin{bmatrix} c\theta_i & -s\theta_i c\alpha_i & s\theta_i s\alpha_i & a_i c\theta_i \\ s\theta_i & c\theta_i c\alpha_i & -c\theta_i s\alpha_i & a_i s\theta_i \\ 0 & s\alpha_i & c\alpha_i & d_i \\ 0 & 0 & 0 & 1 \end{bmatrix} \qquad (2.14)$$

TABLE 2.4: D-H Parameters for the 4-DOF Barrett WAM Manipulator

k	a_k (m)	α_k (rad)	d_k (m)	θ_k
1	0	$-\pi/2$	0	θ_1
2	0	$\pi/2$	0	θ_2
3	0.045	$-\pi/2$	0.55	θ_3
4	-0.045	$\pi/2$	0	θ_4

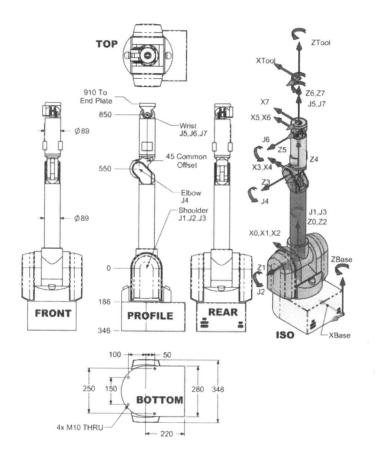

FIGURE 2.11: WAM 7-DOF dimensions and D-H frames.

It should be noted that we follow the convention of [14] in our work. c denotes the cos function and s denotes the sin function. The lengths in the D-H specifications are in meters.

Using equation (2.14) and the D-H parameters of the WAM, we derive the following rotation matrices

$$\mathbf{R}_1^0 = \begin{bmatrix} c\theta_1 & 0 & -s\theta_1 \\ s\theta_1 & 0 & c\theta_1 \\ 0 & -1 & 0 \end{bmatrix} \qquad (2.15)$$

$$\mathbf{R}_2^1 = \begin{bmatrix} c\theta_2 & 0 & s\theta_2 \\ s\theta_2 & 0 & -c\theta_2 \\ 0 & 1 & 0 \end{bmatrix} \qquad (2.16)$$

Barrett Wam Manipulator

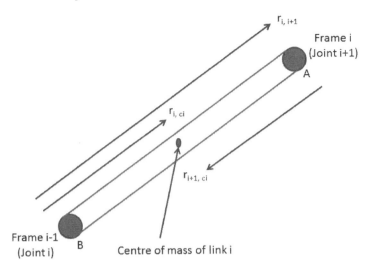

FIGURE 2.12: Vectors associated with link i.

$$\mathbf{R}_3^2 = \begin{bmatrix} c\theta_3 & 0 & -s\theta_3 \\ s\theta_3 & 0 & c\theta_3 \\ 0 & -1 & 0 \end{bmatrix} \quad (2.17)$$

$$\mathbf{R}_4^3 = \begin{bmatrix} c\theta_4 & 0 & s\theta_4 \\ s\theta_4 & 0 & -c\theta_4 \\ 0 & 1 & 0 \end{bmatrix} \quad (2.18)$$

Figure 2.12 shows the different vectors associated with link i. Specifically, we need the link vectors $r^i_{i,i+1}, r^i_{i,c_i}$ and r^i_{i+1,c_i} for $i = 1, 2, 3$ and 4. The superscript i indicates that the vectors need to be expressed in frame i. From the definition of transformation matrices, we know that the vector $\mathbf{v} = [a_i c\theta_i, a_i s\theta_i, d_i]^T$ represents the vector pointing from Joint i to Joint $i+1$ expressed in frame $i-1$. We need to express this vector in frame i. From Figure 2.12, we can see that

$$r^i_{i,i+1} = P^i_A - P^i_B, \quad (2.19)$$

where P^i_A and P^i_B represent the position vectors of points A and B in frame i. Now, we know that any random point P^{i-1}_R in frame $i-1$ is transformed to frame i via the relation

$$\begin{bmatrix} P^i_R \\ 1 \end{bmatrix} = \mathbf{T}^i_{i-1} \begin{bmatrix} P^{i-1}_R \\ 1 \end{bmatrix} \quad (2.20)$$

Now, we have

$$\mathbf{T}^i_{i-1} = (\mathbf{T}^{i-1}_i)^{-1} = \left[\begin{array}{c|c} \mathbf{R}^T & -\mathbf{R}^T \mathbf{v} \\ \hline \mathbf{0} & 1 \end{array} \right], \quad (2.21)$$

TABLE 2.5: Link Vectors

i	r^i_{i+1,c_i} (in mm)	r^i_{i,c_i} (in mm)	$r^i_{i,i+1}$ (in mm)
1	$[0.3506, 132.6795, 0.6286]^T$	$[0.3506, 132.6795, 0.6286]^T$	$[0, 0, 0]^T$
2	$[-0.223, -21.3924, 13.3754]^T$	$[-0.223, -21.3924, 13.3754]^T$	$[0, 0, 0]^T$
3	$[-38.7565, 217.9078, 0.0252]^T$	$[6.2435, -332.0922, 0.0252]^T$	$[45, -550, 0]^T$
4	$[6.2895, -0.001, 111.0633]^T$	$[-38.7105, -0.001, 111.0633]^T$	$[-45, 0, 0]^T$

where $\mathbf{R} = \mathbf{R}_i^{i-1}$. Here, we have used the property that rotation matrices are orthogonal. Clearly, $P_B^{i-1} = [0, 0, 0]^T$ and $P_A^{i-1} = \mathbf{v}$. Using equations (2.20) and (2.21), we get

$$\begin{bmatrix} P_B^i \\ 1 \end{bmatrix} = \begin{bmatrix} \mathbf{R}^T & -\mathbf{R}^T\mathbf{v} \\ \mathbf{0} & 1 \end{bmatrix} \begin{bmatrix} 0 \\ 0 \\ 0 \\ 1 \end{bmatrix} = \begin{bmatrix} -\mathbf{R}^T\mathbf{v} \\ 1 \end{bmatrix} \quad (2.22)$$

and

$$\begin{bmatrix} P_A^i \\ 1 \end{bmatrix} = \begin{bmatrix} \mathbf{R}^T & -\mathbf{R}^T\mathbf{v} \\ \mathbf{0} & 1 \end{bmatrix} \begin{bmatrix} \mathbf{v} \\ 1 \end{bmatrix} = \begin{bmatrix} \mathbf{0} \\ 1 \end{bmatrix} \quad (2.23)$$

Using equations (2.19), (2.22) and (2.23), we get the desired result

$$r^i_{i,i+1} = \mathbf{R}^T \mathbf{v} = [a_i, d_i s\alpha_i, d_i c\alpha_i]^T \quad (2.24)$$

The inertial specifications of the WAM provides the r^i_{i+1,c_i} vector, i.e., the position vector of the center of mass of link i w.r.t to frame i. From Figure 2.12, we see that the following relation holds

$$r^i_{i,c_i} = r^i_{i,i+1} + r^i_{i+1,c_i} \quad (2.25)$$

Equations (2.24) and (2.25) can be used to find the necessary link vectors. Using these equations, the link vectors so obtained are tabulated in Table 2.5. The link masses are - $m_1 = 8.3936$ kg, $m_2 = 4.8487$ kg, $m_3 = 1.7251$ kg and $m_4 = 1.0912$ kg. It should be noted that the masses of the electrical and mechanical cables are not included in the inertial specifications. This data is obtained from the inertial specifications of the Barrett WAM. We also obtain the following inertia matrices from the same source-

$$I_1 = 10^{-6} \begin{bmatrix} 95157.4294 & 246.1404 & -95.0183 \\ 246.1404 & 92032.3524 & -962.6725 \\ -95.0183 & -962.6725 & 59290.5997 \end{bmatrix} \quad (2.26)$$

$$I_2 = 10^{-6} \begin{bmatrix} 29326.8098 & -43.3994 & -129.2942 \\ -43.3994 & 20781.5826 & 1348.6924 \\ -129.2942 & 1348.6924 & 22807.3271 \end{bmatrix} \quad (2.27)$$

$$I_3 = 10^{-6} \begin{bmatrix} 56662.2970 & -2321.6892 & 8.2125 \\ -2321.6892 & 3158.0509 & -16.6307 \\ 8.2125 & -16.6307 & 56806.6024 \end{bmatrix} \quad (2.28)$$

$$I_4 = 10^{-6} \begin{bmatrix} 18890.7885 & -0.8092 & -1721.2915 \\ -0.8092 & 19340.5969 & 17.8241 \\ -1721.2915 & 17.8241 & 2026.8453 \end{bmatrix} \quad (2.29)$$

It should be noted that the unit of each entry of the above matrices is $kg\text{-}mm^2$. This completes the necessary system description. Armed with this data, one can implement Algorithm 3. Two other points need to be mentioned in this regard. For forward recursion, the vector $z_0 = [0,0,1]^T$ and for backward recursion, the vector $g_0 = [0,0,-g]^T$, where g is the acceleration due to gravity. These results follow directly from the way in which the frames are assigned.

The implementation of the Newton-Euler algorithm has been done using the software Maple. The choice was justified by the ability of Maple to carry out heavy symbolic calculations. The Maple code used for deriving the model can be accessed through the following link: *https://drive.google.com/file/d/0B1SCfVjLdPjZekFMLXJpUnRJTzg/view?usp=sharing*. The M, C and G matrices necessary for state-space representation are also derived using this code. For testing our control laws in a simulation environment, we have used the Matlab platform. Thus, the model derived in Maple has been imported to Matlab. The Maple to Matlab conversion has also been demonstrated in the Maple code. It should be noted that this code can be used to derive rigid body models for a generalized n-link manipulator with revolute joints.

Note

- The D-H specifications can be obtained from the following link:
  http://www.me.unm.edu/~starr/research/WAM_UsersGuide_AE-00.pdf.

- The Barrett Arm Inertial Specifications are available from the following *https://www.cs.rpi.edu/twiki/pub/RoboticsWeb/WamTrackingSystem/Arm_InertiaSpecifications.pdf*.

2.6.5 State Space Representation

The standard model representing the dynamics of the robotic system as obtained via the Newton Euler technique is

$$\mathbf{M(q)\ddot{q}} + \mathbf{C(q,\dot{q})} + \mathbf{G(q)} = \boldsymbol{\tau} \quad (2.30)$$

where $\mathbf{q} \in \Re^4$ represents the joint position vector, $\dot{\mathbf{q}} \in \Re^4$ is the joint velocity vector, $\ddot{\mathbf{q}} \in \Re^4$ is the joint acceleration vector, $\mathbf{M(q)} \in \Re^{4\times 4}$ is the symmetric, positive definite inertia matrix, $\mathbf{C(q,\dot{q})} \in \Re^4$ is the Coriolis and Centrifugal vector and $\mathbf{G(q)} \in \Re^4$ is the Gravity vector. $\boldsymbol{\tau}$ represents the vector of applied joint torques to the system.

For applying the control techniques, we need to express the model given by (2.30) in the standard nonlinear control affine form. To this end, we define the following

$$\mathbf{x} \triangleq [q_1, q_2, q_3, q_4]^T$$
$$\mathbf{z} \triangleq [\dot{q}_1, \dot{q}_2, \dot{q}_3, \dot{q}_4]^T$$
$$\mathbf{f}(\mathbf{x}, \mathbf{z}) \triangleq -\mathbf{M}^{-1}(\mathbf{C} + \mathbf{G})$$
$$\mathbf{g}(\mathbf{x}) \triangleq \mathbf{M}^{-1}$$
$$\mathbf{u} \triangleq \boldsymbol{\tau}$$

Based on these notations, the state space model for (2.30) becomes

$$\begin{aligned} \dot{\mathbf{x}} &= \mathbf{z} \\ \dot{\mathbf{z}} &= \mathbf{f}(\mathbf{x}, \mathbf{z}) + \mathbf{g}(\mathbf{x})\mathbf{u} \end{aligned} \quad (2.31)$$

2.7 Summary

Models of PoweCUBE 7 DOF manipulator, Barrett 7 DOF WAM, and UR10 6 DOF manipulator have been introduced. The integration of camera with the system and the corresponding kinematics have been presented. The experimental setup along with these manipulators have been introduced in this chapter.

3

Hand-eye Coordination of a Robotic Arm using KSOM Network

Hand-eye coordination is a process by which biological organisms manipulate objects of interests. Some of the interesting examples of hand-eye coordination are as follows:

- Eagles spot their potential prey from a very long distance and catch the prey with amazing visual control of their speed.

- The great football player Diego Maradona had the amazing ability to pass the ball with accuracy; he had the almost perfect control of the ball as he dribbled past multiple opposing players on a run; and his amazing reaction time to score a goal given an opportunity astounded every one. The key here is again the eye-leg coordination - the precise visual control that he exerted.

- The sand artist Mr. Patel depicts the pastimes of Lord Krishna on the sand by dexterous manipulation of his fingers.

- Such hand eye coordinations are visible when one plays computer games, instruments, and when one is typing or cooking.

In a human brain there are billions of motor neurons that actuate muscles which in turn helps one to manipulate one's hands or legs through visual feedback. We cannot possibly make a mathematical model of human visual motor control mechanisms. However, persons like Maradona and Sudarshan Patnaik have excelled in this through learning and practice. This is the motivation that drives us to present this chapter where we will show many examples of robotic systems that learn to manipulate using visual feedback. We call this as visual motor coordination.

Although humans deal with billions degrees of freedom, a robot manipulator has a considerably lower number of degrees of freedom. A redundant manipulator has a minimum of seven degrees of freedom. To manipulate an object, a robot needs three degrees of freedom in Cartesian task space and three degrees of freedom for orientation - pitch, yaw and roll. With six degrees of freedom, a robot arm can manipulate an object properly. But if one increases this by one, i.e., a seven-degrees of freedom robot manipulator becomes redundant because this can reach a reachable target in theoretically infinite possible kinematic configurations. Thus with redundancy, comes the challenge of dealing with infinite choices. This chapter will explain the process of learning hand-eye coordination using Kohonen's Self Organizing Map for robots with different degrees of freedom. We will start with a simple 2-d planar manipulator to illustrate the learning principle lucidly. Then we will deal with a seven-degrees of freedom Robot Manipulator.

3.1 Kohonen Self Organizing Map

Kohonen [136] proposed an unsupervised learning algorithm that can form clusters for a given data set while preserving topology. A simple configuration of Kohonen self-organizing feature map is illustrated in Fig. 3.1(a). The prominent feature of this network is a lattice that can be m dimensional. Although the dimension of the lattice is a priori fixed, this dimension usually refers to the topology of the real-world data. Another prominent feature is the concept of excitatory learning with a neighborhood around the winning neuron. The size of the neighborhood slowly decreases as learning progresses as shown in Figure 3.1(a). To be precise, in the initial phase, almost all neurons participate in the learning as the network is excited by an input pattern x. But there

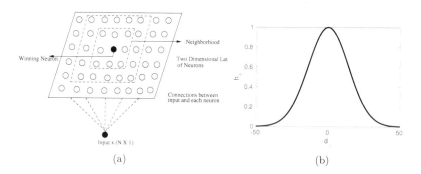

(a) (b)

FIGURE 3.1: (a) A two dimensional self organizing feature map. By updating all the weight connecting to a neighborhood of the target neurons, it enables the neighboring neuron to become more responsive to the same input pattern. Consequently, the correlation between neighboring nodes can be enhanced. Once such a correlation is established, the size of a neighborhood can be decreased gradually based on the desire for having a stronger identity of individual nodes. (b) The Neighborhood Function h_γ. This function value decreases as the lattice distance d_γ of the γth neuron from the winning neuron increases.

Kohonen Self Organizing Map

is a unique winning neuron associated with this input pattern x, which will have a maximum say in the decision making. Other neurons will contribute to the decision making according to their lattice distance d_γ from the winning neuron. This function is shown in Figure 3.1(a)(b).

The basic idea is to discover patterns in the input data in a self-organizing way while similar data are represented by a weight vector w_γ associated with the γth neuron. This clustering takes place in following three steps:

- Competition: For each input pattern, the neurons in the network compute their respective values of a discriminant function. The neuron with the largest value of that function is declared the winner. This discriminant function is usually a measure of Euclidean distance.

- Cooperation: The winning neuron determines the spatial location of a topological neighborhood of excited neurons, i.e., cooperative neighboring neurons.

- Synaptic Adaptation: The excited neurons which are situated in the neighborhood of the winning neuron adjust their synaptic weights in relation to the input pattern.

3.1.1 Competitive Process

Let n be the dimension of the input (data) space and weight vector. Let a randomly chosen input pattern (vector) be

$$x = [x_1, x_2, ..., x_n]^T$$

Let the synaptic weight vector of neuron γ be denoted by

$$w_\gamma = [w_{\gamma 1}, w_{\gamma 2}, ..., w_{\gamma n}]^T, \quad \gamma = 1, 2, ..., N$$

where N=total number of neurons in the network.

Finding the best match of the input vector x with the synaptic weight vectors w_γ is mathematically equivalent to minimizing the Euclidean distance between the vectors x and w_j.

Let $i(x)$ = index to identify the neuron that best matches x,

$$i(x) = \arg \min_\gamma ||x - w_\gamma||, \quad \gamma = 1, 2, ..., N \tag{3.1}$$

3.1.2 Cooperative Process

The winner neuron tends to excite the neurons in its immediate neighborhood more than those farther away from it. Let h_γ denote the topological neighborhood centered on winning neuron $i(x)$ and d_γ denote the lattice distance between winning neuron $i(x)$ and the excited neuron γ.

- The topological neighborhood h_γ is symmetric about the maximum point defined by $d_\gamma = 0$. In other words, it attains its maximum value at the winning neuron $i(x)$ for which the distance d_γ is zero. For the winning neuron $\gamma = i(x)$.

- The amplitude of the topological neighborhood h_γ decreases monotonically with increasing lattice distance d_γ.

A typical choice of h_γ that satisfies these requirements is the Gaussian function as shown in Figure 3.1(a)(b). The expression of a Gaussian neighborhood function is given as:

$$h_\gamma = \exp\left(-\frac{d_\gamma^2}{2\sigma^2}\right), \quad (3.2)$$

where σ is the width of neighborhood function. This width is varied in such a manner so that all neurons participate in the weight update process in the beginning and the width significantly reduced as the training gets completed.

3.1.3 Adaptive Process

Weights associated with the winning neuron and its neighbors are updated as per a neighborhood index h_γ. The winning neuron is allowed to be maximally benefited from this weight update while the neuron that is farthest from the winner is minimally benefited. The Kohonen law by which weights are updated is given as

$$w_\gamma = w_\gamma + \eta_w h_\gamma (x - w_\gamma) \quad (3.3)$$

where η_w is the learning rate. The width of the neighborhood function σ and the learning rate η_w are updated as:

$$\eta = \eta_i \left(\frac{\eta_f}{\eta_i}\right)^{t/t_{max}} \quad (3.4)$$

where $\eta \in \{\eta_w, \sigma\}$. The width of the neighborhood σ is kept usually large initially. This would imply that all neurons in the lattice will be covered by the neighborhood function h_γ in the beginning, allowing all neurons to get excited. They will all participate in the decision making. As learning progresses, this width gets reduced until the neighborhood of the winning neuron shrinks substantially. The learning rate η_w is assigned a large value - usually 1.0 - in the beginning. This value decreases as the learning progresses almost to zero. This implies that once the learning is over, this parameter becomes inactive.

Kohonen Self Organizing Map

Example 3.1. $1 - D$ SOM learns $2 - D$ topology: Select a 1-d lattice and excite the neurons with the data coming from a $2 - d$ plane. Show that the network preserves the topology of the data.

Solution 3.1. In the simulation a neural network is chosen with 100 neurons organized in one dimensional lattice. The network is trained with a two-dimensional input vector x.

- Input data are generated randomly from a $2 - D$ topology.
- Since each data point is two-dimensional, $x = [x_1 \ x_2]^T$, where x_1 represent x coordinate, x_2 represent y coordinate
- w_γ associated with each neuron is also two-dimensional.

The training is done for 6,000 iterations.

$$n = 2$$
$$x = [x_1, x_2]^T$$
$$w_\gamma = [w_\gamma, w_\gamma]^T; \ \gamma = 1, 2, ..., 100$$

The weight vectors of the network are initialized from a random set ($-0.04 < w_{j,1} < 0.04$ and $-0.04 < w_{j,2} < 0.04$). The input x is uniformly distributed in the region ($0 < x_1 < 1$ and $0 < x_2 < 1$). Figure 3.2.a shows the input data space. Figure 3.2.b shows the network weights before training and Figure 3.2.c shows that the weights of the network preserve the topology of the input space. As can be seen in this Figure 3.2, although initial weights have no correlation with the input space, the final weight vector as plotted shows that the data are coming from a $2 - d$ space although neurons are assigned to a $1 - d$ lattice. This example illustrates that KSOM network learns the clusters in the data while preserving the topology.

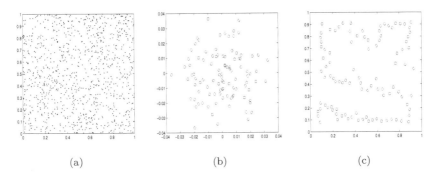

FIGURE 3.2: (a)Input space for training; (b)Initialization of weight for training; (c) Weights after the completion of training

3.2 System Identification using KSOM

We learned how a Kohonen self-organizing map works. In this section, we will learn how this network can be used for learning any arbitrary map $f : x \to y$. That is, given data $\{x, y\}$, we can build a neural architecture around KSOM that will learn the unknown map $f()$. The KSOM network that learns this map $f(.)$ is shown in Figure 3.3.

Let's assume that the following nonlinear map is given:

$$y = f(x), \quad x \in \Re^n, \quad y \in \Re^m$$

We will express this nonlinear function as aggregation of linear functions using first order Taylor series expansion. Given any input vector x_0,

$$y_0 = f(x_0)$$

Using first order Taylor series expansion, the output y can be expressed linearly around x_0 as follows:

$$y = y_0 + \frac{\partial f}{\partial x}\big|_{x=x_0}(x - x_0)$$

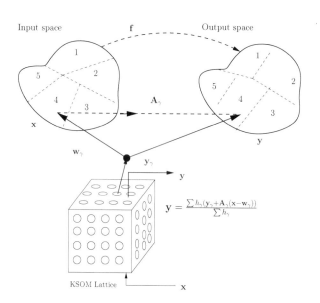

FIGURE 3.3: KSOM network for system identification.

System Identification using KSOM

Let's consider the following Kohonen lattice where each neuron is associated with the following linear model:

$$y^\gamma = y_\gamma + A_\gamma(x - w_\gamma)$$

where given x, y^γ is the linear response of the γth neuron. This neuron is associated with three parameters: w_γ, the natural weight vector; y_γ which should converge to $f(w_\gamma)$; and A_γ which is equivalent of $\frac{\partial f}{\partial x}|_{x=w_\gamma}$.

The linear response of each neuron given x has a weight of h_γ where h_γ is the neighborhood function with respect to the winning neuron. Thus the nonlinear map $y = f(x)$ can be approximated as:

$$y = \frac{\sum h_\gamma y^\gamma}{\sum h_\gamma}$$

where $h_\gamma = e^{-d_\gamma^2/2\sigma^2}$ and d_γ is the lattice distance between the winning neuron i and the γth neuron.

The final expression for the network response can be given as:

$$y = \frac{\sum h_\gamma(y_\gamma + A_\gamma(x - w_\gamma))}{\sum h_\gamma}$$

As shown in Figure 3.3, the network has a collective response y when excited by the input pattern x where each neuron computes its own response linearly. The readers must know that parameters associated with each neuron w_γ, A_γ and y_γ are unknown and are randomly initialized with very small values. We will now derive the update laws for these parameters. Given that w_γ is the natural weight vector, its update will follow the same Kohonen weight update algorithm:

$$w_\gamma = w_\gamma + \eta h_\gamma(x - w_\gamma) \tag{3.5}$$

Let the cost function be $E = \frac{1}{2}\tilde{y}^T\tilde{y}$, $\tilde{y} = y^d - y$ and y^d is the desired response given x while y is the network response. The update law for the y_γ can be derived using gradient descent:

$$\frac{\partial E}{\partial y_\gamma} = -\tilde{y}^T \frac{\partial y}{\partial y_\gamma}$$

$$= -\tilde{y}^T \left(\frac{h_\gamma}{\sum h_\gamma}\right)$$

Thus the update law for y_γ becomes:

$$y_\gamma \leftarrow y_\gamma + \eta \left(\frac{h_\gamma}{\sum h_\gamma}\right)\tilde{y} \tag{3.6}$$

For the update law of A_γ, the gradient term is derived as:

$$\frac{\partial E}{\partial A_\gamma} = -\tilde{y}^T \frac{\partial y}{\partial A_\gamma}$$

$$= -\tilde{y}^T \frac{h_\gamma}{\sum h_\gamma}(x - w_\gamma)$$

$$= -\frac{h_\gamma}{\sum h_\gamma}(x - w_\gamma)\tilde{y}^T$$

Thus the update law becomes:

$$A_\gamma \leftarrow A_\gamma + \eta \tilde{y}(x - w)^T \left(\frac{h_\gamma}{\sum h_\gamma}\right) \quad (3.7)$$

It is important to learn that KSOM based system identification makes use of both unsupervised and supervised learning, which we will call a type of hybrid learning. The following example will demonstrate this idea.

Example 3.2. *Let's consider the following map:*

$$y_1 = e^{x_1^2 + x_2^2} \quad (3.8)$$

$$y_2 = e^{\|x_1 + x_2\|^2} \quad (3.9)$$

Generate input data $x = [x_1 \; x_2]^T$ uniformly distributed in [0,1]. Compute the corresponding output $y = [y_1 \; y_2]^T$. Take a $2-d$ lattice of size 5×5. Update weights w_γ, y_γ and A_γ as given in equations (3.5), (3.6) and (3.7) respectively.

Solution 3.2. *The KSOM network has 25 sets of parameters - each set of parameters associated with each neuron is w_γ, y_γ and A_γ. These parameters are initially uniformly randomly distributed in $[0, 1]$. The network is excited by x which is uniformly randomly generated in [0,1]. Using the corresponding desired response y^d and the network response y, weights are updated.*

The plot of the functional map is given in Figure 3.4.

Five hundred training data sets are generated from this map and these data are used to train the network over 200 epochs. The error convergence over epochs is shown in Figure 3.5(a). During the testing, the input data is generated as $x_1 = 0.5 + 0.5 * \cos(\frac{k\pi}{40})$ and $x_2 = 0.5 + 0.5 * \sin(\frac{k\pi}{40})$. The test

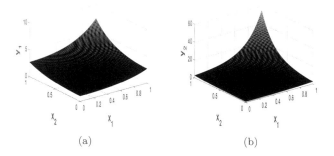

FIGURE 3.4: (a) The plot of y_1 versus input x; (b) The plot of y_2 versus x.

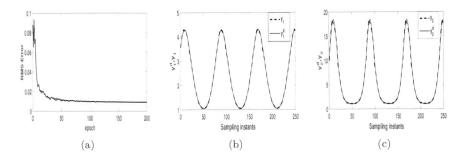

FIGURE 3.5: (a) Error convergence during the training; (b) y_1 versus y_1^d during the testing; (c) y_2 versus y_2^d during the testing.

results y^d versus y are plotted in Figures 3.5 (b) and (c) respectively. One can see from these figures that the actual network response is very accurately following the desired response. The rms tracking error for y_1 and y_2 are 3.5.b and 3.5.c respectively. These results confirm that the KSOM network can be used for learning any unknown map.

```
MATLAB CODE FOR SYSTEM IDENTIFICATION
-------------------------------------
%% Definitions
clc; clear all; close all; format long;
n=100; % No. of dataset
Xt = rand(n,2); nip=size(Xt,2);
for i=1:n
    Ydt(i,1)=exp((Xt(i,1)^2)+(Xt(i,2)^2));
    Ydt(i,2)=exp((Xt(i,1)+Xt(i,2))^2);
end
nop=size(Ydt,2); N=5; %No. of neurons/side in the 2-
    ↪ Dlattice
Ylambda=0.1*rand(nop,N,N); Alambda=rand(nop,nip,N,N);
Wlambda=0.1*rand(nip,N,N); Ylambdanew=rand(nop,N,N);
Alambdanew=rand(nop,nip,N,N); Wlambdanew=rand(nip,N,N);
% Parameters
etah0=10; tauh=0.70; etaw0=0.1; tauw=5;
etaa0=2; taua=50; etay0=2; tauy=50;

%% KSOM Training
N_epoch=20;
for epoch=1:N_epoch
etah=etah0*exp(-epoch/tauh); etaw=etaw0*exp(-epoch/tauw);
etaa=etaa0*exp(-epoch/taua); etay=etay0*exp(-epoch/tauy);
RDPM=randperm(n,n);
```

```
for shuffle=1:n
Xtr(shuffle,:)=Xt(RDPM(shuffle),:);
Ydtr(shuffle,:)=Ydt(RDPM(shuffle),:);
end
for ndata=1:n
    a=100;
    for i=1:N
        for j=1:N
            Dx(i,j) = norm(Xtr(ndata,:)'-Wlambda(:,i,j));
            if(a>Dx(i,j))
                a=Dx(i,j);
                ai=i;
                aj=j;
            end
        end
    end
    for i=1:N
        for j=1:N
            H(i,j) = exp(-((ai-i)^2+(aj-j)^2)/(2*etah*etah));
        end
    end
    s=sum(sum(H));
    Ytemp=zeros(2,1);
    for i=1:N
        for j=1:N
Y(:,i,j)=H(i,j)*(Ylambda(:,i,j)+Alambda(:,:,i,j)
*((Xtr(ndata,:)')-Wlambda(:,i,j)));
            Ytemp=Ytemp+Y(:,i,j);
        end
    end
Ypred(ndata,:)=Ytemp/s; Ytilde(ndata,:)=Ydtr(ndata,:)
-Ypred(ndata,:);
    for i=1:N
        for j=1:N
Ylambda(:,i,j)=Ylambda(:,i,j)+etay*H(i,j)
*(Ytilde(ndata,:)')/s;
Alambda(:,:,i,j)=Alambda(:,:,i,j)+(1/s)*etaa*H(i,j)
*(Ytilde(ndata,:)')*(((Xtr(ndata,:)')-Wlambda(:,i,j))');
Wlambda(:,i,j)=Wlambda(:,i,j)+etaw*H(i,j)
*((Xtr(ndata,:)'-Wlambda(:,i,j)));
        end
    end
end
Error(epoch)=norm(Ytilde)/(size(Ytilde,1)*size(Ytilde,2));
end
%% Error Plot
figure(1)
plot(1:N_epoch,Error,'LineWidth',2);
```

System Identification using KSOM

```
%% Testing
Xte=zeros(1,1); Ydte=zeros(1,1); Y=zeros(nop,N,N); n2 =
    ↪ 250;
for i=1:n2
    Xte(i,1)=0.5*cos(i*pi/40)+0.5;
    Xte(i,2)=0.5*sin(i*pi/40)+0.5;
end
for i=1:n2
    Ydte(i,1)=exp((Xte(i,1)^2)+(Xte(i,2)^2));
    Ydte(i,2)=exp((Xte(i,1)+Xte(i,2))^2);
end
Ypredte=zeros(n2,2);
for ndata=1:n2
    a=100;
    for i=1:N
        for j=1:N
            Dx(i,j) = norm(Xte(ndata,:)'-Wlambda(:,i,j));
            if(a>Dx(i,j))
                a=Dx(i,j);
                ai=i;
                aj=j;
            end
        end
    end
    for i=1:N
       for j=1:N
      H(i,j) = exp(-((ai-i)^2+(aj-j)^2)/(2*etah*etah));
        end
    end
    s=sum(sum(H)); Ytemp=zeros(nop,1);
    for i=1:N
        for j=1:N
Y(:,i,j)=H(i,j)*(Ylambda(:,i,j)+Alambda(:,:,i,j)
*((Xte(ndata,:)')-Wlambda(:,i,j)));
Ytemp=Ytemp+Y(:,i,j);
        end
    end
Ypredte(ndata,:)=(Ytemp/s)'; Ytilde(ndata,:)
=Ydte(ndata,:)-Ypredte(ndata,:);
end

figure(5)
plot(1:n2,Ypredte(:,1),'--','Color',[0,0,0],'LineWidth',4);
hold on;
plot(1:n2,Ydte(:,1),'Color',[0,0,0],'LineWidth',2);

figure(6)
plot(1:n2,Ypredte(:,2),'--','Color',[0,0,0],'LineWidth',4);
hold on;
```

```
plot(1:n2,Ydte(:,2),'Color',[0,0,0],'LineWidth',2);

%% Function Approx Plot
X1funt=0:0.01:1; X2funt=0:0.01:1;
for i=1:size(X1funt,2)
    for j=1:size(X2funt,2)
        Y1funt(i,j)= exp((X1funt(i)^2)+(X2funt(j)^2));
        Y2funt(i,j)= exp((X1funt(i)+X2funt(j))^2);
    end
end
figure(7); surf(X1funt,X2funt,Y1funt)
figure(8); surf(X1funt,X2funt,Y2funt)
```

3.3 Introduction to Learning-Based Inverse Kinematic Control

We will learn how to use KSOM network to learn inverse kinematics of a robot manipulator. The human arm has 7 degrees of freedom (DOF). The five fingers have 21 degrees of freedom. If we consider two arms with two hands, then we are talking of 56 degrees of freedom. These degrees of freedom help a sculpturer to give an aesthetic shape to a statue. Although with degrees of freedom, the dexterity of manipulation enhances as choices in terms of kinematic configurations also increase manifold times. A sculpturer is in general ignorant of scientific characterization of forward and inverse kinematics, but he is expert in manipulation simply through learning. This section will teach you how to learn inverse kinematics of a robot manipulator. To make it easy for the learners, we will start with a planar two link manipulator. Referring to Figure 3.6, it has two links of link lengths l_1 and l_2 respectively. The angle of the first link with respect to the horizontal x axis is θ_1 and the angle of the second link with respect to the link 1 axis is θ_2 as shown in the figure. From the figure, we can derive the following forward kinematics:

$$x = l_1 cos(\theta_1) + l_2 cos(\theta_1 + \theta_2); \quad y = l_1 sin(\theta_1) + l_2 sin(\theta_1 + \theta_2) \qquad (3.10)$$

In general, for a robot manipulator, it is very easy to find out the forward kinematics. Given a forward kinematic equation, we can find out the tip position of the robot manipulator in the Cartesian space if we are given joint angles of each link. However, in practice, the robot has to know its own joint angles given a target defined in the Cartesian space. It is usual for a robot to reach a target, i.e., the desired tip position is known and the robot has to find out its own joint angles. For the above planar manipulator, let's find out the

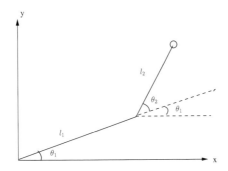

FIGURE 3.6: Two link planar manipulator.

inverse kinematics:

$$cos\theta_2 = \frac{x^2 + y^2 - l_1^2 - l_2^2}{2l_1 l_2} \qquad (3.11)$$

$$\theta_2 = cos^{-1}\frac{x^2 + y^2 - l_1^2 - l_2^2}{2l_1 l_2} \qquad (3.12)$$

$$\theta_1 = tan^{-1}\frac{y}{x} - tan^{-1}\frac{l_2 sin\theta_2}{l_1 + l_2 cos\theta_2} \qquad (3.13)$$

You can note that although forward kinematics has simple expressions, the inverse kinematics is not that simple. In fact, as the degrees of freedom will increase, the inverse kinematic solution cannot be obtained in closed form equations as derived above. We will show you how to learn this inverse kinematic solutions using KSOM network.

In general the forward kinematics is given as

$$x = f(\theta); \quad x \in R^m; \theta \in R^n \qquad (3.14)$$

The inverse kinematic relation can be expressed as

$$\theta = g(x); \quad x \in R^m; \theta \in R^n; g = f^{-1} \qquad (3.15)$$

The above inverse kinematic equation can be linearized around θ_0 using first order Taylor series expansion:

$$\theta = \theta_0 + A_0(x - x_0)$$

where $A_0 = \frac{g}{x}|_{x=x_0}$ which is called inverse Jacobian at x_0. It should be noted that $\theta_0 = g(x_0)$.

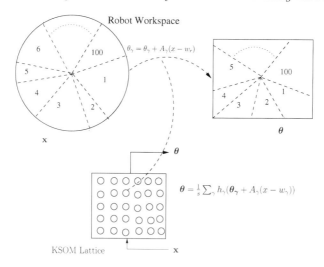

FIGURE 3.7: KSOM network for kinematic control of 2-d planar manipulator.

3.3.1 The Network

Please refer to the KSOM network as shown in Fig 3.7. Here the input is the Cartesian space coordinate vector x. Since it is a planar manipulator, the 2-d lattice of size 10×10 has been taken. It is assumed that $l_1 = 1m; l_2 = 1m$. If you look at the planar manipulator, you will see that its workspace covers a circle of radius $2m$. The lattice has 100 neurons and each neuron represents a discrete cell in this circle of 2m radius. We associate a linear model with each of these neurons:

$$\theta = \theta_\gamma + A_\gamma(x - w_\gamma)$$

where w_γ is the weight associated with the γth neuron where $\gamma \in [1, 100]$. A_γ is the inverse Jacobian at $x = w_\gamma$. $\theta_\gamma = g(w_\gamma)$ is the local joint angle vector around which the above liner model is valid.

Just like in the previous section, the network response to input x is given in terms of collective response model:

$$\theta = \frac{1}{s} \sum_\gamma h_\gamma(\theta_\gamma + A_\gamma(x - w_\gamma)) \qquad (3.16)$$

where $s = \frac{1}{\sum_\gamma h_\gamma}$. This is a normalizing factor. As said earlier, each neuron contributes according to its lattice distance from the winning neuron - this is represented by the neighborhood function h_γ.

Introduction to Learning-Based Inverse Kinematic Control 69

3.3.2 The Learning Problem

In the previous section on system identification, we are given the vector pairs $\{x, y\}$, i.e., input vector x and desired vector y. It is desired to learn the network parameters given both input and output vectors. But in the inverse kinematic learning problem, we are given only the input vector x, the target at which the manipulator is desired to reach. With reference to equation (3.16), all parameters θ_γ, A_γ, w_γ are unknown. Since there are 100 neurons as shown in Figure, there are 300 sets of parameters that have to be learned based on a given set of random targets within the circular robot workspace of 2m radius. This is even more difficult than the previous system identification problem.

3.3.3 The Approach

We need to generate error corrector terms to be able to update the parameters. In the beginning, all 100 sets of parameters - θ_γ, A_γ, w_γ are randomly initialized and these random values are made as small as possible to test the efficacy of the update algorithms. The coarse action is initiated first. This action is the network response with the current set of network parameters. Given x, the network response is

$$\theta_0^{out} = \frac{1}{s} \sum_\gamma h_\gamma (\theta_\gamma + A_\gamma (x - w_\gamma)) \tag{3.17}$$

where θ_0^{out} is the coarse action that is actuated to take the robot to some point in Cartesian space. If we feed these angles to equation (3.10), we will get $v_0 = [x \ y]^T$. This course action θ_0^{out} will take the robot to Cartesian point in the task space v_0. This pair θ_0^{out}, v_0 will help us to generate the error corrector term. But this v_0 is absolutely random. Hence we need to create a fine action that will take the robot toward the actual target x. The fine action is generated as:

$$\theta_1^{out} = \theta_0^{out} + s^{-1} \sum_\gamma h_\gamma A_\gamma (x - v_0) \tag{3.18}$$

In this expression, the pair θ_0^{out}, v_0 is an inverse kinematic pair, i.e., $\theta_0^{out} = g(v_0)$. Hence this fine action will surely take the manipulator tip position closer to the target position x. If we feed this angle vector θ_1^{out} to equation (3.10), we will get $v_1 = [x \ y]^T$. That is, the fine action θ_1^{out} will take the robot to Cartesian position in the task space v_1. Through the fine action, we get another inverse kinematic pair θ_1^{out}, v_1, i.e. $\theta_1^{out} = g(v_1)$. These informations will help us to formulate the cost function.

3.3.4 The Formulation of Cost Function

Since w_γ will be updated as per the Kohonen law (3.5), we need to formulate the cost function for the update of parameters A_γ and θ_γ. During the coarse and fine control actions, we obtained the following two inverse kinematic pairs: $\theta_0^{out} = g(v_0)$ and θ_1^{out}, v_1. Using these pairs, following two expressions can be written as:

$$\theta_0^{out} = \frac{1}{s}\sum_{\gamma} h_\gamma(\theta_\gamma + A_\gamma(v_0 - w_\gamma)) \tag{3.19}$$

$$\theta_1^{out} = \frac{1}{s}\sum_{\gamma} h_\gamma(\theta_\gamma + A_\gamma(v_1 - w_\gamma)) \tag{3.20}$$

If we assume that all network parameters are exact, then equations (3.19) and (3.20) are correct as v_0 will initiate the joint actuation θ_0^{out} and v_1 will initiate the joint actuation θ_1^{out}.

By subtracting (3.19) from (3.20), we obtain:

$$\Delta\theta^{out} = \frac{1}{s}\sum_{\gamma} h_\gamma A_\gamma \Delta v \tag{3.21}$$

where $\Delta\theta^{out} = \theta_1^{out} - \theta_0^{out}$ and $\Delta v = v_1 - v_0$. The equation (3.21) is valid if the network parameters are exact. However, they are not exact. Hence the cost function for the update of A_γ naturally becomes:

$$E_A = \frac{1}{2}\|\Delta\theta^{out} - \frac{1}{s}\sum_{\gamma} h_\gamma A_\gamma \Delta v\|^2 \tag{3.22}$$

The cost function for the update of θ_γ can be obtained using the same logic from equation (3.19) as

$$E_\theta = \frac{1}{2}\|\theta_0^{out} - \frac{1}{s}\sum_{\gamma} h_\gamma(\theta_\gamma + A_\gamma(v_0 - w_\gamma))\|^2 \tag{3.23}$$

3.3.5 Weight Update Laws

Using gradient descent, we can update the parameter matrices as

$$\theta_\gamma \leftarrow \theta_\gamma - \eta_\theta \frac{\partial E_\theta}{\partial \theta_\gamma} \tag{3.24}$$

$$A_\gamma \leftarrow A_\gamma - \eta_A \frac{\partial E_A}{\partial A_\gamma} \tag{3.25}$$

We can thus write the weight update algorithms as

$$w_\gamma \leftarrow w_\gamma + \eta_w h_\gamma(x - w_\gamma) \tag{3.26}$$
$$\theta_\gamma \leftarrow \theta_\gamma + \eta_\theta \Delta\theta_\gamma \tag{3.27}$$
$$A_\gamma \leftarrow A_\gamma + \eta_A \Delta A_\gamma \tag{3.28}$$

Introduction to Learning-Based Inverse Kinematic Control 71

where

$$\Delta\theta_\gamma = \frac{h_\gamma}{s}\left[\theta_0^{out} - s^{-1}\sum_\gamma h_\gamma(\theta_\gamma + A_\gamma(v_0 - w_\gamma))\right] \quad (3.29)$$

$$\Delta A_\gamma = \frac{h_\gamma}{s\|\Delta v\|^2}\left[\Delta\theta^{out} - s^{-1}\sum_\gamma h_\gamma A_\gamma \Delta v\right]\Delta v^T \quad (3.30)$$

Although we started with the description of the two link planar manipulator, the update laws are valid for any n-link manipulator. The only change will be the network lattice. Since the workspace of any n-link manipulator will have 3-D structure, the lattice has to be that 3-d size.

Example 3.3. Let's consider the two-link planar manipulator as shown in Fig 3.6. The link lengths are $l_1 = 1m$ and $l_2 = 1m$. As shown in the corresponding Kohonen network (Figure 3.7), to learn its inverse kinematics, the workspace of the manipulator is a planar circle with radius 2m. The network lattice is a 2d lattice of size 10×10. This network is presented with as a random target position from this circle. Through coarse and fine learning, we construct the cost functions E_A and E_θ as given in equations (3.22), and (3.23) respectively. Write a MATLAB code to update weights as given in (3.26)-(3.28). After training is over, draw the following graphs:

- Draw the final weight vectors associated with 100 neurons in the same plot.
- Test your trained network for five random target positions and verify using exact kinematic inverse equations.
- Make the manipulator to track a straight line.
- Make the manipulator to track a circle.

Solution 3.3. Please look at the MATLAB code. All initial weights are assigned a random number uniformly distributed in $[0, 0.1]$. That is, the robot is absolutely ignorant about its own kinematics as well as its own workspace. The objective is to start from no knowledge to complete knowledge by updating weights of KSOM network as shown in Figure 3.7. To compute h_γ, σ is updated as

$$\sigma = \sigma_i \left(\frac{\sigma_f}{\sigma_i}\right)^{(t/t_{max})} \quad (3.31)$$

where $\sigma_i = 2.5$ and $\sigma_f = 0.01$. The large initial value of σ implies that as weights w_γ are updated as per the Kohonen law, all neurons participate in the decision making in the beginning. This value is gradually reduced to 0.01 which implies that mostly the winning neuron makes the decision as training comes to an end. The learning rate η_w of the Kohonen law starts with a value 1.0 and ends with a value 0.05 following the similar tuning law for σ. The

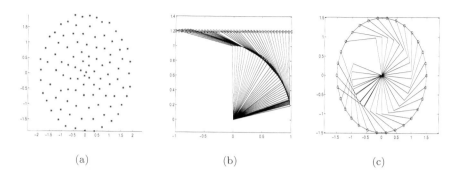

FIGURE 3.8: (a) Final weights, (b) Tracking a line, (c) Tracking a circle.

learning rate η_A for A_γ and the learning rate η_θ for θ_γ are both assigned to 0.9. The number of maximum iterations is fixed at 3000. To train the network, we use 2-d lattice of dimension 10×10 where 3000 samples (random target positions) are used for training which are generated using a forward kinematic model. After the training is over, the weights of each neuron are plotted in Figure 3.8a.

We can see from Figure 3.8(a) that each neural weight has taken position within the robot workspace which is a circle of radius 2m. This signifies that the network has captured the topology of the input data x as expected.

The trained network is tested for three cases. First, this trained network was given some random target positions. Five such target positions in the task space are $[x_1 \ x_2] = [0 \ 1.414]$, $[1.414 \ 0]$, $[1 \ 1]$, $[-1 \ -1]$ and $[0.8 \ 1.2]$. The network response θ is given in Table 3.1. For a given target position $(x_1 = x, x_2 = y)$ in task space, the joint angles θ_1 and θ_2 are calculated using the inverse kinematic equations (3.11). We find that the network responses are consistent with the exact inverse kinematics results.

Next, the KSOM network is tested to track a line $y = 1.2$ while $x \in [-1, +1]$ and a circle of radius 1.5m. The results are shown in Figure 3.8(b)

TABLE 3.1: Two-link manipulator reaches five target positions

KSOM based inverse kinematics results				Exact inverse kinematics results			
x	y	θ_1	θ_2	x	y	θ_1	θ_2
0.0167	1.4161	45.76°	89.84°	0.0167	1.4161	44.41°	89.863°
1.4186	-0.0394	-46.39°	89.60°	1.4186	-0.0394	-46.39°	89.60°
1.0210	1.0371	2.14°	86.61°	1.0210	1.0371	2.14°	86.62°
-0.9995	-0.9980	-180.11°	90.14°	-0.9995	-0.9980	-180.11°	90.14
0.7898	1.1894	11.96°	88.90°	0.7898	1.1894	11.97°	88.90°

Introduction to Learning-Based Inverse Kinematic Control 73

and (c) respectively which show that two-link manipulator properly follows the desired paths.

Figures 3.8(b) and (c) show the tracking of a line and circle with respective kinematic configurations as well. The rms error for the line tracking is 0.0175m and the rms error for the tracking a circle is 0.0223m.

The complete MATLAB code - for training, for testing on random targets, for tracking a line and a circle are provided:

```
MATLAB CODE FOR TRAINING
------------------------
clear all;
%% Initialization of Model parameters
sig_i=2.5; sig_f=0.01;
etaw_i=1;etaw_f=0.05;
etaA_i=0.9;etaA_f=0.9;
l1=1;l2=1; % link length
A_g=0.1*rand(2,2,100)
w_g=0.1*rand(2,1,100)
th_g=0.1*rand(2,1,100)
%% 2d Lattice formation of size 10x10
[lx,ly]=ind2sub([10,10],1:100);
lattice=[lx;ly];iterations=6000;
%% Iterations and update
for i=1:iterations
th1=(rand-0.5)*2*pi;th2=(rand-0.5)*2*pi;
x=l1*cos(th1)+l2*cos(th2+th1);
y=l1*sin(th1)+l2*sin(th2+th1);
u=[x;y];
    for j=1:100
        dist(j)=norm(u-w_g(:,:,j));
    end
[~,win_val]=min(dist);
win=[lx(win_val),ly(win_val)];% Winning Neuron
sig(i)=sig_i*((sig_f/sig_i)^(i/iterations));
eta_wg(i)=etaw_i*((etaw_f/etaw_i)^(i/iterations));
eta_Ag(i)=etaA_i*((etaA_f/etaA_i)^(i/iterations));
d=repmat(win',1,100)-lattice;
H_g=exp(-(sum(d.^2))/(2*(sig(i)^2)));
% Coarse action:
s=sum(H_g);s2=0;s3=0;
    for k=1:100
        s1=H_g(k)*(th_g(:,:,k)+A_g(:,:,k)*(u-w_g(:,:,k)));
        s2=s2+s1;
    end
th_o=s2/s;
x_o=l1*cos(th_o(1))+l2*cos(th_o(2)+th_o(1));
y_o=l1*sin(th_o(1))+l2*sin(th_o(2)+th_o(1));
v_o=[x_o;y_o];
% Fine action
```

```
    for k=1:100
        s4=H_g(k)*(A_g(:,:,k)*(u-v_o));
        s3=s3+s4;
    end
th_1=th_o+s3/s;
x_1=l1*cos(th_1(1))+l2*cos(th_1(2)+th_1(1));
y_1=l1*sin(th_1(1))+l2*sin(th_1(2)+th_1(1));
v_1=[x_1;y_1];
% Update equations
del_v = v_1-v_o;del_th = th_1-th_o;s5=0;s7=0;
for k=1:100
    s6=H_g(k)*(th_g(:,:,k)+A_g(:,:,k)*(v_o-w_g(:,:,k)));
    s5=s5+s6;
end
    for t=1:100
        deltheta_g(:,:,t) = (H_g(t)/s)*(th_o-(s5/s));
    end
    for k=1:100
        s8=H_g(k)*(A_g(:,:,k)*del_v);s7=s8+s7;
    end
for t=1:100
    deltaA_g(:,:,t)=(H_g(t)/(s*norm(del_v)^2))
    *(del_th-s7/s)*(del_v');
    w_g(:,:,t)=w_g(:,:,t)+eta_wg(i)*H_g(t)
    *(u-w_g(:,:,t)); % Update Weights
    th_g(:,:,t)=th_g(:,:,t)+eta_Ag(i)
    *deltheta_g(:,:,t); % Update Theta_g
    A_g(:,:,t)=A_g(:,:,t)+eta_Ag(i)
    *deltaA_g(:,:,t); % Update A_g
end
end
% Plot final Weights
figure(1); hold on;
for t = 1:100
plot(w_g(1,1,t),w_g(2,1,t),'*')
end;
```

```
MATLAB CODE FOR POINT TRACKING
------------------------------
% Tracking of given five points
u1=[0 1.414 ; 1.414 0 ; 1 1 ; -1 -1 ; 0.8 1.2]';

for m=1:size(u1,2)

u=u1(:,m)

    for j=1:100

        dist(j)=norm(u-w_g(:,:,j));
```

```
    end
[~,win_val]=min(dist);
win=[lx(win_val),ly(win_val)];
d=repmat(win',1,100)-lattice;
H_g=exp(-(sum(d.^2)/(2*(sig_f^2))));
% Corse action:
s=sum(H_g);s2=0;s3=0;
for k=1:100
    s1=H_g(k)*(th_g(:,:,k)+A_g(:,:,k)*(u-w_g(:,:,k)));
    s2=s2+s1;
end
theta=s2/s
th_degree=theta*180/(pi)
x=l1*cos(theta(1))+l2*cos(theta(2)+theta(1));
y=l1*sin(theta(1))+l2*sin(theta(2)+theta(1));
v=[x;y] % Tracked point


end
```

```
MATLAB CODE FOR LINE TRACKING
-------------------------
% Track the line
x = linspace(-1,1,41);y = 1.2*ones(size(x));
test2 = [x;y];t=size(x,2);
for m=1:t
    u1=test2(:,m);
```

```
    for j=1:100
        dist(j)=norm(u1-w_g(:,:,j));
    end
    [~,win_val]=min(dist);
    win=[lx(win_val),ly(win_val)];
    d=repmat(win',1,100)-lattice;
    H_g=exp(-(sum(d.^2))/(2*(sig_f^2)));
    s=sum(H_g);s2=0;
for k=1:100
    s1=H_g(k)*(th_g(:,:,k)+A_g(:,:,k)*(u1-w_g(:,:,k)));
    s2=s2+s1;
end
    theta=s2/s;
    x_o=l1*cos(theta(1))+l2*cos(theta(2)+theta(1));
    y_o=l1*sin(theta(1))+l2*sin(theta(2)+theta(1));
    v_o=[x_o;y_o];
    th(:,m) = theta;
end
for i = 1:t
    x_Position(i,:) = [0 l1*cos(th(1,i))
    l1*cos(th(1,i))+l2*cos(th(2,i)+th(1,i))];

    y_Position(i,:) = [0 l1*sin(th(1,i))
    l1*sin(th(1,i))+l2*sin(th(2,i)+th(1,i))];
end
figure; plot(test2(1,:),test2(2,:),'-ok');hold on;
for i = 1:t
```

```
        plot(x_Position(i,:),y_Position(i,:),'k')
end
axis equal
```

MATLAB CODE FOR CIRCLE TRACKING

```
% Track the Circle

t=0:pi/15:2*pi;test1=[1.5*cos(t);1.5*sin(t)];

for m=1:length(t)

u2=test1(:,m);

    for j=1:100

        dist(j)=norm(u2-w_g(:,:,j));

    end

    [~,win_val]=min(dist);

    win=[lx(win_val),ly(win_val)];

    d=repmat(win',1,100)-lattice;

    H_g=exp(-(sum(d.^2))/(2*(sig_f^2)));

    s=sum(H_g);s2=0;

for k=1:100

    s1=H_g(k)*(th_g(:,:,k)+A_g(:,:,k)*(u2-w_g(:,:,k)));

    s2=s2+s1;

end

    theta=s2/s;

    x_o=l1*cos(theta(1))+l2*cos(theta(2)+theta(1));

    y_o=l1*sin(theta(1))+l2*sin(theta(2)+theta(1));

    v_o=[x_o;y_o];
```

```
        th(:,m) = theta;

end

for i = 1:length(t)

    x_Position(i,:) = [0 l1*cos(th(1,i)) l1*cos(th(1,i))
    +l2*cos(th(2,i)+th(1,i))];

    y_Position(i,:) = [0 l1*sin(th(1,i)) l1*sin(th(1,i))
    +l2*sin(th(2,i)+th(1,i))];

end

figure; plot(test1(1,:),test1(2,:),'-ok');hold on;

for i = 1:length(t)

    plot(x_Position(i,:),y_Position(i,:),'k')

end

axis equal
```

Example 3.4. Let's consider a CRS PLUS manipulator as shown in Figure 3.9. The link lengths are taken from its service manual. The kinematics are simplified by assuming the wrist (joint 4) to be rigid. Actually, the wrist is servoed to maintain its pose with respect to the base plane. This ensures that the manipulator behaves as a 3 DOF manipulator.

The forward kinematics are given as:

$$z = l_2 sin(\theta_2) + l_3 sin(\theta_3) + l_1 \qquad (3.32)$$
$$x = R cos(\theta_1) \qquad (3.33)$$
$$y = R sin(\theta_1) \qquad (3.34)$$

where $R = l_2 cos(\theta_2) + l_3 cos(\theta_3) + t$ and l_1, l_2, l_3 are the respective link lengths, and t is the length of the rigid portion of the wrist.

For the CRS PLUS Manipulator, $l_1 = l_2 = l_3 = 254mm$, $t = 50mm$. It is assumed that the wrist of the manipulator is rigid, assured by locking the joint for our purpose, and always holds the end-effector parallel to the work table (the xy plane).

The exact inverse kinematics of this manipulator is given as:

$$\theta_1 = tan^{-1}\left(\frac{y}{x}\right) \qquad (3.35)$$
$$\theta_2 = cos^{-1}(p) \qquad (3.36)$$

Introduction to Learning-Based Inverse Kinematic Control

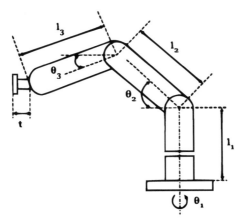

FIGURE 3.9: CRS Plus manipulator.

$$\theta_3 = sin^{-1}\left(\frac{b - l_2 sin(\theta_2)}{l_3}\right) \quad (3.37)$$

Here p is the solution of $Ap^2 + Bp + C = 0$, where $a = \sqrt{x^2 + y^2} - t$, $b = z - l_1$, $K = b + l_2^2 - l_3^2 + a^2$, $A = 4a^2 l_2^2 + 4b^2 l_2^2$, $B = -4al_2 K$, and $C = K^2 - 4b^2 l_2^2$.

It is desired that these inverse kinematic solutions are derived using the KSOM network. Since the robot task-space is 3 dimensional, a 3-d lattice of size $12 \times 7 \times 4$ is selected. The maximum training iterations are fixed at 30000. This implies that the network is presented with 30000 random target positions sample from its workspace. Train the KSOM network and test the network for tracking a straight line and circle.

Solution 3.4. *Please look at the MATLAB code. All initial weights are assigned a random number uniformly distributed in [0, 0.1]. That is, the robot is absolutely ignorant about its own kinematics as well as its own workspace. The objective is to start from no knowledge to complete knowledge by updating weights of KSOM network as shown in Fig 3.7. The learning parameters are updated during the training as:*

$$\eta = \eta_i \left(\frac{\eta_f}{\eta_i}\right)^{(t/t_{max})} \quad (3.38)$$

where $\eta \in \{\eta_w, \eta_\theta, \eta_A, \sigma\}$. $\sigma_i = 2.5$ and $\sigma_f = 0.01$. The large initial value of σ implies that as weights w_γ are updated as per the Kohonen law, all neurons participate in the decision making in the beginning. This value is gradually reduced to 0.01 which implies that mostly the winning neuron makes the decision as training comes to an end. The learning rate η_w of the Kohonen law (3.5) starts with a value 1.0 and ends with a value 0.05. The learning rate

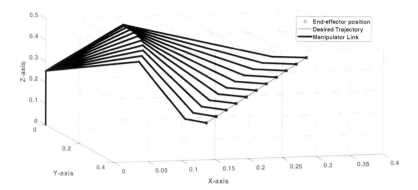

FIGURE 3.10: CRS manipulator tracks a line.

η_A for A_γ and the learning rate η_θ for θ_γ are fixed at 0.9. The number of maximum iterations is fixed at 30000. To train the network, we use $3-d$ lattice of dimension $12 \times 7 \times 4$ where 30000 samples (random target positions) are used for training which are generated using the forward kinematic model (3.32). After the training, the robot is asked to follow a straight line. The tracking results are shown in Figure 3.10 along with the kinematic configurations. Readers should note that the network has effectively learned the inverse kinematics where all network parameters are initialized randomly and the network is presented with the random target positions without the corresponding joint space solutions. The coarse and fine actions are used to build the effective cost functions gradients of which helped to derive the weight update laws. The MATLAB code will help you to understand these concepts even better.

```
MATLAB CODE-KSOM TRAINING-3 LINK MANIPULATOR
----------------------------------
clear all;
format long; kstar=1000; Xmin=0.15; Xmax=0.35;
Ymin=0.15; Ymax=0.35; Zmin=0.15; Zmax=0.35;
Xmiddle=(Xmax+Xmin)/2; Ymiddle=(Ymax+Ymin)/2;
Zmiddle=(Zmax+Zmin)/2; ndsD=200; %No. of datasets
l1=0.254; l2=0.254; l3=0.254; t=0.05;
% Lattice dimensions
p=12; q=7; r=4; nip=3;%Input dimension
nop=3;%Output dimension
W(1,:,:,:)=(Xmax-Xmin)*rand(1,p,q,r)+Xmin;
W(2,:,:,:)=(Ymax-Ymin)*rand(1,p,q,r)+Ymin;
W(3,:,:,:)=(Zmax-Zmin)*rand(1,p,q,r)+Zmin;
A=rand(nop,nip,p,q,r); Y_lambda=zeros(nop,p,q,r);
Theta_lambda=2*pi*(rand(nop,p,q,r)-0.5*ones(nop,p,q,r));
D=zeros(p,q,r); sigini=2.5; sigfin=0.01;
etaini=0.5; etafin=0.9; etaWini=0.5; etaWfin=0.05;
epoch=300; a=0;
```

```
for ep=1:epoch
    ep
    if(a>1)
fprintf('More than one min distance values');
        break;
    end
Xd(1,:)=(Xmax-Xmin)*rand(1,ndsD)+Xmin;
Xd(2,:)=(Ymax-Ymin)*rand(1,ndsD)+Ymin;
Xd(3,:)=(Zmax-Zmin)*rand(1,ndsD)+Zmin;
for nds=1:ndsD
    etaW=etaWini*(etaWfin/etaWini)^((ep/(epoch)));
    eta=etaini*(etafin/etaini)^((ep)/(epoch));
    etaT=eta;
sigma=sigini*(sigfin/sigini)^(ep/epoch);
    for i=1:p
        for j=1:q
            for k=1:r
                Dis(i,j,k)=norm(W(:,i,j,k)-Xd(:,nds));
            end
        end
    end
    Dis_min=min(min(min(Dis)));
    a=0;
    for i=1:p
        for j=1:q
            for k=1:r
                if(Dis(i,j,k)==Dis_min)
                    a=a+1;ai=i;aj=j;ak=k;
                end
            end
        end
    end
    if(a>1)
        fprintf('More than one min distance values');
        break;
    end
    for i=1:p
        for j=1:q
            for k=1:r
H(i,j,k)=exp(-((ai-i)^2+(aj-j)^2+(ak-k)^2)/(2*sigma*sigma))
    ↪ ;
            end
        end
    end
    Y_temp=zeros(3,1);
    for i=1:p
        for j=1:q
            for k=1:r
Y_lambda(:,i,j,k)=H(i,j,k)*(Theta_lambda(:,i,j,k)
```

```
+A(:,:,i,j,k)*(Xd(:,nds)-W(:,i,j,k)));
Y_temp=Y_temp+Y_lambda(:,i,j,k);
            end
        end
    end
s=(sum(sum(sum(H))));Theta_0=(Y_temp)/s;
R=(l2*cos(Theta_0(2,1)))+(l3*cos(Theta_0(3,1)))+t;
V_0(1,1)= R*cos(Theta_0(1,1));V_0(2,1)= R*sin(Theta_0(1,1))
    ↪ ;
V_0(3,1)=(l2*sin(Theta_0(2,1)))+(l3*sin(Theta_0(3,1)))+l1;
Y_temp_Theta_0=zeros(nop,1);
Y_lambda_Theta_0=zeros(nop,p,q,r);
for i=1:p
for j=1:q
for k=1:r
Y_lambda_Theta_0(:,i,j,k)=H(i,j,k)*(Theta_lambda(:,i,j,k)
+A(:,:,i,j,k)*(V_0-W(:,i,j,k)));
Y_temp_Theta_0=Y_temp_Theta_0+Y_lambda_Theta_0(:,i,j,k);
            end
        end
    end
    Y_temp_Theta_0=Y_temp_Theta_0/s;
    Co_act=zeros(nop,p,q,r);
Corr_action=zeros(nop,1);
    for i=1:p
        for j=1:q
            for k=1:r
Co_act(:,i,j,k)=H(i,j,k)*A(:,:,i,j,k)*(Xd(:,nds)-V_0);
Corr_action=Corr_action+Co_act(:,i,j,k);
            end
        end
    end
Theta_1=Theta_0+(Corr_action/s);
R=(l2*cos(Theta_1(2,1)))+(l3*cos(Theta_1(3,1)))+t;
V_1(1,1)= R*cos(Theta_1(1,1));
V_1(2,1)= R*sin(Theta_1(1,1));
V_1(3,1)=(l2*sin(Theta_1(2,1)))+(l3*sin(Theta_1(3,1)))+l1;
del_V=V_1-V_0;
del_Theta=Theta_1-Theta_0;
YA_up=zeros(nop,p,q,r);
YA_update=zeros(nop,1);
    for i=1:p
        for j=1:q
            for k=1:r
YA_up(:,i,j,k)=H(i,j,k)*A(:,:,i,j,k)*(V_1-V_0);
YA_update=YA_update+YA_up(:,i,j,k);
            end
        end
    end
```

```
       for i=1:p
           for j=1:q
               for k=1:r
Theta_lambda(:,i,j,k)=Theta_lambda(:,i,j,k)+(etaT*(Theta_0
-Y_temp_Theta_0)*H(i,j,k)/s);
W(:,i,j,k)=W(:,i,j,k)+etaW*H(i,j,k)*(Xd(:,nds)-W(:,i,j,k));
               end
           end
       end
       for i=1:p
           for j=1:q
               for k=1:r
A(:,:,i,j,k)=A(:,:,i,j,k)+eta*H(i,j,k)*(del_Theta-
(YA_update/s))*(del_V')/(s*norm(del_V)*norm(del_V));
               end
           end
       end
       valA(nds)=A(1,1,5,6,4);
  valW(nds)=W(1,5,6,4);
       valT(nds)=Theta_lambda(1,5,6,4);
end
end
%% Testing 1
a=0;
Xmiddle=(Xmax+Xmin)/2;
Ymiddle=(Ymax+Ymin)/2;
Zmiddle=(Zmax+Zmin)/2;
Rp=0.08;
Xt=0;
Xdt=0;
for j=0:0.1:1
    a=a+1;
Xdt(1,a)=Xmin*(1-j)+Xmax*j;
Xdt(2,a)=Ymax*(1-j)+Ymin*j;
Xdt(3,a)=Zmin*(1-j)+Zmax*j;
end
ndsT=a;
for nds=1:ndsT
    nds
    for i=1:p
        for j=1:q
            for k=1:r
Dis(i,j,k)=norm(W(:,i,j,k)-Xdt(:,nds));
            end
        end
    end
    Dis_min=min(min(min(Dis)));
    a=0;
    for i=1:p
```

```
            for j=1:q
                for k=1:r
                    if(Dis(i,j,k)==Dis_min)
                        a=a+1;
ai(a)=i;
aj(a)=j;
                        ak(a)=k;
                    end
                end
            end
        end
        if(a>1)
            fprintf('More than one min distance values');
            break;
        end
        for i=1:p
            for j=1:q
                for k=1:r
H(i,j,k)=exp(-((ai-i)^2+(aj-j)^2+(ak-k)^2)/
(2*sigfin*sigfin));
                end
            end
        end
        Y_temp=zeros(3,1);
        for i=1:p
            for j=1:q
                for k=1:r
Y_lambda_Theta_0(:,i,j,k)=H(i,j,k)*(Theta_lambda(:,i,j,k)
+A(:,:,i,j,k)*(Xdt(:,nds)-W(:,i,j,k)));
Y_temp=Y_temp+Y_lambda_Theta_0(:,i,j,k);
                end
            end
        end
s=(sum(sum(sum(H))));
Theta(:,nds)=(Y_temp)/s;
R=(12*cos(Theta(2,nds)))+(13*cos(Theta(3,nds)))+t;
V_0= R*cos(Theta(1,nds));
V_0= R*sin(Theta(1,nds));
V_0=(12*sin(Theta(2,nds)))+(13*sin(Theta(3,nds)))+l1;
        lol=0;
        error=10;
        while(error>0.001)
            lol=lol+1;
Co_act=zeros(nop,p,q,r);
Corr_action=zeros(nop,1);
        for i=1:p
            for j=1:q
                for k=1:r
Co_act(:,i,j,k)=H(i,j,k)*A(:,:,i,j,k)*(Xdt(:,nds)-V_0);
```

```
            Corr_action=Corr_action+Co_act(:,i,j,k);
                end
            end
        end
Theta_1=Theta_0+(Corr_action/s);
R=(l2*cos(Theta_1(2,1)))+(l3*cos(Theta_1(3,1)))+t;
Xt(1,nds)= R*cos(Theta_1(1,1));
Xt(2,nds)= R*sin(Theta_1(1,1));
Xt(3,nds)=(l2*sin(Theta_1(2,1)))+(l3*sin(Theta_1(3,1)))+l1;
    error=norm(Xt(:,nds)-Xdt(:,nds));
    Theta_0=Theta_1;
V_0=Xt(:,nds);
    if(lol>100)
        break;
    end
    end
Teeta(:,nds)=Theta_1;
ptsl(:,nds)=[0;0;0];
pts2(:,nds)=[0;0;l1];
pts3(:,nds)=[l2*cos(Theta_1(2,1))*cos(Theta_1(1,1));
l2*cos(Theta_1(2,1))*sin(Theta_1(1,1));
l2*sin(Theta_1(2,1))+l1];
pts4(:,nds)=[(R-t)*cos(Theta_1(1,1));
(R-t)*sin(Theta_1(1,1));Xt(3,nds)];
pts5(:,nds)=[Xt(1,nds);Xt(2,nds);Xt(3,nds)];
end
figure(1)
grid on;
scatter3(Xt(1,:)',Xt(2,:)',Xt(3,:)',50,'b','o');
grid on;
hold on;
legend('End effector position','Desired Trajectory',
'Manipulator Link')
plot3(Xdt(1,:),Xdt(2,:),Xdt(3,:),'k','lineWidth',2);
hold on;
xlabel('X-axis');
ylabel('Y-axis');
zlabel('Z-axis');
for ii=1:nds
% scatter3(pts1(1,ii),pts1(2,ii),pts1(3,ii));
line([pts1(1,ii),pts2(1,ii)],[pts1(2,ii),pts2(2,ii)]
,[pts1(3,ii),pts2(3,ii)],'LineWidth',4)
% scatter3(pts2(1,ii),pts2(2,ii),pts2(3,ii));
line([pts3(1,ii),pts2(1,ii)],[pts3(2,ii),pts2(2,ii)]
,[pts3(3,ii),pts2(3,ii)],'LineWidth',4)
% scatter3(pts3(1,ii),pts3(2,ii),pts3(3,ii));
line([pts3(1,ii),pts4(1,ii)],[pts3(2,ii),pts4(2,ii)]
,[pts3(3,ii),pts4(3,ii)],'LineWidth',4)
% scatter3(pts4(1,ii),pts4(2,ii),pts4(3,ii));
```

```
line([pts5(1,ii),pts4(1,ii)],[pts5(2,ii),pts4(2,ii)]
,[pts5(3,ii),pts4(3,ii)],'LineWidth',4)
% scatter3(pts5(1,ii),pts5(2,ii),pts5(3,ii));
end
hold on;
% legend('Manipulator Link')
e=norm(Xt-Xdt)/size(Xt,2);
for i=1:p
    for j=1:q
        for k=1:r
            figure(3)
scatter3(W(1,i,j,k),W(2,i,j,k),W(3,i,j,k))
            hold on;
        end
    end
end
```

```
MATLAB COODE FOR TESTING-3 LINK MANIPULATOR
-------------------------
%% Testing 1
load('ksom_3dof_weights');
a=0;
Xmiddle=(Xmax+Xmin)/2;
Ymiddle=(Ymax+Ymin)/2;
Zmiddle=(Zmax+Zmin)/2;
Rp=0.08;
Xt=0;
Xdt=0;
for j=0:0.1:1
    a=a+1;
Xdt(1,a)=Xmin*(1-j)+Xmax*j;
Xdt(2,a)=Ymax*(1-j)+Ymin*j;
Xdt(3,a)=Zmin*(1-j)+Zmax*j;
end
ndsT=a;
for nds=1:ndsT
    nds
    for i=1:p
        for j=1:q
            for k=1:r
Dis(i,j,k)=norm(W(:,i,j,k)-Xdt(:,nds));
            end
        end
    end
    Dis_min=min(min(min(Dis)));
    a=0;
    for i=1:p
        for j=1:q
            for k=1:r
```

Introduction to Learning-Based Inverse Kinematic Control

```
                    if(Dis(i,j,k)==Dis_min)
 a=a+1;
ai(a)=i;
aj(a)=j;
ak(a)=k;
                    end
                end
            end
        end
        if(a>1)
            fprintf('More than one min distance values');
            break;
        end
        for i=1:p
            for j=1:q
                for k=1:r
H(i,j,k)=exp(-((ai-i)^2+(aj-j)^2+(ak-k)^2)/
(2*sigfin*sigfin));
                end
            end
        end
        Y_temp=zeros(3,1);
        for i=1:p
            for j=1:q
                for k=1:r
Y_lambda_Theta_0(:,i,j,k)=H(i,j,k)*(Theta_lambda(:,i,j,k)
+A(:,:,i,j,k)*(Xdt(:,nds)-W(:,i,j,k)));
Y_temp=Y_temp+Y_lambda_Theta_0(:,i,j,k);
                end
            end
        end
        s=(sum(sum(sum(H))));
Theta(:,nds)=(Y_temp)/s;
    R=(l2*cos(Theta(2,nds)))+(l3*cos(Theta(3,nds)))+t;
    V_0= R*cos(Theta(1,nds));
 V_0= R*sin(Theta(1,nds));
    V_0=(l2*sin(Theta(2,nds)))+(l3*sin(Theta(3,nds)))+l1;
    lol=0;
 error=10;
    while(error>0.001)
        lol=lol+1;
    Co_act=zeros(nop,p,q,r);
 Corr_action=zeros(nop,1);
    for i=1:p
        for j=1:q
            for k=1:r
Co_act(:,i,j,k)=H(i,j,k)*A(:,:,i,j,k)*(Xdt(:,nds)-V_0);
Corr_action=Corr_action+Co_act(:,i,j,k);
            end
```

```
            end
        end
Theta_1=Theta_0+(Corr_action/s);
R=(l2*cos(Theta_1(2,1)))+(l3*cos(Theta_1(3,1)))+t;
Xt(1,nds)= R*cos(Theta_1(1,1));
Xt(2,nds)= R*sin(Theta_1(1,1));
Xt(3,nds)=(l2*sin(Theta_1(2,1)))+(l3*sin(Theta_1(3,1)))+l1;
error=norm(Xt(:,nds)-Xdt(:,nds));
Theta_0=Theta_1;
V_0=Xt(:,nds);
    if(lol>100)
        break;
    end
    end
Teeta(:,nds)=Theta_1;
pts1(:,nds)=[0;0;0];
pts2(:,nds)=[0;0;l1];
pts3(:,nds)=[l2*cos(Theta_1(2,1))*cos(Theta_1(1,1));
l2*cos(Theta_1(2,1))*sin(Theta_1(1,1));
l2*sin(Theta_1(2,1))+l1];
pts4(:,nds)=[(R-t)*cos(Theta_1(1,1));(R-t)*sin(Theta_1(1,1)
    ↪ )
;Xt(3,nds)];
pts5(:,nds)=[Xt(1,nds);Xt(2,nds);Xt(3,nds)];
end
figure(1)
grid on;
scatter3(Xt(1,:)',Xt(2,:)',Xt(3,:)',50,'b','o');
grid on;
hold on;
legend('End effector position','Desired Trajectory',
'Manipulator Link')
plot3(Xdt(1,:),Xdt(2,:),Xdt(3,:),'k','lineWidth',2);
hold on;
xlabel('X-axis');
ylabel('Y-axis');
zlabel('Z-axis');
for ii=1:nds
% scatter3(pts1(1,ii),pts1(2,ii),pts1(3,ii));
line([pts1(1,ii),pts2(1,ii)],[pts1(2,ii),pts2(2,ii)],
[pts1(3,ii),pts2(3,ii)],'LineWidth',4)
% scatter3(pts2(1,ii),pts2(2,ii),pts2(3,ii));
line([pts3(1,ii),pts2(1,ii)],[pts3(2,ii),pts2(2,ii)],
[pts3(3,ii),pts2(3,ii)],'LineWidth',4)
% scatter3(pts3(1,ii),pts3(2,ii),pts3(3,ii));
line([pts3(1,ii),pts4(1,ii)],[pts3(2,ii),pts4(2,ii)],
[pts3(3,ii),pts4(3,ii)],'LineWidth',4)
% scatter3(pts4(1,ii),pts4(2,ii),pts4(3,ii));
line([pts5(1,ii),pts4(1,ii)],[pts5(2,ii),pts4(2,ii)],
```

```
[pts5(3,ii),pts4(3,ii)],'LineWidth',4)
% scatter3(pts5(1,ii),pts5(2,ii),pts5(3,ii));
end
hold on;
e=norm(Xt-Xdt)/size(Xt,2);
for i=1:p
    for j=1:q
        for k=1:r
            figure(3)
scatter3(W(1,i,j,k),W(2,i,j,k),W(3,i,j,k))
            hold on;
        end
    end
end
```

3.4 Visual Motor Control of a Redundant Manipulator using KSOM Network

In the previous section, it is discussed how to make use of KSOM based network to learn the inverse kinematics of a robot manipulator. In this section we consider a 7 degrees of freedom manipulator as shown in figure 3.11. This robot manipulator is a modular power PowerCube manufactured by Schunk. The kinematic model of this manipulator is given in Chapter 2. The visual feedback is provided by the two overhead cameras. The integration of the camera model with the kinematic model is also provided in Chapter 2.

The KSOM network for visual motor control (VMC) using the standard model given in the previous section is given in Figure 3.12.

The input to this network is u_t which is the target position in the visual space. A 3-d lattice has been taken as the task space is always that of 3-d even if the manipulator considered here has 7 DOF. Each camera gives the centroid of the target object as x_c, y_c. Thus $u_t = [x_c^r y_c^r x_c^l y_c^l]^T$. r and l refer to right and left overhead cameras. In this network the coarse and fine actions are given as:

$$\theta_0^{out} = \frac{1}{s}\sum_\gamma h_\gamma(\theta_\gamma + A_\gamma(u_t - w_\gamma)) \tag{3.39}$$

$$\theta_1^{out} = \theta_0^{out} + s^{-1}\sum_\gamma h_\gamma A_\gamma(u_t - v_0) \tag{3.40}$$

One can notice that the only change that has happened here is that x has been replaced by u_t. The weight update laws for the visual motor control are thus given as

$$w_\gamma \leftarrow w_\gamma + \eta_w h_\gamma(u_t - w_\gamma) \tag{3.41}$$

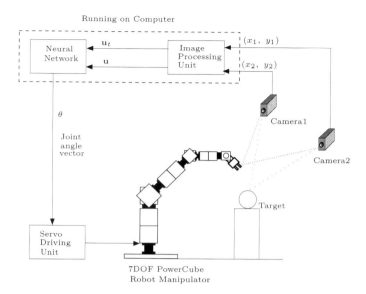

FIGURE 3.11: Schematic of a visual motor control system. $\mathbf{u}_t$ and $\mathbf{u}_r$ are the 4-dimensional image coordinate vectors for target point and robot end-effector respectively.

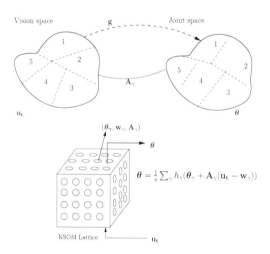

FIGURE 3.12: KSOM network for visual motor control.

$$\theta_\gamma \leftarrow \theta_\gamma + \eta_\theta \Delta\theta_\gamma \tag{3.42}$$

$$A_\gamma \leftarrow A_\gamma + \eta_A \Delta A_\gamma \tag{3.43}$$

where

$$\Delta\theta_\gamma = \frac{h_\gamma}{s}\left[\theta_0^{out} - s^{-1}\sum_\gamma h_\gamma(\theta_\gamma + A_\gamma(v_0 - w_\gamma))\right] \tag{3.44}$$

$$\Delta A_\gamma = \frac{h_\gamma}{s\|\Delta v\|^2}\left[\Delta\theta^{out} - s^{-1}\sum_\gamma h_\gamma A_\gamma \Delta v\right]\Delta v^T \tag{3.45}$$

The above standard SOM-based VMC scheme has following limitations which restrict its applicability to redundant manipulators:

- It is found that for a redundant manipulator with 6 or higher degrees of freedom, although the SOM lattice neurons preserve topology of the input space as shown in Fig. 3.13, the lattice fails to preserve the topology of output (joint angle) space as shown in Fig. 3.14. In Fig. 3.13, it can be seen that the weight vectors ($\mathbf{w}_\gamma$) represented by square 'boxes' are spread out uniformly over the input space. On the other hand, in Fig. 3.14, we find that the clusters in joint angle space ($\boldsymbol{\theta}_\gamma$) represented by square 'boxes' are concentrated at one location. It would be shown in the simulation section that because of the fact that the network fails to capture the output topology, the positioning accuracy attained using standard SOM algorithm is sensitive to initial conditions.

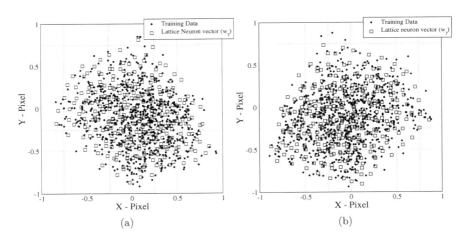

FIGURE 3.13: Clustering in image-coordinate (input) space. Lattice neurons capture the topology of input space during training. The circular dots denote the actual input data generated during training and the square represent the cluster centers $\mathbf{w}_\gamma$.

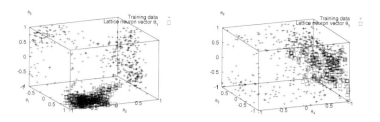

(a) Space formed by first 3 joint angles (b) Space formed by last 3 joint angles

FIGURE 3.14: The topology of output space is not captured by the original VMC algorithm during the evolution of parameters (training phase). While the training data shown by '+' signs are distributed across the entire volume, the cluster centers $\boldsymbol{\theta}_\gamma$ are collected at one location.

- The standard SOM algorithm returns a unique inverse kinematic solution for any target in the manipulator workspace. This might not be desirable in case of redundant manipulators where one would like to choose a different configuration to satisfy some additional requirements. Even though the training data sets are replete with redundant solutions, there is no provision to *preserve* this redundancy during the evolution of parameters.

3.4.1 The Problem

In case of 2-d and 3-d manipulators, during the training, the robot is asked to reach 3000 and 6000 random target positions. But in case of 7 DOF manipulator, the number of training examples required may increase significantly. During the training, we actuate a random joint angle vector within given joint constraints to the forward kinematics. This will make the robot tip position to reach a random target position. This random position as seen by the two overhead cameras becomes the input for the KSOM network to actuate a joint angle vector. If we select ten random angles for each joint, then for seven joints, the total number of target positions generated will be 10^7. This is surely a very large number of training data set. In stead we generate 50,000 random joint angle vectors. Using the integrated kinematic-camera model, 50,000 target positions in the visual space are obtained. It will be shown in this section that these many training examples are sufficient to train the network.

There is another problem associated with redundancy. A redundant manipulator can reach a target position in infinite possible kinematic configurations. In practice, many joint angle vectors will make the robot tip position which is also called as the end-effector to reach the same target positions. So the robot will reach same target position in many possible kinematic configurations.

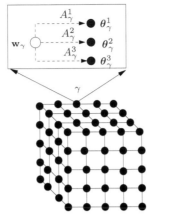

3-dimensional KSOM lattice

FIGURE 3.15: Sub-clustering in joint angle space: Each node γ is associated with one weight vector $\mathbf{w}_\gamma$ and several $\boldsymbol{\theta}$ vectors.

Earlier we associated only one linear model with each neuron. But each neuron has to be associated with multiple linear models for a redundant manipulator as shown in figure 3.15.

The idea is to associate each lattice neuron with several joint angle vectors. Each joint angle vector is a linear expression in terms of θ_γ, A_gamma and w_γ. The advantage is that for every target position, it is possible to have several configurations, and one can choose a suitable configuration based on some task oriented criterion. In this approach, clustering is carried out independently in the task space as well as in the configuration space and a linearized inverse kinematic relationship is learned between each pair of input-output clusters. Since each cluster in task space is associated with more than one cluster in configuration space, redundancy is resolved in real time using different criteria. This approach is different from PSOM based methods proposed by Walter and Ritter [137–139], where constraints are included in forming a map manifold over which the training is carried out. This in turn, necessitates a priori knowledge of the task at hand.

Another problem associated with the redundant manipulator is that the number of kinematic configurations in which the robot can reach a target position will get reduced as the target position varies from the interior of the workspace toward the external boundary of the workspace. So we cannot assign a fixed number of linear models with each neuron. Thus following modifications have been done to the network architecture.

- Instead of fixing the number of sub-clusters a priori, the number is decided adaptively during on-line training process. A new sub-cluster is created whenever the incoming data vector is far away from the currently existing sub-clusters. Through simulations, it is shown that this scheme helps in preserving the topology in joint angle space by avoiding creation of outliers.

- The smoothness of joint angle trajectories can be preserved by using the neighborhood concept where the network output is taken as the weighted average of individual neuron outputs. The weighting coefficients are obtained from a neighborhood function. Since each input cluster is associated with more than one output cluster, the conventional neighborhood concept as used by Martinetz et al. [140, 141] cannot be used. A modified neighborhood concept is proposed to preserve the conservative property of the inverse kinematic solution. The concept explained in detail later in this chapter.

3.5 KSOM with Sub-Clustering in Joint Angle Space

Any point within the 3-dimensional Cartesian workspace of the manipulator may be reached using only 3 degrees of freedom. The presence of higher degrees of freedom provide dexterity in performing the task at hand. In other words, apart from reaching the point, the extra degrees of freedom may be used to perform some additional tasks like avoiding obstacles, meeting joint angle limits or satisfying other motion constraints. The proper utilization of available degrees of freedom has been an interesting problem for researchers.

In the context of visual motor control using KSOM networks, redundancy resolution has been dealt with by many authors. For instance, Martinetz et al. [142] applied KSOM algorithm to a 5 DOF manipulator and argued that because of neighborhood function, the redundancy is resolved 'naturally' by using 'lazy arm method.' Han et al. [143] and Zha et al. [144] used SOM for avoiding obstacles. Zheng et al. [145] resolved redundancy by optimizing some task oriented criteria. Most of these methods have been applied to 3, 4 or 5 DOF manipulators and each method resolves redundancy in only one way. The redundancy resolution scheme is learned during the training phase itself. Hence once the network is trained, it is not possible to change the redundancy resolution criterion while computing the inverse kinematic solution. Since most of the current VMC schemes involve time consuming training process, retraining the network for a new redundancy resolution criterion is not desirable.

In case of redundant manipulators, several sets of joint angle vectors may lead to same end-effector position. Thus, the data generated during training phase consists of redundant data sets. The current VMC algorithms explained in the beginning of the Section 3.4 do not have provisions to preserve this redundancy in the solution space. In the previous chapter, a concept called

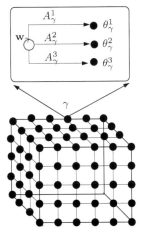

FIGURE 3.16: Sub-clustering in joint-angle space. Each lattice neuron γ is associated with one weight vector $\mathbf{w}_\gamma$ and several joint angle vectors $\boldsymbol{\theta}_\gamma^\lambda$, $\lambda = 1, 2, \ldots, N_\gamma$. Here $N_\gamma = 3$.

'*sub-clustering in joint angle space*' is introduced to preserve this redundancy in a useful manner. The idea is to associate multiple angle vectors with each lattice neuron as shown in Fig. 3.16. Since the number of redundant solutions available for a target point varies across the manipulator workspace, the number of sub-clusters to be associated with each neuron is decided on-line based on the distribution of generated data points.

3.5.1 Network Architecture

The network architecture for sub-clustering is reproduced here for convenience. It consists of a 3-dimensional SOM lattice where each lattice neuron γ is associated with a 4-dimensional weight vector $\mathbf{w}_\gamma$ and several 6-dimensional joint angle vectors as shown in Fig. 3.16. Let us assume that each lattice neuron γ is associated with N_γ numbers of angle vectors given by $\boldsymbol{\theta}_\gamma^j$, $j = 1, 2 \ldots, N_\gamma$ and an equal number of Jacobian matrices A_γ^j $j = 1, 2, \ldots, N_\gamma$ of dimension 6×4. The number N_γ varies with each γ and is decided on-line based on the actual data distribution. When lattice neuron γ becomes a winner for a given input vector $\mathbf{u}_t$, this network architecture can actuate N_γ kinematic configurations by which the robot manipulator can reach the same target as per following relation:

$$\theta^j = \theta_\gamma^j + A_\gamma^j(\mathbf{u}_t - w_\gamma); \quad j = 1, 2 \ldots, N_\gamma \qquad (3.46)$$

For this network architecture, the parameters $\boldsymbol{\theta}_\gamma^j$ and A_γ^j cannot be learned using standard SOM algorithm. Unlike standard SOM algorithm, we propose an on-line clustering algorithm to learn θ_γ^j while the error-correcting gradient learning for A_γ in standard SOM algorithm has been adapted to learn A_γ^j.

3.5.2 Training Algorithm

The training phase consists of following steps:

1. Set the iteration counter $k = 1$.

2. *Data generation*: A training data set $(\boldsymbol{\theta}_t, \mathbf{u}_t)$ is generated using robot and camera models during simulation. The robot manipulator is commanded a movement in joint angle space by generating a random vector $\boldsymbol{\theta}_t$ within physical limits while the input vector $\mathbf{u}_t$ is recorded from camera output.

3. *Clustering in input space*: For each 4-dimensional target input $\mathbf{u}_t$, a winner neuron μ is selected based on minimum Euclidean distance as shown in equation (3.1). The weight vectors corresponding to the winner neuron μ and the neighboring neurons are updated as per equation (3.41) for the given target vector $\mathbf{u}_t$.

4. *Clustering in output space*: Let's assume that this winner neuron μ is associated with N_μ number of θ vectors given by $\boldsymbol{\theta}_\mu^j, j = 1, 2, \ldots, N_\mu$. The incoming target angle vector $\boldsymbol{\theta}_t$ is used to create a new angle vector or update the existing angle vectors as per following conditions:

    - Case I: If $N_\mu = 0$, i.e., there is no θ vector associated with this neuron, then assign the target joint angle vector $\boldsymbol{\theta}_t$ as its first center. That is, $\boldsymbol{\theta}_\mu^{N_\mu+1} = \boldsymbol{\theta}_\mu^1 = \boldsymbol{\theta}_t$.
    - Case II: If $N_\mu > 0$, following steps are followed:
        - Find the angle vector $\boldsymbol{\theta}_\mu^j$ which is nearest to the incoming angle vector $\boldsymbol{\theta}_t$. Let's call the winner among these angle vectors be $\boldsymbol{\theta}_\mu^\beta$ where
        $$\beta = \arg\min_j \|\boldsymbol{\theta}_\mu^j - \boldsymbol{\theta}_t\|, \; j = 1, 2, \ldots, N_\mu \quad (3.47)$$
        - If the minimum distance $d_{min} = \|\boldsymbol{\theta}_\mu^\beta - \boldsymbol{\theta}_t\| < K$, where K is a user-defined threshold, the angle vectors are updated using a competitive rule given by
        $$\boldsymbol{\theta}_\mu^j(k+1) = \boldsymbol{\theta}_\mu^j(k) + \eta h_{\beta j}(\boldsymbol{\theta}_t - \boldsymbol{\theta}_\mu^j(k)) \quad (3.48)$$
        where $h_{\beta j} = e^{\frac{-(\beta-j)^2}{2\sigma_t^2}}$ is the neighborhood function used for subclustering. A suitable value of spread of Gaussian function σ_t is selected for this purpose.
        - If $d_{min} > K$, create a new centre and assign the incoming $\boldsymbol{\theta}_t$ vector to it and increment the count of angle centers associated with this winner neuron μ from N_μ to $N_\mu + 1$. In other words, $\boldsymbol{\theta}_\mu^{N_\mu+1} = \boldsymbol{\theta}_t$.

5. *Coarse Movement*: Because of sub-clustering, the winner neuron μ is associated with N_μ sub-clusters in joint angle space given by $\boldsymbol{\theta}_\mu^j$, $j = 1, 2, \ldots, N_\mu$. For the given target point $\mathbf{u}_t$, the network has N_μ outputs given by

$$\boldsymbol{\theta}_0^j = \boldsymbol{\theta}_\mu^j + A_\mu^j(\mathbf{u}_t - \mathbf{w}_\mu); \quad j = 1, 2, \ldots, N_\mu. \tag{3.49}$$

These are called coarse movements. These coarse movements lead to end-effector positions $\mathbf{v}_0^j$, $j = 1, 2, \ldots, N_\mu$ as recorded by the cameras.

6. *Fine movement*: Based on the current positioning accuracy for each end-effector position $\mathbf{v}_0^j$, a fine movement may also be carried out as follows:

$$\boldsymbol{\theta}_1^j = \boldsymbol{\theta}_0^j + A_\mu^j(\mathbf{u}_t - \mathbf{v}_0^j) \tag{3.50}$$

The new end-effector positions are recorded as $\mathbf{v}_1^j$, $j = 1, 2, \ldots, N_\mu$.

7. The difference $\Delta \mathbf{v}^j = \mathbf{v}_1^j - \mathbf{v}_0^j$ is used to update the corresponding Jacobian matrix A_μ^j so as to minimize the error

$$E_j = \frac{1}{2}(\Delta \boldsymbol{\theta}^j - A_\mu^j \Delta \mathbf{v}^j)^2 \tag{3.51}$$

This gives following update law for Jacobian matrices:

$$A_\mu^j(k+1) = A_\mu^j(k) + \frac{\eta}{\|\Delta \mathbf{v}^j\|^2}(\Delta \boldsymbol{\theta}^j - A_\mu^j \Delta \mathbf{v}^j)\Delta \mathbf{v}^{j^T} \tag{3.52}$$

8. Increment the iteration counter $k = k + 1$ and go to step 1.

Note that during training phase, the output of the network is not an weighted average of all the neurons as was previously done. It is because of the fact that the number of $\boldsymbol{\theta}$ clusters associated with neurons are not same and hence the usual neighborhood concept cannot be used in this case.

3.5.3 Testing Phase

- For a given target point $\mathbf{u}_t$, the winner neuron μ is computed based on its minimum Euclidean distance from the target in input space as given by (3.1). This winner neuron is associated with several, say, N_μ sub-clusters in joint angle space.

- One can choose among these sub-clusters based on some criterion. The redundancy is resolved using three criteria namely, lazy arm movement, minimum angle norm and minimum condition number of Jacobian matrix. Let the winning joint angle sub-cluster be β. Once the winner indices μ and β are computed, coarse and fine joint angle outputs are given by

$$\boldsymbol{\theta}_0 = \boldsymbol{\theta}_0^\beta = \boldsymbol{\theta}_\mu^\beta + A_\mu^\beta(\mathbf{u}_t - \mathbf{w}_\mu) \tag{3.53}$$
$$\boldsymbol{\theta}_1 = \boldsymbol{\theta}_1^\beta = \boldsymbol{\theta}_0 + A_\mu^\beta(\mathbf{u}_t - \mathbf{v}_0) \tag{3.54}$$

where $\mathbf{v}_0$ is the end-effector position recorded after coarse movement. Multiple steps may be taken to improve the positioning accuracy further.

3.5.4 Redundancy Resolution

Sub-clustering gives rise to multiple configurations for every target position. Let the index of winner neuron in input space be μ and this winner neuron is associated with N_μ sub-clusters. One can select a suitable configuration based on different criteria. In this chapter, the following three criteria are used for resolving redundancy:

- *Lazy arm movement:* The angle sub-cluster which is closest to the current robot configuration is selected as the winner. The winning sub-cluster for these criteria is given by

$$\beta = \arg\min_j \|\boldsymbol{\theta}_\mu^j - \boldsymbol{\theta}_c\| \tag{3.55}$$

where $\boldsymbol{\theta}_c$ is the current robot configuration.

- *Minimum angle norm:* The angle sub-cluster whose norm is minimum is selected as the winner. The winning sub-cluster for this criteria is given by

$$\beta = \arg\min_j \|\boldsymbol{\theta}_\mu^j\| \tag{3.56}$$

- *Minimum condition number:* The matrix A_μ^j represents a local inverse image Jacobian matrix associated with each joint angle vector $\boldsymbol{\theta}_\mu^j$. For visual motor control, it is desirable to have low condition number for image Jacobian matrices to improve the robustness and numerical stability of the system [146]. Sometimes the condition number of image Jacobian matrix is used as measure of *perceptibility* of motion [147, 148]. The *perceptibility* is a quantitative measure of the ability of a camera setup to observe the changes in image feature due to motion of robot end-effector. It is used to evaluate the ease of achieving vision-based control and steering away from singular configurations [147].

Since several joint angle configurations are available for a given winner neuron, each associated with an inverse Jacobian matrix, one can choose a particular configuration based on the minimum condition number of these matrices. The winning sub-cluster based on minimum condition number is given by

$$\beta = \arg\min_j [cond(A_\mu^j)] \tag{3.57}$$

where $cond(A_\mu^j)$ is the condition number of the matrix A_μ^j.

3.5.5 Tracking a Continuous Trajectory

Unlike original VMC algorithm, the neighborhood function is not used during the training phase. This might not be of concern when the task is to reach isolated points in the workspace. However, one would like to have a continuous trajectory in joint angle space for a continuous trajectory in image coordinate space. In other words, the conservative property [149] of inverse kinematic solution is desirable and needs to be preserved.

Use of neighborhood helps in maintaining a continuous trajectory in the joint angle space by avoiding abrupt changes in the joint motion. This happens because the network output is obtained by taking the weighted average of individual neuron outputs within a neighborhood around the winner neuron. In order to facilitate further discussion, we divide the available configurations into following two classes:

- Each joint angle sub-cluster represents a particular robot configuration. Two configurations, $\boldsymbol{\theta}_\gamma^i$ and $\boldsymbol{\theta}_\lambda^j$, are said to be *similar* if $\|\boldsymbol{\theta}_\gamma^i - \boldsymbol{\theta}_\lambda^j\| < K$, $i \neq j$, $\gamma \neq \lambda$ and K is an arbitrarily small constant. Otherwise, they would be called *dissimilar* configurations.

- All angle sub-clusters associated with each neuron γ are *dissimilar*. In other words, $\|\boldsymbol{\theta}_\gamma^i - \boldsymbol{\theta}_\gamma^j\| > K$ for $i \neq j$. Note that K is the threshold that was used for creating a new sub-cluster during training phase. Refer to the discussion under section "*clustering in output space*".

The concept of neighborhood in a sub-clustered environment employed in this paper is explained in Fig. 3.17. In this figure, $\mathbf{w}_\gamma$ and $\mathbf{w}_\lambda$ are two neighboring neurons, represented by their respective weight vectors. The angle sub-clusters associated with them are represented by $\boldsymbol{\theta}_\gamma^i$, $i = 1, 2, \ldots, N_\gamma$ and $\boldsymbol{\theta}_\lambda^j$, $j = 1, 2, \ldots, N_\lambda$ respectively. For simplicity, it is assumed that $N_\gamma = N_\lambda = 5$. Let's assume that the lazy-arm criterion (LA) selects the angle vector '2' in case of γ neuron and '5' in case of λ neuron (refer to (3.55)). On the other hand, the minimum-angle norm criterion (MA) selects angle vector '4' in case of γ and '3' in case of λ neuron (refer to (3.56)). The dotted lines show that the configurations corresponding to these sub-clusters are *similar*. In other words, the configurations $(\gamma, 2)$ and $(\lambda, 5)$ are *similar* and so are the configurations $(\gamma, 4)$ and $(\lambda, 3)$.

The network output is obtained as a weighted average over all *similar* joint angle vectors. The coarse and fine joint angle movements, after incorporating neighborhood, are given by following two equations

$$\boldsymbol{\theta}_0 = \frac{\sum_\gamma h_\gamma (\boldsymbol{\theta}_\gamma^{\beta(\gamma)} + A_\gamma^{\beta(\gamma)} (\mathbf{u}_t - \mathbf{w}_\gamma))}{\sum_\gamma h_\gamma} \qquad (3.58)$$

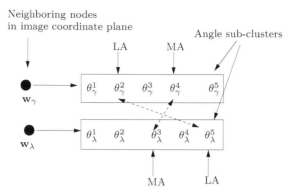

FIGURE 3.17: Defining neighborhood in sub-clustered environment. $(\mathbf{w}_\gamma, \boldsymbol{\theta}_\gamma^2)$ and $(\mathbf{w}_\lambda, \boldsymbol{\theta}_\lambda^5)$ are similar configurations for lazy-arm criterion when γ and λ are neighboring neurons in input space.

$$\boldsymbol{\theta}_1 = \boldsymbol{\theta}_0 + \frac{\sum_\gamma h_\gamma A_\gamma^{\beta(\gamma)}(\mathbf{u}_t - \mathbf{v}_0)}{\sum_\gamma h_\gamma} \qquad (3.59)$$

where $h_\gamma = e^{-\frac{\|\gamma - \mu\|^2}{\sigma^2}}$ is the neighborhood function defined in input space. μ is the winner neuron in input (image coordinate) space. $\beta(\gamma)$ is the index of the winning angle sub-cluster associated with the neuron γ. The winning sub-cluster for each neuron γ is obtained using the same criterion which was used for winner neuron μ. In the simulation section, it would be shown that the use of neighborhood results in a smooth trajectory in the joint angle space.

3.6 Simulation and Results

3.6.1 Network Architecture and Workspace Dimensions

A 3-dimensional neural lattice with $7 \times 7 \times 7$ neurons is selected for the task. Note that $10 \times 10 \times 10$ nodes were used for least square based method in the previous chapter. With the current scheme, this smaller network is found to be adequate for obtaining better accuracy. Training data is generated using forward kinematic model (2.1) and camera model (4.12). A Cartesian workspace of dimension of $600\ mm \times 500\ mm \times 500\ mm$ is considered for both simulation as well as experiment. All points within this workspace are visible through both the cameras of the stereo-vision system. Joint angle values are generated

Simulation and Results 101

randomly within the physical limits of the manipulator and only those input-output pairs are retained where the end-effector positions are visible by both the cameras simultaneously. The ranges of input and output spaces are given in Table 2.2. Since end-effector positions in camera plane and joint angles have different range of values, data points are normalized within ± 1.

3.6.2 Training

The network is trained offline using 50,000 data generated using forward kinematic model (2.1) and (4.12). Again, it is to be noted that this data size is one-tenth the size of the training set used in the least-square based method discussed in the previous chapter. The training can be carried out 'on-line' which would necessitate generating data by moving the robot continuously. Generating such a large number of data on a real system might not be convenient. Hence we follow the hybrid approach proposed by Behera et al. [150] where, a network is trained offline using approximate models and then it is fine-tuned during online operation.

A new θ sub-cluster is formed whenever the distance of incoming θ_t from the existing nearest sub-cluster exceeds the threshold $K = 1.0$. The distribution of sub-clusters for lattice neurons is shown in Fig. 3.18(a). The number of sub-clusters associated with each neuron varies between 10 to 35. The distribution of joint angle sub-clusters in 3-dimensional manipulator workspace is shown in Fig. 3.18(b). It is seen that the points with less number of redundant solutions lie toward the boundary of the workspace as shown by '+' symbols (no. of solutions < 19). The number of points with very large number of redundant solutions is also less as shown by circles (no. of solutions > 30). The square symbols represent points with number of solutions in between 19 and 30. It is a common observation that the number of inverse kinematic solutions for a given target position varies across the manipulator workspace. This distribution of redundant solutions is captured effectively by the proposed architecture as shown in the Figure 3.18(b).

The discretization of input and output spaces by the lattice neurons is shown in Fig. 3.19. It is seen that the topology is captured by the lattice neurons both in input and output spaces. Since the number of sub-clusters for each neuron is decided based on the input data distribution, the outliers are automatically avoided. Outliers are the neurons which do not represent input data. In Fig. 3.19(a)-3.19(b), it is seen that all sub-clusters are surrounded by training data points and they do not lie in an empty region. These results are in contrast to that of a standard SOM algorithm as shown in figure 3.14 where clusters are localized and some of them are outliers.

3.6.3 Testing

The following tasks were performed to demonstrate the efficacy and usefulness of the proposed schemes.

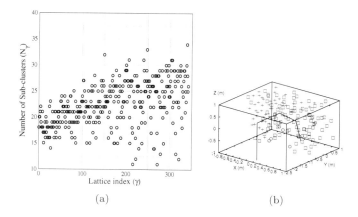

(a) (b)

FIGURE 3.18: (a) The number of sub-clusters for lattice neurons. The number of sub-clusters vary from one neuron to another. The number of sub-clusters for a neuron is decided on-line based on the training data distribution. (b) Distribution of joint angle sub-clusters in the 3-dimensional manipulator workspace. The number of solutions available for a given target point varies across the workspace. There are very few points where the number of available solutions are too high. The number of solutions decreases toward boundary of workspace. In this figure, '+' represents the points where the number of sub-clusters $N_\gamma < 18$, squares represent the points with N_γ is between 19 to 30 and circles represent points with $N_\gamma > 30$. A typical robot configuration is shown in solid line.

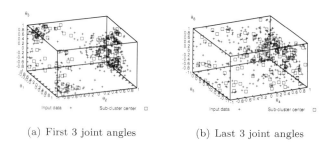

(a) First 3 joint angles   (b) Last 3 joint angles

FIGURE 3.19: KSOM-SC architecture captures the topology in the output space thereby eliminating the limitations of standard KSOM-based architectures. Since the number of joint angle sub-clusters are decided on-line based on the actual data distribution, the outliers are automatically avoided. It's because a new sub-cluster is created only when the new training data is far away from current sub-clusters.

Simulation and Results 103

3.6.3.1 Reaching Isolated Target Positions in the Workspace

The joint angle vectors were computed for 20,000 target positions located randomly within the manipulator workspace. Only one step is used to compute the necessary joint angles. The performance of the proposed sub-clustering based scheme is compared with the standard SOM-based scheme [141, 150]. The performance comparison is provided in Table 3.2. Since the performance of standard SOM-based schemes is found to depend on initial values of network parameters, the test results are averaged over twenty different runs. Each run starts with a different random initialization.

As discussed earlier, sub-clustering based scheme preserves the redundancy available in training data and provides a finite number of joint angle vectors for every target position in the manipulator workspace. The redundancy is resolved using three criteria namely, "lazy-arm method (LA)," "minimum angle norm (MA)," and "minimum condition number (MC)." Various solutions obtained for a given target point after resolving redundancy is shown in Fig. 3.20.

From the Table 3.2, it is clear that sub-clustering based methods give better positioning accuracy than the standard SOM-based schemes proposed by Martinetz [140] and Schulten [141]. It is seen in Table 3.2 that standard SOM algorithm gives rise to very large joint angle variation as reflected in the magnitude of joint angle norm.

It is said earlier that the joint angle vectors associated with lattice neurons do not capture the topology of output space. This makes the convergence of standard SOM-based schemes sensitive to initial values of network parameters as shown in Fig. 3.21(a). In this figure, average positioning error over 20,000 isolated target positions are shown for different runs, where each run starts with a different random initialization and includes a training and a testing phase. It is observed in Fig. 3.21(a) that the average positioning error varies widely across different runs and can be as high as 100 mm. On the contrary, the performance of sub-clustering based scheme using lazy-arm redundancy resolution technique is comparatively less sensitive to initial conditions and

TABLE 3.2: Performance comparison for reaching isolated points

Scheme	Average positioning error		Angle norm (normalized)	learning parameters $(\eta_w, \eta_t, \eta_a, \sigma)$
	Cartesian space (mm)	Image space (pixels)		
Standard SOM	24.93	7.87	1.54	0.1, 0.2, 0.9, 0.1
SC + LA	3.83	1.22	0.83	0.1, 0.5, 0.9, 1.5
SC + MA	3.0	0.95	0.76	0.1, 0.5, 0.9, 1.5
SC + MC	18.67	6.33	1.38	0.1, 0.9, 0.9, 0.2

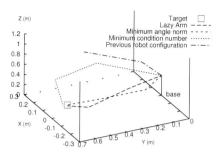

FIGURE 3.20: Redundancy resolution using various criteria. Different criteria give different configurations for the same target point.

performance remains constant across different runs. The average positioning accuracy over all runs is below 4 mm implying that the proposed scheme is very accurate in position tracking.

The advantage of an on-line incremental learning scheme is that one can execute multiple fine movements (refer (3.54)) to improve the positioning accuracy as shown in Fig. 3.21(b). This figure plots the number of fine movement steps required to achieve a given average positioning error over 1000

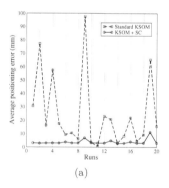
(a)

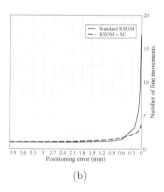
(b)

FIGURE 3.21: (a) Dependence of convergence on initial conditions. Standard SOM algorithm is sensitive to initial conditions and hence varies widely for different runs. The performance of sub-clustering based scheme is independent of initial conditions. (b) Improving positioning accuracy using multiple fine movements. The results are obtained by averaging over 1000 test points. Standard SOM algorithm requires very large number of fine movements to attain accuracy below 1 mm as compared to sub-clustering based method.

Simulation and Results 105

test points. Since the standard SOM algorithm is sensitive to initial network parameters, we have taken initial network parameters corresponding to run 14 where both standard SOM and proposed algorithm have same level of performance. This figure shows that the standard SOM algorithm requires seventeen fine movement steps to attain the average positioning accuracy of 0.1 mm while the proposed sub-clustering based method takes three steps to attain the same accuracy. The result shown in this figure is in sharp contrast to those reported by Angulo et al. [151] where authors take more than 1000 steps to attain that level of accuracy even when they use orientation information to compute the inverse kinematic solution. Note that in Fig. 3.21(b), the result for standard SOM is shown for those initial conditions for which the network performance is comparable with the proposed scheme. Otherwise, the performance of the standard SOM will further deteriorate for other initial conditions.

3.6.3.2 Tracking a Straight Line Trajectory

The desired straight line trajectory in Cartesian space is given by

$$y = \frac{5}{6}x + \frac{11}{20}$$
$$z = \frac{5}{6}(x + 0.3) \tag{3.60}$$

where $-0.3\ m \leq x \leq 0.3\ m$. Six hundred points are generated sequentially on this line and the joint angles for each point are computed using only one fine movement. Since learning algorithm has been designed so that it can be used on-line, the parameters are updated even during the testing phase. This helps in improving tracking accuracy. A typical tracking result obtained using sub-clustering scheme is shown in Fig. 3.22(a). In this case, the redundancy is resolved using lazy-arm criterion. The corresponding joint angle trajectory is shown in Fig. 3.22(b). It is seen that for a continuous trajectory in Cartesian space, the joint angle movement is continuous and hence the inverse kinematic solution is conservative in nature. Performance comparison among various redundancy resolution schemes is provided in Table 3.3. The solution based on minimum condition number (MC) criterion does not give rise to continuous joint angle movement. While the joint angle trajectory obtained using MA criterion is unique for a given task space trajectory, the joint angle trajectory obtained using LA criterion depends on current robot configuration. If all the joint angles of the manipulator are reset to zero prior to robot movement, then the solution for lazy arm movement converges to that of minimum angle norm movement. Different criteria give rise to different trajectories in the configuration space as shown in Fig. 3.23. This figure shows the solutions obtained using two LA and MA criteria. The purpose of this figure is to show that different criteria lead to different solution trajectories.

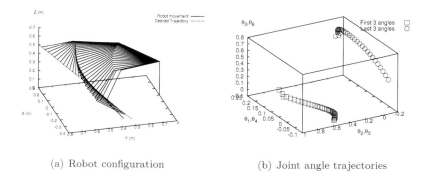

(a) Robot configuration   (b) Joint angle trajectories

FIGURE 3.22: Tracking a straight line using lazy-arm criterion. A continuous trajectory in task-space gives rise to a continuous trajectory in joint angle space.

TABLE 3.3: Performance comparison for tracking a straight line trajectory

Criterion	Average positioning error		Angle norm (normalized)	learning parameters (η_w, η_t, η_a, σ)
	Cartesian space (mm)	Image space (pixels)		
SC + LA	2.91	0.75	0.89	0.9, 0.9, 0.9, 1.4
SC + MA	2.70	0.63	0.77	0.9, 0.9, 0.9, 1.4
SC + MC	34.78	9.74	1.361	0.9, 0.9, 0.9, 1.4

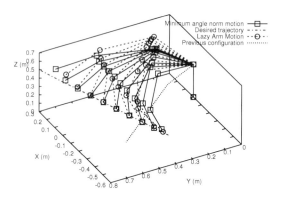

FIGURE 3.23: Redundancy resolution while tracking a straight line trajectory. The minimum angle norm (MA) solution is independent of initial configuration. The lazy arm (LA) solution depends on initial condition.

3.6.3.3 Tracking an Elliptical Trajectory

The desired trajectory to be traversed is given by

$$x = 0.2\sin t; \qquad y = 0.5 + 0.2\cos t; \qquad z = \frac{5}{6}(x + 0.3) \qquad (3.61)$$

where t varies from 0 to 2π. A total of 628 points are generated sequentially on this trajectory, the joint angles are computed in one step for each point. A typical trajectory obtained using lazy-arm criterion and sub-clustering technique is shown in Fig. 3.24(a). The corresponding trajectories for joint angles is shown in Fig. 3.24(b). This reaffirms our previous assertion that the inverse kinematic solution obtained is conservative. The performance comparison for different schemes is provided in Table 3.4. Similar references can be drawn from this table as it was done in case of a straight line trajectory.

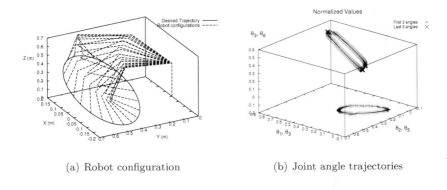

(a) Robot configuration   (b) Joint angle trajectories

FIGURE 3.24: Tracking an elliptical trajectory using lazy arm movement. The inverse kinematic solution is conservative in the sense that a closed loop trajectory in task space gives rise to a closed trajectory in configuration space.

TABLE 3.4: Performance comparison for tracking an elliptical trajectory

Criterion	Average positioning error		Angle norm (normalized)	learning parameters $(\eta_w, \eta_t, \eta_a, \sigma)$
	Cartesian space (mm)	Image space (pixels)		
SC + LA	1.44	0.42	1.0	0.9, 0.9, 0.9, 2.0
SC + MA	1.28	0.42	0.82	0.9, 0.9, 0.9, 2.0
SC + MC	22.78	6.82	1.55	0.9, 0.9, 0.9, 2.0

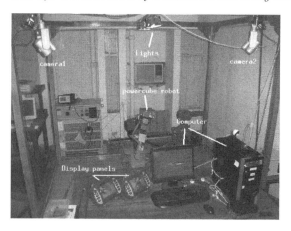

FIGURE 3.25: Experimental setup for VMC experiment.

3.6.4 Real-Time Experiment

The actual setup used for experiment is shown in Figure 3.25. The Cartesian workspace visible by both cameras has a dimension of 600 mm × 600 mm × 500 mm. The image frame has a dimension of 320 × 240 pixels. A yellow ball is taken as a target and robot tip is identified using pink color. The regions of interest are extracted using thresholding and filtering operations. The centroid of the region is used by the VMC algorithm to compute necessary joint angles. All image processing tasks are carried out using OpenCV library [135]. The algorithm is implemented using C/C++ on a computer with Intel Pentium 4 1.8 GHz processor. The cameras are calibrated using Reg Wilson's C implementation of Tsai algorithm [134]. In order to reduce positioning error in real-time experiment, LEDs are used to detect end-effector position as well as target position in a dark environment. The time required for manipulator to execute a given joint angle command is approximately 80 milliseconds. The image processing unit must provide the coordinates of the target within this time interval. During closed loop operation, the synchronization between the image processing unit and robot arm motion is carried out using software timers.

The Cartesian workspace visible by both cameras has a dimension of 600 mm × 600 mm × 500 mm. The image frame has a dimension of 320 × 240 pixels. The target and the robot tip are identified with yellow and pink colors respectively. The initial location of robot end-effector and target in the image plane is shown in Figure 3.26(a). The regions of interest are extracted using thresholding and filtering operations. The centroid of the region is used by the VMC algorithm to compute necessary joint angles. These joint angles are applied to the robot and it moves to a position as shown in Figure 3.26(b). This figure shows the final state of manipulator obtained after robot movement. The

Simulation and Results 109

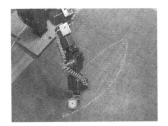

(a) Initial state   (b) Final state

FIGURE 3.26: Extraction of pixel coordinates for robot end-effector and target. Initial state is the state before robot movement. The final state refers to state obtained after robot movement.

error is computed after making corrections for the pixel width of the robot end-effector as well as the target object. The accuracy in detecting tip position is further improved by using LEDs against a dark background.

The proposed scheme can be implemented on-line as was done by Schulten [140, 141]. However, we used a hybrid approach as suggested by Behera et al. [150] where the SOM network is trained offline by generating data from the model rather than from the actual system. This reduces the demand on on-line data generation.

The trained network was used on-line to compute joint angle vectors for 20 random locations in the manipulator workspace. Since it was not possible to accurately measure the manipulator tip positions in world coordinate, the distance error was measured directly in pixel coordinates. Only one fine step was used to compute the necessary joint angle vector for each point. The average distance error in the image plane is computed to be 12 pixels. This error can be reduced by taking multiple fine steps. It takes fifteen steps, on an average, for reaching an accuracy of about 1 pixel for a given target point in the image plane.

3.6.4.1 Redundant Solutions

Although a redundant manipulator can reach a target point using more than one joint angle configuration, existing learning algorithms can provide only a unique inverse kinematic solution. However, the proposed redundancy preserving network provides multiple solutions simultaneously for any given target position. Fig. 3.27 shows some of these inverse kinematic solutions for a given target position represented by a yellow ball. Readers should be able to appreciate that the yellow ball has been reached by many kinematic configurations. Although the network provides multiple solutions, a particular configuration can be selected based on the task requirement.

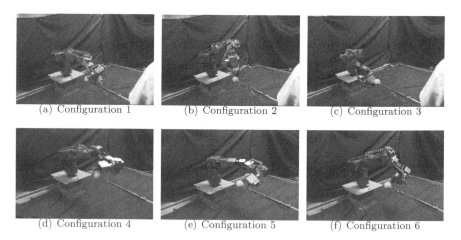

(a) Configuration 1 (b) Configuration 2 (c) Configuration 3

(d) Configuration 4 (e) Configuration 5 (f) Configuration 6

FIGURE 3.27: Redundant solutions for a given target position: The target is a yellow ball which is reached by the manipulator using various kinematic configurations. Six such kinematic configurations are shown in this figure.

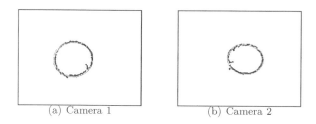

(a) Camera 1 (b) Camera 2

FIGURE 3.28: Tracking a circular trajectory as observed in the image plane. Thin line represents the desired trajectory and the thick line is the trajectory of actual end-effector position. The end-effector is detected using an LED in a dark environment. The image has been processed to ensure visibility on paper.

3.6.4.2 Tracking a Circular and a Straight Line Trajectory

The real-time experimental results for tracking a circular trajectory are shown in Fig. 3.28 and the results for tracking a straight line trajectory are shown in Fig. 3.29. The desired trajectories are specified directly on the image plane using a camera model. In Fig. 3.28, the desired trajectory is shown as a thin line and the actual end-effector trajectory is shown in thick points. In Fig. 3.29, the desired and actual positions are shown in red and blue colors respectively. Due to the overlap of colors, the actual trajectory appears discontinuous. The end-effector position is detected using an LED against a dark background. The average positioning error in both cases is 8 pixels.

Summary

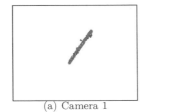
(a) Camera 1

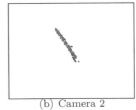
(b) Camera 2

FIGURE 3.29: Tracking a straight line trajectory as observed in the image plane. The actual and the desired trajectories of the end-effector position are demonstrated. The end-effector is detected using an LED against a dark background.

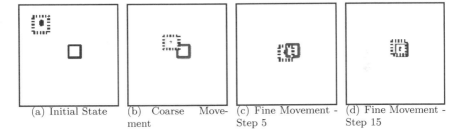

(a) Initial State (b) Coarse Movement (c) Fine Movement - Step 5 (d) Fine Movement - Step 15

FIGURE 3.30: Multi-step Movement: The rectangular box with solid boundary is the target specified on image plane that is to be reached. The current end-effector position is shown with a box with dashed boundary. (a) shows the *initial state*. The end-effector LED appears as black dot. (b) shows the state after coarse movement. (c) shows the end-effector position after five fine movement steps, and (d) shows the final end-effector position obtained after 15 fine movement steps. The final positioning error is less than 1 pixel. The images have been processed to ensure visibility on paper.

3.6.4.3 Multi-Step Movement

In order to improve positioning accuracy of end-effector, multiple fine movement steps are applied to the manipulator, and the corresponding results are shown in Fig. 3.30. The first step is a coarse movement as shown in Fig. 3.30(b). The positioning error is approximately 15 pixels after the coarse movement. This error has been further reduced to 7 pixels and 1 pixels respectively after five and fifteen fine movements as shown in Fig. 3.30(c) and Fig. 3.30(d).

3.7 Summary

This chapter shows that existing KSOM-based visual motor control algorithms are inefficient for applications in redundant manipulators. The existing learning architectures do not preserve topology of the output space (refer Fig. 3.14). Thus such algorithms become sensitive to initial network parameters as shown in Fig. 3.21(a). Since existing learning architectures do not preserve redundancy, the redundant manipulator cannot perform dexterous tasks using these VMC algorithms.

Thus a KSOM based redundancy preserving network proposed in the previous chapter is used to provide several kinematic configurations for a given target position. A real-time algorithm to learn network parameters has been proposed. Since each lattice neuron is associated with multiple solutions in joint angle space, an online adaptive clustering algorithm has been proposed to learn these joint angle vectors. It is shown that this adaptive sub-clustering in output space leads to the preservation of topology in both input and output spaces by the KSOM lattice neurons. It is also shown that the proposed KSOM network is insensitive to initial network parameters unlike the standard KSOM network. The smoothness of joint angle trajectories is maintained through a modified neighborhood concept thereby preserving the conservative property of the inverse kinematic solution.

These modifications lead to following improvements over existing methods:

- It is possible to attain a positioning error less than 1 mm in real-time experiment. In simulation, it is even possible to reach less than 0.1 mm error.

- It leads to a ten-fold reduction in the amount of data needed for training the network. A smaller sized network (with $7 \times 7 \times 7$ nodes) gives rise to an accuracy of 1 mm when it is trained with only 50,000 data points. The training can be carried out on-line.

- The inverse kinematic solution is conservative. In other words, the joint angle trajectories are smooth and continuous for a continuous task space trajectory.

- The positioning accuracy attained is less dependent on initial conditions of the network as compared to the standard SOM-based algorithms.

- In a multi-step movement, the number of steps needed for obtaining lesser positioning error increases very slowly as compared the existing SOM-based algorithms.

Three criteria namely, lazy-arm movement, minimum angle-norm movement, and minimum condition number of Jacobian matrices, are used to resolve redundancy. Apart from providing dexterity, it is shown that the proposed scheme provides the best positioning accuracy as compared to standard SOM-based schemes. Finally, simulation results are validated through experiments on a 7 DOF PowerCube ™ robot manipulator.

4

Model-based Visual Servoing of a 7 DOF Manipulator

4.1 Introduction

This chapter presents the theoretical development of the MBVS control law using image moments. It is focused on determining the analytical form of the interaction matrix related to selected image moment features from a segmented image. A real-time experimental results using a 7 DOF PowerCube robot manipulator are presented to validate the system convergence.

This chapter starts the discussion by reviewing the basic image moment definitions and computations as presented Section 4.5. The following Section 4.6 gives detail derivation of the first order of the image moments with respect to the time using Green's theorem. A fundamental pinhole camera model is described in Section 4.7 which then is used for further development to determine the interaction matrix. Section 4.8 presents the interaction matrix derivation of the selected image moment features. The real-time experimental results are presented in Section 4.9 and followed by the summary of this chapter as described in Section 4.10.

4.2 Kinematic Control of a Manipulator

The position of the end-effector and the associated joint angle configuration are coupled with forward and inverse kinematic relationship of the manipulator. The forward kinematic map expresses the Cartesian space position of the end-effector $\mathbf{x}$ for the given joint angle configuration $\boldsymbol{\theta}$, as

$$\mathbf{x} = \mathbf{f_x}(\boldsymbol{\theta}), \qquad (4.1)$$

where the dimension of the task space $\mathbf{x}$ is n, and that of the joint angle space $\boldsymbol{\theta}$ is m. In the case of redundant manipulators $n < m$ and the degree of redundancy is given by $m - n$. $\mathbf{f_x}(\boldsymbol{\theta})$ is highly nonlinear and is obtained from the geometry of the manipulator using Denavit-Hartenberg (D-H) parameters [152].

The inverse kinematic relation computes the joint angle space configuration θ which is required to reach the desired position $\mathbf{x}_d$. The closed form inverse kinematic relationship exists only for simple manipulator configurations. In general inverse kinematic control is achieved with forward differential kinematic relationship, since it expresses a linear relationship between the joint angular velocity $\dot{\theta}$ and the Cartesian space velocity $\dot{\mathbf{x}}$. The forward differential kinematic relationship between $\dot{\theta}$ and $\dot{\mathbf{x}}$ is represented as

$$\dot{\mathbf{x}} = \mathbf{J}\dot{\theta}, \qquad (4.2)$$

where $\mathbf{J} = \frac{\partial \mathbf{f}_x}{\partial \theta}$ is the kinematic Jacobian of the manipulator. The joint angular velocity required for the given end-effector velocity is computed using inverse Jacobian. The pseudo-inverse method computes the value of $\dot{\theta}$ as

$$\dot{\theta} = \mathbf{J}^{-1}\dot{\mathbf{x}}_d, \qquad (4.3)$$

where $\mathbf{J}^{-1}$ is the pseudo-inverse of the kinematic Jacobian, and $\dot{\mathbf{x}}_d$ is the desired end-effector velocity. The open-loop solution obtained using the above equation, unavoidably leads to solution drift due to numerical integration and, hence, results in task space error $\mathbf{e} = \mathbf{x}_d - \mathbf{x}$. To overcome this drawback in open-loop control, the closed loop kinematic control is developed with the task space error $\mathbf{e}$. In closed loop kinematic control the joint velocity is computed as

$$\dot{\theta} = k_p \mathbf{J}^{-1} \mathbf{e}, \qquad (4.4)$$

where $k_p > 0$ is proportional gain which control the speed of the convergence to the desired position $\mathbf{x}_d$.

4.2.1 Kinematic Control of Redundant Manipulator

Kinematic control is difficult for a redundant manipulator since infinite number of solutions exist to reach the given workspace position. The control of a kinematically redundant manipulator to reach the object is a highly challenging task owing to the one-to-many inverse kinematic relationship.

In the case of redundant manipulators, $\mathbf{J}$ is not a square matrix and theoretically infinite joint angular velocity $\dot{\theta}$ exist to generate the given end-effector velocity. Inverse Jacobian does not exist in case of redundant manipulators since the associated Jacobian is not square and, hence, the pseudo-inverse has been employed. Inverse kinematic control of the redundant manipulator using generalized pseudo-inverse. Joint velocity is computed for redundant manipulator in closed loop as

$$\dot{\theta} = k_p \mathbf{J}^+ \mathbf{e}, \qquad (4.5)$$

where $\mathbf{J}^+$ is the pseudo-inverse of the kinematic Jacobian, and $\dot{\mathbf{x}}_d$ is the desired end-effector velocity. Henceforth, the notation $(.)^+$ will be used to indicate the generalized pseudo-inverse of $(.)$. The pseudo-inverse based solution results in lazy-arm movement, i.e., it minimizes the joint angular velocity in least square sense.

Pseudo-inverse based kinematic control is widely popular since the relationship between the various joint angular velocities, which can generate the desired end-effector velocity can be established using the pseudo-inverse of Jacobian $\mathbf{J}$. $\mathbf{J}^+$ obeys the property that the matrix $(\mathbf{I} - \mathbf{J}^+\mathbf{J})$ projects onto the null space of $\mathbf{J}$ and, hence, the vector $\mathbf{J}(\mathbf{I} - \mathbf{J}^+\mathbf{J})\boldsymbol{\phi} = \mathbf{0}$ for all vectors $\boldsymbol{\phi}$. A joint angular velocity computed as $\dot{\boldsymbol{\theta}} = (\mathbf{I} - \mathbf{J}^+\mathbf{J})\boldsymbol{\phi}$ for any vector $\boldsymbol{\phi} \in R^m$ does not generate any end-effector motion but only changes the internal joint angle configuration of the manipulator. The internal reconfiguration of the manipulator is popularly known as self-motion of the manipulator. The different joint angular velocities which can generate the given end-effector velocity are given by the relationship,

$$\dot{\boldsymbol{\theta}} = \mathbf{J}^+\dot{\mathbf{x}} + k_n(\mathbf{I} - \mathbf{J}^+\mathbf{J})\boldsymbol{\phi}, \tag{4.6}$$

where $\mathbf{I}$ is the identity matrix of order m, and k_n is the gain which determines the magnitude of the self-motion.

4.3 Visual Servoing

Vision is employed in robotics owing to its flexibility during manipulation. Visual feedback gives dynamic information about the environment and the object. Typically, vision-based manipulator control is executed in open loop fashion, "looking" and then "moving." This results in poor positioning accuracy due to the model inaccuracies. An alternative approach is to use a visual control loop which is generally referred as visual servoing. Vision-based manipulator control uses either single camera or multiple cameras to give visual feedback to the manipulator system. Visual servoing systems use one of two camera configurations: eye-in-hand or eye-to-hand. In eye-in-hand configuration, a camera is mounted on the end-effector while in eye-to-hand configuration, the cameras are fixed in the workspace. Eye-to-hand configuration is also known as stand-alone camera system.

Visual servoing schemes use the image features $\mathbf{u}$ to represent the position of the end-effector and the object in the vision space. The desired position $\mathbf{x}_d$ and the current position $\mathbf{x}$ of the end-effector are observed through the camera as the desired image feature vector $\mathbf{u}_d$ and the current image feature vector $\mathbf{u}$, respectively. In general, visual servoing uses the linear relationship between the change in the image feature vector $\mathbf{u}$ and the change in the Cartesian space position of the end-effector $\mathbf{x}$ for controlling the manipulator. The image Jacobian $\mathbf{L}$, represents the relationship between the end-effector motion and the image feature motion as

$$\dot{\mathbf{u}} = \mathbf{L}\dot{\mathbf{x}}, \tag{4.7}$$

where $\mathbf{u}, \dot{\mathbf{u}} \in R^p$. $\mathbf{L}$ is a $p \times n$ matrix and is also referred as interaction matrix in literatures. The control task is to compute the necessary Cartesian space velocity motion such that the end-effector will reach the desired position in vision space asymptotically.

The simple proportional control law which results in asymptotic stabilization is expressed as

$$\dot{\mathbf{x}}_d = k_p \, \mathbf{L}^+ (\mathbf{u}_d - \mathbf{u}), \tag{4.8}$$

where k_p is a proportional gain, $\mathbf{L}^+$ is the pseudo-inverse of $\mathbf{L}$ and $\mathbf{u}_d - \mathbf{u}$ is the error between the desired and the current image features. Here afterwards, the error vector in the visions space is expressed as $\mathbf{e}_u$ and, hence, $\mathbf{e}_u = \mathbf{u}_d - \mathbf{u}$: $\in R^p$. The above controller requires the exact knowledge of $\mathbf{L}$ and its pseudo-inverse though it ensures global stability. Hence, the image Jacobian $\mathbf{L}$ is to be estimated at each instant. To reduce the computational complexity, the image Jacobian is estimated at the desired position and then the pseudo-inverse is evaluated for the estimated image Jacobian to implement the controller. This eliminates the continuous estimation of $\mathbf{L}$ and the computation of the pseudo-inverse $\mathbf{L}^+$ in real-time. But, this results in a locally stabilizing controller since the sufficient positivity condition of stability is valid in local region only.

4.3.1 Estimating the Vision Space Motion with Camera Model

The camera acts as a sensor and gives visual feedback about the position of the object and the environment. The feature extracted from the camera must give enough information such that the manipulator can be controlled to reach the object located in a 3-D workspace.

Two cameras fixed in the workspace can be used to give visual feedback about the environment. The position of the centroid of the object in the image plane of the stereo-vision is used as the feedback to locate the position of the object. If a single fixed camera is used instead of two cameras, the depth information about the environment will not be available while identifying only the centroid of the object. To control the manipulator in 3-D workspace, more features are to be extracted with single camera. Typically, area of the object gives a good information about the distance between the camera and the object. Hence, the manipulator can be controlled to reach the object in a 3-dimensional workspace with the image features area, and position of the object in the image plane. Alternatively, multiple non-coplanar points can be identified in the object to get the information about the position of the object. But such an approach demands the knowledge of the relative position of the feature points in the 3-D workspace, for effective control.

The position of the object is estimated by extracting the image feature as follows: the positional coordinates of the end-effector in the Cartesian space get projected as pixel coordinates in the frame buffer of the image plane. The position $\mathbf{x} = [x \ y \ z]^T$ in the Cartesian space gets projected into the camera frame buffer as (x_f, y_f), which corresponds to the $x - y$ coordinates of the

Visual Servoing

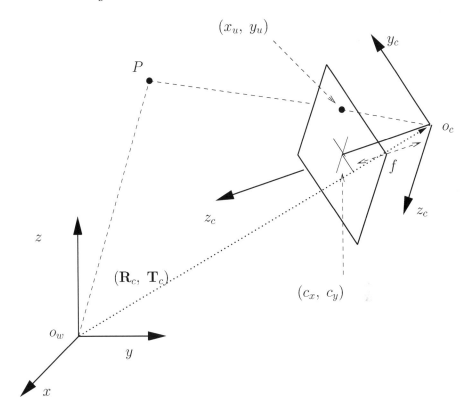

FIGURE 4.1: Transformation from Cartesian space to vision space.

camera frame buffer, respectively. The position of the end-effector in the vision space is obtained through series of transformations. These transformations are computed as a camera model, which computes the position of the point in the vision space, from the point's position in the Cartesian space. This necessitates a camera model to compute **u** from **x** in simulations.

4.3.2 Transformation from Cartesian Space to Vision Space

The transformation associated with computing a point's position in the vision from the Cartesian space is shown in Figure 4.1. The origin of the world coordinate frame and that of the camera coordinate frame are shown as O_w and O_c, respectively. The origin of the camera coordinate frame is located at $[T_x \; T_y \; T_z]^T$ in the world coordinate frame and the orientation is represented using $\mathbf{R}_c$. The origin of the camera coordinate frame (x_c, y_c, z_c) coincides with front nodal point of the camera and the z_c axis coincides with the camera's optical axis. The image plane is assumed to be parallel to the (x_c, y_c) plane at a distance of f from the origin, where f is the effective focal length of the camera. The position (x_f, y_f) of a point P in the camera plane is obtained from the point's position $\mathbf{x} = [x\ y\ z]^T$ in the world coordinate frame as follows:

The position $\mathbf{x} = [x\ y\ z]^T$ is transformed from the world coordinate frame to the position $\mathbf{x}_c = [x_c\ y_c\ z_c]^T$ in the camera coordinate frame through rotation $\mathbf{R}_c$ and translation $\mathbf{T}_c$. The transformation is expressed in the form of equation as

$$\begin{bmatrix} x_c \\ y_c \\ z_c \end{bmatrix} = \mathbf{R}_c \begin{bmatrix} x \\ y \\ z \end{bmatrix} + \begin{bmatrix} T_x \\ T_y \\ T_z \end{bmatrix}, \quad \text{where } \mathbf{R}_c = \begin{bmatrix} r_1 & r_2 & r_3 \\ r_4 & r_5 & r_6 \\ r_7 & r_8 & r_9 \end{bmatrix}. \quad (4.9)$$

In the above equation, $\mathbf{R}_c$ describes the orientation of the camera in the world coordinate frame, and $\mathbf{T}_c = [T_x\ T_y\ T_z]^T$ is the translational position of the camera in the world coordinate frame. The projected position of the point P in the image plane is computed using an ideal pinhole camera model. This transformation is obtained by perspective projection as follows:

$$\begin{aligned} x_u &= f\frac{x_c}{z_c} \\ y_u &= f\frac{y_c}{z_c}. \end{aligned} \quad (4.10)$$

The position (x_u, y_u) is computed with the assumption of an ideal pinhole camera. But, there exists distortion and the position obtained with ideal pinhole model is not accurate. This is compensated for by using the lens distortion coefficient κ. The true position of the point's image (x_d, y_d) in the sensor plane is computed from the ideal undistorted position as

$$\begin{aligned} x_u &= x_d(1 + \kappa\rho^2) \\ y_u &= y_d(1 + \kappa\rho^2), \end{aligned} \quad (4.11)$$

where $\rho = \sqrt{x_d^2 + y_d^2}$. Finally the image of the point is transformed from the sensor plane to its coordinates in the camera's frame buffer (x_f, y_f) as

$$\begin{aligned} x_f &= \frac{s_x x_d}{d_x} + c_x \\ y_f &= \frac{y_d}{d_y} + c_y, \end{aligned} \quad (4.12)$$

where

c_x, c_y : Pixel coordinates of optical center;

s_x : Scale factor to account for any uncertainty due to imperfections in hardware timing for scanning and digitization;

d_x : Dimension of camera's sensor element along x coordinate direction (in mm/sel);

d_y : Dimension of camera's sensor element along y coordinate direction (in mm/sel).

Visual Servoing

The computation of a point's position in the frame buffer requires the geometric and camera parameters used in the aforementioned transformations. The parameters which specify the position and the orientation of the camera relative to the world coordinate frame is commonly known as extrinsic or external parameters. The camera parameters which project the point from the camera coordinate frame to the frame buffer are known as intrinsic or internal parameters. The camera calibration is the process of estimation of a model for camera overlooking a workspace.

4.3.3 The Camera Model

Tsai's algorithm is a popularly known camera calibration technique which is based on the pinhole perspective projection discussed above and estimates eleven parameters: f, κ, c_x, c_y, s_x, T_x, T_y, T_z, R_x, R_y, and R_z. The Tsai model represents the rotation angles for the transformation between the world and camera coordinates with (R_x, R_y, R_z). The elements of the rotation matrix $\mathbf{R}_c$ is computed from $[R_x, R_y, R_z]$ as follows:

$$\begin{aligned} r_1 &= c_\beta c_\gamma \\ r_2 &= c_\gamma s_\alpha s_\beta - c_\alpha s_\gamma \\ r_3 &= s_\alpha s_\gamma + c_\alpha c_\gamma s_\beta \\ r_4 &= c_\beta s_\gamma \\ r_5 &= s_\alpha s_\beta s_\gamma + c_\alpha c_\gamma \\ r_6 &= c_\alpha s_\beta s_\gamma - c_\gamma s_\alpha \\ r_7 &= -s_\beta \\ r_8 &= c_\beta s_\alpha \\ r_9 &= c_\alpha c_\beta, \end{aligned}$$

(4.13)

where $c_\alpha = \cos(R_x)$, $c_\beta = \cos(R_y)$, $c_\gamma = \cos(R_z)$, $s_\alpha = \sin(R_x)$, $s_\beta = \sin(R_y)$, $s_\gamma = \sin(R_z)$.

In addition to the above eleven variable camera parameters, Tsai's model uses the following six fixed intrinsic camera constants:

d_x : Size of camera's sensor element in x coordinate direction (in mm/sel),

d_y : Size of camera's sensor element in y coordinate direction (in mm/sel),

N_{cx} : Number of sensor elements in camera's x direction (in sels),

N_{fx} : Number of pixels in frame grabber's x direction (in pixels),

d_{px} : Effective number of pixel in y coordinate direction of the frame buffer (in mm/pixel), and

d_{py} : Effective number of pixel in y coordinate direction of the frame buffer (in mm/pixel).

These six parameters can be obtained from the manufacturer's data sheet.

The image feature vector $\mathbf{u}$ is obtained using the estimated model for stereo-vision as $\mathbf{u} = [u_1 \ u_2 \ u_3 \ u_4]^T$ where (u_1, u_2) and (u_3, u_4) are the $x - y$ coordinates of the first and the second camera, respectively. Hence, (u_1, u_2) is the (x_f, y_f) of the first camera, and (u_3, u_4) corresponds to the (x_f, y_f) of the second camera, respectively. Hence, the control vectors $\mathbf{u}_d$ and $\mathbf{u}$ belong to R^4 in the all the experiments presented in this chapter.

4.3.4 Computation of Image Feature Velocity in the Vision Space

The image Jacobian which represents the motion of the image features with respect to the motion in the Cartesian space is given by

$$\mathbf{L} = \begin{bmatrix} \frac{k_{cx}}{z_c} & 0 & \frac{-k_{cx}(x_f - c_x)}{z_c} \\ 0 & \frac{k_{cy}}{z_c} & \frac{-k_{cy}(y_f - c_y)}{z_c} \end{bmatrix}, \quad (4.14)$$

where (k_{cx}, k_{cy}) are the gains associated to transform the Cartesian space position to the x-y coordinate of the vision space, and z_c is the distance between the image plane and the object in the camera coordinate frame. The camera gains (k_{cx}, k_{cy}) are computed using the camera parameters as follows:

$$k_{cx} = \frac{fs_x}{d_x}$$
$$k_{cx} = \frac{f}{d_y}. \quad (4.15)$$

The vision space velocity is computed from the Cartesian space velocity as

$$\begin{bmatrix} \dot{u}_1 \\ \dot{u}_2 \\ \dot{u}_3 \\ \dot{u}_4 \end{bmatrix} = \begin{bmatrix} \mathbf{L}_1 & \mathbf{0} \\ \mathbf{0} & \mathbf{L}_2 \end{bmatrix} \begin{bmatrix} \dot{\mathbf{x}}_{C_1} \\ \dot{\mathbf{x}}_{C_2} \end{bmatrix}, \quad (4.16)$$

where $\mathbf{L}_i$ is the image Jacobian of the ith camera, $\dot{\mathbf{x}}_{C_i} = \begin{bmatrix} \dot{x}_{c_i} & \dot{y}_{c_i} & \dot{z}_{c_i} \end{bmatrix}^T$ represents the velocity of the end-effector in the coordinate frame of the ith camera.

The velocity of the end-effector in the coordinate frame of the ith camera is

$$\begin{bmatrix} \dot{x}_{c_i} \\ \dot{y}_{c_i} \\ \dot{z}_{c_i} \end{bmatrix} = \mathbf{R}_{c_i} \begin{bmatrix} \dot{x} \\ \dot{y} \\ \dot{z} \end{bmatrix}$$

$$\dot{\mathbf{x}}_{c_i} = \mathbf{R}_{c_i} \dot{\mathbf{x}}, \tag{4.17}$$

where $\mathbf{R}_{c_i}$ is the rotational transformation between the robot coordinate frame and the camera coordinate frame. The parameters (k_{cx}, k_{cy}), (c_x, c_y), and $\mathbf{R}_c$ are obtained from the camera model estimated with Tsai algorithm.

4.4 Kinematic Control of a Manipulator Directly from Vision Space

As discussed in previous sections, classical approaches estimate the Cartesian space velocity of the end-effector from the vision space with the visual servoing schemes, and then the joint velocity is computed to follow the Cartesian space trajectory generated by the visual servoing scheme. Visual servoing scheme uses the image features $\mathbf{u}_d$, $\mathbf{u}$, and the Cartesian space information $\mathbf{x}$ to compute the end-effector velocity $\dot{\mathbf{x}}$. Inverse kinematic schemes compute the joint angular velocity from the end-effector velocity $\dot{\mathbf{x}}$, using the current joint angle $\boldsymbol{\theta}$ and the environmental constraints.

Alternatively, the vision space trajectories can be directly controlled from the joint angle space by combining visual servoing with redundancy resolution in a single framework. The redundancy is achieved for the trajectories specified in the vision space while satisfying the additional constraints introduced by the environment. The controller computes the joint angle configuration directly from the visual feedback resulting in a direct and efficient control over the vision space.

The relationship between image feature velocity and joint angular velocity is obtained by combining equations (4.2) and (4.7) as

$$\begin{aligned} \dot{\mathbf{u}} &= \mathbf{LPJ}\dot{\boldsymbol{\theta}} \\ &= \mathbb{J}\dot{\boldsymbol{\theta}}, \end{aligned} \tag{4.18}$$

where $\mathbf{P}$ is the transformation matrix representing the coordinate transformation between the world coordinate frame and the camera coordinate frame, and $\mathbb{J} = \mathbf{LPJ}$ is a $p \times m$ Jacobian matrix from the joint angle space to the vision space. Here afterwards, the notation $\mathbb{J}$ will be used to represent the Jacobian from the joint angle space to the vision space.

The closed loop proportional controller resulting in asymptotic stabilization is given as

$$\dot{\boldsymbol{\theta}} = k_p \, \mathbb{J}^+ \mathbf{e}_u. \tag{4.19}$$

Since the pseudo-inverse of the Jacobian $\mathbb{J}$ is used in the control law, the controller would result in "lazy-arm movement." The null space of $\mathbb{J}$ can be used to satisfy the additional constraints required in the dynamic environment.

4.5 Image Moments

One of the basic problems in the design of an imagery pattern recognition system relates to the selection of a set of appropriate numerical attributes of features to be extracted from the object of interest for the purpose of classification. The recognition of objects from imagery may be achieved with many methods by identifying an unknown object as a member of a set of known objects. Efficient object recognition techniques abstracting characterizations uniquely from objects for representation and comparison are crucially important for a given pattern recognition system. One of the popular techniques to characterise an object in the image space is to use image moments.

Image moments are mathematical entities whose inferred values can describe objects in the image space defined either from closed contour or a set of points in a segment or within the boundary. Image moment descriptors are related to geometrical properties of a segment, e.g., position, orientation or size of a segment in the image space. The mathematical concept of moments has been around for many years and has been used in many diverse fields ranging from mechanics and statistics to pattern recognition and image understanding. In [153] the mathematical concept of image moments was first introduced in 1962, it also proved that the image moment functions are insensitive to a particular segment's changes, such as translation, rotation and scaling, based on the theory of algebraic invariant. Since then, the moment invariants theory has been applied in several applications, such as character recognition [154], pose estimation [155], [156], [157], and object matching [158]. At the same time of widespread use of image moments for such applications, the theoretical developments of the moment invariants have also been presented. In [159] detecting objects using n-fold rotation symmetry was presented to resolve the limitation of the moment invariant to detect objects which present symmetries. A formulation of image moments which reduces the computational time was proposed in [160], by defining image moments as a function of image coordinates of the points lying on the boundary of the considered segment, instead of taking into account the whole points of an object's image. A survey of the theoretical developments of image moments was presented in [161].

The objective of this section is to introduce image moments and to illustrate the usefulness in visual servoing. The usefulness of the image moments approach in the visual servoing is because of their key feature. The key feature of the image moments is that they can be used as a generic representation of an object in the image space in the form of an image segment. The image moments' descriptors are computed using information of all image points lying on a segmented image. Image moments approach does not consider an individual point as a descriptor but a global information of a segmented image, therefore, they are robust in the presence of image noise. Image moments are potential entities that can be used as a feedback signal vector in visual

Image Moments

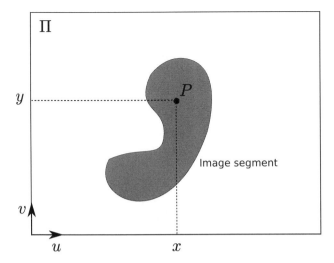

FIGURE 4.2: Image segment on an image plane Π.

servoing systems since they consider the shape of the object which has more geometric meaning compared with the image points.

It is well known that the key solution for visual servoing controller development is to construct the interaction matrix associated with selected visual features. A visual servoing controller is developed to have smooth and continues velocity control trajectories to follow the movement of the target object. Therefore, the interaction matrix that maps the camera kinematic screw into visual feature velocity is needed. To develop the interaction matrix, the visual feature properties have to be modeled and to be derived. Firstly the geometric moments and its definitions are presented in the following discussion.

The 2D geometric moment of order $i+j$ of a density distribution function $f(x,y)$ is defined in terms of the surface integral as

$$m_{ij} = \int_{-\infty}^{\infty} \int_{-\infty}^{\infty} x^i y^j f(x,y) dx dy \qquad (4.20)$$

where $f(x,y)$ could be related to the pixel intensity value, color or other pixel image properties, e.g., $f(x,y)$ is the grey level of a point P at coordinate (x,y) lying on the image segment in the orthogonal frame (u,v) (see Figure 4.2). Therefore, it can be deduced that moments are strongly correlated to the shape of the image segment, as it is formulated in the product term of $x^i y^j$. In [153], Hu stated that $f(x,y)$ is piecewise continous and has nonzero values only in a finite region of Π plane, then the moment sequence m_{ij} is uniquely determined by $f(x,y)$, and the other way around that $f(x,y)$ is uniquely determined by the moment sequence m_{ij}. Therefore, complete moments can be computed and used to uniquely describe an object image if moments of all orders exist and the image segment has finite area. But, it requires an infinite number of moment descriptors to obtain all information contained in an image segment. Thus, it becomes very important to select a meaningful subset of the moment descriptors which contains minimal information to characterize an object image uniquely for a specific application.

The fundamental geometric properties contained in the image segment can be represented by the lower order moment functions. Those geometric properties are called: area, center of gravity, centered moments, and orientation. The definition of the zeroth order moment m_{00} of the function $f(x,y)$

$$m_{00} = \int_{-\infty}^{\infty} \int_{-\infty}^{\infty} f(x,y) dx dy \tag{4.21}$$

represents the total mass of the given function or image $f(x,y)$. The zeroth order moment is basically the total area for the case of a binary image. The first two order moments determine the position of the center of gravity or the center of the area, defined as

$$m_{10} = \int_{-\infty}^{\infty} \int_{-\infty}^{\infty} x f(x,y) dx dy \tag{4.22}$$

$$m_{01} = \int_{-\infty}^{\infty} \int_{-\infty}^{\infty} y f(x,y) dx dy \tag{4.23}$$

The center of gravity is the point where all the mass of the image $f(x,y)$ concentrated without changing the first moment of the image about any axis. In the 2D case, the moment values of the center of gravity coordinates are denoted as

$$x_g = \frac{m_{10}}{m_{00}}, \quad y_g = \frac{m_{01}}{m_{00}} \tag{4.24}$$

Generally, it is common practice that the center of gravity is chosen to represent the position of an object image in the field of view, since it defines a unique location of an image segment $f(x,y)$ that can be used as a reference point.

The centered moments of $f(x,y)$ are defined as

$$\mu_{ij} = \int_{-\infty}^{\infty} \int_{-\infty}^{\infty} (x-x_g)^i (y-y_g)^j f(x,y) dx dy \tag{4.25}$$

The centered moments μ_{ij} are invariant under the translation of coordinates [153].

$$x' = x + c_x \tag{4.26}$$
$$y' = y + c_y \tag{4.27}$$

where x', y' are new position coordinates of centered moments after those are translated by constants c_x and c_y. Given the center of the gravity and the

Image Moments

moments m_{kl} the centered moments can be described as a binomial function as

$$\mu_{ij} = \sum_{k=0}^{i}\sum_{l=0}^{j} \binom{i}{k}\binom{j}{l}(-x_g)^{i-k}(y_g)^{j-l} m_{kl} \qquad (4.28)$$

where the binomial coefficients

$$\binom{i}{k} = \frac{i!}{k!(i-k)!} \qquad (4.29)$$

The reciprocal relationship to (4.28) which describes m_{kl} by giving the centered moments is computed as

$$m_{kl} = \sum_{i=0}^{k}\sum_{j=0}^{l} \binom{k}{i}\binom{l}{j}(x_g)^{k-i}(y_g)^{l-j} \mu_{ij} \qquad (4.30)$$

The second order moments m_{02}, m_{11}, m_{20} are known as the moment of inertia. These descriptors are used to define another important geometric image feature, the orientation of an image segment. The description of the image segment orientation is generally measured in the direction of the principal axes, describing how the image lies in the field of view. In terms of moments, the orientation of the principal axis α (see Figure 4.3) is described as

$$\alpha = \frac{1}{2}\tan^{-1}\left(\frac{2\mu_{11}}{\mu_{20}-\mu_{02}}\right) \qquad (4.31)$$

where

$$\mu_{11} = \int_{-\infty}^{\infty}\int_{-\infty}^{\infty}(x-x_g)(y-y_g)f(x,y)dxdy \qquad (4.32)$$

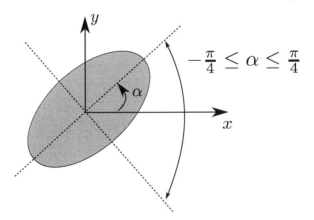

FIGURE 4.3: Image segment of an ellipse-shaped object.

$$\mu_{20} = \int_{-\infty}^{\infty}\int_{-\infty}^{\infty}(x-x_g)^2 f(x,y)dxdy \qquad (4.33)$$

$$\mu_{02} = \int_{-\infty}^{\infty}\int_{-\infty}^{\infty}(y-y_g)^2 f(x,y)dxdy \qquad (4.34)$$

In (4.31) the principal axis angle α is in the range of $-\frac{\pi}{4} \leq \alpha \leq \frac{\pi}{4}$. The following discussion will be focused on the determination of the relationship between the camera movement and the time variation of the image moment m_{ij}.

4.6 Image Moment Velocity

The system modeling objective is to determine the linear relationship between the camera movement $\dot{\mathbf{x}}_c = (\mathbf{v}_c, \boldsymbol{\omega}_c)$ and the time variation $\dot{m}_{ij}$ of image moment m_{ij}. Recall equation (2.2) for the case of the image moments as a image feature set

$$\dot{m}_{ij} = \mathbf{L}_{m_{ij}}\dot{\mathbf{x}}_c \qquad (4.35)$$

where $\mathbf{L}_{m_{ij}}$ is the interaction matrix determined by the image moments. Let's recall Green's theorem which gives the relationship between a line integral around a simple closed contour $\mathcal{C}$ and a double integral over the segment region $\mathcal{D}$ [162]. Figure 4.4 shows an image segment $\mathcal{D}$ and its contour $\mathcal{C}$, the line integral of the vertical component of the tangential component of vector field $\mathbf{F}$ along $\mathcal{C}$ as the double integral of the vertical component curl $\mathbf{F}$ over the region $\mathcal{D}$ enclosed by $\mathcal{C}$.

$$\oint_{\mathcal{C}} \mathbf{F} \cdot d\mathbf{r} = \iint_{\mathcal{D}} (\mathrm{curl}\mathbf{F}) \cdot \mathbf{k}\, dA \qquad (4.36)$$

where $\mathbf{k}$ is a unit vector perpendicular to the segment $\mathcal{D}$. The contour $\mathcal{C}$ can be described by vector equation

$$\mathbf{x}(t) = x(t)\mathbf{i} + y(t)\mathbf{j} \qquad (4.37)$$

while the unit tangent vector $\mathbf{t}$ is given by

$$\mathbf{t}(t) = \frac{x(t)}{|\mathbf{r}(t)|}\mathbf{i} + \frac{y(t)}{|\mathbf{r}(t)|}\mathbf{j} \qquad (4.38)$$

then the normal vector $\mathbf{n}(t)$ is described by

$$\mathbf{n}(t) = \frac{y(t)}{|\mathbf{r}(t)|}\mathbf{i} - \frac{x(t)}{|\mathbf{r}(t)|}\mathbf{j} \qquad (4.39)$$

Developing (4.36), another version of Green's theorem can be derived which says that the line integral of the normal component of $\mathbf{F}$ along $\mathcal{C}$ is equal to

Image Moment Velocity

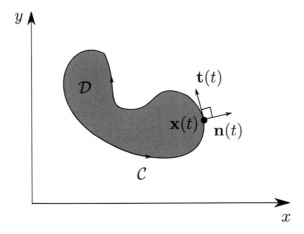

FIGURE 4.4: Image segment $\mathcal{D}$ enclosed by contour $\mathcal{C}$.

double integral of the divergence of $\mathbf{F}$ over the segment $\mathcal{D}$ enclosed by $\mathcal{C}$ [162].

$$\oint_{\mathcal{C}} \mathbf{F} \cdot \mathbf{n}\, ds = \iint_{\mathcal{D}} \mathrm{div}\mathbf{F}(x,y)\, dxdy \qquad (4.40)$$

In relation to the definition stated in Green's theorem, the image moment function can be defined as

$$m_{ij}(t) = \iint_{\mathcal{D}(t)} h(x,y)\, dxdy \qquad (4.41)$$

where $h(x,y) = x^i y^j f(x,y)$. It can be seen in the right side of (4.41) that the only part that varies on t is the image segment $\mathcal{D}(t)$. Therefore the image moment velocity $\dot{m}_{ij}$ can be obtained by analyzing the variation of $\mathcal{C}(t)$, since there is a relationship between a segment and its contour (4.40). Figure 4.5 shows the variation of contour $\mathcal{C}(t)$, the variation of m_{ij} is computed using m_{ij} on segmented area between $\mathcal{C}(t+1)$ and $\mathcal{C}(t)$. The segmented area between $\mathcal{C}(t+1)$ and $\mathcal{C}(t)$ can be obtained by integrating every point $\mathbf{x} = (x,y)$ along $\mathcal{C}(t)$ until it reaches $\mathcal{C}(t+1)$ by the scalar product between point's velocity $\dot{\mathbf{x}}$ and the normal vector $\mathbf{n}$. Thus using (4.40), the image moment velocity can be written as [163, 164]

$$\begin{aligned}
\dot{m}_{ij} &= \iint_{\mathcal{D}(t)} \mathrm{div}[h(x,y)\dot{\mathbf{x}}]\, dxdy \\
&= \iint_{\mathcal{D}(t)} \left(\frac{\partial h(x,y)\dot{\mathbf{x}}}{\partial x} + \frac{\partial h(x,y)\dot{\mathbf{x}}}{\partial y} \right) dxdy \\
&= \iint_{\mathcal{D}(t)} \left(\frac{\partial h(x,y)}{\partial x}\dot{x} + \frac{\partial h(x,y)}{\partial y}\dot{y} + h(x,y)\left(\frac{\partial \dot{x}}{\partial x} + \frac{\partial \dot{y}}{\partial y} \right) \right) dxdy
\end{aligned} \qquad (4.42)$$

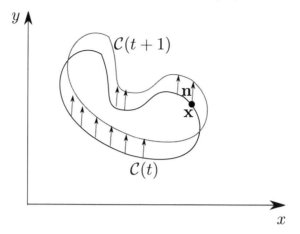

FIGURE 4.5: Variation of contour $\mathcal{C}(t)$.

In (4.42), the term $\dot{x}$, $\dot{y}$, $\frac{\partial \dot{x}}{\partial x}$ and $\frac{\partial \dot{y}}{\partial y}$ can be described as the kinematic screw of the camera. Similarly, by applying Green's theorem, the variation of the centered image moments can be defined as

$$\dot{\mu}_{ij} = \iint_{\mathcal{D}(t)} \frac{\partial h_\mu(x,y)}{\partial x}(\dot{x} - \dot{x}_g) + \frac{\partial h_\mu(x,y)}{\partial y}(\dot{y} - \dot{y}_g)$$
$$+ h_\mu(x,y)\left(\frac{\partial \dot{x}}{\partial x} + \frac{\partial \dot{y}}{\partial y}\right) dx dy \qquad (4.43)$$

where $h_\mu(x,y) = (x - x_g)^i (y - y_g)^j f(x,y)$. To continue the development of the visual servoing model using image moment m_{ij} and μ_{ij}, the relationship between the camera kinematic screw and the corresponding point velocity expressed in the camera frame has to be known. This relationship can be obtained using a pinhole camera model as explained in the following section.

4.7 A Pinhole Camera Projection

To model the visual servoing system, the projection of the object with respect to the pinhole camera system must be described [165, 166]. A pin hole camera projection is shown in Figure 4.6 as the position of a point p with respect to the camera frame $o_c x_c y_c z_c$ and p_π is a projection point of p on the image plane. The coordinate $p_\pi(x_\pi, y_\pi, f_c)$ is expressed relative to the camera frame $o_c x_c y_c z_c$. The origin of a 2D local coordinate of the image plane is denoted as (u_0, v_0) and it is also called the principal point, measured in pixels. The pixel coordinate of p_π is represented by (x_i, y_i). The transformation of the

A Pinhole Camera Projection

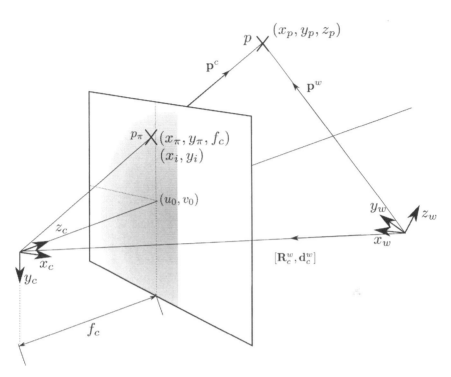

FIGURE 4.6: A pinhole camera projection.

camera frame $o_c x_c y_c z_c$ with respect to the reference frame $o_w x_w y_w z_w$ is given by $[\mathbf{R}_c^w, \mathbf{d}_c^w]$. The normalized projection coordinate p into p_π can be expressed as

$$x_\pi = f_c \frac{x_p}{z_p}, \; y_\pi = f_c \frac{y_p}{z_p} \qquad (4.44)$$

In order to relate the 3D coordinate into 2D pixel coordinate we determine the projection relationship between (x_π, y_π) and (x_i, y_i) as follows

$$x_\pi = \frac{(x_i - u_0)}{k_u}, \; y_\pi = \frac{(y_i - v_0)}{k_v} \qquad (4.45)$$

where (k_u, k_v) are the conversion factors from meters to pixels for the horizontal and vertical camera axis, respectively.

Substituting (4.44) into (4.45), we have

$$x_i = f_c k_u \frac{x_p}{z_p} + u_0, \; y_i = f_c k_v \frac{y_p}{z_p} + v_0 \qquad (4.46)$$

Thus, once the values of the intrinsic camera parameters f_c, k_u, k_v, u_0, v_0 are determined the mapping of a point in a 3D coordinate system to image the coordinate system can be computed. In the matrix form, we have

$$\begin{bmatrix} x_i \\ y_i \\ 1 \end{bmatrix} = \begin{bmatrix} f_c k_u & 0 & u_0 \\ 0 & f_c k_v & v_0 \\ 0 & 0 & 1 \end{bmatrix} \begin{bmatrix} X_p \\ Y_p \\ 1 \end{bmatrix} \quad (4.47)$$

$$\mathbf{x}_i = \mathbf{A}\mathbf{x}_p \quad (4.48)$$

where $X_p = \frac{x_p}{z_p}$ and $Y_p = \frac{y_p}{z_p}$ denote normalized coordinates.

To continue the derivation of the velocity relationship of a point p between coordinate frames in Figure 4.6, we need to introduce a Skew symmetric matrix $\mathbf{S}(\boldsymbol{\omega})$ [152]. It is expressed as

$$\mathbf{S}(\boldsymbol{\omega}) = \begin{bmatrix} 0 & -\omega_z & \omega_y \\ \omega_z & 0 & -\omega_x \\ -\omega_y & \omega_x & 0 \end{bmatrix} \quad (4.49)$$

where $\boldsymbol{\omega}$ is the angular velocity vector of the rotating frame with respect to the reference frame. The transpose of $\mathbf{S}(\boldsymbol{\omega})$ can be easily derived as

$$\mathbf{S}^T(\boldsymbol{\omega}) = \mathbf{S}(-\boldsymbol{\omega}) \quad (4.50)$$

In relation with the first derivative of a rotation matrix $\mathbf{R}$, the Skew symmetric matrix can be defined as

$$\mathbf{S}(\boldsymbol{\omega}) = \dot{\mathbf{R}}\,\mathbf{R}^T \quad (4.51)$$

furthermore, the transpose of $\mathbf{S}(\boldsymbol{\omega})$ can also be derived as

$$\mathbf{S}^T(\boldsymbol{\omega}) = (\dot{\mathbf{R}}\,\mathbf{R}^T)^T = \dot{\mathbf{R}}^T\,\mathbf{R} \quad (4.52)$$

By multiplying the right term and the left term of (4.52) with $\mathbf{R}^T$, we have

$$\mathbf{S}^T(\boldsymbol{\omega})\mathbf{R}^T = \dot{\mathbf{R}}^T\mathbf{R}\mathbf{R}^T \quad (4.53)$$

Let's remark the orthogonality property of a rotation matrix $\mathbf{R}$

$$\mathbf{R}\,\mathbf{R}^T = \mathbf{I} \quad (4.54)$$

where $\mathbf{I} \in \mathbb{R}^3$ is an identity matrix. Substituting (4.54) into (4.53)

$$\dot{\mathbf{R}}^T = \mathbf{S}^T(\boldsymbol{\omega})\mathbf{R}^T \quad (4.55)$$

The Skew symmetric matrix properties in (4.51) and (4.55) are important properties for the next derivation of the velocity relationship of a point p between coordinate frames in Figure 4.6. The representation of the coordinate point p with respect to reference frame $o_w x_w y_w z_w$ is given as

$$\mathbf{p}^w = \mathbf{R}_c^w \mathbf{p}^c + \mathbf{d}_c^w \quad (4.56)$$

A Pinhole Camera Projection

Since the relative position of a point p with respect to the camera frame is considered, we have

$$\mathbf{p}^c = (\mathbf{R}_c^w)^T (\mathbf{p}^w - \mathbf{d}_c^w) \tag{4.57}$$

The velocity vector of a point p is then derived as

$$\dot{\mathbf{p}}^c = (\dot{\mathbf{R}}_c^w)^T (\mathbf{p}^w - \mathbf{d}_c^w) - (\mathbf{R}_c^w)^T \dot{\mathbf{d}}_c^w \tag{4.58}$$

Note, a point p is static and the camera frame is moving, thus $\dot{\mathbf{p}}^w = 0$. Rearranging (4.57) and substituting into (4.58)

$$\dot{\mathbf{p}}^c = (\dot{\mathbf{R}}_c^w)^T \mathbf{R}_c^w \mathbf{p}^c - (\mathbf{R}_c^w)^T \dot{\mathbf{d}}_c^w \tag{4.59}$$

Using Skew symmetric properties (4.54) and (4.55) yields

$$\dot{\mathbf{p}}^c = \mathbf{S}^T(\boldsymbol{\omega}) (\mathbf{R}_c^w)^T \mathbf{R}_c^w \mathbf{p}^c - (\mathbf{R}_c^w)^T \dot{\mathbf{d}}_c^w \tag{4.60}$$
$$= \mathbf{S}(-\boldsymbol{\omega}) \mathbf{I} \mathbf{p}^c - \mathbf{R}_w^c \dot{\mathbf{d}}_c^w \tag{4.61}$$
$$= \mathbf{S}(-\boldsymbol{\omega}) \mathbf{p}^c - \dot{\mathbf{d}}_w^c \tag{4.62}$$

where $\dot{\mathbf{d}}_w^c$ is the camera frame translational velocity. In the matrix form, it is expressed as

$$\begin{bmatrix} \dot{x}_p \\ \dot{y}_p \\ \dot{z}_p \end{bmatrix} = \begin{bmatrix} 0 & \omega_z & -\omega_y \\ -\omega_z & 0 & \omega_x \\ \omega_y & -\omega_x & 0 \end{bmatrix} \begin{bmatrix} x_p \\ y_p \\ z_p \end{bmatrix} - \begin{bmatrix} \dot{x}_c \\ \dot{y}_c \\ \dot{z}_c \end{bmatrix} \tag{4.63}$$

then we have

$$\dot{x}_p = y_p \omega_z - z_p \omega_y - \dot{x}_c \tag{4.64}$$
$$\dot{y}_p = z_p \omega_x - x_p \omega_z - \dot{y}_c \tag{4.65}$$
$$\dot{z}_p = x_p \omega_y - y_p \omega_x - \dot{z}_c \tag{4.66}$$

The first derivative of (4.46) is derived as

$$\dot{x}_i = f_c k_u \frac{\dot{x}_p z_p - x_p \dot{z}_p}{(z_p)^2} \tag{4.67}$$

$$\dot{y}_i = f_c k_v \frac{\dot{y}_p z_p - y_p \dot{z}_p}{(z_p)^2} \tag{4.68}$$

Note, we assume an ideal camera model where the principal point (u_0, v_0) is centered $(0,0)$ and the image is undistorted $\frac{k_u}{k_v} = 1$. The velocity of the projected point on the image plane associated with the camera movement is then derived by substituting (4.64), (4.65), (4.66) into (4.67) and (4.68)

$$\dot{x}_i = f_c k_u \frac{(y_p \omega_z - z_p \omega_y - \dot{x}_c) z_p - x_p (x_p \omega_y - y_p \omega_x - \dot{z}_c)}{(z_p)^2}$$

$$= f_c k_u \frac{\left(\left(\frac{y_i z_p}{f_c k_v}\right)\omega_z - z_p\omega_y - \dot{x}_c\right)z_p - \left(\frac{x_i z_p}{f_c k_u}\right)\left(\left(\frac{x_i z_p}{f_c k_u}\right)\omega_y - \left(\frac{y_i z_p}{f_c k_v}\right)\omega_x - \dot{z}_c\right)}{(z_p)^2}$$

$$= \frac{-f_c k_u}{z_p}\dot{x}_c + \frac{x_i}{z_p}\dot{z}_c + \frac{x_i y_i}{f_c k_v}\omega_x - \frac{f_c^2 k_u^2 + x_i^2}{f_c k_u}\omega_y + y_i\omega_z \tag{4.69}$$

$$\dot{y}_i = f_c k_v \frac{(z_p\omega_x - x_p\omega_z - \dot{y}_c)z_p - y_p(x_p\omega_y - y_p\omega_x - \dot{z}_c)}{(z_p)^2}$$

$$= f_c k_v \frac{\left(z_p\omega_x - \left(\frac{x_i z_p}{f_c k_u}\right)\omega_z - \dot{y}_c\right)z_p - \left(\frac{y_i z_p}{f_c k_v}\right)\left(\left(\frac{x_i z_p}{f_c k_u}\right)\omega_y - \left(\frac{y_i z_p}{f_c k_v}\right)\omega_x - \dot{z}_c\right)}{(z_p)^2}$$

$$= \frac{-f_c k_v}{z_p}\dot{y}_c + \frac{y_i}{z_p}\dot{z}_c + \frac{f_c^2 k_v^2 + y_i^2}{f_c k_v}\omega_x - \frac{x_i y_i}{f_c k_u}\omega_y - x_i\omega_z \tag{4.70}$$

It can be composed in the matrix form as follows

$$\begin{bmatrix}\dot{x}_i\\\dot{y}_i\end{bmatrix} = \begin{bmatrix}\frac{-f_c k_u}{z_p} & 0 & \frac{x_i}{z_p} & \frac{x_i y_i}{f_c k_v} & -\frac{f_c^2 k_u^2 + x_i^2}{f_c k_u} & y_i \\ 0 & \frac{-f_c k_v}{z_p} & \frac{y_i}{z_p} & \frac{f_c^2 k_v^2 + y_i^2}{f_c k_v} & -\frac{x_i y_i}{f_c k_u} & -x_i\end{bmatrix}\begin{bmatrix}\dot{x}_c\\\dot{y}_c\\\dot{z}_c\\\omega_x\\\omega_y\\\omega_z\end{bmatrix} \tag{4.71}$$

At this step, without loss generality and to simplify the model, (4.71) can be expressed as

$$\dot{\mathbf{x}} = \begin{bmatrix}-\frac{1}{Z} & 0 & \frac{x}{Z} & xy & -1-x^2 & y \\ 0 & -\frac{1}{Z} & \frac{y}{Z} & 1+y^2 & -xy & -x\end{bmatrix}\dot{\mathbf{x}}_c \tag{4.72}$$

where the depth $Z = z_p$, $f_c = k_u = k_v = 1$, $\dot{\mathbf{x}} = [\dot{x}_i, \dot{y}_i]$ and $\dot{\mathbf{x}}_c = [\mathbf{v}_c, \boldsymbol{\omega}_c]$. Note that the camera parameters obtained from camera calibration can be easily replugged into the final model of the IBVS using image moments. However, the camera parameters described in (4.72) are sufficient to be applied in practice. The robustness of the IBVS in the presence of camera calibration errors is described in Appendix C.

4.8 Image Moment Interaction Matrix

In [167], the projection analysis of the 3D geometric primitive (lines, cylindrical, spherical, etc.) parameters into the image plane has been discussed. Figure 4.7 shows the projection of a 3D object into the planar limb surface and the image plane [168]. The projection of 3D geometric parameters of an object into its planar limb surface is expressed as

$$\mathbf{h}(\mathbf{X}, \mathbf{P}) = 0 \tag{4.73}$$

Image Moment Interaction Matrix

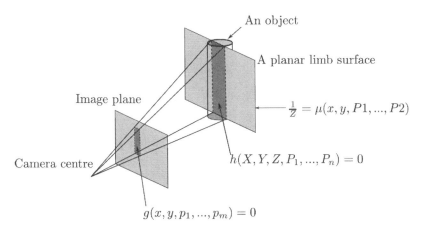

FIGURE 4.7: Projection of the geometric primitive points and parameters into the image plane and the planar limb surface.

and its projection into the image plane is expressed as

$$\mathbf{g}(\mathbf{x}, \mathbf{p}) = 0 \qquad (4.74)$$

where $\mathbf{P}$ and $\mathbf{p}$ are the geometric primitive parameters. The link between the 3D geometric parameters and the 2D image is described as

$$\frac{1}{Z} = \mu(\mathbf{x}, \mathbf{P}) \qquad (4.75)$$

As an example, if a line in space is represented by the intersection of two planes, then

$$\mathbf{h}(\mathbf{X}, \mathbf{P}) = \{ \begin{array}{rcl} h_1 & = & A_1 X + B_1 Y + C_1 Z + D_1 = 0 \\ h_2 & = & A_2 X + B_2 Y + C_2 Z = 0 \end{array} \qquad (4.76)$$

Using the perspective projection (a pinhole camera projection), the function $\mu(\mathbf{x}, \mathbf{P})$ can be described as

$$\mu(\mathbf{x}, \mathbf{P}) = \frac{1}{Z} = Ax + By + C \qquad (4.77)$$

For the case of h_1 the projected geometric primitive parameters are defined as $A = -\frac{A_1}{D_1}$, $B = -\frac{B_1}{D_1}$ and $C = -\frac{C_1}{D_1}$. The geometric primitive parameter of h_2 projected in the image plane can be expressed as

$$ax + by + c = 0 \quad \text{with} \quad a = A_2, b = B_2, c = C_2$$

In this case, the depth any 3D points belonging to an object is in the continous form which expressed as [163]:

$$\frac{1}{Z} = \sum_{q \geq 0, r \geq 0} A_{qr} x^q y^r \tag{4.78}$$

In this study, a planar object is used. The degenerated analysis of the 3D geometric of a non planar object's properties from the projected image is beyond the scope of this chapter. For a planar object $A_{00} = C$, $A_{10} = A$, $A_{01} = B$ and all other terms A_{qr} equal to 0. Using (4.72) and (4.77), the relationship between the camera kinematic screw and the velocity of the image feature in the image plane involving the 3D geometric primitive can be expressed as

$$\begin{aligned}
\dot{x} &= -(Ax + By + C)v_x + x(Ax + By + C)v_z \\
&\quad + xy\omega_x - (1 + x^2)\omega_y + y\omega_z
\end{aligned} \tag{4.79}$$

$$\begin{aligned}
\dot{y} &= -(Ax + By + C)v_y + y(Ax + By + C)v_z \\
&\quad + (1 + y^2)\omega_x - xy\omega_y - x\omega_z
\end{aligned} \tag{4.80}$$

The partial derivative of (4.79) and (4.80) can be derived as

$$\frac{\partial \dot{x}}{\partial x} = -Av_x + (2Ax + By + C)v_z + y\omega_x - 2x\omega_y \tag{4.81}$$

$$\frac{\partial \dot{y}}{\partial y} = -Bv_y + (A_x + 2By + C)v_z + 2y\omega_x - x\omega_y \tag{4.82}$$

Substituting (4.79), (4.80), (4.81) and (4.82) into the developed image moment velocity using Green's theorem (4.42) and arrange the derivation in the form of

$$\dot{m}_{ij} = \mathbf{L}_{m_{ij}} \dot{\mathbf{x}}_c \tag{4.83}$$

and the interaction matrix components are denoted as

$$\mathbf{L}_{m_{ij}} = [m_{vx}\ m_{vy}\ m_{vz}\ m_{\omega x}\ m_{\omega y}\ m_{\omega z}] \tag{4.84}$$

Knowing that $h(x,y) = x^i y^j f(x,y)$, $\frac{\partial h(x,y)}{\partial x} = ix^{i-1}y^j f(x,y)$, $\frac{\partial h(x,y)}{\partial x} = jx^i y^{j-1} f(x,y)$ the result of the derivation can be written as

$$\begin{aligned}
m_{vx} &= -i(Am_{ij} + Bm_{i-1,j+1} + Cm_{i-1,j}) - Am_{ij} & (4.85) \\
m_{vy} &= -j(Am_{i+1,j-1} + Bm_{ij} + Cm_{i,j-1}) - Bm_{ij} & (4.86) \\
m_{vz} &= (i+j+3)(Am_{i+1,j} + Bm_{i,j+1} + Cm_{ij}) - Cm_{ij} & (4.87) \\
m_{\omega x} &= (i+j+3)m_{i,j+1} + jm_{i,j-1} & (4.88) \\
m_{\omega y} &= -(i+j+3)m_{i+1,j} - im_{i-1,j} & (4.89) \\
m_{\omega z} &= im_{i-1,j+1} - jm_{i+1,j-1} & (4.90)
\end{aligned}$$

Equations (4.85) - (4.90) are general forms for further derivation of any particular image moment m_{ij} defined by the value of i and j. Let's start with

Image Moment Interaction Matrix

image moment area $a = m_{00}$. The interaction matrix for image moment area can be derived from the general forms using $i = j = 0$.

$$m_{vx} = -Am_{ij} \tag{4.91}$$
$$m_{vy} = -Bm_{ij} \tag{4.92}$$
$$m_{vz} = 3(Am_{10} + Bm_{01} + Cm_{00}) - Cm_{00} \tag{4.93}$$
$$m_{\omega x} = 3m_{01} \tag{4.94}$$
$$m_{\omega y} = -3m_{10} \tag{4.95}$$
$$m_{\omega z} = 0 \tag{4.96}$$

Considering that $\frac{1}{Z_g} = Ax_g + By_g + C$, $x_g = \frac{m_{10}}{m_{00}}$ and $y_g = \frac{m_{01}}{m_{00}}$ (see equation (4.24)), the interaction matrix of the image moment area can be simplified as

$$\mathbf{L}_a = \begin{bmatrix} -aA & -aB & a\left(\frac{3}{Z_g} - C\right) & 3ay_g & -3ax_g & 0 \end{bmatrix} \tag{4.97}$$

A special case of the camera-object position configuration is used to know the behavior of the camera kinematic screw in relation with the given set of image features. This special case simplifies the parameters described in the system by assuming that the object plane is parallel to the image plane and the object center (x_g, y_g) is in the optical axis of the camera frame. It can be seen from (4.97) that when the object image is centered and parallel to the image plane where $A = B = x_g = y_g = 0$, $\dot{m}_{00}$ is only affected by the translation movement along z axis, v_z (expressed in the camera frame).

Image points are other geometric properties that are useful in the visual servoing. In term of the image moment, the variables that can be computed by image moment descriptors and represent a point coordinates are the image moment centroid $x_g = \frac{m_{10}}{m_{00}}$ and $y_g = \frac{m_{01}}{m_{00}}$. Similar to the previous development of $\mathbf{L}_a$, the interaction matrix of the image moment centroid coordinates x_g and y_g can be obtained by knowing that

$$\dot{x}_g = \frac{\dot{m}_{10}m_{00} - \dot{m}_{00}m_{10}}{m_{00}^2} \tag{4.98}$$

$$\dot{y}_g = \frac{\dot{m}_{01}m_{00} - \dot{m}_{00}m_{01}}{m_{00}^2} \tag{4.99}$$

The final result of the interaction matrix of the image centroid coordinates are expressed as

$$\mathbf{L}_{x_g} = \begin{bmatrix} -\frac{1}{Z_g} & 0 & x_{g_{vz}} & x_{g_{\omega x}} & x_{g_{\omega y}} & y_g \end{bmatrix} \tag{4.100}$$

$$\mathbf{L}_{y_g} = \begin{bmatrix} 0 & -\frac{1}{Z_g} & y_{g_{vz}} & y_{g_{\omega x}} & y_{g_{\omega y}} & -x_g \end{bmatrix} \tag{4.101}$$

where

$$x_{g_{vz}} = \frac{x_g}{Z_g} + 4(An_{20} + Bn_{11}) \tag{4.102}$$

$$y_{g_{vz}} = \frac{y_g}{Z_g} + 4(An_{11} + Bn_{02}) \tag{4.103}$$

$$x_{g_{\omega x}} = -y_{g_{\omega y}} = x_g y_g + 4n_{11} \tag{4.104}$$

$$x_{g_{\omega y}} = -(1 + x_g^2 + 4n_{20}) \tag{4.105}$$

$$y_{g_{\omega x}} = 1 + y_g^2 + 4n_{20} \tag{4.106}$$

and the normalized centered moments n_{11}, n_{02} and n_{20} are expressed as

$$n_{11} = \frac{m_{11} - ax_g y_g}{a} \tag{4.107}$$

$$n_{02} = \frac{m_{02} - ay_g^2}{a} \tag{4.108}$$

$$n_{20} = \frac{m_{20} - ax_g^2}{a} \tag{4.109}$$

Again by assuming that the object image is centered and parallel to the image plane where $A = B = x_g = y_g = 0$, the interaction matrix $\mathbf{L}_{x_g}$ and $\mathbf{L}_{y_g}$ obtained as

$$\mathbf{L}_{x_g} = \begin{bmatrix} -\frac{1}{Z_g} & 0 & 0 & 0 & -1 & 0 \end{bmatrix} \tag{4.110}$$

$$\mathbf{L}_{y_g} = \begin{bmatrix} 0 & -\frac{1}{Z_g} & 0 & 1 & 0 & 0 \end{bmatrix} \tag{4.111}$$

which deducing that the motion of x_g is mostly affected by the camera translational motion along camera x axis (v_x) and the rotational motion about camera y axis (ω_y). Similarly, v_y and ω_x are the main variables that cause the motion of y_g.

Using (4.43) and the relationship between m_{ij} and μ_{ij} which has been described in (4.28) and (4.30), the interaction matrix associated with μ_{ij} can be derived in the following form

$$\mathbf{L}_{\mu_{ij}} = \begin{bmatrix} \mu_{vx} & \mu_{vy} & \mu_{vz} & \mu_{\omega x} & \mu_{\omega y} & \mu_{\omega z} \end{bmatrix} \tag{4.112}$$

where matrix components of the general form interaction matrix $\mathbf{L}_{\mu_{ij}}$ are denoted as

$$\mu_{vx} = -(i+1)A\mu_{ij} - iB\mu_{i-1,j+1} \tag{4.113}$$

$$\mu_{vy} = -jA\mu_{i+1,j-1} - (j+1)B\mu_{ij} \tag{4.114}$$

$$\mu_{vz} = -A\mu\omega_y + B\mu\omega_x + (i+j+2)C\mu_{ij} \tag{4.115}$$

$$\mu_{\omega x} = (i+j+3)\mu_{i,j+1} + ix_g\mu_{i-1,j+1}$$
$$+(i+2j+3)y_g\mu_{ij} - 4in_{11}\mu_{i-1,j} - 4jn_{02}\mu_{i,j-1} \tag{4.116}$$

$$\mu_{\omega y} = -(i+j+3)\mu_{i+1,j} - (2i+j+3)x_g\mu_{ij}$$
$$-jy_g\mu_{i+1,j} + 4in_{20}\mu_{i-1,j} - 4jn_{11}\mu_{i,j-1} \tag{4.117}$$

$$\mu_{\omega z} = i\mu_{i-1,j+1} - j\mu_{i+1,j-1} \tag{4.118}$$

Image Moment Interaction Matrix

An interesting image moment feature that can be selected using image moment in the second order is the object orientation α as described in (4.31), for convenient analysis, it is rewritten as follows

$$\alpha = \frac{1}{2}\tan^{-1}\left(\frac{2\mu_{11}}{\mu_{20}-\mu_{02}}\right) \tag{4.119}$$

Using $\frac{d(1/2)\tan^{-1}(2u)}{dt} = \frac{\dot{u}}{1+4u^2}$, the first derivative of α can be derived as

$$\begin{aligned}\dot{\alpha} &= \frac{\frac{\dot{\mu}_{11}(\mu_{20}-\mu_{02})-\mu_{11}(\dot{\mu}_{20}-\dot{\mu}_{02})}{(\mu_{20}-\mu_{02})^2}}{1+4(\frac{\mu_{11}}{\mu_{20}-\mu_{02}})^2} \\ &= \frac{\dot{\mu}_{11}(\mu_{20}-\mu_{02})-\mu_{11}(\dot{\mu}_{20}-\dot{\mu}_{02})}{\Delta}\end{aligned} \tag{4.120}$$

where $\Delta = (\mu_{20}-\mu_{02})^2 + 4\mu_{11}^2$. By deriving $\dot{\mu}_{11}, \dot{\mu}_{20}$ and $\dot{\mu}_{02}$ using the general form of the interaction matrix associated with the image centered moments (substituting i, j combinations using $i = j = 1$, $i = 2; j = 0$ and $i = 0; j = 2$ into equation (4.112)), the interaction matrix that relates the camera kinematic screw $\dot{\mathbf{x}}_c$ with the velocity of α can be obtained as

$$\mathbf{L}_\alpha = \begin{bmatrix} \alpha_{vx} & \alpha_{vy} & \alpha_{vz} & \alpha_{wx} & \alpha_{wy} & -1 \end{bmatrix} \tag{4.121}$$

where

$$\begin{aligned}\alpha_{vx} &= a_\alpha A + b_\alpha B & (4.122)\\ \alpha_{vy} &= -c_\alpha A - a_\alpha B & (4.123)\\ \alpha_{vz} &= -A\alpha_{wy} + B\alpha_{wx} & (4.124)\\ \alpha_{wx} &= -b_\alpha x_g + a_\alpha y_g + d_\alpha & (4.125)\\ \alpha_{wy} &= a_\alpha x_g - c_\alpha y_g + e_\alpha & (4.126)\end{aligned}$$

and

$$\begin{aligned}a_\alpha &= \frac{\mu_{11}(\mu_{20}+\mu_{02})}{\Delta} & (4.127)\\ b_\alpha &= \frac{2\mu_{11}^2 + \mu_{02}(\mu_{02}-\mu_{20})}{\Delta} & (4.128)\\ c_\alpha &= \frac{2\mu_{11}^2 + \mu_{20}(\mu_{20}-\mu_{02})}{\Delta} & (4.129)\\ d_\alpha &= \frac{5(\mu_{12}(\mu_{20}-\mu_{02})+\mu_{11}(\mu_{03}-\mu_{21}))}{\Delta} & (4.130)\\ e_\alpha &= \frac{5(\mu_{21}(\mu_{02}-\mu_{20})+\mu_{11}(\mu_{30}-\mu_{12}))}{\Delta} & (4.131)\end{aligned}$$

Let's assume a special case of the camera-object position configuration where the object plane is parallel and centered along the camera optical axis ($A = 0$,

$B = 0$). When the special case is considered, it can be noted that from (4.121), the motion of α does not depend on any translational camera motion (v_x, v_y, v_z). It also can be deduced that there is a strong association between the camera rotation motion ω_z and the motion of α, which can be seen from the last column's value of $\mathbf{L}_\alpha$.

The visual features x_g, y_g, a, and α have been derived using the image moments up to second order. As it has been discussed, the selected visual features $(x_g, y_g, a, \text{ and } \alpha)$ have a strong relationship with particular component of the camera kinematic screw and invariant to the rest. For example, the motion of a is strongly affected by the camera motion along the camera optical axis (v_z) and invariant to the other camera kinematic screw components $(v_x, v_y, \omega_x, \omega_y, \omega_z)$. Using image moment features (x_g, y_g, a, α), 4-DOF movement of the robot in the task space can be controlled, specifically the translational movements (v_x, v_y, v_z) and the rotational movement about the camera optical axis (ω_z). A full 6-DOF task space robot control requires the image moments derivation up to third order and the computation of the image moment invariants [164]. In this chapter, 4-DOF robot movement in the task space is considered, focusing the discussion on how to use the learning algorithm to reduce the computational complexity of the traditional visual servoing scheme in next chapters.

Let's consider developing visual servoing algorithm using four visual features (x_g, y_g, a, α). A new form of the combined interaction matrix $\mathbf{L_s}$ can be obtained by stacking together the obtained interaction matrices of $\mathbf{L}_a, \mathbf{L}_{x_g}, \mathbf{L}_{y_g}$ and $\mathbf{L}_\alpha$ (represented in (4.97) (4.100) (4.101) (4.121), respectively), it is described as

$$\mathbf{L_s} = \begin{bmatrix} \mathbf{L}_{x_g} & \mathbf{L}_{y_g} & \mathbf{L}_a & \mathbf{L}_\alpha \end{bmatrix}^T$$

$$= \begin{bmatrix} -\frac{1}{Z_g} & 0 & x_{g_{vz}} & x_{g_{\omega x}} & x_{g_{\omega y}} & y_g \\ 0 & -\frac{1}{Z_g} & y_{g_{vz}} & y_{g_{\omega x}} & y_{g_{\omega y}} & -x_g \\ -aA & -aB & a\left(\frac{3}{Z_g} - C\right) & 3ay_g & -3ax_g & 0 \\ \alpha_{vx} & \alpha_{vy} & \alpha_{vz} & \alpha_{\omega x} & \alpha_{\omega y} & -1 \end{bmatrix} \quad (4.132)$$

Since the visual servoing controller is designed to control $(v_x, v_y, v_z, \omega_z)$ and by making ω_x and ω_y are equal to zero, the fourth and fifth column of the $\mathbf{L_s}$ can be cancelled. Thus, the interaction matrix $\mathbf{L_s}$ can be simplified as

$$\mathbf{L_s} = \begin{bmatrix} -\frac{1}{Z_g} & 0 & x_{g_{vz}} & y_g \\ 0 & -\frac{1}{Z_g} & y_{g_{vz}} & -x_g \\ -aA & -aB & a\left(\frac{3}{Z_g} - C\right) & 0 \\ \alpha_{vx} & \alpha_{vy} & \alpha_{vz} & -1 \end{bmatrix} \quad (4.133)$$

In the following section, the experimental results of the MBVS of a 7 DOF manipulator using the selected image moment features and the developed interaction matrix are presented.

4.9 Experimental Results using a 7 DOF Manipulator

The experimental set up consists of a 7 DOF PowerCube robot manipulator as shown in Figure 4.8. A specific detail about the kinematic derivation of the 7 DOF PowerCube robot manipulator is given in Appendix A. In this experiment, a firewire CCD camera is mounted at the robot end-effector so that the camera frame coincided with the robot end-effector frame, $\mathbf{x}_c = \mathbf{x}_e$. In the presented experimental results, the visual servoing control law is described as

$$\dot{\mathbf{x}}_c = -\kappa \mathbf{L}_{\mathbf{s}^*}^{-1} \mathbf{e}$$

where $\dot{\mathbf{x}}_c$ is the camera velocity, κ is a positive gain and $\mathbf{L}_{\mathbf{s}^*}^{-1}$ is the inverse of the interaction matrix $\mathbf{L}_{\mathbf{s}^*}$ and $\mathbf{e} = (\mathbf{s} - \mathbf{s}^*)$. $\mathbf{L}_{\mathbf{s}^*}$ is the interaction matrix which is computed using the desired value of $\mathbf{s}$. The relationship between the image feature velocity vector $\dot{\mathbf{s}}$ and the joint angle velocity vector $\dot{\boldsymbol{\theta}}$ is obtained as

$$\dot{\mathbf{s}} = \mathbf{L}_{\mathbf{s}^*} \mathbf{J}_e \dot{\boldsymbol{\theta}} \quad (4.134)$$

Ensuring an exponential decrease of the error $\dot{\mathbf{e}} = -\kappa \mathbf{e}$, the final eye-in-hand MBVS control law is obtained as

$$\dot{\boldsymbol{\theta}} = \kappa \mathbf{J}_e^\dagger \mathbf{L}_{\mathbf{s}^*}^{-1} (\mathbf{s}^* - \mathbf{s}) \quad (4.135)$$

where $\mathbf{J}_e^\dagger$ is the pseudo-inverse of the robot kinematic Jacobian $\mathbf{J}_e$ expressed in the robot end-effector frame. In this experiment scenario, the desired camera

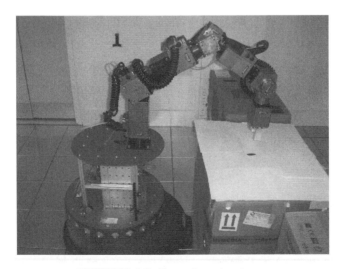

FIGURE 4.8: Experiment setup.

pose is parallel with an offset (Z^*) to the target object plane where the object segment center is on the camera optical axis. Therefore, the object segment geometric parameters A and B are set to zero. Immediately, the interaction matrix can be expressed as

$$\mathbf{L_{s^*}} = \begin{bmatrix} -\frac{f_c}{Z_g^*} & 0 & \frac{x_g^*}{Z_g^*} & y_g^* \\ 0 & -\frac{f_c}{Z_g^*} & \frac{y_g^*}{Z_g^*} & -x_g^* \\ 0 & 0 & a^*\left(\frac{2}{Z_g^*}\right) & 0 \\ 0 & 0 & 0 & -1 \end{bmatrix} \quad (4.136)$$

where $f_c = 0.0053$ is the camera focal length given by the camera manufacturer's specification. The image feature $\mathbf{s}$ and the corresponding components of $\mathbf{L_{s^*}}$ are expressed in meter. The desired centroid coordinates were set as the center coordinates of the camera image view. A firewire camera with the resolution of 320×240 was used in this experiment, therefore, the image view center coordinates $(x_{gi}^*, y_{gi}^*) = (160, 120)$. The corresponding coordinates point on the camera frame are $(x_g^*, y_g^*) = (0, 0)$. The desired depth Z_g^* and area a^* can be obtained by bringing the camera into the desired pose with respect to the target object position.

The final pose of the robot end-effector was set to be 20 cm of distance from the object, $Z_g^* = 20\,cm$, with $\alpha^* = 0$. In this final pose the desired image area was computed as $a_{image}^* = 5046$ pixels^2, using (4.45), it can be deduced that $a = f^2 a_{image}$. The numerical value of the interaction matrix from the desired image feature values was computed as

$$\mathbf{L_{s^*}} = \begin{bmatrix} -0.0265 & 0 & 0 & 0 \\ 0 & -0.0265 & 0 & 0 \\ 0 & 0 & 0.1417 & 0 \\ 0 & 0 & 0 & -1 \end{bmatrix} \quad (4.137)$$

At the desired image feature values $\mathbf{s}^*$, the interaction matrix $\mathbf{L_{s^*}}$ has a perfect decoupling property since it is a diagonal matrix.

The initial robot joint configuration of $\boldsymbol{\theta} = [0°, 60°, 0, 30°, 0°, 90°, 30°]$ displaced the camera pose from its desired pose which was approximately composed of the translation of 50, 15,10 cm along x, y, z axes, and the rotation of $0°$, $0°$, $30°$ rotation about x, y, z axes, respectively. The result realized from the MBVS control law measured in the joint space is depicted in Figure 4.9. The joint velocities converged to zero in 2 s when the desired camera pose was reached. At $t > 2\,s$, the MBVS controller maintained the camera pose at the desired values by giving the error signal between the desired and the current image feature sets. Small joint velocity fluctuations as shown in the Figure 4.9 were caused by the small image noise captured by the camera. The corresponding image feature error trajectories are shown in Figure 4.10.

It can be seen from Figure 4.10(a), the error of the image centroid coordinate is more sensitive to the image noise compared to the other image feature

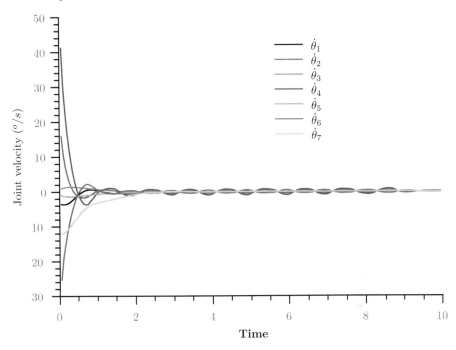

FIGURE 4.9: Joint velocity results of the MBVS.

errors, since small pixels' noise would affect less in the computation result of the image area a and α (see Figure 4.10(b) and 4.10(c)). As an example, 2 pixels noise will not significantly affect the area of a segmented image of 70×70 pixels. This condition also applies to the computation of α which involves the computation of the second order centered image moments. Figure 4.11 shows segmented target images captured at initial and desired camera positions. OpenCV 2.0 [135, 169] library was used to preprocess images from the camera's raw data to the obtained binary segmented images which included image color converter (RGB to greyscale image), blurring, Canny edge detector and contour finder functions.

4.10 Summary

This chapter has presented a detail development of the MBVS of a 7 DOF manipulator using image moments. The visual features x_g, y_g, a, and α have been derived using the image moments up to second order. These image moment features were chosen to have nice decoupling property of the

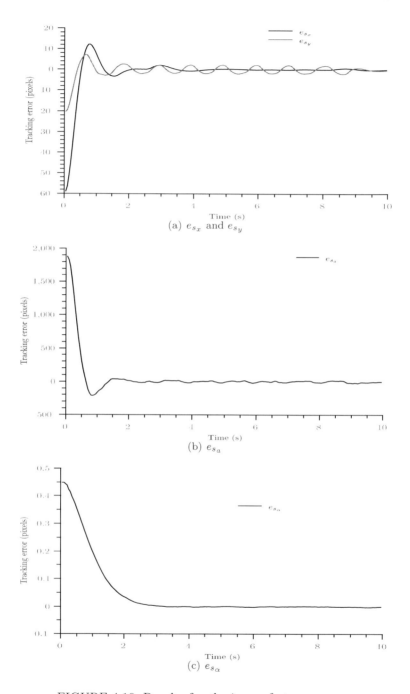

FIGURE 4.10: Results for the image feature errors.

Summary

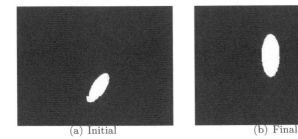

(a) Initial (b) Final

FIGURE 4.11: Segmented target images at initial and desired camera position.

interaction matrix, e.g., the movement of the centroid coordinates (x_g, y_g) are significantly affected by the movement of the camera in x and y camera axes, the changes of the segmented image area a is caused by the movement of the camera along the camera optical axis, and the orientation α of the segmented image is significantly affected by the orientation of the camera about the camera optical axis. As a result, the interaction matrix of the desired image is a diagonal matrix which has perfect decoupling property.

The presented MBVS control law has been validated in a real-time experiment using 7 DOF PowerCube robot manipulator. By giving four image moment features, the MBVS controlled each joint of the robot manipulator using velocity command, to position the attached camera on the robot end-effector from the initial pose to the desired pose. The system convergence was reached when the current image moment feature set $\mathbf{s}$ was approximately the same with the desired image moment feature set $\mathbf{s}^*$. In the real-time experiment, image noises cannot be avoided; as a result the joint velocity trajectories fluctuated in a small region near zero, in order to keep the camera pose at the desired position. The development of the presented MBVS in this chapter is important for further analysis and comparison in the next following contribution chapters.

Readers may refer [15] for inverse kinematic control of the redundant manipulator using generalized pseudo-inverse. Vision-based manipulator control in open loop fashion is discussed in [51]. A detailed survey on visual servoing can be found in [10], [11] and [52]. Vision-based control in eye-in-hand configuration is discussed in [53, 54]. Visual servoing in eye-to-hand configuration is discussed in [52, 55, 56]. Model-based redundancy resolution for a visually controlled manipulator is discussed in [81] for trajectories defined in vision space. The trajectories are defined in vision space from a single camera in eye-in-hand configuration, and then task sequencing is used to prioritize the task for achieving kinematic limit avoidance. Mansard and Chaumette [82] achieved obstacle avoidance by task sequencing while following vision space trajectories in eye-in-hand configuration. Later the approach is extended for multiple-task considering occlusion and kinematic limit avoidance together in [83]. The learning-based servoing scheme proposed in [61] for a non-redundant manipulator focuses on learning the inverse Jacobian at the chosen operation point only. Tsai's algorithm to calibrate the camera is discussed in [133], and an online implementation of the Tsai calibration algorithm is available by R. Willson [134].

5

Learning-Based Visual Servoing

Kohonen's self-organizing map has been used to kinematically control the redundant manipulator but the type of solution learned with the associated map is not discussed in the literature. This chapter analyzes the map learned with the KSOM based kinematic control algorithm. It is experimentally shown that the learned KSOM actually approximates the pseudo-inverse of the Jacobian with a linear map in every local zone. A globally asymptotically stable visual servoing method is proposed with the learned map, and it is shown that the proposed scheme is Lyapunov stable, if the approximation is accurate. A KSOM based global positioning scheme is further generalized for redundancy resolution using a weighted norm solution method [170].

5.1 Introduction

Real-life implementation of the redundant manipulator control requires the ability to control the manipulator over the entire workspace to reach the objects scattered in the environment. Model-based schemes compute the joint angular velocity from the vision space as,

$$\dot{\boldsymbol{\theta}} = k_p \, \mathbb{J}^+ \mathbf{e}_u. \qquad (5.1)$$

Such model-based approaches are inefficient while implementing in a dynamic environment in the following aspects:

- The model-based visual control schemes require the exact Cartesian depth information between the camera and the environment for the computation of the interaction matrix $\mathbf{L}$, which may not be available over the entire workspace in a dynamic environment.

- The pseudo-inverse of Jacobian $\mathbb{J}$ is required at each instant to control the manipulator. The computation of pseudo-inverse is computationally intensive. The method may lead to instability in dynamic environment due to sensor and model inaccuracies since the method is sensitive to parameter variations.

These problems are circumvented in model-based paradigm, by computing $\mathbb{L}$ only at a given operating point during the control process. In general the desired position $\mathbf{u}_d$ is chosen to estimate the image Jacobian, which is denoted as $\mathbf{L}_{\mathbf{u}_d}$. Then the pseudo-inverse is computed for the Jacobian estimated at $\mathbf{u}_d$ as $\mathbb{J}_{\mathbf{u}_d} = \mathbf{L}_{\mathbf{u}_d}\mathbf{PJ}$. If the camera and the kinematic model are not available, then $\mathbf{P}$ and $\mathbf{J}$ are also estimated. Such methodology is computationally cost effective but results in local stabilization. Model-based locally stabilizing schemes cannot be used in the real-world since the objects are scattered over the environment. The image Jacobian $\mathbf{L}$ has to be estimated for every object and the position of the objects will be continuously changing in dynamic environment. Hence a local estimation of the Jacobian is ineffective. The global stabilization can be achieved by estimating $\mathbb{J}$ over the entire workspace and then computing the pseudo-inverse at each operating point. The Jacobian from the joint space to the vision space is estimated at every instant and then an adaptive control strategy is proposed visual servoing in [171,172]. The global Jacobian from the joint space to the vision space is estimated using a K-nearest neighbor network in [173] and receptive field weighted regression neural network in [174, 175]. The learned map is used to compute the pseudo-inverse at each instant to control the manipulator. All the above discussed approaches require the computation of the pseudo-inverse at each instant and, hence, the approaches are computationally intensive. An alternative approach is to estimate the pseudo-inverse $\mathbb{J}^+$ directly over the entire workspace, which reduces the computation complexity associated with the pseudo-inverse. The learning-based visual servoing scheme discussed in [61] computes the pseudo-inverse of $\mathbf{L}$ at a chosen operating point and achieves local stabilization. The pseudo-inverse of the image Jacobian $\mathbf{L}$ is estimated with an online update algorithm in [176] while realizing task sequencing. Hence, the control is a two stage process, and the redundancy is resolved while following the Cartesian space trajectory generated from the estimate of the image Jacobian's pseudo-inverse.

On the other hand, model-free control of the redundant manipulator from vision space has been addressed using KSOM based kinematic control schemes [177]. KSOM based learning schemes compute the joint angles directly from the vision space but mostly tested on non-redundant manipulators [51]. In [178], it has been shown through experimentation that KSOM learns a smooth map for redundant manipulators owing to its topology preserving nature, and yet a detailed analysis about the type of solution is not studied. KSOM has been used for obstacle avoidance in [179]. The algorithm presumes that the KSOM approximates the pseudo-inverse of the Jacobian and the Jacobian is estimated by computing the pseudo-inverse of the learned map. The null space of the Jacobian is then used to achieve obstacle avoidance, but the assumption is not confirmed with any analysis. The discussed experimental results also show that the redundant manipulator is tested for end-effector collision avoidance only and the method suffers from positioning inaccuracy due to open loop mode of operation. Asuni et al. [180] used a growing neural gas architecture to learn the inverse kinematics of a redundant manipulator. The approach does not resolve the redundancy for any particular task. It has been shown through the experimental results that the learned

Introduction

map is robust to model inaccuracies and it can adapted to environmental changes such as clamped links and extended tool tips. A detailed survey of KSOM based kinematic control schemes is discussed in [181]. The implementation of the learning-based control schemes is constrained by the number of data generated to train the network. Angulo and Torras [182] suggested function decomposition for manipulators with last three joints crossing at a point to improve the speed of the learning process. This method can be used for both non-redundant and redundant manipulators provided that the last three joints cross at a point. Kumar et al. [183] suggested an inverse-forward adaptive scheme to reduce the required number of training data during the learning stage. The approach approximates the forward map with a radial basis function network and then the inverse kinematic solution is obtained with a KSOM based hint generator for redundancy resolution. The learned forward map is updated online while controlling the manipulator to improve the positioning accuracy. Alternatively, Behera and Kirubanandan [51] suggested learning an approximate inverse kinematic map with the available kinematic and camera models and then the learned map is improved in real-time. Such an approach uses the model to train the network and the learned model is adapted during the operational phase.

Kumar et al. [184], proposed a KSOM network with joint angle space sub-clustering which allows to learn multiple solutions for each end-effector position. The network acts as a look-up table for redundant solutions and it works based on the principle of "look and move." Though KSOM based approaches control the manipulator over the entire workspace, there is no proper study associated with the relationship between the learned KSOM and the type of resulting solution. This thesis work analyzes the learned map with experimental studies. It is empirically proved that the learned map approximates the inverse Jacobian as a linear map in each operating zone. With such experimental verification, a globally asymptotically stable visual control scheme is proposed for redundant manipulators. In addition, KSOM based kinematic control scheme is generalized to learn a particular solution to resolve the redundancy for the chosen additional task.

Initially a KSOM based neural network is used to learn the inverse kinematics of the redundant manipulator offline. The input to the network is 4 dimensional image coordinate vector viewed from two cameras while the output is 7 dimensional joint angle vector. Each neuron in KSOM approximates the inverse kinematics relation from the vision space to the joint angle space within a local operating zone. The output of KSOM neuron lattice consists of a joint angle configuration required to reach near the corresponding input vision space position, and a local first order map to move closer to the desired position. This thesis work mainly focuses on the learned local linear model of the KSOM network. It is shown experimentally that KSOM approximates the pseudo-inverse of the Jacobian matrix with the local linear map. This observation motivated us to use the learned KSOM for closed loop visual servoing. The classical proportional feedback [164] is chosen for the closed loop control. Further experiments revealed that a globally stabilizing controller can be obtained by using conventional proportional feedback in conjunction with the inverse kinematic map learned using KSOM. Since the approximate inverse kinematic relationship is learned offline over the entire workspace, a simple proportional controller results in global stability. Lyapunov analysis shows that the global stability can be achieved if the learned map accurately approximates the local inverse Jacobian. The obtained inverse Jacobian approximation also eliminates the necessity of online pseudo-inverse computation required in visual servoing and makes the proposed scheme computationally efficient. With the empirical observations of convergence to pseudo-inverse, the KSOM based kinematic control is extended for redundancy resolution under weighted norm formulation. KSOM is learned to resolve the redundancy directly from the vision space while minimizing an instantaneous cost function.

The remaining portion of this chapter is organized as follows. The following section briefly introduces the KSOM based kinematic control scheme. The problem is defined in Section 5.3, and the proposed control strategy is presented in Section 5.4. The simulations and the experiments performed for controlling the robotic system discussed in Chapter 2 are presented in Section 5.7. The contributions made in this chapter are finally summarized in Section 5.8.

5.2 Kinematic Control using KSOM

The forward map from 7 dimensional joint angle space to 4 dimensional image coordinate space can be derived using manipulator forward kinematic model (2.1) and camera model obtained through the Tsai algorithm. This forward mapping is represented as,

$$\mathbf{u} = \mathbf{f_{ux}}(\boldsymbol{\theta}) \tag{5.2}$$

where $\mathbf{f_{ux}}$ represents the nonlinear map from the joint angle space to the vision space. In robotic manipulation, the inverse relationship plays a key role, since the knowledge of the joint angle configuration which can reach the desired position $\mathbf{u}_d$ is necessary, for manipulating the objects scattered in the workspace. The inverse kinematic relationship is given by,

$$\boldsymbol{\theta} = \mathbf{f_{ux}^{-1}}(\mathbf{u}_d) = \mathbf{r}(\mathbf{u}_d). \tag{5.3}$$

In KSOM based visual control, a smooth solution is learned over the entire workspace and the learned map is used to reach any desired position in the workspace. A brief discussion about KSOM based NN architecture for kinematic control is presented in the following subsection to aid understanding.

5.2.1 KSOM Architecture

The inverse kinematic relationship of the redundant manipulator (5.3) is a nonlinear relationship and, hence, it is difficult to learn. One easier approach to this problem involves the discretization of both the input as well as the output spaces into several small cells so that a linear map from the input to output space holds good within each cell. KSOM discretizes the input vision space into number of cells and associates a vector and a linear map in the output joint angle space for each region.

In this thesis work, a 3 dimensional KSOM lattice is used to discretize the input and output spaces. Lattice node indices are represented by γ and each such node is associated with a vision space vector $\mathbf{w}_\gamma \in R^p$, a joint angle vector $\boldsymbol{\theta}_\gamma \in R^m$, and a linear map $\mathbf{A}_\gamma : R^p \to R^m$. The vectors $\mathbf{w}_\gamma$ and $\boldsymbol{\theta}_\gamma$ discretize the input and output space respectively. $\mathbf{A}_\gamma$ approximates the inverse kinematic relationship in each region with a linear map. The joint angle required to reach any desired position is computed using KSOM as follows:

Given a desired position $\mathbf{u}_d$, a winner neuron μ is selected based on its Euclidean distance metric in the input space. The neuron whose weight vector is closest to the desired position is declared winner as shown below.

$$\mu = \min_\gamma \|\mathbf{u}_d - \mathbf{w}_\gamma\|_2. \tag{5.4}$$

The arm is given a coarse movement $\boldsymbol{\theta}_0^{out}$ given by,

$$\boldsymbol{\theta}_0^{out} = s^{-1} \sum_{\gamma=1}^{N_n} h_\gamma (\boldsymbol{\theta}_\gamma + \mathbf{A}_\gamma (\mathbf{u}_d - \mathbf{w}_\gamma)) \tag{5.5}$$

where $s = \sum_{\gamma=1}^{N_n} h_\gamma$, $h_\gamma = e^{(\frac{-\|\mu-\gamma\|}{2\sigma^2})}$, and N_n is the number of neurons located in the KSOM lattice. Because of this coarse movement, the end-effector reaches a position $\mathbf{u}_0$ in vision space. A correcting fine movement $\boldsymbol{\theta}_1^{out}$ is evaluated as follows:

$$\boldsymbol{\theta}_1^{out} = \boldsymbol{\theta}_0^{out} + s^{-1} \sum_{\gamma=1}^{N_n} h_\gamma \mathbf{A}_\gamma (\mathbf{u}_d - \mathbf{u}_0). \tag{5.6}$$

This corrective movement results in a final movement of the end-effector to $\mathbf{u}_1$. Although one can use several such corrective movements to increase the accuracy of tracking, usually one corrective movement is used.

5.2.2 KSOM: Weight Update

The parameters of the KSOM network are updated as,

$$\mathbf{A}_\gamma^{new} = \mathbf{A}_\gamma^{old} + s^{-1} \eta\, h_\gamma\, \Delta \mathbf{A}_\gamma \tag{5.7}$$

$$\mathbf{w}_\gamma^{new} = \mathbf{w}_\gamma^{old} + s^{-1} \eta\, h_\gamma\, \Delta \mathbf{w}_\gamma \tag{5.8}$$

$$\boldsymbol{\theta}_\gamma^{new} = \boldsymbol{\theta}_\gamma^{old} + s^{-1}\, \eta\, h_\gamma\, \boldsymbol{\Delta\theta}_\gamma \,. \tag{5.9}$$

The change in the network parameters $\boldsymbol{\Delta A}_\gamma$, $\boldsymbol{\Delta\theta}_\gamma$ and $\boldsymbol{\Delta w}_\gamma$ are computed as follows:

The local linear map $\mathbf{A}_\gamma$ is updated similar to gradient descent rule, by minimizing the function,

$$E = \frac{1}{2} \parallel \boldsymbol{\Delta\theta}_{01} - \mathbf{A}_\gamma \boldsymbol{\Delta u}_{01} \parallel^2 \tag{5.10}$$

where $\boldsymbol{\Delta\theta}_{01} = \boldsymbol{\theta}_1 - \boldsymbol{\theta}_0$ and $\boldsymbol{\Delta u}_{01} = \mathbf{u}_1 - \mathbf{u}_0$. The value of $\boldsymbol{\Delta A}_\gamma$ is obtained from equation (5.10) as,

$$\boldsymbol{\Delta A}_\gamma = \parallel \boldsymbol{\Delta u}_{01} \parallel^{-2} (\boldsymbol{\Delta\theta}_{01} - \mathbf{A}_\gamma \boldsymbol{\Delta u}_{01}) \boldsymbol{\Delta u}_{01}^T. \tag{5.11}$$

The change in the value of $\boldsymbol{\theta}_\gamma$ is computed as,

$$\boldsymbol{\Delta\theta}_\gamma = \boldsymbol{\theta}_0 - \boldsymbol{\theta}_\gamma - \mathbf{A}_\gamma^{new}(\mathbf{u}_0 - \mathbf{w}_\gamma), \tag{5.12}$$

such that $\boldsymbol{\theta}_\gamma \to \boldsymbol{\theta}_0$.

The value of $\boldsymbol{\Delta w}_\gamma$ is computed with the basic KSOM based clustering algorithm to identify a center around the desired position $\mathbf{u}_d$ as,

$$\boldsymbol{\Delta w}_\gamma = \mathbf{u}_d - \mathbf{w}_\gamma. \tag{5.13}$$

5.2.3 Comments on Existing KSOM Based Kinematic Control Schemes

The above approach has been used for visual motor coordination of non-redundant manipulators [51] as well as redundant manipulators [178], [185]. While the application of KSOM to kinematic control of the non-redundant manipulators has been analyzed extensively, it has not been applied much to the redundant manipulators, since the redundancy is lost in the learning phase. It is demonstrated in [178] that, the above control algorithm is capable of resolving the redundancy by minimizing the variations of joint angles, in the case of manipulators with higher degrees of freedom. Han et al. [179] used KSOM to avoid obstacles with multiple camera setup for a 4 DOF manipulator, but the approach involves the computation of the pseudo-inverse during learning phase.

The learned map is generally used in open loop mode which suffers from positioning inaccuracy. KSOM based kinematic control algorithm considers only the desired position and the current end-effector position is ignored during the coarse movement. Hence, the path traversed during coarse movement from the current position to the desired position is not controlled. Since the manipulator is controlled with joint angle reference, it is difficult to resolve redundancy with existing approaches for different subtasks.

5.3 Problem Definition

As discussed in previous sections, existing visual servoing techniques are model-dependant and computationally intensive. Though model-free strategies are analyzed for position level control, they are inaccurate and not suitable for redundancy resolution. Considering these challenges associated in the visual control of the redundant manipulators, the problem is formulated as follows:

"Given a redundant manipulator with stereo vision overlooking the workspace in eye-to-hand configuration, develop a model-free visual control technique which can control the redundant manipulator over the entire workspace while resolving the redundancy for the chosen additional task. With any initial manipulator configuration $\boldsymbol{\theta}_0$ resulting in end-effector position $\mathbf{u}$, and the desired end-effector position $\mathbf{u}_d$ in vision space, identify the control law $\dot{\boldsymbol{\theta}} = \mathbf{f}(\boldsymbol{\theta}, \mathbf{e}_u)$, where $\mathbf{e}_u = \mathbf{u}_d - \mathbf{u}$, such that the manipulator end-effector asymptotically reaches the desired position from the initial position."

The main focus of the proposed approach is to achieve global positioning of the end-effector through visual servoing, and it is achieved by analyzing the linear map learned using the KSOM based kinematic control algorithm. Following are the prime issues addressed in this thesis work:

- A computationally less intensive model-free architecture for visual servoing.

- Global positioning of the redundant manipulator without the computation of pseudo-inverse at each instant.

- Redundancy resolution from vision space while minimizing an instantaneous cost function under learning paradigm.

5.4 Analysis of Solution Learned Using KSOM

As discussed in Section 5.2, KSOM learns to control the redundant manipulator with a linear map in each operating zone. In case of redundant manipulators, it is shown through simulation [178] that the KSOM resolves the redundancy by learning a smooth movement in the workspace. A smooth solution is learned since it tries to minimize the joint angle variation due to its topology conserving nature. In this thesis work, the solution learned with KSOM is analyzed using eigenvalue approach, and it is experimentally shown that the pseudo-inverse of Jacobian matrix is learned locally. In such a case it is argued that the KSOM can be considered as an approximation of the pseudo-inverse of the Jacobian matrix for the learned joint angle configuration θ_γ. In the

5.4.1 KSOM: An Estimate of Inverse Jacobian

The correcting fine movement (5.6) can be rewritten as,

$$\boldsymbol{\theta}_1^{out} - \boldsymbol{\theta}_0^{out} = s^{-1} \sum_{\gamma=1}^{N_n} h_\gamma \mathbf{A}_\gamma (\mathbf{u}_d - \mathbf{u}_0)$$

$$\Delta \boldsymbol{\theta}^{out} = s^{-1} \sum_{\gamma=1}^{N_n} h_\gamma \mathbf{A}_\gamma (\Delta \mathbf{u}) \tag{5.14}$$

where $\Delta \boldsymbol{\theta}^{out}$ represents the estimated change in the joint angle to generate the end-effector position change of $\Delta \mathbf{u}$ in the vision space. The above equation can be represented in velocity form by actuating a joint angular velocity, $\dot{\boldsymbol{\theta}}^{out}$ for a duration of Δt as follows,

$$\frac{\Delta \boldsymbol{\theta}^{out}}{\Delta t} = s^{-1} \sum_{\gamma=1}^{N_n} h_\gamma \mathbf{A}_\gamma \left(\frac{\Delta \mathbf{u}}{\Delta t} \right)$$

$$\dot{\boldsymbol{\theta}}^{out} = s^{-1} \sum_{\gamma=1}^{N_n} h_\gamma \mathbf{A}_\gamma \dot{\mathbf{u}}. \tag{5.15}$$

By comparing equations (5.1) and (5.15), it is easy to infer that KSOM may approximate the inverse of the Jacobian from the joint angle space to the vision space as,

$$\mathbb{J}^+ \simeq s^{-1} \sum_{\gamma=1}^{N_n} h_\gamma \mathbf{A}_\gamma. \tag{5.16}$$

This thesis work proposes that the KSOM approximates the inverse Jacobian as a linear map in each operating zones. The linear map is valid within its local zone and the global nonlinear inverse is obtained by clustering in the lattice space. To verify the proposition, empirical experiments are performed. For simplicity, the simulations are performed for inverse kinematic relation from the Cartesian space to the joint angle space. The same experiments can also be extended to the vision space which also requires the computation of image Jacobian at every point in the visible workspace.

5.4.2 Empirical Verification

If KSOM approximates the pseudo-inverse of the kinematic Jacobian $\mathbf{J}$ while controlling from the Cartesian space, then the following relationships are valid.

Analysis of solution learned using KSOM 153

- Around non-singular points,

$$\mathbf{J}\sum_{\gamma=1}^{N_n} h_\gamma \mathbf{A}_\gamma \approx \mathbf{I} \tag{5.17}$$

- Around singular points,

$$\mathbf{J}\sum_{\gamma=1}^{N_n} h_\gamma \mathbf{A}_\gamma \approx \mathbf{I}' \tag{5.18}$$

where $\sum_{\gamma=1}^{N_n} h_\gamma \mathbf{A}_\gamma$ is the linear approximation of the inverse Jacobian learned by KSOM. $\mathbf{I}$ is the identity matrix of order n and $\mathbf{I}'$ is a positive definite matrix of order n. The $\mathbf{I}'$ matrix of rank r will have $n-r$ eigenvalues as 0. Considering these relationships, following simulations are performed to check whether above properties are satisfied with KSOM. The simulations are performed with the kinematic model of the PowerCube™ manipulator discussed in Section 2.2. The parameters are taken the same as an actual setup in simulation so that it matches with the experimental result.

A 3 dimensional neural lattice with $7 \times 7 \times 7$ neurons is selected to learn the inverse kinematics. The inverse kinematic relation from the Cartesian space to the joint angle space is learned with $5,00,000$ training patterns. The input to KSOM network is 3 dimensional Cartesian position of the end-effector and the output is 6 dimensional joint angle coordinates.

5.4.2.1 Inverse Jacobian Evolution in Learning Phase

It is observed from the inverse kinematic solutions that KSOM learns a smooth motion. This learned mapping improves as the number of patterns increases. It is easier to infer then, that, as the learning progresses, KSOM approaches the pseudo-inverse. To validate this assumption, a typical neuron is selected and $\mathbf{I}' = \mathbf{J}\sum_{\gamma=1}^{N_n} h_\gamma \mathbf{A}_\gamma$ is computed in regular intervals of learning. If KSOM learns the generalized pseudo-inverse, then eigenvalues of $\mathbf{I}'$ converge to 1.

The neuron located at $(4, 4, 4)$ of neuron lattice is considered to check the eigenvalue evolution in the learning phase. The eigenvalues are computed at regular interval of 200 data points. The simulation results are shown in Figure 5.1. It is clear from the figure that the eigenvalues approach 1 with the learning, which confirms the proposition.

5.4.2.2 Testing Phase: Inverse Jacobian Estimation at each Operating Zone

The inverse Jacobian relationship at every nodes of KSOM network is checked after learning. The results are shown in Figure 5.2. It is clear from the figure

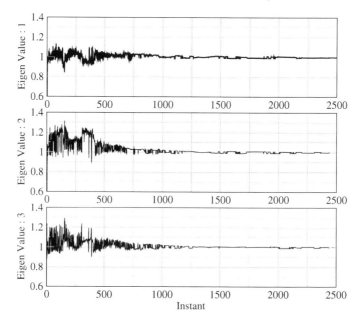

FIGURE 5.1: Evolution of eigenvalue.

that KSOM approximates the pseudo-inverse in most of the centers of networks, and in some of the nodes the eigenvalues have not yet converged to 1 which belongs to the neurons located at the corner of lattice. To conclude further, the positioning accuracy at each center is checked and the result is shown in Figure 5.3, which clearly shows that learning is not accurate at the corresponding centers where eigenvalues have not yet converged to 1. Hence, the eigenvalues may converge to 1, if the learning is extended further.

5.4.2.3 Inference

It is clear from the above two experiments that KSOM approximates the pseudo-inverse of Jacobian and the approximation improves with learning. Though the above experiments are performed from the Cartesian space to the joint angle space for simplicity, it can be extended to the vision space too.

It is claimed that the KSOM learns the pseudo-inverse of kinematic relationship as a cluster of locally valid inverse maps, and the claim is corroborated with the empirical results. KSOM reaches the pseudo-inverse, since the linear approximation of inverse Jacobian is learned by minimizing equation (5.10), which is equivalent to,

$$\dot{\boldsymbol{\theta}} = \mathbb{J}^+ \dot{\mathbf{x}}. \tag{5.19}$$

Analysis of Solution Learned Using KSOM

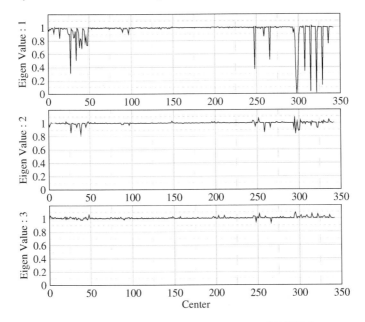

FIGURE 5.2: Eigenvalues at centers of KSOM.

Hence, KSOM based kinematic control algorithm estimates the inverse Jacobian with average value of locally valid linear inverse Jacobian maps in each operating zone. This inverse Jacobian is required in the visual servoing control algorithm (5.1). This is one of the major contribution achieved in this thesis work.

With these observations, the learned KSOM is considered to be an approximation of the inverse Jacobian from the vision space to the joint angle space. These observations play a significant role in the visual control of a kinematically redundant manipulator as follows:

- The map learned with KSOM based kinematic control algorithm is an estimate of the inverse Jacobian for the global workspace. A globally stable visual servoing algorithm can be formulated with this estimated map.

- The KSOM based kinematic control algorithm can be generalized to approximate a particular solution which satisfies the desired additional task by learning an appropriate inverse Jacobian map.

It will be shown in subsequent sections of this chapter that it is indeed possible to achieve global visual servoing while satisfying the desired additional task with the KSOM based kinematic control scheme. Global positioning scheme is achieved by using the KSOM in conjunction with the proportional gain, while redundancy is resolved by expressing the instantaneous cost function in weighted norm formulation.

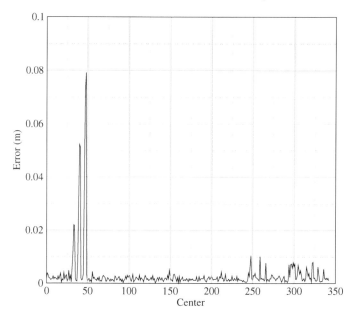

FIGURE 5.3: Positioning error at centers of KSOM.

5.5 KSOM in Closed Loop Visual Servoing

Through experimental analysis, it is shown that the inverse Jacobian is approximated as a linear map in each operation zone with KSOM based kinematic control algorithm. This learned map can be used as an approximate inverse Jacobian for visual servoing. With the learned KSOM based approximation of the inverse Jacobian, image based visual servoing can be performed from the joint angle space. The KSOM based visual servoing simplifies the following issues:

- With the learned KSOM map, the approximate pseudo-inverse from the vision space to the joint angle space is known over the entire space. This eliminates the computation of the pseudo-inverse during servoing. It is known that only the winner neuron contributes to the learned map after learning phase. Hence, the computation cost of real-time servoing reduces to a simple matrix multiplication.

- Since KSOM learns a unique relationship between the vision space and the joint angle space, it resolves the redundancy in learning phase itself. This facilitates to analyze the visual servoing and redundancy resolution

KSOM in Closed Loop Visual Servoing

in a simple integrated framework with direct computation of joint angle space trajectories from vision space.

The conventional proportional controller with pseudo-inverse computation at the desired location ensures local asymptotic stability only. The global asymptotic stability can be achieved by either estimating the forward Jacobian at each instant and then computing the pseudo-inverse of the forward Jacobian or estimating the pseudo-inverse of the Jacobian over the entire workspace. As discussed in the previous section, KSOM approximates the inverse kinematic Jacobian from the vision space to the joint angle space at discrete operating points. Hence, KSOM based learning approach is a holistic methodology to learn the inverse kinematic relationship over the entire workspace.

In this thesis work, this KSOM based approximation of the inverse Jacobian is used to achieve global stabilization with conventional proportional controller. KSOM eliminates the computation of pseudo-inverse along the trajectory, since the inverse kinematic relationship is learned offline. With this control scheme, the input to KSOM network is given as,

$$\Delta \mathbf{u} = k_p\, \mathbf{e}_u. \tag{5.20}$$

The global stabilizing controller can be obtained only if the inverse adaptively changes along the path. In conventional KSOM algorithm, the winner neuron is selected based on the position of the object and, hence, the inverse will be fixed for a given desired location. This approach results in a local stabilizing controller. In this work, the winner neuron is selected based on the current end-effector position such that the inverse Jacobian changes as the end-effector traverses along the path. The desired joint angular velocity is then computed with the above input as,

$$\dot{\boldsymbol{\theta}}^{out} = k_p\, s^{-1} \sum_{\gamma=1}^{N_n} h_\gamma \mathbf{A}_\gamma \mathbf{e}_u. \tag{5.21}$$

After training, the winner neuron is the major contributor to the joint angular velocity. Hence, the computation reduces to a simple matrix multiplication in real-time which makes the algorithm computationally efficient. This is a major improvement in case of visual servoing, where currently the computation poses a constraint in real-time implementation due to the computation cost associated with the image processing techniques.

5.5.1 Stability Analysis

It is clear from the empirical observation that the KSOM approximates the pseudo-inverse of the Jacobian. In this section, Lyapunov stability of the proposed control scheme is analyzed. Let's consider the Lyapunov candidate as the quadratic position error,

$$V = \mathbf{e}_u^T \mathbf{e}_u \tag{5.22}$$

where the error, $\mathbf{e}_u = \mathbf{u}_d - \mathbf{u}$ and $\mathbf{u}_d$ is constant for positioning task. The time derivative of the Lyapunov function is given by,

$$\begin{aligned}
\dot{V} &= \mathbf{e}_u^T \dot{\mathbf{e}}_u \\
&= -\mathbf{e}_u^T \dot{\mathbf{u}} \\
&= -\mathbf{e}_u^T \mathbb{J} \dot{\boldsymbol{\theta}} \\
&= -\mathbf{e}_u^T \mathbb{J} \, k_p s^{-1} \sum_{\gamma=1}^{N_n} h_\gamma \mathbf{A}_\gamma \mathbf{e}_u \\
&= -k_p s^{-1} \mathbf{e}_u^T \mathbb{J}. \sum_{\gamma=1}^{N_n} (h_\gamma \mathbf{A}_\gamma) \mathbf{e}_u \\
&= -k_p s^{-1} \mathbf{e}_u^T \mathbb{J} \{ \mathbb{J}^+ - \mathbb{J}^+ + (\sum_{\gamma=1}^{N_n} h_\gamma \mathbf{A}_\gamma) \} \mathbf{e}_u \\
&= -k_p s^{-1} \mathbf{e}_u^T \mathbb{J} \mathbb{J}^+ \mathbf{e}_u - k_p s^{-1} \mathbf{e}_u^T \mathbb{J} \{ \mathbb{J}^+ - (\sum_{\gamma=1}^{N_n} h_\gamma \mathbf{A}_\gamma) \} \mathbf{e}_u \\
&= -k_p s^{-1} \mathbf{e}_u^T \mathbf{I}' \mathbf{e}_u - k_p s^{-1} \mathbf{e}_u^T \mathbb{J}(\tilde{\mathbf{A}}) \mathbf{e}_u \\
&= -k_p s^{-1} \mathbf{e}_u^T \mathbf{I}' \mathbf{e}_u - k_p s^{-1} \mathbf{e}_u^T \tilde{\mathbf{I}} \mathbf{e}_u
\end{aligned} \quad (5.23)$$

where $\mathbf{I}' = \mathbb{J}\mathbb{J}^+$, $\tilde{\mathbf{A}} = \{ \mathbb{J}^+ - (\sum_{\gamma=1}^{N_n} h_\gamma \mathbf{A}_\gamma) \}$ is the approximation error of KSOM network and $\tilde{\mathbf{I}} = \mathbb{J}\tilde{\mathbf{A}}$. It is well known that, $\mathbf{I}' > 0$ and $\tilde{\mathbf{I}}$ is sign indefinite.

The above equation can be further simplified as,

$$\dot{V} < -k_p s^{-1} \mathbf{e}_u^T \mathbf{I}' \mathbf{e}_u + k_p s^{-1} \| \tilde{\mathbf{I}} \| \| \mathbf{e}_u \|. \quad (5.24)$$

It is clear from the above equation that $\dot{V}$ is negative definite, if,

$$\mathbf{e}_u^T \mathbf{I}' \mathbf{e}_u > \| \tilde{\mathbf{I}} \| \| \mathbf{e}_u \|. \quad (5.25)$$

Empirical observation has clearly shown that the linear map of KSOM approaches the local pseudo-inverse of the kinematic Jacobian with training and, hence, $\tilde{\mathbf{I}} \approx 0$. Thus equation (5.25) is true, which implies that the stability condition given by equation (5.24) is also satisfied. To make the algorithm robust, one can increase the number of neurons which would increase the discretization of the workspace and, hence, $\| \tilde{\mathbf{I}} \|$ will be bounded. The global Lyapunov stability of the proposed scheme is thus guaranteed, with accurate offline learning of KSOM network. *A globally asymptotically stable visual servoing scheme can be designed with the inverse kinematic map learned using the KSOM based kinematic control algorithm. Visual servoing does not require the estimation of the Jacobian and the computation of its pseudo-inverse at each instant, which makes the proposed KSOM based visual servoing scheme computationally efficient. The KSOM based visual servoing simplifies the development of global visual servoing scheme with offline learning process.*

5.6 Redundancy Resolution

The proposed KSOM based closed loop control strategy is further extended for resolving the redundancy. As discussed in section 5.4.2.3, KSOM generates a smooth minimum joint angle space motion since the linear map is updated by minimizing the equation (5.10). KSOM based kinematic control algorithm is generalized to resolve the redundancy by minimizing weighted norm as discussed in [18]. Weighted norm solution penalizes the joint angle space motion for achieving the desired additional task. The joint angular velocity $\dot{\boldsymbol{\theta}}$ which minimizes the weighted norm, $\| \dot{\boldsymbol{\theta}}^T \mathbf{W}_R \dot{\boldsymbol{\theta}} \|$ is given as,

$$\dot{\boldsymbol{\theta}} = \mathbf{W}_R^{-1/2} \mathbb{J}_w^+ \dot{\mathbf{u}} \tag{5.26}$$

where $\mathbf{W}_R \in R^{m \times m}$ is the weight matrix which penalizes the joint angle space motion to achieve the additional task, $\mathbb{J}_w = \mathbb{J} \mathbf{W}_R^{-1/2}$ and $\mathbb{J}_w^+ = \mathbf{W}_R^{-T/2} \mathbb{J}^T (\mathbb{J} \mathbf{W}_R^{-1} \mathbb{J}^T)^{-1}$. The detailed discussion about weighted least norm solution is available in [18]. Comparing equations (5.19) and (5.26), the Jacobian matrix of KSOM is updated to minimize,

$$\frac{1}{2} \| \Delta \boldsymbol{\theta}_{01} - \mathbf{W}_R^{-1/2} \mathbf{A}_i \Delta \mathbf{u}_{01} \|^2 . \tag{5.27}$$

The above equation is analogous to equation (5.10), which converges to the minimum norm solution (5.19). The above cost function, is same as equation (5.10), if $\mathbf{W}_R = \mathbf{I}$, where $\mathbf{I}$ is the identity matrix. Hence, the existing KSOM based learning method [51] is a particular case of the proposed generalized update law. The change in the value of $\mathbf{A}_\gamma$ with the proposed generalization is computed to be,

$$\Delta \mathbf{A}_\gamma = \| \Delta \mathbf{u}_{01} \|^{-2} \left(\Delta \boldsymbol{\theta}_{01} - \mathbf{W}_R^{-1/2} \mathbf{A}_\gamma \Delta \mathbf{u}_{01} \right) \Delta \mathbf{u}_{01}^T . \tag{5.28}$$

The change in the joint angle vector $\boldsymbol{\theta}_\gamma$ is evaluated as,

$$\Delta \boldsymbol{\theta}_\gamma = \boldsymbol{\theta}_0 - \boldsymbol{\theta}_\gamma - \mathbf{W}_R^{-1/2} \mathbf{A}_\gamma^{new} (\mathbf{u}_0 - \mathbf{w}_\gamma), \tag{5.29}$$

such that $\boldsymbol{\theta}_\gamma \to \boldsymbol{\theta}_0$ while minimizing $\| \dot{\boldsymbol{\theta}}^T \mathbf{W}_R \dot{\boldsymbol{\theta}} \|$. The update law for $\mathbf{w}_\gamma$ is chosen as same as equation (5.13), since the weights are penalizing the joint angle space only.

The manipulator has to be actuated with generalized KSOM based kinematic control scheme to move near the desired position $\mathbf{u}_d$ with a coarse movement. The coarse movement is generalized as,

$$\boldsymbol{\theta}_0^{out} = s^{-1} \sum_{\gamma=1}^{N_n} h_\gamma (\boldsymbol{\theta}_\gamma + \mathbf{W}_R^{-1/2} \mathbf{A}_\gamma (\mathbf{u}_d - \mathbf{w}_\gamma)). \tag{5.30}$$

The manipulator moves to the position $\mathbf{u}_0$ in the vision space, with the joint angle configuration $\boldsymbol{\theta}_0^{out}$ computed for the joint penalization $\mathbf{W}_R$. The positioning error in the vision space after the coarse movement is given by $\mathbf{u}_d - \mathbf{u}_0$.

The positioning accuracy can be further improved with a fine movement defined as,

$$\boldsymbol{\theta}_1^{out} = \boldsymbol{\theta}_0^{out} + s^{-1}\mathbf{W}_R^{-1/2}\sum_{\gamma=1}^{N_n} h_\gamma \mathbf{A}_\gamma (\mathbf{u}_d - \mathbf{u}_0). \quad (5.31)$$

The manipulator reaches the position $\mathbf{u}_1$, when it is commanded with the joint angle vector $\boldsymbol{\theta}_1^{out}$. The positioning accuracy can be further improved with multi-steps similar to the existing KSOM based kinematic control schemes.

The above algorithm penalizes the joint angle space motion based on the additional task and, hence, it is expected to resolve the redundancy during learning phase. The learned map is expected to converge to $\mathbb{J}_w^+ = \mathbf{W}_R^{-T/2}\mathbb{J}^T(\mathbb{J}\mathbf{W}_R^{-1}\mathbb{J}^T)^{-1}$ with the proposed generalization of the KSOM based learning scheme. It will be further corroborated with empirical results that the proposed generalized scheme indeed resolves the redundancy by penalizing the joint angle space motion with weight matrix $\mathbf{W}_R$. *Thus, the proposed generalization of KSOM based kinematic control algorithm resolves the redundancy during learning phase, provided the additional task is expressed in weighted norm formulation. Such generalization is highly desirable in vision-based redundant manipulator control, since the accurate model of the system is not known, and the redundancy of the manipulator can be resolved only by estimating the Jacobian with a learning algorithm.*

5.7 Results

The performance of the proposed controller scheme is tested from the vision space to control the PowerCube manipulator, in both simulation and real time. The controller is tested for positioning task first within the learned workspace. The performance of the redundancy resolution scheme is then analyzed for tracking a straight line and elliptical trajectory in simulation.

5.7.1 Learning Inverse Kinematic Relationship using KSOM

A 3 dimensional neural lattice with $7 \times 7 \times 7$ nodes is selected for learning the inverse kinematic map from the vision space to the joint angle space. Each node in the KSOM lattice is associated with an input weight vector, $\mathbf{w}_\gamma$ of dimension 4×1 which represents the pixel coordinates of the object in the stereo vision system.

TABLE 5.1: Cartesian workspace limit

Cartesian workspace
$-0.4\ m \leq x \leq 0.4\ m$
$0.3\ m \leq y \leq 0.8\ m$
$-0.15\ m \leq z \leq 0.38\ m$

Training data is generated using the forward kinematic model (2.1), and the camera model obtained using Tsai algorithm [134] discussed in section 2.3.2. The dimension of the workspace, visible through the stereo-vision is tabulated in Table 5.1. Random end-effector positions are estimated from the randomly generated joint angle values within the manipulator kinematic bounds, and only those end-effector positions are retained which lie within the visible workspace volume.

The inverse kinematic relationship is learned with 50,000 random points, which resulted in an average positioning accuracy of 0.12m over the entire workspace.

5.7.2 Visual Servoing

The learned map is used for closed loop control of the redundant manipulator. The initial and the final positions are considered same in both simulation and real-time experiment. The chosen initial and final positions of the object in Cartesian space are tabulated in Table 5.2. The learned map is used as an approximation of the inverse Jacobian in closed loop visual servoing. The joint angular velocity computed using equation (5.21) are applied to the robot manipulator. The proportional gain is chosen as $k_p = 0.05$. The sampling rate is chosen as 0.1s in simulation, to match the experimental setup. The end-effector motion in the vision space is shown in Figure 5.4. The trajectory is smooth in simulation, however it is noisy in real-time. There is noise in the trajectory measurement during real-time implementation, due to the inaccuracies in image processing method. It is clear from the figure that the real-time and the simulation results are much similar, indicating that the controller is performing good though the learning is approximate.

The joint angular velocity is shown in Figure 5.5. In real-time, joint angular velocity is noisy due to noisy measurement of the end-effector position and numerical differentiation. It is clear from Figure 5.5 and Table 2.2, that in both

TABLE 5.2: Initial and final end-effector positions

Position	x(m)	y(m)	z(m)
Initial	-0.1	0.55	0.15
Final	0.1	0.75	0.35

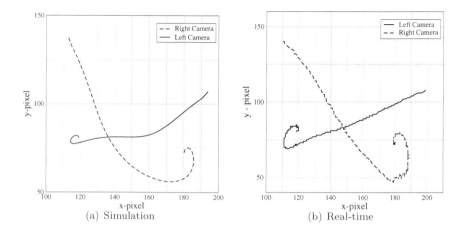

FIGURE 5.4: End-effector motion in the vision space: Both right and left camera views are shown (pixel). Smooth end-effector motion is generated from initial position to final position.

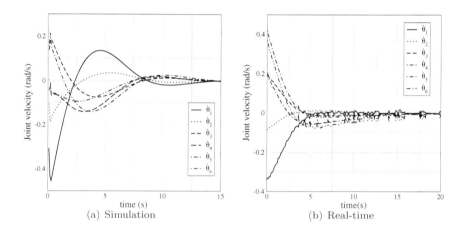

FIGURE 5.5: Joint angular velocity of all links (rad/s). The joint angular velocity is within the limit and smooth resulting a smooth motion. Joint angular velocity converges to zero as the end-effector reaches the desired position.

simulation and real-time the velocity of each joints are within their physical limit and finally go to zero as the end-effector reaches the desired position, indicating the stability of the proposed algorithm. The joint angle trajectory of each links, while moving toward the desired position, are shown in Figure 5.6

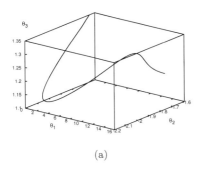
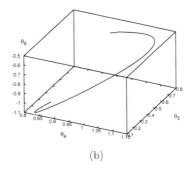

FIGURE 5.6: Joint angle space motion for positioning task in simulation: (a) Joint angle: Link 1,2 and 3 (rad), (b) Joint angle: Link 4,5 and 6 (rad).

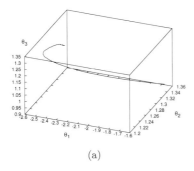
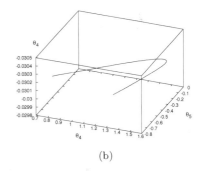

FIGURE 5.7: Joint angle space motion for positioning task in real-time: (a) Joint angle: Link 1, 2 and 3 (rad), (b) Joint angle: Link 4, 5 and 6 (rad).

and 5.7 respectively for simulation and real-time. The joint angle variation is smooth and the angles are within the limit.

The end-effector motion in Cartesian space in a real-time experiment is presented in Figure 5.8. It is observed that the end-effector reaches the final position with 2mm accuracy. In simulation, it is observed that the desired position can be reached with an accuracy of 0.24 pixel error. This accuracy can be further increased by executing the simulation for longer intervals. In real-time a minimum of 1 pixel error can be achieved due to the image processing limitation, which resulted in 2mm error. The real-time performance is influenced by the measurement noise, which affects the positioning accuracy. The response is slightly sluggish in real-time compared to simulations. The sluggish response is due to the image processing noise which would be comparable as the end-effector approaches the desired position.

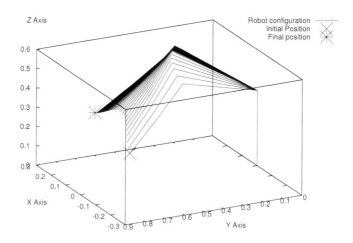

FIGURE 5.8: End-effector motion in Cartesian space from the initial position to the final position (m).

5.7.3 Redundancy Resolution

This section discusses about the simulations performed to analyze the proposed redundancy resolution scheme. The simulation is performed with the end-effector trajectories defined in Cartesian space. The trajectories are defined in Cartesian space, and are projected to the vision space with the available camera model. The resulting vision space position is given as the input to the learned Kohonen's self-organizing map. The main purpose of this simulation is to show that KSOM network learns the inverse kinematic map over the entire workspace and, hence, it guarantees global stability. To confirm the proposition that a particular solution can be learned with the proposed generalization of the KSOM based kinematic control scheme, the learned controller is tested to track two trajectories in Cartesian space: (i) a straight line and (2) an ellipse.

The trajectories are tracked with KSOMs learned with the existing algorithm and the proposed generalized learning algorithm. The weighted norm solution is learned with the weight matrix, $\mathbf{W}_R = diag(1, 1, 100, 1, 1, 1, 1)$, which constrains the motion of the third joint of the manipulator. In case of the weighted norm, the manipulator is expected to track the trajectory with constrained motion of the third joint.

5.7.3.1 Tracking a Straight Line

The proposed scheme is tested first for tracking a straight line in Cartesian space. The straight line is particularly chosen since it is well known that tracking a straight line is much more difficult than a smooth curved trajectory, with a revolute joint manipulator. The desired end-effector position in the vision space is obtained for the straight line, using the camera model and is given as input to the controller. A straight line passing across the entire workspace can be tracked only if the inverse Jacobian $\mathbb{J}^+$ is learned accurately around each operating point.

The line connecting the points $[0.3, 0.7, 0.05]^T$ and $[-0.2, 0.6, 0.28]^T$ is considered for tracking. The end-effector trajectory while moving along the line is shown in Figure 5.9. The corresponding vision space trajectory is shown in Figure 5.10. The r.m.s. tracking errors with the minimum norm and the weighted minimum norm solution are 0.7mm and 1.3mm respectively. The corresponding errors in the vision space are observed as 0.067 and 0.27 pixels, respectively.

The instantaneous tracking error in the vision space is shown in Figure 5.11 which shows that controller tracks the trajectory with an accuracy of ±4 pixel. The tracking error is lesser than ±1 pixel along the major portion of the trajectory. The large deviation is observed in a small section of the line indicating that the learning is not complete at those locations.

The joint angle trajectory of each links are shown in Figure 5.12 and it is clear that the trajectory of the individual link is smooth while moving along

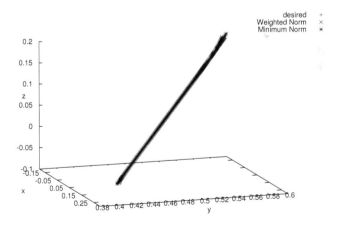

FIGURE 5.9: End-effector motion in Cartesian space while tracking the line (m).

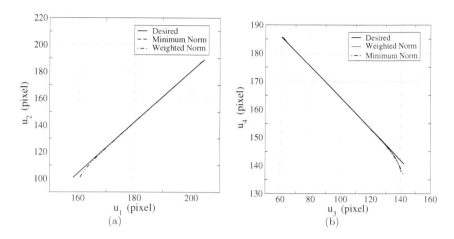

FIGURE 5.10: Manipulator end-effector position in the vision space while tracking the line: (a) Camera: 1, (b) Camera: 2.

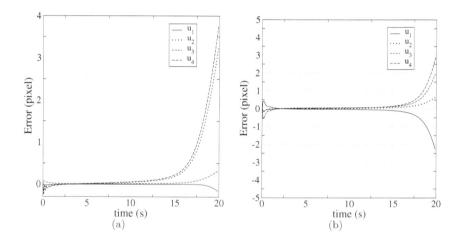

FIGURE 5.11: Vision space error while tracking the line: (a) Minimum norm, (b) Weighted norm.

the line due to topology preserving nature of the KSOM network. The angular configuration of the third joint is less in case of the weighted norm solution. To analyze the effect of redundancy resolution on the third joint, its trajectory is shown separately in Figure 5.13. It is clear from the figure that the motion of the third joint is constrained in the case of a weighted norm solution.

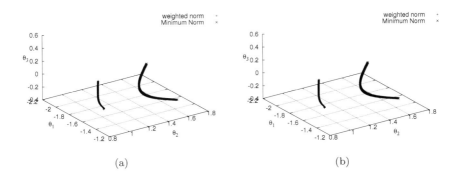

FIGURE 5.12: Joint angle space motion while tracking the line: (a) Joint angle configuration: Link 1, 2, and 3 (rad), (b) Joint angle configuration: Link 4, 5, and 6 (rad).

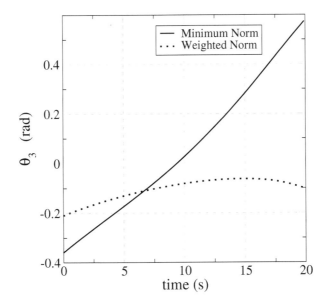

FIGURE 5.13: Motion of third joint while tracking the line. Minimum change is observed in the case of a weighted norm solution.

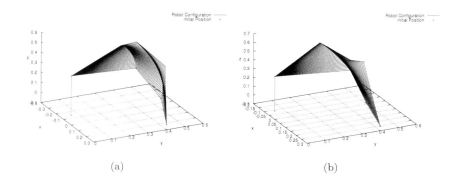

FIGURE 5.14: Manipulator configuration while tracking the line: (a) Minimum norm solution (m), (b) Weighted norm solution (m).

The manipulator configuration while moving along the straight line is shown in Figure 5.14 which shows the effect of weighted norm on each joints.

5.7.3.2 Tracking an Elliptical Trajectory

An elliptical trajectory is further tested to check the performance along the closed path. The tracking result is shown in Figure 5.15. The demonstrated result shows the controller performance while tracking the elliptical trajectory

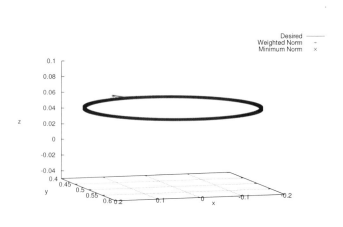

FIGURE 5.15: End-effector motion in Cartesian space from initial position to final position (m).

Results

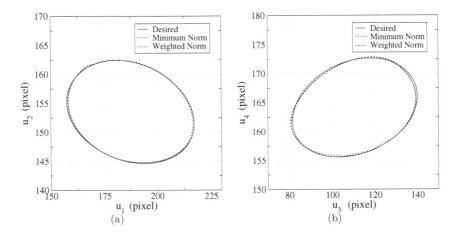

FIGURE 5.16: Manipulator end-effector position in the vision space, while tracking the ellipse: (a) Camera: 1, (b) Camera: 2.

given by,

$$x = 0.2\sin(t)$$
$$y = 0.5 + 0.1\cos(t)$$
$$z = 0.05.$$

The r.m.s. tracking errors in Cartesian space are observed as 0.68mm and 0.62mm for minimum norm and weighted minimum norm solutions respectively. The end-effector trajectory in the vision space is shown in Figure 5.16. The r.m.s. tracking errors in the vision space are observed as 0.165 pixels in case of minimum norm and 0.164 pixels for weighed minimum norm solution.

The instantaneous tracking error in the vision space is shown in Figure 5.17 which shows that controller tracks the trajectory with an accuracy of ± 1 pixel.

The joint angle trajectories are shown in Figure 5.18 and the motion of the third joint is presented in Figure 5.19. It is clear from the figures that the weighed norm constrains the motion of the third joint and the joint angle trajectory is also following the closed path.

The manipulator configuration while tracking the ellipse is shown in Figure 5.20 which clearly shows that the weighted norm solution constrains the motion of the third joint which in turn effects the larger movement of the other joints.

170   *Learning-Based Visual Servoing*

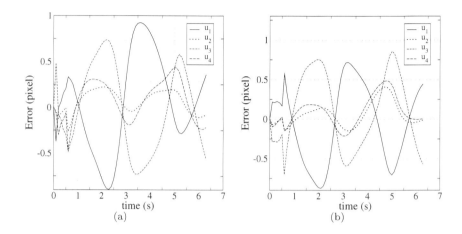

FIGURE 5.17: Vision space error while tracking the elliptical path: (a) Minimum norm, (b) Weighted norm.

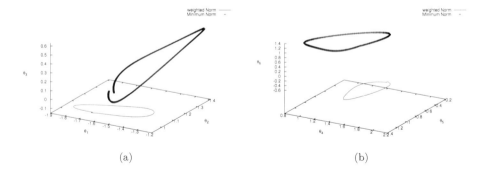

FIGURE 5.18: Joint angle space motion in simulation: (a) Joint angle configuration: Link 1, 2, and 3 (rad), (b) Joint angle configuration: Link 4, 5, and 6 (rad).

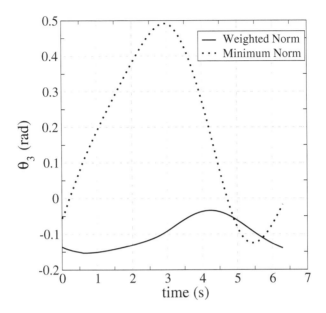

FIGURE 5.19: Motion of third joint while tracking the ellipse. Minimum change is observed in case of weighted norm solution.

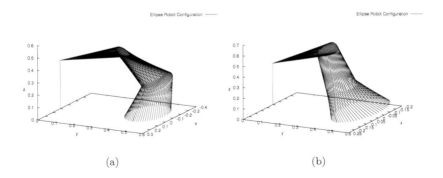

FIGURE 5.20: Manipulator configuration while tracking the ellipse: (a) Minimum norm solution (m), (b) Weighted norm solution (m).

5.8 Summary

A learning-based scheme to estimate the inverse Jacobian from the joint angle space to the vision space has been proposed. It has been shown experimentally that the KSOM approximates the pseudo-inverse of the Jacobian over the entire workspace as a cluster of locally valid linear inverse Jacobian maps. This eliminates the computation of the pseudo-inverse over the entire workspace during visual servoing. With this observation, a generalized learning algorithm is proposed for KSOM based kinematic control to resolve the redundancy in the learning phase by minimizing an instantaneous cost function. The learned KSOM is used in conjunction with the proportional controller in real-time, for closed loop visual servoing. It is shown through the Lyapunov stability analysis that the proposed controller guarantees global stability, if the learned map is sufficiently accurate.

5.9 Reinforcement Learning-Based Optimal Redundancy Resolution Directly from the Vision Space

The adaptive critic based redundancy resolution is proposed for the trajectories defined in the Cartesian space in the previous chapter. In real-world applications, the trajectories are defined in the vision space and the robot manipulation becomes efficient if the joint angle space motion is deduced directly from the vision space so that the effect of the sensor noise can be directly controlled in the closed loop operation. This chapter proposes SNAC based redundancy resolution for the trajectories defined in the vision space.

5.10 Introduction

The single network adaptive critic based redundancy resolution scheme has been proposed in the previous chapter for the Cartesian space trajectories. In a dynamic environment the manipulator is guided with the visual feedback and, hence, it is desirable to define the trajectories in the vision space compared to the Cartesian space. Hence, the SNAC based redundancy resolution approach is extended to the vision space, and the redundancy resolution is achieved while tracking the trajectory defined in the vision space. The advantage of resolving the redundancy for visual trajectory is that the inaccuracies in the

Introduction

vision system are incorporated during the redundancy resolution which makes the approach more robust to the model inaccuracies. The state space is defined in the vision space and the change in the joint angle is considered as the input. Similar to the Cartesian space approach proposed in Section 7.4, the primary and the additional tasks are expressed as an integral cost function which facilitates to achieve optimal solution with single network adaptive critic architecture.

The optimal redundancy resolution from the vision space with the critic based approach is not as simple as that from the Cartesian space since the associated forward Jacobian depends on the joint angle configuration of the manipulator and the position of the end-effector in both the Cartesian space and vision space. The end-effector positioning accuracy is poor, if the given end-effector position is not associated with a unique joint angle configuration. The redundancy can be effectively resolved from the vision space only if the exact correspondence between the joint angle space and the vision space is known a priori and the control process is initiated with a unique joint angle configuration.

To circumvent these challenges, two novel neural network architectures are proposed to learn the costate vector from the vision space, while the correspondence between the vision space and the joint angle space is also identified. The first neural architecture is a T-S fuzzy model-based critic network. The vision space is clustered with fuzzy boundaries and a linear relationship is learned between the costate vector and the positioning error in the vision space. In addition, a joint angle configuration is learned in each fuzzy zone for initializing the control process with a unique joint angle. The second critic network is based on KSOM, which spatially orders the vision space in a 3-D lattice, and thereby learns the relationship between the costate vector and the positioning error. It is observed in our studies, that the KSOM based critic network performs similarly to the human way of operation by clustering and ordering the input space on a lattice space, with fewer parameters to be tuned during the training phase. Since the input space is clustered with the neighborhood defined in the lattice space the training algorithm is robust to the changes occurring in the topology of the workspace. The algorithm is tested for grasping the ball in real-time with the PowerCube manipulator integrated with Barrett Hand [186]. The manipulator is visually guided to grasp the ball located in the workspace, with the optimal control policy obtained from the adaptive critic.

This chapter is organized as follows: The redundancy resolution problem from the vision space is discussed in the next section. The optimal SNAC based redundancy resolution scheme from the vision space and the associated challenges in learning the costate vector are discussed in Section 5.12. The proposed T-S fuzzy model-based critic network architecture and KSOM based critic network architecture are discussed in Sections 5.13 and 5.14 respectively. The simulation results are presented in Section 5.15. The real-time performance analyses for moving along the simulated trajectory and the grasping of a ball with the learned critic based controller are discussed Section 5.16. The discussion is finally summarized in Section 5.17.

5.11 Redundancy Resolution Problem from the Vision Space

Consider the forward kinematic relationship from the joint angle space to the vision space,

$$\dot{\mathbf{u}} = \mathbb{J}(\mathbf{u}, \boldsymbol{\theta})\dot{\boldsymbol{\theta}} \tag{5.32}$$

where $\mathbb{J}(\mathbf{u}, \boldsymbol{\theta}) = \mathbf{LPJ}$ is a 4×7 Jacobian matrix from the joint angle space to the vision space.

Adaptive critic based redundancy resolution from the vision space is achieved by formulating the positioning task as a discrete-time input affine system in vision space. The approach follows the Cartesian space formulation discussed in Section 7.3.

The forward difference kinematics in the vision space is represented as,

$$\boldsymbol{\Delta}\mathbf{u} = \mathbb{J}\boldsymbol{\Delta}\boldsymbol{\theta} \tag{5.33}$$

where $\boldsymbol{\Delta}\mathbf{u} = \begin{bmatrix} \Delta u_1 & \Delta u_2 & \Delta u_3 & \Delta u_4 \end{bmatrix}^T$, represents the change in the position of the end-effector in the vision space due to the change in the joint angle $\boldsymbol{\Delta}\boldsymbol{\theta}$.

Following the Cartesian space formalism, the forward difference kinematics is expressed as a set of discrete-step motion of the end-effector in the vision space at different instants as,

$$\begin{aligned} \boldsymbol{\Delta}\mathbf{u} &= \mathbb{J}\boldsymbol{\Delta}\boldsymbol{\theta} \\ \mathbf{u}(\mathbf{k}+1) - \mathbf{u}(\mathbf{k}) &= \mathbb{J}\boldsymbol{\Delta}\boldsymbol{\theta}(k) \end{aligned} \tag{5.34}$$

where $\boldsymbol{\Delta}\boldsymbol{\theta}(k)$ is the change in the joint angle at the kth instant, $\mathbf{u}(k+1)$ and $\mathbf{u}(k)$ are the end-effector positions at $(k+1)$th and kth instants respectively. The aforementioned discrete motion results in the dynamic evolution of the end-effector's position in the vision space as,

$$\mathbf{u}(k+1) = \mathbf{u}(k) + \mathbb{J}\boldsymbol{\Delta}\boldsymbol{\theta}(k). \tag{5.35}$$

The closed loop error dynamics which moves the end-effector from the current position $\mathbf{u}$ to the desired position $\mathbf{u}_d$ is obtained as,

$$\mathbf{e}_u(k+1) = \mathbf{e}_u(k) - \mathbb{J}\boldsymbol{\Delta}\boldsymbol{\theta}(k) \tag{5.36}$$

where $\mathbf{e}_u(k) = \mathbf{u}_d(k) - \mathbf{u}(k)$, and $\mathbf{u}_d(k+1) = \mathbf{u}_d(k)$. The assumption $\mathbf{u}_d(k+1) = \mathbf{u}_d(k)$ means that the position of the object is fixed in the vision space.

The manipulator reaches the object located at the desired position in multi-step movement.

The aforementioned dynamical system representation of the positioning task is in nonlinear input affine form with state vector $x(k) = \mathbf{e}_u(k)$, input vector $u(k) = \Delta\theta(k)$, $\mathbf{f}(x) = \mathbf{I}$ and $\mathbf{g}(x) = -\mathbb{J}$. The control task is to compute the optimal input $\Delta\theta^*(k)$ such that the desired position defined in the vision space is reached, while performing the chosen additional task. The major challenge associated with the redundancy resolution from the vision space is that the input matrix $\mathbf{g}(x)$ depends on the image Jacobian $\mathbf{L}(\mathbf{u},\mathbf{x})$, the projection matrix $\mathbf{P}$, and the kinematic Jacobian $\mathbf{J}(\theta)$. The current joint angle configuration θ is not known and the optimal joint configuration has to be evaluated from the redundancy resolution scheme. The projection matrix $\mathbf{P}$ depends on the pose of the cameras in the world coordinate frame.

The above formalism defines the state space in the task space, i.e., the vision space. Hence, the SNAC based redundancy resolution scheme proposed in Section 7.5 can be directly extended to the trajectories defined in the vision space with this formalism.

5.12 SNAC Based Optimal Redundancy Resolution from Vision Space

The discrete-time input affine formulation of the closed loop positioning task discussed in equation (5.36) is suitable for SNAC based redundancy resolution if the primary and the additional tasks are modeled in the form of quadratic cost functions as discussed in Section 7.1.3. It will be shown in further discussions that such formulation is indeed possible by following the strategy discussed in Section 7.5.

The dynamical system representation of the closed loop positioning task considers the positioning error in the vision space as the system state, i.e., $x(k) = \mathbf{e}_u(k)$. The state $\mathbf{e}_u(k)$ of the closed loop positioning task depends on both the desired and the current positions of the end-effector. The input $u(k)$ to the dynamical system is $\Delta\theta(k)$.

The joint angle input $\Delta\theta$ is computed in SNAC based control methodology as,

$$\Delta\theta(k) = \mathbf{R}^{-1}\mathbb{J}^T\hat{\lambda}(k+1) \tag{5.37}$$

where $\hat{\lambda}(k+1) \in R^4$ is the costate vector defined from the vision space, for the chosen cost function and is estimated using a critic network. The input to the manipulator depends on the current joint angle θ and the position of the end-effector in the vision space $\mathbf{u}$. The schematic of the optimal redundancy resolution scheme with SNAC based reinforcement learning is shown in Figure 5.21. The current position and the positioning error in the vision space is

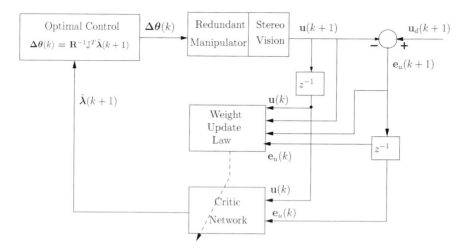

FIGURE 5.21: Optimal redundancy resolution from the vision space with adaptive critic.

given as input to the critic network to estimate the costate vector. The input, i.e., the change in the joint angle is computed using equation (5.37). The cost function has to be appropriately chosen to achieve the additional task so that the redundancy can be resolved optimally in SNAC framework. The formalism discussed in the Cartesian space is used to model the cost function for the vision space too. Hence, the primary positioning task is achieved with the state weight matrix, and the joint angle space motion is penalized based on the additional task requirement.

5.12.1 Selection of Cost Function

As discussed in previous section, the redundancy can be resolved in real-time using SNAC, if the primary positioning task and the additional task can be specified in the form of quadratic cost. The cost function is chosen as,

$$J_c = \frac{1}{2}\sum_{k=0}^{\infty} (\ \mathbf{e}_u^T(k)\mathbf{Q}\mathbf{e}_u(k) + \mathbf{\Delta\theta}^T(k)\mathbf{R}\mathbf{\Delta\theta}(k)\ ) \tag{5.38}$$

where the primary positioning task is defined in the vision space. Hence $\mathbf{Q}$ is a 4×4 positive semi-definite matrix. The input weight matrix $\mathbf{R}$ has to be chosen to penalize the individual joints based on the additional task.

As discussed in the previous chapter, the state weight matrix $\mathbf{Q}$ is always chosen as an identity matrix to ensure uniform convergence toward the desired position in every coordinate direction. The input weight matrix $\mathbf{R}$ is chosen

SNAC Based Optimal Redundancy Resolution from Vision Space

to penalize individual joints based on the desired additional task. The redundancy is resolved directly from the vision space by satisfying the additional tasks:

- Global weighted norm minimization
- Kinematic limit avoidance

The cost function is defined similarly to the previous chapter and the redundancy is resolved for the trajectories defined directly in the vision space. Such formalism extends all the merits discussed for the Cartesian space to the vision space too.

5.12.2 Control Challenges

The T-S fuzzy model-based critic network was used for SNAC based redundancy resolution from the Cartesian space. The critic network discussed in section 7.5.1 can be extended to the vision space as follows:

The workspace spanned by the end-effector in the vision space is to be fuzzified, similar to the Cartesian space control. Then, the ith rule of the critic network to model the costate vector from the vision space is defined as,

IF $u_1(k)$ is F_1^i AND $u_2(k)$ is F_2^i AND $u_2(k)$ is F_3^i AND $u_4(k)$ is F_4^i THEN,

$$\hat{\boldsymbol{\lambda}}_i(k+1) = \mathbf{W}_i \mathbf{e}_u(k)$$

where $\mathbf{W}_i \in R^{4 \times 4}$ is the linear map to approximate the costate vector in the ith fuzzy zone. The fuzzy space is defined along individual coordinate direction of the vision space as,

$$\mu_{u_1}^i(u_1) = e^{\left(\frac{-(u_1 - c_{u_1}^i)^2}{2(\sigma_{u_1}^i)^2}\right)}$$

$$\mu_{u_2}^i(u_2) = e^{\left(\frac{-(u_2 - c_{u_2}^i)^2}{2(\sigma_{u_2}^i)^2}\right)}$$

$$\mu_{u_3}^i(u_3) = e^{\left(\frac{-(u_3 - c_{u_3}^i)^2}{2(\sigma_{u_3}^i)^2}\right)}$$

$$\mu_{u_4}^i(u_4) = e^{\left(\frac{-(u_4 - c_{u_4}^i)^2}{2(\sigma_{u_4}^i)^2}\right)}. \tag{5.39}$$

The fuzzy membership value $\mu_i(\mathbf{u})$ is computed from the product rule as,

$$\mu_i(\mathbf{u}) = \mu_{u_1}^i \mu_{u_2}^i \mu_{u_3}^i \mu_{u_4}^i. \tag{5.40}$$

The nonlinear costate vector can be computed similar to the Cartesian space with the weighted average method. The critic model similar to the Cartesian space suffers with following shortcomings in the vision space:

- The positioning accuracy achieved with the critic network is poor. In [187], the costate vector is learned from an initial position with a joint angle vector $\boldsymbol{\theta}_0$ in the chosen fuzzy zone $\mathbf{x}_{f0}$. It is observed in the experiments that such initialization results in accurate positioning, only around the operating point $\mathbf{x}_{f0}$, in case of visual control. As the operating zone increases, the closed loop system may become unstable. The discussion in Section 5.11 shows that the input matrix of the dynamical system (5.36) depends on $\boldsymbol{\theta}$, $\mathbf{x}$ and $\mathbf{u}$. Initially $\boldsymbol{\theta}$, and $\mathbf{x}$ are unknown, and $\mathbf{u}$ is received from the visual feedback. The same end-effector position can be reached with many joint angle configurations in case of redundant manipulators. Without initial knowledge of the joint angle configuration, the same end-effector position may get represented by different $\boldsymbol{\theta}$ while training the critic. This results in the computation of different Jacobians $\mathbf{J}$ for the same vision space position. This effect becomes predominant as the system moves away from the initial $\boldsymbol{\theta}_0$. The experimental results presented in sections 7.7.1 and 7.7.2 have shown that the same end-effector position is reached with different joint angles based on the initial joint configuration. Hence, the network may get trained for different Jacobians for the same end-effector position as the initial operating point changes.

  Experimental analysis shows that the inaccuracies in the computation of $\mathbf{L}$ and $\mathbf{J}$ affect the stability during training phase and, hence, the critic network does not move toward the optimality. The variation in $\mathbf{J}$ due to the availability of multiple joint configurations may affect the stability during the training phase. This instability affects the convergence of the network. The change in the value of $\mathbf{J}$ is not deteriorating the positioning accuracy in case of the Cartesian space control. But it plays a significant role while learning the costate vector from the vision space.

  The visual control requires a unique initial joint angle to compute $\mathbb{J}$ for effective control over the entire vision space. Hence, the correspondence between the joint angle and the end-effector position in the vision space is necessary in visual control. It will be shown through experiments, that the critic actually moves toward the optimality by learning the correspondence between the joint angle space and the vision space.

- The Cartesian space was fuzzified into equally spaced fuzzy zones in the previous chapter. The fuzzy space was pre-initialized, and the network was trained to update only the linear map of the costate vector. This approach simplified the learning process, since the critic network is linear relative to the parameter $\mathbf{W}_i$. The vision space cannot be fuzzified with equi-distant fuzzy zones similar to the Cartesian space, since there is a nonlinear transformation from the Cartesian space to the vision space. If the fuzzy centers are placed at equal distance in the vision space, then the corresponding discretization in the Cartesian space is not uniform. The number of fuzzy zones will be more in some portion of the Cartesian space and the remaining portion will be represented with fewer numbers of fuzzy zones, if fuzzy zones are created by partitioning the vision space with equal intervals. Hence, the fuzzification of the workspace is not direct in case of the visual control. The vision space can be fuzzified by either choosing the fuzzy centers from equally spaced Cartesian points or using clustering schemes such as Fuzzy c-means clustering algorithm [188]. If the fuzzy zones are chosen from the equally spaced points in the Cartesian space, then the fuzzy zones are not equi-distant in the vision space, but a uniform convergence over the entire workspace can be achieved. Clustering techniques create fuzzy zones based on the data distribution and, hence, it is expected to give better approximation over the entire workspace. In addition, the spread of the fuzzy zone σ has to be initialized such that at least one fuzzy zone is effective at every operating point during the training phase.

In these contexts, two novel critic network architectures are discussed to learn the costate vector which learns the correspondence between the joint angle space and the vision space, while learning the costate vector. The first network architecture extends the T-S fuzzy model proposed for the Cartesian space. The second network is based on KSOM. The structure of the individual neuron is same for both T-S fuzzy model and KSOM based critic networks. The individual neuron structures and the proposed T-S fuzzy model and KSOM based critic networks are presented in subsequent sections.

5.13 T-S Fuzzy Model-Based Critic Neural Network for Redundancy Resolution from Vision Space

The T-S fuzzy model-based critic network proposed for the redundancy resolution from the Cartesian space in Section 7.5.1 is extended to the vision space. The network includes a joint angle vector in the output section to incorporate the correspondence between the joint angle space and the vision space.

5.13.1 Fuzzy Critic Model

The T-S fuzzy model-based critic network for redundancy resolution from the vision space is shown Figure 5.22. The input space is discretized into local operating zones by fuzzification. Each fuzzy zone is defined using a Gaussian function with mean $c_i \in R^4$ and standard deviation $\sigma_i \in R^4$.

The output section of every fuzzy zone is associated with a joint angle vector $\theta_i \in R^m$ and a linear matrix $\mathbf{W}_i \in R^{4 \times 4}$ to compute the local costate vector. The joint angle vector θ_i is used to initialize the control process, with an initial guess of joint angle $\theta(0)$. The output of the ith fuzzy zone while

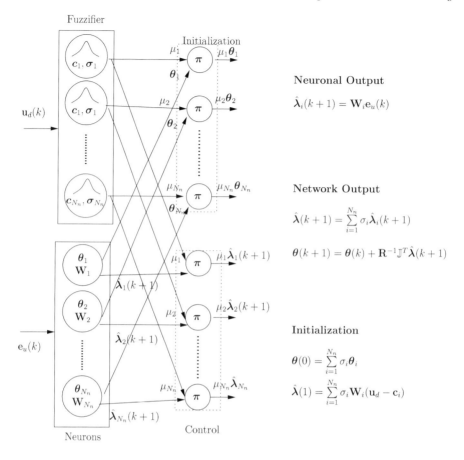

FIGURE 5.22: T-S Fuzzy critic network.

controlling the manipulator at kth instant is given by,

$$\hat{\boldsymbol{\lambda}}_i(k+1) = \mathbf{W}_i \mathbf{e}_u(k). \tag{5.41}$$

The costate vector is computed from the input vision space as,

$$\begin{aligned}
\hat{\boldsymbol{\lambda}}(k+1) &= \frac{\sum_{i=1}^{N_n} \mu_i \hat{\boldsymbol{\lambda}}_i(k+1)}{\sum_{i=1}^{N_n} \mu_i} \\
&= \sum_{i=1}^{N_n} \sigma_i \hat{\boldsymbol{\lambda}}_i(k+1) \tag{5.42}
\end{aligned}$$

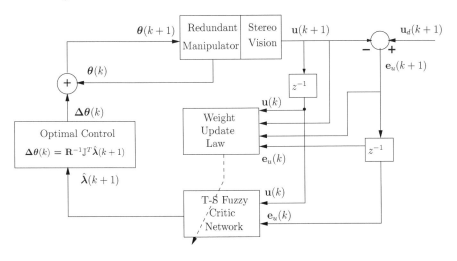

FIGURE 5.23: Visual control using fuzzy critic network.

where μ_i is computed for the current end-effector position $\mathbf{u}$ and $\sigma_i = \frac{\mu_i}{\sum_{i=1}^{N_n} \mu_i}$. The change in the joint angle is computed using equation (5.37), and the manipulator is actuated with,

$$\boldsymbol{\theta}(k+1) = \boldsymbol{\theta}(k) + \mathbf{R}^{-1}\mathbb{J}^T\hat{\boldsymbol{\lambda}}(k+1) \qquad (5.43)$$

where Jacobian $\mathbb{J}$ is computed for the current end-effector position $(\mathbf{u}(k), \boldsymbol{\theta}(k))$.

The schematic diagram of visual control of the redundant manipulator with the proposed T-S fuzzy model-based critic network is shown in Figure 5.23. The positioning error is given as input to the critic network to compute the costate vector with the fuzzy membership value computed for the current position $\mathbf{u}(k)$. The required change in the joint angle is computed using equation (5.37), and the manipulator is commanded to move to $\boldsymbol{\theta}(k+1)$.

5.13.2 Weight Update Law

The fuzzy zones are pre-initialized using the available camera model. After initialization, the fuzzy zones are not updated, and only the output parameters $(\boldsymbol{\theta}_i, \mathbf{W}_i)$ are updated during the training process. The update law for the output parameters is defined as,

$$\mathbf{W}_i^{new} = \mathbf{W}_i^{old} + \eta\,\sigma_i\,\boldsymbol{\Delta}\mathbf{W}_i \qquad (5.44)$$
$$\boldsymbol{\theta}_i^{new} = \boldsymbol{\theta}_i^{old} + \eta\,\sigma_i\boldsymbol{\Delta}\boldsymbol{\theta}_i. \qquad (5.45)$$

The linear map of the local costate vector $\mathbf{W}_i$ is updated by minimizing,

$$E = \frac{1}{2} \parallel \boldsymbol{\lambda}_d(k+1) - \hat{\boldsymbol{\lambda}}_i(k+1) \parallel^2 \tag{5.46}$$

where $\boldsymbol{\lambda}_d(k+1)$ is computed by substituting $\hat{\boldsymbol{\lambda}}(k+2)$ and $\mathbf{e}_u(k+1)$ in the costate vector (7.12). $\hat{\boldsymbol{\lambda}}(k+2)$ is estimated from the critic as,

$$\hat{\boldsymbol{\lambda}}(k+2) = \sum_{i=1}^{N_n} \sigma_i \mathbf{W}_i \mathbf{e}_u(k+1). \tag{5.47}$$

$\boldsymbol{\theta}_i$ is updated to represent the current joint configuration of the manipulator. It is updated by minimizing,

$$E = \frac{1}{2} \parallel \boldsymbol{\theta}(k) - \boldsymbol{\theta}_i \parallel^2 . \tag{5.48}$$

The generalized update algorithm for multi-step movement is obtained as,

$$\boldsymbol{\Delta}\mathbf{W}_i(k) = \parallel \boldsymbol{\Delta}\mathbf{u}(k) \parallel^{-2} \left(\boldsymbol{\lambda}_d(k+1) - \hat{\boldsymbol{\lambda}}_i(k+1) \right) \boldsymbol{\Delta}\mathbf{u}^T(k) \tag{5.49}$$

$$\begin{aligned}\boldsymbol{\Delta}\boldsymbol{\theta}_i &= \boldsymbol{\theta}(k) - \boldsymbol{\theta}_i \\&\quad - \mathbf{R}^{-1}\mathbb{J}^T(\boldsymbol{\theta}(k), \mathbf{u}(k)) \mathbf{W}_i^{new}(\mathbf{u}(k) - \mathbf{c}_i).\end{aligned} \tag{5.50}$$

5.13.3 Selection of Fuzzy Zones

The network is pre-initialized with the fuzzy zones such that the fuzzy workspace spans over the vision space. The fuzzy zones are selected such that the Cartesian space is discretized in equal intervals. The Cartesian space is divided into equal intervals in all the coordinate direction, and then, the centers of the fuzzy zone $\mathbf{c}$ are computed using the camera model with the Cartesian space centers. Though the fuzzy zones are not equi-distant in the vision space, but they are arranged in regular interval in the Cartesian space. The spread of the fuzzy zone $\boldsymbol{\sigma}$ is to be initialized such that at least one fuzzy zone is effective at every operating point.

The spatial distance between each fuzzy zones varies in the vision space, since the Cartesian to vision space mapping is nonlinear. In general, the fuzzy membership value of each zone is computed either using minimum fuzzification method or the product method. Hence, the fuzzy membership value depends on the entity u^j of the vision space which is farther from the given fuzzy center c_i.

The distance d_i between the ith fuzzy zone and its closest neighbor is computed as,

$$d_i = \min_{k, k \neq i} \max_{j} d_j^{ik} \tag{5.51}$$

where $d_j^{ik} = |x_j^i - x_j^k|$ where $j = 1, 2, 3, 4$. The maximum overlap of the closest neighbor at the fuzzy center of ith fuzzy zone is assumed, and then the standard deviation of the Gaussian function is computed from the distance d_i as,

$$\sigma_i = \sqrt{\frac{d_i^2}{2\log(\mu_i^{cen})}} \tag{5.52}$$

where μ_i^{cen} is the overlap of the closest neighbor at the fuzzy center c_i. The standard deviation of the fuzzy sets are chosen as same value in all coordinate directions for the ith fuzzy zone in the experiments for simplicity, i.e, $\boldsymbol{\sigma}_i = [\sigma_i\ \sigma_i\ \sigma_i\sigma_i]^T$.

Initially, the contribution of closest fuzzy zone is taken as a large value during the training phase, and then gradually reduced with training, i.e., the variance is gradually decreased from a larger value. A larger variance is chosen initially so that the entire network will get trained together which ensures the learning of a smoother joint configuration map over the entire workspace. The maximum overlap at the fuzzy center is varied as,

$$\mu^{cen}(i) = \mu_i^{cen}\left(\frac{\mu_f^{cen}}{\mu_i^{cen}}\right)^{(i/N_e)} \tag{5.53}$$

where μ_i^{cen} and μ_f^{cen} are the initial and the final contribution of the closest neighbors at the fuzzy centers respectively, and then the standard deviation is computed using equation (5.52). N_e represents the maximum number of training patterns used to update the critic during the learning phase. As the variance of the Gaussian function reduces the local nature of the costate vector is captured better.

5.13.4 Initialization of the Fuzzy Network Control

As discussed in previous section, the knowledge of the current joint configuration is not available initially and, hence, it is to be estimated. The initial joint configuration is estimated with the proposed T-S fuzzy model-based critic network as follows:

Given the desired position $\mathbf{u}_d$, the critic network internally estimates the joint angle vector, $\boldsymbol{\theta}(0)$ as,

$$\boldsymbol{\theta}(0) = \frac{\sum_{i=1}^{N_n} \mu_i \boldsymbol{\theta}_i}{\sum_{i=1}^{N_n} \mu_i} \tag{5.54}$$

where μ_i is computed with fuzzifier for the position $\mathbf{u}_d$. The end-effector is actuated to move near to the desired position with the joint angle $\boldsymbol{\theta}(1)$ as,

$$\boldsymbol{\theta}(1) = \boldsymbol{\theta}(0) + \mathbf{R}^{-1}\mathbf{J}^T\hat{\boldsymbol{\lambda}}(1). \tag{5.55}$$

Since, the error vector $\mathbf{e}_u$ is not available initially, $\hat{\boldsymbol{\lambda}}(1)$ is computed as,

$$\hat{\boldsymbol{\lambda}}(1) = \frac{\sum_{i=1}^{N_n} \mu_i \mathbf{W}_i (\mathbf{u}_d - \mathbf{c}_i)}{\sum_{i=1}^{N_n} \mu_i}$$

$$= \sum_{i=1}^{N_n} \sigma_i \mathbf{W}_i (\mathbf{u}_d - \mathbf{c}_i). \qquad (5.56)$$

Jacobian $\mathbb{J}$ is computed as a function of $(\mathbf{u}_d, \boldsymbol{\theta}(0))$, and $\sigma_i = \frac{\mu_i}{\sum_{i=1}^{N_n} \mu_i}$. The end-effector reaches the position $\mathbf{u}(1)$ with the joint angle $\boldsymbol{\theta}(1)$. After the initial movement $\boldsymbol{\theta}(1)$, the knowledge about a corresponding pair from the joint space to the vision space $(\mathbf{u}(1), \boldsymbol{\theta}(1))$ is available, which gives the complete information about the current configuration of the manipulator. The position of the manipulator end-effector in the Cartesian space $\mathbf{x}(1)$, and the kinematic Jacobian $\mathbf{J}$ can be computed from $\boldsymbol{\theta}(1)$ using the forward kinematic relationship discussed in equation (2.1). The positioning error corresponding to the initial movement is given as $\mathbf{e}_u(1) = \mathbf{u}_d - \mathbf{u}(1)$.

The manipulator is further actuated with the critic network by computing the costate vector as,

$$\boldsymbol{\lambda}(2) = \sum_{i=1}^{N_n} \sigma_i \mathbf{W}_i \mathbf{e}_u(1) \qquad (5.57)$$

where σ_i is computed from the membership value obtained for the current end-effector position $\mathbf{u}(1)$. The input to the manipulator is computed as $\boldsymbol{\Delta\theta}(1) = \mathbf{R}^{-1}\mathbb{J}^T\boldsymbol{\lambda}(2)$, where $\mathbb{J}$ is computed using $(\boldsymbol{\theta}(1), \mathbf{u}(1))$. The end-effector reaches the position $\mathbf{u}(2)$ with $\boldsymbol{\theta}(2) = \boldsymbol{\theta}(1) + \boldsymbol{\Delta\theta}(1)$. The end-effector can be guided to reach the desired position with arbitrary accuracy by computing the input using equations (5.42) and (5.43).

Hence, the initialization differs from the normal control process as follows:

- Fuzzy membership value is computed for $\mathbf{u}_d$, while it is computed for the current position $\mathbf{u}$ during the control process.

- Initial value of the Jacobian $\mathbb{J}$ is computed for $(\mathbf{u}_d, \boldsymbol{\theta}(0))$, since the corresponding joint angle and vision space positions are not available.

5.13.4.1 Remark

The pre-initialization of the fuzzy zones plays a key role in learning the critic network, and it requires the camera model for initializing the fuzzy centers. The procedure is not complex for critic based redundancy resolution from the Cartesian space since the fuzzy zones were equally spaced in the task space [187]. In contrast, the fuzzy zones are not equally spaced in the vision, since the map from Cartesian space to the vision space is nonlinear, and equally spaced Cartesian centers are not equi-distant is the vision space. The pose of the camera in the world frame determines the topology of the vision space. It is observed in experimental analysis that the pose of the camera in the world coordinate frame affects the choice of $\boldsymbol{\sigma}_i$ during the training phase, since the topology of the input space changes.

It is observed in brain cortex that the visual features closer to each other are spatially ordered [189] for efficient processing. The KSOM proposed by Kohonen [190] spatially orders the input vectors by defining the neighborhood in the lattice space. The neurons are ordered in the lattice space, in spite of the nonlinearities in the input space. The topology of the input space is learned with KSOM using the neighborhood defined in the lattice space. Considering the constraints discussed above in learning the T-S fuzzy model-based critic network, a KSOM based critic network is proposed for parameter-free optimal learning of the costate vector.

5.14 KSOM Based Critic Network for Redundancy Resolution from Vision Space

KSOM critic network architecture is similar to the T-S fuzzy model-based critic network proposed in previous section. The outputs of the individual neurons are same in both the cases. The neighborhood is defined in the lattice space in KSOM, while it is defined through fuzzification of the vision space in the T-S fuzzy model-based critic network.

5.14.1 KSOM Critic Model

The architecture of the proposed KSOM based critic network to resolve the redundancy from the vision space is shown in Figure 5.24. In this work, a 3 dimensional KSOM lattice is considered to define the neighborhood. Each neuron consists of an input weight vector $\mathbf{w}_i \in R^4$, which discretizes the input vision space. The output side of the individual neurons consists of a joint angle vector $\boldsymbol{\theta}_i \in R^m$, and a linear map $\mathbf{W}_i \in R^{4\times 4}$ which learns the relationship between the costate vector and the input vision space within the local zone defined by $\mathbf{w}_i$. Hence, the structure of the individual neuron is same for both T-S fuzzy model and KSOM based network. In case of T-S fuzzy model the input space is fuzzified around the center $\boldsymbol{c}_i$, while it is discretized around $\mathbf{w}_i$ in KSOM based critic network.

KSOM based critic network discretizes the vision space with the neighborhood defined in the lattice space, and associates a joint angle vector and local costate vector in every operating zone. The neighborhood is defined in the lattice space by identifying a winner neuron μ for the end-effector position $\mathbf{u}$. Winner neuron μ is selected based on the Euclidean distance metric between

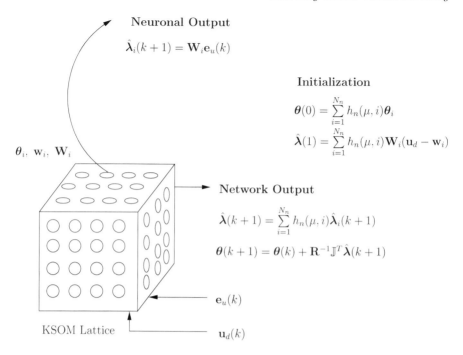

FIGURE 5.24: KSOM based critic network.

the vision space position $\mathbf{u}$ and the input weight vector $\mathbf{w}_i$. The winner neuron is declared as follows:
$$\mu = \min_i \|\mathbf{u} - \mathbf{w}_i\|_2. \tag{5.58}$$
The neighborhood between neuron i and the winner neuron μ is defined as
$$h(\mu, i) = e^{\frac{(-\|\mu - i\|)}{2\sigma^2}} \tag{5.59}$$
where σ defines the spread of the neighborhood in the lattice space which is chosen same for all the neurons at each instant. The individual neuron in KSOM lattice computes the costate vector similar to the neurons in T-S fuzzy model. The local costate vector is computed with the ith neuron, at kth instant as,
$$\hat{\boldsymbol{\lambda}}_i(k+1) = \mathbf{W}_i \mathbf{e}_u(k). \tag{5.60}$$
The overall nonlinear costate vector is computed as,
$$\begin{aligned}\hat{\boldsymbol{\lambda}}(k+1) &= s^{-1} \sum_{i=1}^{N_n} h(\mu, i) \hat{\boldsymbol{\lambda}}_i(k+1) \\ &= \sum_{i=1}^{N_n} h_n(\mu, i) \hat{\boldsymbol{\lambda}}_i(k+1)\end{aligned} \tag{5.61}$$

where the neighborhood $h(\mu, i)$ is defined for the current end-effector position $\mathbf{u}$, $s = \sum_{i=1}^{N_n} h(\mu, i)$, and $h_n(\mu, i) = s^{-1} h(\mu, i)$, is the normalized neighborhood between the ith neuron and the winner μ. The manipulator is moved to $\boldsymbol{\theta}(k+1)$ from $\boldsymbol{\theta}(k)$, by computing the control input $\Delta\boldsymbol{\theta}(k)$ using equation (5.37) as

$$\boldsymbol{\theta}(k+1) = \boldsymbol{\theta}(k) + \mathbf{R}^{-1} \mathbb{J}^T \hat{\boldsymbol{\lambda}}(k+1). \tag{5.62}$$

Similar to the T-S fuzzy model, the Jacobian $\mathbb{J}$ is computed for the current end-effector position. Joint angle vector $\boldsymbol{\theta}_i$ will be used to guess the initial joint angle configuration, just like the T-S fuzzy model. Hence, the individual neurons behave in same manner for both T-S fuzzy model and KSOM. The nonlinear costate vector is computed using the fuzzy clusters in case of T-S fuzzy model, while it is computed in KSOM using the neighborhood defined in lattice space.

The schematic diagram of SNAC based redundancy resolution scheme while following the trajectories defined in the vision space, using the proposed KSOM based critic network is shown in Figure 5.25. A KSOM based critic network behaves similarly to the T-S fuzzy model as shown in Figure 5.23, except that the neighborhood is defined in the lattice space. The critic network computes the necessary change in the joint angle input $\Delta\boldsymbol{\theta}(k)$ using the estimated value of the costate vector $\hat{\boldsymbol{\lambda}}(k+1)$. The costate vector is computed for the current positioning error $\mathbf{e}_u(k)$ in the vision space with the neighborhood defined in the lattice space for the current position $\mathbf{u}(k)$.

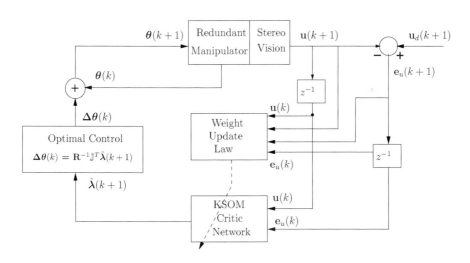

FIGURE 5.25: Visual control using KSOM based critic network.

5.14.2 KSOM: Weight Update

The network weight update law is derived similar to the T-S fuzzy model-based critic network, since the individual neurons behave in identical manner in both the networks. The input vector is initialized randomly in KSOM, and has to be updated in contrast to the fuzzy network where the fuzzy centers are pre-initialized. The pre-initialization of the input weight is not need in KSOM network and hence, the camera model is not required to learn the critic with KSOM based network.

The update law for the parameters of the KSOM based critic network is given by,

$$\mathbf{w}_i^{new} = \mathbf{w}_i^{old} + \eta\, h_n(\mu, i)\, \Delta\mathbf{w}_i \tag{5.63}$$
$$\mathbf{W}_i^{new} = \mathbf{W}_i^{old} + \eta\, h_n(\mu, i)\, \Delta\mathbf{W}_i \tag{5.64}$$
$$\boldsymbol{\theta}_i^{new} = \boldsymbol{\theta}_i^{old} + \eta\, h_n(\mu, i)\, \Delta\boldsymbol{\theta}_i. \tag{5.65}$$

The above update law is similar to the update proposed for the T-S fuzzy model except that the parameters are updated based on the neighborhood defined in the lattice space. The width of the neighborhood is initially taken large and gradually reduced with training to learn a smoother map. The initial and the final value of the width depends on the size of the KSOM lattice and it is independent of the input space.

Similar to T-S fuzzy model $\mathbf{W}_i$ is updated to minimize,

$$E = \frac{1}{2} \parallel \boldsymbol{\lambda}_d(k+1) - \hat{\boldsymbol{\lambda}}_i(k+1) \parallel^2$$

and $\boldsymbol{\theta}_i$ is updated by minimizing,

$$E = \frac{1}{2} \parallel \boldsymbol{\theta}(k) - \boldsymbol{\theta}_i \parallel^2 .$$

The cost function of the T-S fuzzy model itself is chosen to update the network since the parameters serve the same purpose in both the networks. Hence, $\Delta\boldsymbol{\theta}_i$ and $\Delta\mathbf{W}_i$ are computed using equation (5.50) and equation (5.49) respectively.

The update law for the output joint angle vector $\boldsymbol{\theta}_i$ and the costate vector $\mathbf{W}_i$ is similar to the fuzzy network with the fuzzy membership function replaced by the neighborhood in the lattice space. $\Delta\mathbf{w}_i$ is defined to cluster the vision space as,

$$\Delta\mathbf{w}_i = \mathbf{u} - \mathbf{w}_i. \tag{5.66}$$

5.14.3 Initialization of KSOM Network Control

The initialization of the control process with unique joint angle configuration poses a challenge in case of KSOM based visual control also. The optimal control is initialized with KSOM as follows:

Given the desired position $\mathbf{u}_d$, the winner μ is selected for the desired position using equation 5.58. Then, the critic network estimates the initial joint angle vector $\boldsymbol{\theta}(0)$ as,

$$\begin{aligned}\boldsymbol{\theta}(0) &= s^{-1} \sum_{i=1}^{N_n} h(\mu, i) \boldsymbol{\theta}_i \\ &= \sum_{i=1}^{N_n} h_n(\mu, i) \boldsymbol{\theta}_i \end{aligned} \quad (5.67)$$

where $h(\mu, i)$ represents the neighborhood between the ith neuron and the winner μ in the lattice space for the desired position $\mathbf{u}_d$.

The end-effector is moved to the position $\mathbf{u}(1)$ with the joint angle $\boldsymbol{\theta}(1)$ computed as,

$$\boldsymbol{\theta}(1) = \boldsymbol{\theta}(0) + \mathbf{R}^{-1} \mathbb{J}^T \boldsymbol{\lambda}(1) \quad (5.68)$$

where

$$\boldsymbol{\lambda}(1) = \sum_{i=1}^{N_n} h_n(\mu, i) \mathbf{W}_i (\mathbf{u}_d - \mathbf{w}_i). \quad (5.69)$$

Jacobian $\mathbb{J}$ is computed as a function of $(\mathbf{u}_d, \boldsymbol{\theta}(0))^T$ similar to the T-S fuzzy model-based critic network. After the initial movement with $\boldsymbol{\theta}(1)$, we get corresponding pairs $(\mathbf{u}(1), \boldsymbol{\theta}(1))$ from joint space to vision space. Then the manipulator can be controlled by generating a series of $\boldsymbol{\theta}(k+1)$, by computing $\mathbb{J}$ with the corresponding pair $(\mathbf{u}(k), \boldsymbol{\theta}(k))$ and defining the neighborhood using the current position $\mathbf{u}(k)$, where $k = 1, 2, \ldots$. The arbitrary positioning accuracy can be achieved similar to the T-S fuzzy model.

The functioning of the KSOM based critic network differs from the T-S fuzzy model in the following aspects:

- The neighborhood is defined in the lattice space in contrast to the input vision space as in the T-S fuzzy model. This makes the network training robust to the variations in the topology of the input space. Hence, a network designed with N_n neurons on a KSOM lattice can be easily retrained for various camera positions without any change in the training parameters (neighborhood).

- The input vector $\mathbf{w}_i$ plays a role similar to the fuzzy center c_i. But the pre-initialization is needed in case of fuzzy centers, while the input weight vector is learned through self-organization in case of KSOM during the learning stage.

5.15 Simulation Results

The redundancy resolution is tested with the proposed neural architectures on the experimental setup discussed in Chapter 2. The kinematic model discussed in Section 2.2, and the camera model discussed in Section 4.3.1 are used to learn the critic network initially. The networks are trained for 10,000 random positions and the learning rate is varied from 1 to 0.005 for both the networks during training phase as,

$$\eta(i) = \eta_i \left(\frac{\eta_f}{\eta_i}\right)^{(i/N_e)}$$

where i corresponds to the training instant, η_i and η_f are the initial and the final learning rates respectively. The costate vector is learned to resolve the redundancy for three additional tasks,

- Minimum norm movement,
- Weighted norm movement,
- Kinematic limit avoidance.

In weighted norm movement, the entry corresponding to the fourth joint in **R** matrix is chosen as 1.5 and the rest is chosen as 1. The kinematic limit avoidance is implemented on all the joints with the physical constraints tabulated in Table 2.2.

The trained network is tested by following an elliptical trajectory,

$$\begin{aligned} x &= 0.55 + 0.15\cos(0.05 t_k) \\ y &= 0.4\sin(0.05 t_k) \\ z &= 0.2. \end{aligned} \quad (5.70)$$

The trajectories are drawn for a sampling time of 200ms to match with the real-time results. The ellipse is tracked by computing the joint angle input $\Delta\boldsymbol{\theta}(k)$ with the available kinematic model as discussed in section 7.7.1.

5.15.1 T-S Fuzzy Model

The T-S fuzzy model-based critic network is initialized with $7 \times 7 \times 7$ neurons to compare the performance. The Cartesian workspace is fuzzified into seven fuzzy zones in x, y and z directions and the corresponding vision space feature vector is computed using the camera model to initialize the input vector. The initial and final overlap of the closest neighbor at the fuzzy centers are chosen as 0.6 and 0.1 respectively. This contribution of the neighborhood is obtained with trial and error method and it is observed in the simulation that the initial

Simulation Results

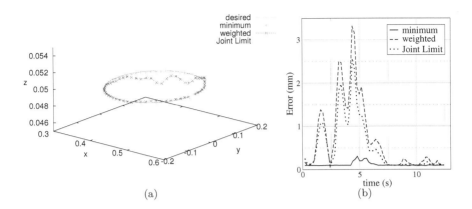

FIGURE 5.26: Simulation: Tracking an elliptical trajectory from vision space using T-S fuzzy model-based critic network (a) Trajectory (m), (b) Tracking error.

and the final value changes with the change in the camera pose respective to the world coordinate frame.

The elliptical trajectory traversed by the end-effector with T-S fuzzy model and the corresponding tracking error are shown in Figure 5.26. The figure clearly shows that the elliptical trajectory is tracked effectively within an accuracy of 3mm while performing the additional tasks too.

The corresponding joint angle trajectories are shown in Figure 5.27 and the vision space trajectories are shown in Figure 5.28. The effect of the weighted norm on the fourth link is clearly visible from Figure 5.27(d) indicating that the effectiveness of added weight 1.5. The weight 1.5 for the fourth link is computed iteratively and the kinematic limits of the other joints are not guaranteed always. The joint angle trajectory of the fourth link is closer to its physical limits in case of kinematic limit avoidance than the weighted norm. But the kinematic limit avoidance scheme ensures that all the links are well within their kinematic limit and the weighted norm may violate the limit for some other trajectory. Figure 5.28 shows that the position is reached with sub-pixel accuracy in simulation which is not possible in real-time implementation. Hence the real-time accuracy will be slightly lower than that in the simulation.

5.15.2 Kohonen's Self-organizing Map

A $7 \times 7 \times 7$ neuron lattice is used in the KSOM so that the number of neurons will be same as the T-S fuzzy model. The initial and final standard deviation, σ of the Gaussian neighborhood function is chosen as 3 and 0.01 respectively. It is observed in our experiments, that the choice of σ depends on the dimension

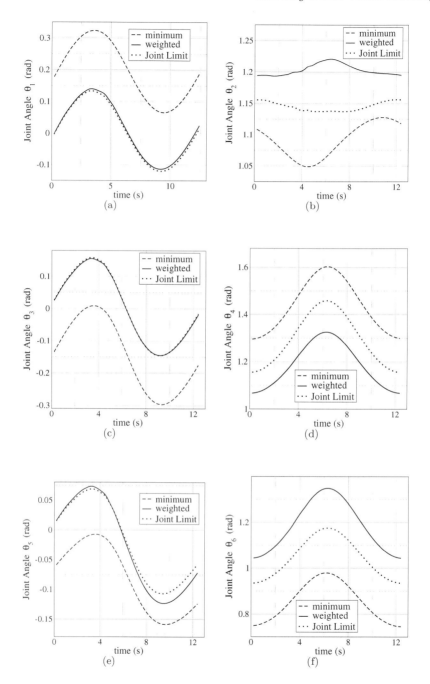

FIGURE 5.27: Simulation: Joint angle configuration while tracking the elliptical trajectory from the vision space using T-S fuzzy model-based critic network.

Simulation Results

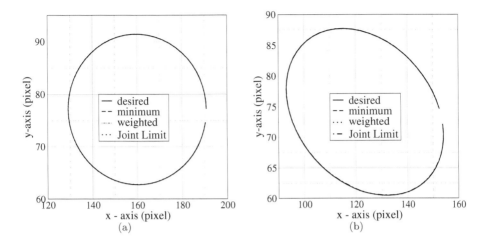

FIGURE 5.28: Simulation: Image space elliptical trajectory from the vision space using T-S fuzzy model-based critic network (a) Left camera, (b) Right camera.

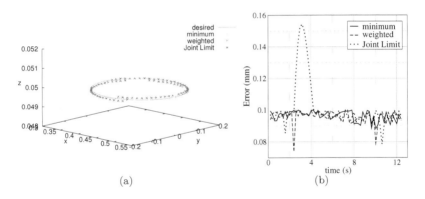

FIGURE 5.29: Simulation: Tracking an elliptical trajectory from vision space using KSOM based critic network (a) Trajectory (m), (b) Tracking error.

of the lattice, and it is independent of the position of the camera. The standard deviation of the neighborhood is varied as,

$$\sigma(i) = \sigma_i \left(\frac{\sigma_f}{\sigma_i}\right)^{(i/N_e)}$$

where σ_i and σ_f are the initial and the final value of the standard deviation of the neighborhood defined on the lattice space respectively. The elliptical trajectory and the tracking error with the SOM are shown in Figure 5.29. The figure clearly shows that the SOM performs better than the T-S fuzzy model with a maximum tracking error of 0.16 mm. The joint angle trajectory

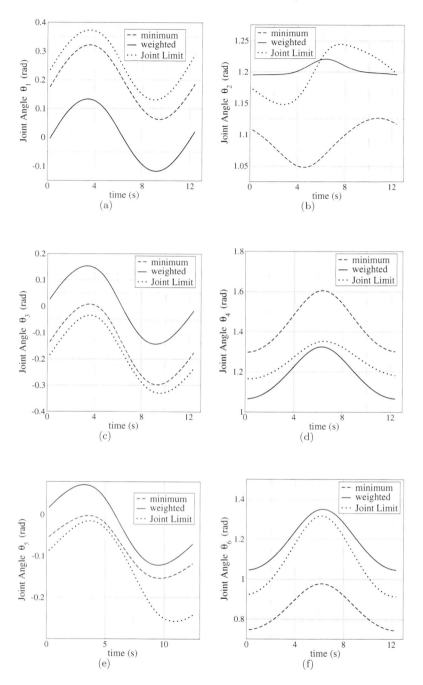

FIGURE 5.30: Simulation: Joint angle configuration while tracking the elliptical trajectory using KSOM.

TABLE 5.3: Simulation: Average number of iterations with SNAC to reach the desired position in the vision space

Critic Network	Minimum Norm	Weighted Norm	Joint Limit
T-S	11	11	18
KSOM	16	17	24

while tracking the ellipse is shown in Figure 5.30, which clearly shows that the learned trajectories shows similar trend with the T-S fuzzy model. It is clear from Figure 5.30(d) that the initial angle of the fourth link is more in case of kinematic limit avoidance than the weighted norm, and it gradually reduces as the joint approaches its limit. The effect of avoidance scheme is predominant as the joint moves near to its limit and it provides an automatic weighing which allows the manipulator to move freely if the joints are away from kinematic limits and gets constrained near the limits. In case of weighted norm, the weight on the joint is constant irrespective of its position relative to its limits. The performance is further compared with the average number of iterations to reach the desired position in 1mm accuracy in Table 5.3. The number of iterations required by the KSOM is typically more than that required by the T-S fuzzy model-based critic network. The costate vector given by the SOM is constant for a winner neuron irrespective of the end-effector position while it varies smoothly from one neuron to the other neuron in T-S fuzzy model-based critic network. The slower performance of the KSOM critic network may be due to the inaccuracy in modelling the costate vector compared to T-S fuzzy model-based critic network. The end-effector position in the vision space is not shown since it exactly follows the desired trajectory, and the performance is similar to that of T-S fuzzy model-based critic network.

5.16 Real-Time Experiment

The networks trained in sections 5.15.1 and 5.15.2 are used in real-time experiment on PowerCube manipulator setup shown in Figure 5.31. The end-effector is identified with the red tape wrapped around it. The centroid of the identified region of the red tape is used as the feedback from the stereo-vision. The proposed critic based neural network controllers are tested in real-time in two phases: (a) Tracking the elliptical trajectory defined in equation (5.70), (b) Ball is grasped with the PowerCube manipulator integrated with Barrett Hand. The elliptical trajectory (5.70) is used to compare the real-time performance with the simulation results and analyze the effect of the sensor noise on positioning accuracy. A learning rate $\eta = 0.01$ is used during real-time experiment to adapt to the inaccuracies in the system model.

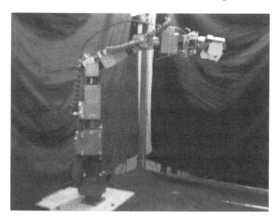

FIGURE 5.31: 7 DOF PowerCube manipulator with Barrett Hand.

5.16.1 Tracking Elliptical Trajectory

The desired position of the end-effector along the trajectory is computed using the camera model at each instant, and is given as input to the network $\mathbf{u}_d(t_k)$. The position of the end-effector $\mathbf{u}(t_k)$ observed through the stereo-vision is used as feedback to the critic network while moving along the trajectory.

5.16.1.1 T-S Fuzzy Model

The position of the end-effector $\mathbf{u}(t_k)$ is used to compute the membership function in real-time. The elliptical trajectory traversed by the end-effector with T-S fuzzy model and the corresponding tracking error are shown in Figure 5.32. The figure clearly shows that the tracking error shows a similar trend as the simulation but with a maximum tracking error of 2.2mm.

There is an increase in the tracking error in real-time due to the inaccuracies associated with the camera model which is predominantly visible in Figure 5.33. The tracking error in the vision space is around 5 pixels which is higher than the simulation. The large vision space error has contributed for the Cartesian space tracking error 2.2mm. The critic based redundancy resolution follows the primary positioning task well, in spite of noisy feedback from the camera which clearly shows the robustness of the proposed scheme to model inaccuracies.

The joint angle trajectories while tracking the ellipse are shown in Figure 5.34 which clearly shows that the real-time trajectory follows a similar trend as the simulation results. The trajectory is not exactly same due to the model inaccuracy but the network could track the desired trajectory by adapting to the model inaccuracies. To check the robustness of the control scheme, the fifth

Real-Time Experiment

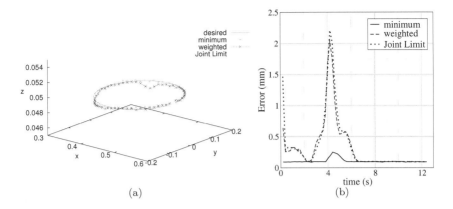

FIGURE 5.32: Tracking of elliptical trajectory from the vision space using T-S fuzzy model-based critic network in real-time experiment (a) Trajectory (m), (b) Tracking error.

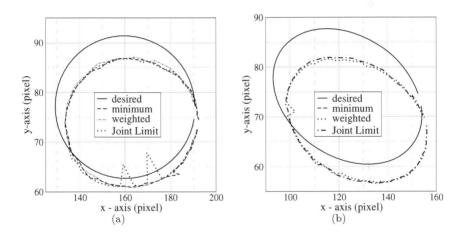

FIGURE 5.33: Image space trajectory while tracking the ellipse using T-S fuzzy model-based critic network in real-time experiment (a) Left camera, (b) Right camera.

link is clamped at the initial position and then the controller is executed. The critic network adapts to this new constraint and tracks the desired trajectory effectively. The manipulator configurations while tracking the trajectory with minimum norm, weighted norm, and kinematic limit avoidance are shown in Figure 5.38(a), Figure 5.38(b) and Figure 5.38(c) respectively.

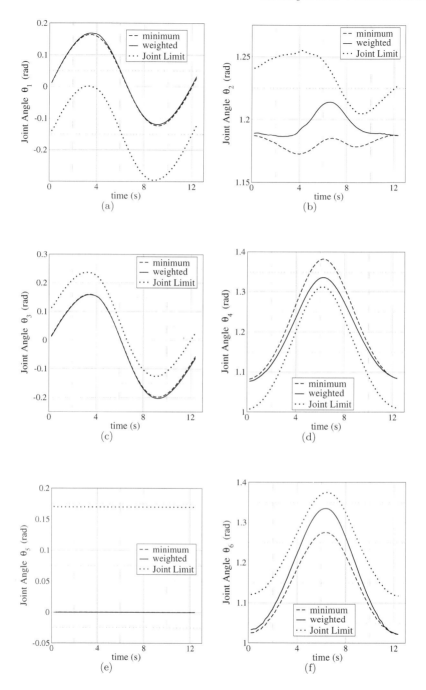

FIGURE 5.34: Joint angle configuration while tracking the elliptical trajectory from the vision space using T-S fuzzy model in real-time experiment.

5.16.1.2 KSOM

The elliptical trajectory and the tracking error with the KSOM are shown in Figure 5.35, which clearly shows that the trajectory is tracked with an error of 1 mm which is slightly more than the simulation. The corresponding vision space trajectory is shown in Figure 5.36 which is similar to the T-S fuzzy model. The joint angle trajectory while tracking the ellipse is shown in Figure 5.37. It is clear from the figure that the trajectory is similar to the T-S fuzzy

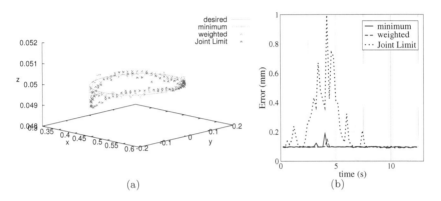

FIGURE 5.35: Tracking an elliptical trajectory from the vision space using KSOM based critic network in real-time experiment (a) Trajectory (m), (b) Tracking error.

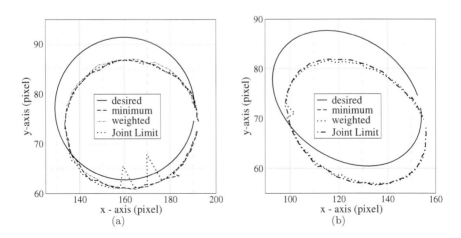

FIGURE 5.36: Image space trajectory while tracking the ellipse from the vision space using KSOM based critic network in real-time experiment (a) Left camera, (b) Right camera.

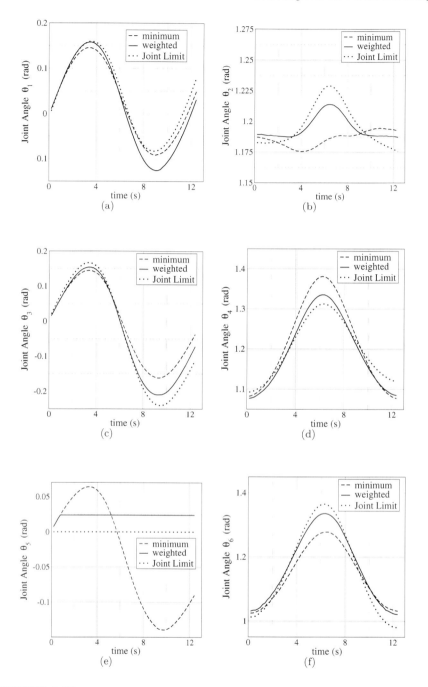

FIGURE 5.37: Joint angle configuration while tracking the elliptical trajectory from the vision space using KSOM based critic network in real-time.

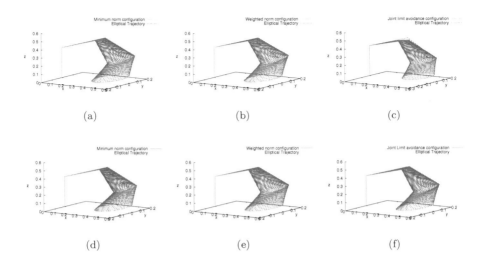

FIGURE 5.38: Real-time experiment: Manipulator configuration while tracking the elliptical trajectory (a)-(c) T-S fuzzy model, (d)-(f) KSOM.

model indicating that both architectures are effective in adapting to the model inaccuracies which occur in real-time.

The manipulator configurations while tracking the trajectory with minimum norm, weighted norm and kinematic limit avoidance are shown in Figure 5.38(d), Figure 5.38(e), and Figure 5.38(f) respectively.

5.16.2 Grasping a Ball with Hand-manipulator Setup

The proposed SNAC based redundancy resolution scheme is further used to grasp a ball in the workspace. The ball is seen through the stereo-vision and is identified based on the color. The centroid of the identified ball is considered as the desired position and is given as the input to the trained network. Ball grasping is achieved with an integrated PowerCube manipulator and Barrett Hand setup shown in Figure 5.31. In the experimental setup the Barrett Hand increases the length of d_7 by 0.18 m. The critic network is retrained considering the increased length with the parameters discussed in section 5.15.2. KSOM based critic network proposed in section 5.14, is used in the experiment for grasping the ball seen through the stereo-vision.

The experimental views of the manipulator while grasping the ball in different positions in the workspace are shown in Figure 5.39. The sub-figures 5.39(a), 5.39(d) and 5.39(g) show the positions of the ball seen through the left camera, and the corresponding views from the right camera are shown

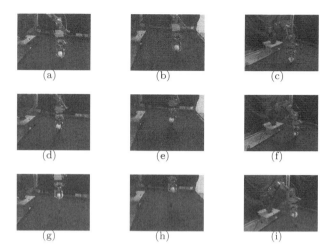

FIGURE 5.39: Real-time experiment: Grasping a ball with integrated manipulator and Barrett Hand Setup from various positions - first column: left camera view, second column: right camera view, third column: manipulator configuration.

in Figures 5.39(b), 5.39(e) and 5.39(h). The manipulator configurations while grasping the ball are shown in figures 5.39(c), 5.39(f) and 5.39(i). It is observed in the experiment that the physical limit of the second joint is violated for positions closer to the base of the manipulator for the global minimum norm and the weighted norm movement. The joint angle configurations generated with the kinematic limit avoidance scheme are observed to be within the kinematic limit. The centroid of the ball as seen from the stereo-vision gives a noisy position information about the position of the ball. It is clear from the figures that the proposed controller is robust to noise and can grasp the ball successfully.

5.17 Summary

This chapter focused on developing an optimal redundancy resolution scheme for visually controlled redundant manipulator. The redundancy is resolved optimally while reaching the positions defined in the vision space. The SNAC based redundancy resolution scheme proposed in the previous chapter has been extended to the vision space with similar dynamical system formulation. The redundancy resolution directly from the vision is difficult, since the exact correspondence between the joint angular space and the vision space is required for accurate global positioning, and also the initialization of the

Summary

fuzzy zones is not simple. Two novel neural network architectures have been proposed to cope with the challenges occurring while defining the control trajectories in the vision space. The first neural network extends the T-S fuzzy model proposed in section 7.5.1, which discretizes the input vision space with fuzzy boundaries. The consequent portion of each fuzzy zone is associated with a joint angle configuration to learn the correspondence between the joint angle space and the vision space, and a linear map to represent the relationship between the costate vector and the positioning error in the vision space. The T-S fuzzy model requires the camera model for the initialization of the center of the fuzzy zones. In addition, the training requires the proper selection of the spread of the fuzzy zones, since it varies with the pose of the camera in the world frame. KSOM based critic network is proposed following the architecture of T-S fuzzy model, to circumvent the initialization problems that occur while training the fuzzy model. KSOM based critic network does not require a priori initialization of the input centers, as they are learned during the training phase by spatially ordering the input in the lattice space. The training parameters of the KSOM based network are independent of the camera position, since neighborhood is defined in the lattice space. The proposed neural network architectures are tested both in simulation and real-time on the PowerCube manipulator for trajectory spanning the entire vision space. Finally the critic based redundancy resolution scheme is successfully used to control the PowerCube manipulator mounted with Barrett Hand, to grasp the ball located in various positions within the workspace.

6

Visual Servoing using an Adaptive Distributed Takagi-Sugeno (T-S) Fuzzy Model

This chapter is concerned with the design and implementation of a distributed proportional-derivative (PD) controller for a 7 degrees of freedom (DOF) robot manipulator using the Takagi-Sugeno (T-S) fuzzy framework. Existing machine learning approaches to visual servoing involve system identification of image and kinematic Jacobians. In contrast, the proposed approach actuates a control signal primarily as a function of the error and derivative of the error in the desired visual feature space. This approach leads to a significant reduction in the computational burden compared with model-based approaches, as well as existing learning approaches to model inverse kinematics. The simplicity of the controller structure will make it attractive in industrial implementations where PD/PID type schemes are in common use. While the initial values of PD gain are learned with the help of a model-based controller, an online adaptation scheme has been proposed that is capable of compensating for local uncertainties associated with the system and its environment.

Firstly, the T-S fuzzy PD parameters are initialized by the offline learning of the MBVS. Thereafter the pseudo-inverse robot Jacobian and the inverse interaction matrix are not computed during servoing implementation using T-S fuzzy PD. This estimated model is applied in a straightforward manner in order to map the image error vector to the joint velocities vector of the robot. This chapter also proposes an online adaptation scheme whereby the PD parameters are updated during servoing such that the local uncertainties associated with the system and its environment can be compensated. In the following, Section 6.1 provides a general review of the T-S fuzzy model. Section 6.2 discusses in detail the development of the adaptive T-S fuzzy PD visual servoing which includes the development of the offline and online learning algorithms as well as the stability analysis of the proposed visual servoing control law. Real-time experiments using a 7 DOF robot manipulator have been carried out to validate and measure the performance of the proposed approach; the experimental results are provided in Section 6.3. The computational complexity analysis of the algorithm is given in Section 6.4. Finally, this work is summarized in Section 6.5.

6.1 T-S Fuzzy Model

Fuzzy inference systems have been widely used in many applications such as robotics, control systems and system modeling. In control systems, fuzzy techniques are used to emulate human deductive thinking to infer conclusions [191]. In contrast, traditional control approaches require formal modeling of the physical reality. Therefore, fuzzy control approaches become more practical to be used than traditional control approaches when the mathematical model representations of the physical systems are too complex to be formalized.

There are two typical types of fuzzy control approaches: Mamdani and Takagi-Suge no types. The Mamdani type fuzzy controller is a direct approach and a language-driven type fuzzy inference system where the controllers are based on the fuzzy rule base. The behavior of the system is adjusted by the expert operator knowledge and experiences. In this approach an explicit system representation cannot be identified. The T-S fuzzy controller is an indirect approach and a data-driven type fuzzy inference system where an explicit system model can be approximated. Therefore the T-S fuzzy model presents systemic understanding [191].

General fuzzy rule representations of Mamdani and T-S types are defined by (6.1) and (6.2), respectively:

$$R_i : \text{if } x_1 \text{ is } A_{i1} \text{ and } x_2 \text{ is } A_{i2} \cdots \text{ and } x_M \text{ is } A_{iM}$$
$$\text{then } y_i \text{ is } Y_i \qquad (6.1)$$

$$R_i : \text{if } x_1 \text{ is } A_{i1} \text{ and } x_2 \text{ is } A_{i2} \cdots \text{ and } x_M \text{ is } A_{iM}$$
$$\text{then } y_i = \sum_{m=1}^{M} c_{im} x_m + c_{i0} \qquad (6.2)$$

where R_i is the i-th fuzzy rule, $\mathbf{x} = [x_1, x_2, \cdots, x_M]^T$ is a fuzzifier input vector which is known as the premise variable or the antecedent variable and y_i is the fuzzy output which is also known as the consequent variable. Y_i represents a linguistic fuzzy set of the i-th Mamdani type rule consequent and $\mathbf{c}_i = [c_{i0}, c_{i1}, \cdots, c_{iM}]^T$ is the consequent parameter vector of the i-th T-S type rule consequent.

T-S fuzzy approach provides a general class representation of a nonlinear dynamical system [191–193]. The consequent variables in each fuzzy rule is a linear function of a nonlinear dynamical local region. Thus, a complete model of nonlinear dynamical system can be approximated by merging different operating linear fuzzy regions. T-S fuzzy models can be catagorized into three types based on their consequent function: affine T-S fuzzy model, homogenous T-S fuzzy model, and singleton T-S fuzzy model.

The affine T-S fuzzy model is the general representation of the T-S fuzzy (6.2) where the consequent part is a linear regression model [194].

T-S Fuzzy Model

A homogenous T-S fuzzy model is defined at the offset $c_{i0} = 0$ and can be represented as

$$R_i : \textbf{if } x_1 \text{ is } A_{i1} \textbf{ and } x_2 \text{ is } A_{i2} \cdots \textbf{ and } x_M \text{ is } A_{iM}$$
$$\textbf{then } y_i = \sum_{m=1}^{M} c_{im} x_m \qquad (6.3)$$

Singleton T-S fuzzy model is also known as the zero-order T-S fuzzy model where the consequent variable in each fuzzy rule is constant. The i-th fuzzy rule of a singleton type can be expressed as

$$R_i : \textbf{if } x_1 \text{ is } A_{i1} \textbf{ and } x_2 \text{ is } A_{i2} \cdots \textbf{ and } x_M \text{ is } A_{iM}$$
$$\textbf{then } y_i = c_{i0} \qquad (6.4)$$

The nonlinear dynamic system is then estimated by finding the approximation of the consequent parameters $\mathbf{c}_i$ of the T-S fuzzy local linear function y_i.

The T-S fuzzy model is constructed based on the selection of some parameters: the membership function shapes and its positions, the distribution of the membership functions, the rule based model, the logical operations and the consequent functions. It is difficult to develop a unique method capable of determining all of the parameters simultaneously. However, in general, at least the type of the membership functions, the logical operations, and the consequent function models have to be selected in advance based on some criteria, e.g., differentiability of the consequent functions and membership functions. The logical "AND" operation is used since it provides an analytical expression which is differentiable based on the given optimization cost function. The remaining T-S fuzzy parameters can be adjusted from the input-output data set. There are several techniques for adjusting the T-S fuzzy parameters; in general those techniques are based on optimization approaches which minimize the error cost function between the real output values from the data set and the approximated values from the T-S fuzzy output. In general, based on the adjusted parameters, T-S fuzzy techniques can be categorized as

1. Lookup table method [195]

2. Gradient descent [191, 194]

3. Clustering [196]

4. Evolutionary [197, 198]

The lookup table method fixes the T-S fuzzy parameters except the consequent function parameters $\mathbf{C} = [\mathbf{c}_1, \mathbf{c}_2, \cdots, \mathbf{c}_r]^T$ is adjusted where r is the number of the T-S fuzzy rule. The T-S fuzzy method using gradient descent defines the type and the number of the membership function in advance whereas the membership function parameters and their position as well as the consequent function parameters $\mathbf{C}$ are tuned. In the clustering method, the membership

TABLE 6.1: T-S fuzzy learning methods

Method	MF's type	MF's number	MF's parameters	C
Table lookup	fixed	fixed	fixed	refined
Gradient Descent	fixed	fixed	refined	refined
Clustering	fixed	refined	refined	refined
Evolutionary	refined	refined	refined	refined

functions parameters are defined by a clustering algorithm in advance, the membership function position and the consequence function parameters **C** are refined during the learning process. Evolutionary strategies can optimize all the T-S fuzzy parameters. However the complexity of the algorithm is increased. A summary of the T-S fuzzy techniques based on its learning is shown in Table 6.1.

This chapter focuses on the development of an uncalibrated visual servoing control where the T-S fuzzy with the gradient descent technique is used. The T-S fuzzy parameters will be optimized online. In this case, the gradient descent technique is more suitable than other T-S fuzzy learning method (see Table 6.1) to be implemented for an online learning approach for the highly nonlinear and realtime control system. The differentiable cost function for online T-S fuzzy learning is designed such that the T-S fuzzy parameters being tuned are optimal. In clustering and evolutionary methods, the membership function type and the membership function number are not differentiable given the cost function for an online T-S fuzzy learning scheme.

6.2 Adaptive Distributed T-S Fuzzy PD Controller

The adaptive distributed T-S fuzzy PD controller is designed such that the visual servoing control system actuates a control signal primarily as a function of the error and the derivative of the error in the visual feature space. The T-S fuzzy offline and online learning algorithms are required in this proposed approach. The T-S fuzzy offline learning is needed to initialize the T-S fuzzy PD parameters with the help of model-based controller. Then, the T-S fuzzy online algorithm is applied to adaptively compensate local uncertainties associated with the system model and its environment. Figure 6.1 shows the closed-loop controller diagrams of the T-S fuzzy PD visual servoing and MBVS.

Using T-S fuzzy approach (see Figure 6.1(a)), the controller output $\widehat{\boldsymbol{\theta}}$ is computed directly from the error signal **e**. In contrast, MBVS requires the computation of the pseudo-inverse of the coupled robot-image Jacobian $\mathbf{J}_e^\dagger \mathbf{L}_{\mathbf{s}^*}^\dagger$ which depends upon both system input states, the image error **e** and the robot

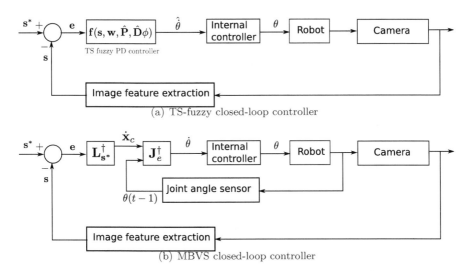

FIGURE 6.1: T-S fuzzy PD visual servoing and MBVS controller approaches.

joint angles $\boldsymbol{\theta}$. The MBVS controller law described in the previous chapters can be rewritten as

$$\dot{\boldsymbol{\theta}} = \kappa \mathbf{J}_e^\dagger(\boldsymbol{\theta}) \mathbf{L}_{\mathbf{s}^*}^\dagger (\mathbf{s}^* - \mathbf{s}) \tag{6.5}$$

The T-S fuzzy PD visual servoing model will be designed and explained in the following sections.

6.2.1 Offline Learning Algorithm

Figure 6.2 depicts the architecture of the T-S fuzzy PD for jth joint. There are four functionality layers in the T-S fuzzy architecture. In layer-1 the current image features vector $\mathbf{s} = [s_1, s_2, \cdots, s_M]^T$ is presented as input to the fuzzifier, where M is the total number of fuzzifier inputs. The error rate in the image space is denoted $\Delta e(k) = e(k) - e(k-1)$ where k indicates the index of the time step. The ith fuzzy rule of the T-S fuzzy PD system is defined as

$$\mathrm{R}_i^j : \text{if } s_1 \text{ is } A_{n_1 1}^j \text{ and } s_2 \text{ is } A_{n_2 2}^j \cdots \text{ and } s_M \text{ is } A_{n_M M}^j$$
$$\textbf{then } y_i^j = \sum_{m=1}^{M} \left(P_{im}^j e_m \right) + \sum_{m=1}^{M} \left(P_{im}^j \Delta e_m \right) \tag{6.6}$$

where $j = 1, 2, \cdots, N$ represents the joint's number, N is the number of robot degrees of freedom (DOF) and $f_i^j = \overline{w}_i^j y_i^j$ represents the output function in the ith fuzzy zone. $\mathbf{A}_{n_m m}$ represents the fuzzy linguistics variable vector, where

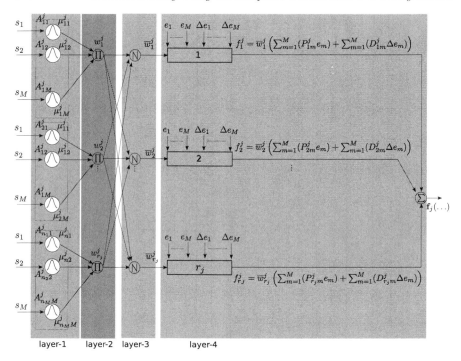

FIGURE 6.2: T-S fuzzy PD architecture for j-th joint.

n_m is the index of the membership function in the m-th fuzzifier input.

$$\sum_{m=1}^{M}\left(P_{im}^{j}e_m\right) \tag{6.7}$$

is the proportional term, and the derivative term can be represented as

$$\sum_{m=1}^{M}\left(P_{im}^{j}\Delta e_m\right) \tag{6.8}$$

where $\mathbf{P}_j$ and $\mathbf{D}_j$ represent the proportional gain vector and the derivative gain vector of the jth joint velocity model, respectively. The number of rules associated with the jth joint is given by $r_j = \prod_{m=1}^{M} N_m^j$, where N_m^j is the number of fuzzy membership functions describing the mth fuzzifier input and M is the total number of image features. In 6.6, it is observed that the control actuation is only a function of the error and the derivative of the error. This way of learning the controller map is novel as well as simple in structure.

The complete expression of this fuzzy PD controller can be given as

$$\widehat{\dot{\theta}} \approx \mathbf{f}(\mathbf{s}, \mathbf{w}, \widehat{\mathbf{P}}, \widehat{\mathbf{D}}\phi)(\mathbf{e}) \tag{6.9}$$

where ϕ is the differential operator $\frac{d}{dt}$. The term $\phi\mathbf{e}$ is approximated as a normalized term $\Delta\mathbf{e}$. $\widehat{\mathbf{P}}$ and $\widehat{\mathbf{D}}$ are some estimates of the T-S fuzzy PD parameters. This equation shows that the control action is only a function of error in the visual feature space. Since the map is not a function of model dynamics such as inverse kinematic and image Jacobians, the structure is simple and easy to compute. As shown in Figure 6.2, for layer-1 the computation of the Gaussian membership function values associated with every node corresponding to individual inputs are calculated using

$$\mu_{n_m m}^j = e^{-\frac{1}{2}\left(\frac{s_m - c_{n_m m}^j}{\sigma_{n_m m}^j}\right)^2} \tag{6.10}$$

where $\mathbf{c}_{n_m m}^j$ and $\boldsymbol{\sigma}_{n_m m}^j$ denote the jth vector of the mean and the variance of the n_mth Gaussian membership function in the mth fuzzifier input. In layer-2, the product of membership values of all input variables is used to make the model differentiable with respect to the parameters of the fuzzifier. For simplicity we assume the number of fuzzy memberships in each input are all the same and the product can be described as

$$w_i^j = \prod_{m=1}^{M} \mu_{n_m m}^j \tag{6.11}$$

where

$$n_m = \lfloor \frac{(i-1)}{(N_m)^{m-1}} \rfloor \mod N_m + 1 \tag{6.12}$$

In layer-3, the ith node calculates the normalized firing strengths:

$$\overline{w}_i^j = \frac{w_i^j}{\sum_{i=1}^{r_j} w_i^j} \tag{6.13}$$

In layer-4, every node i is calculated as the product of the normalized firing strength and the function of the input and the consequent parameter set, $\overline{w}_i^j y_i^j, i = 1, 2, .., r_j$. Finally, the modeled joint velocity $\widehat{\theta}_j$ can be obtained from the summation of all the outputs from layer-4:

$$\widehat{\theta}_j = \mathbf{f}_j = \sum_{i=1}^{r_j} f_i^j \tag{6.14}$$

The gradient descent adaptation algorithm is implemented to tune the parameters used in the T-S fuzzy PD control system. This algorithm seeks to decrease the value of the cost function of the error:

$$E_j = \frac{1}{2}(\dot{\theta}_j^* - \widehat{\dot{\theta}}_j)^2$$

where $\dot{\theta}_j^*$ is the desired velocity output of the jth joint. Thus the iterative gradient descent learning algorithm to update the consequent parameter set and the Gaussian function parameters can be described as

$$\begin{bmatrix} P_{im}^j(k+1) \\ D_{im}^j(k+1) \\ c_{n_m m}^j(k+1) \\ \sigma_{n_m m}^j(k+1) \end{bmatrix} = \begin{bmatrix} P_{im}^j(k) \\ D_{im}^j(k) \\ c_{n_m m}^j(k) \\ \sigma_{n_m m}^j(k) \end{bmatrix} - \eta \begin{bmatrix} \partial E_j/\partial P_{im}^j \\ \partial E_j/\partial D_{im}^j \\ \partial E_j/\partial c_{n_m m}^j \\ \partial E_j/\partial \sigma_{n_m m}^j \end{bmatrix} \quad (6.15)$$

where η is the learning rate parameter. The partial derivatives of E_j with respect to the change of the consequent parameters are described as follows

$$\frac{\partial E_j}{\partial P_{im}^j} = -(\dot{\theta}_j^* - \widehat{\dot{\theta}}_j)\overline{w}_i^j e_m \quad (6.16)$$

$$\frac{\partial E_j}{\partial D_{im}^j} = -(\dot{\theta}_j^* - \widehat{\dot{\theta}}_j)\overline{w}_i^j \Delta e_m \quad (6.17)$$

The partial derivatives of E_j with respect to the change of the Gaussian function parameters are derived as

$$\frac{\partial E_j}{\partial c_{n_m m}^j} = B\frac{\left(s_m - c_{n_m m}^j\right)}{(\sigma_{n_m m}^j)^2} \quad (6.18)$$

$$\frac{\partial E_j}{\partial \sigma_{n_m m}^j} = B\frac{\left(s_m - c_{n_m m}^j\right)^2}{(\sigma_{n_m m}^j)^3} \quad (6.19)$$

where

$$B = -(\dot{\theta}_j^* - \widehat{\dot{\theta}}_j)f_i^j \left(\frac{\frac{w_i^j}{\mu_{n_m m}^j}\sum_{i=1}^{r_j} w_i^j - w_i^j C}{(\sum_{i=1}^{r_j} w_i^j)^2} \right) \mu_{n_m m}^j \quad (6.20)$$

and

$$C = \sum_{i=1}^{r_j} \prod_{a=1}^{M} \mu_{n_m a}^j, \quad \text{for} \quad a \neq m \quad (6.21)$$

6.2.2 Online Adaptation Algorithm

The T-S fuzzy PD controller architecture is given in Figure 6.1. The estimated controller parameters $\widehat{\mathbf{P}}$ and $\widehat{\mathbf{D}}$ were used to control the robot using the estimated joint velocity $\dot{\boldsymbol{\theta}} = \widehat{\dot{\boldsymbol{\theta}}}$. To make the controller more generic in terms of adaptability in a locally changing environment, these parameters must be adapted online. In this section, a simple yet elegant structure of an online parameter adaptation scheme is presented. For further analysis following notations are used: $\mathbf{f}$ to represent $\mathbf{f}(\mathbf{s}, \mathbf{w}, \mathbf{P}, \mathbf{D}\boldsymbol{\phi})$ and $\widehat{\mathbf{f}}$ to represent $\mathbf{f}(\mathbf{s}, \mathbf{w}, \widehat{\mathbf{P}}, \widehat{\mathbf{D}}\boldsymbol{\phi})$.

Adaptive Distributed T-S Fuzzy PD Controller

In online learning, the distributed T-S fuzzy PD parameters should thus be updated such that $\hat{\mathbf{f}} \to \mathbf{f}$. Given Figure 6.1, during the realtime operation, the controller parameterized by $\mathbf{P}$ and $\mathbf{D}$ actuates joint velocity vector $\dot{\boldsymbol{\theta}}$. This joint velocity vector guides the robot to a new position, $\mathbf{s}^{\ddagger}$, in the image space.

This data observation can be interpreted as follows: if $\mathbf{P}$ and are actual parameters of the controller $\mathbf{f}$ then $\dot{\boldsymbol{\theta}} = \mathbf{f}(\mathbf{s}^{\ddagger} - \mathbf{s}) = \mathbf{fe}$, i.e is $\mathbf{s}^{\ddagger}$. It should be noted that in the presence of the ideal controller $\mathbf{f}$, $\mathbf{s}^* = \mathbf{s}^{\ddagger}$ will make the robot end-effector move from $\mathbf{s}$ to $\mathbf{s}^{\ddagger}$ as shown in Figure 6.1. Since the controller is $\hat{\mathbf{f}}$, it can be seen in retrospective that $\mathbf{e} = \mathbf{s}^{\ddagger} - \mathbf{s}$ is the input and $\dot{\boldsymbol{\theta}}$ is the desired output for the fuzzy controller $\hat{\mathbf{f}}$. In essence, $\dot{\boldsymbol{\theta}} = \mathbf{f}(\mathbf{s}^{\ddagger} - \mathbf{s})$ is the observed and $\hat{\dot{\boldsymbol{\theta}}} = \hat{\mathbf{f}}(\mathbf{s}^{\ddagger} - \mathbf{s})$ is the estimated in real-time. This requires that the online update algorithm be derived in such a way that the instantaneous cost function $\|\dot{\boldsymbol{\theta}} - \mathbf{fe}\|$ be minimized. It is interesting to note that the instantaneous cost function is a function of data observed in real-time, i.e., $\dot{\boldsymbol{\theta}}$ is the actuated joint velocity vector and $\mathbf{s}^{\ddagger}$ is the position reached by the robot in the image space. The adaptation algorithm is derived based on the Lyapunov stability condition. The Lyapunov candidate function is given as

$$V = \sum_{j=1}^{7} \frac{1}{2}(\dot{\theta}_j - \mathbf{f}_j \mathbf{e})^2 \qquad (6.22)$$

where $\mathbf{f}_j = \mathbf{f}(\mathbf{s}, \mathbf{w}, \hat{\mathbf{P}}, \hat{\mathbf{D}} \phi)$, $\mathbf{e} = \mathbf{s}^{\ddagger} - \mathbf{s}$ and $\dot{\theta}_j = \mathbf{f}_j(\mathbf{s}^* - \mathbf{s})$. The derivative of the Lyapunov function for the j-th manipulator joint can be written as

$$\dot{V} = -(\dot{\theta}_j - \hat{\mathbf{f}}_j \mathbf{e}) \left(\left[\frac{\partial \mathbf{f}_j \mathbf{e}}{\partial \mathbf{P}_j} \right]^T \dot{\mathbf{P}}_j + \left[\frac{\partial \mathbf{f}_j \mathbf{e}}{\partial \mathbf{D}_j} \right]^T \dot{\mathbf{D}}_j \right) \qquad (6.23)$$

where

$$\frac{\partial \mathbf{f}_j \mathbf{e}}{\partial \mathbf{P}_j} = \left[\frac{\partial f_1^j \mathbf{e}}{\partial \mathbf{P}_1^j} \frac{\partial f_2^j \mathbf{e}}{\partial \mathbf{P}_2^j} \cdots \frac{\partial f_i^j \mathbf{e}}{\partial \mathbf{P}_i^j} \right]^T$$

$$= \begin{bmatrix} \left[\frac{\partial f_1^j \mathbf{e}}{\partial P_{11}^j} \frac{\partial f_1^j \mathbf{e}}{\partial P_{12}^j} \frac{\partial f_1^j \mathbf{e}}{\partial P_{13}^j} \right]^T \\ \left[\frac{\partial f_2^j \mathbf{e}}{\partial P_{21}^j} \frac{\partial f_2^j \mathbf{e}}{\partial P_{22}^j} \frac{\partial f_2^j \mathbf{e}}{\partial P_{23}^j} \right]^T \\ \vdots \\ \left[\frac{\partial f_i^j \mathbf{e}}{\partial P_{i1}^j} \frac{\partial f_i^j \mathbf{e}}{\partial P_{i2}^j} \frac{\partial f_i^j \mathbf{e}}{\partial P_{i3}^j} \right]^T \end{bmatrix} = \begin{bmatrix} \overline{w}_1^j \mathbf{e}^T \\ \overline{w}_2^j \mathbf{e}^T \\ \vdots \\ \overline{w}_i^j \mathbf{e}^T \end{bmatrix} \qquad (6.24)$$

and

$$\frac{\partial \mathbf{f}_j \mathbf{e}}{\partial \mathbf{D}_j} = \left[\frac{\partial f_1^j \mathbf{e}}{\partial \mathbf{D}_1^j} \frac{\partial f_2^j \mathbf{e}}{\partial \mathbf{D}_2^j} \cdots \frac{\partial f_i^j \mathbf{e}}{\partial \mathbf{D}_i^j} \right]$$

$$= \begin{bmatrix} \left[\dfrac{\partial f_1^j \mathbf{e}}{\partial D_{11}^j} \dfrac{\partial f_1^j \mathbf{e}}{\partial D_{12}^j} \dfrac{\partial f_1^j \mathbf{e}}{\partial D_{13}^j}\right]^T \\ \left[\dfrac{\partial f_2^j \mathbf{e}}{\partial D_{21}^j} \dfrac{\partial f_2^j \mathbf{e}}{\partial D_{22}^j} \dfrac{\partial f_2^j \mathbf{e}}{\partial D_{23}^j}\right]^T \\ \vdots \\ \left[\dfrac{\partial f_i^j \mathbf{e}}{\partial D_{i1}^j} \dfrac{\partial f_i^j \mathbf{e}}{\partial D_{i2}^j} \dfrac{\partial f_i^j \mathbf{e}}{\partial D_{i3}^j}\right]^T \end{bmatrix} = \begin{bmatrix} \overline{w}_1^j \Delta \mathbf{e}^T \\ \overline{w}_2^j \Delta \mathbf{e}^T \\ \vdots \\ \overline{w}_i^j \Delta \mathbf{e}^T \end{bmatrix} \qquad (6.25)$$

The adaptive laws of P_{im}^j and D_{im}^j are designed as

$$\dot{P}_{im}^j = \eta_p(\dot{\theta}_j - \mathbf{f}_j \mathbf{e})\overline{w}_i^j e_m \qquad (6.26)$$
$$\dot{D}_{im}^j = \eta_d(\dot{\theta}_j - \mathbf{f}_j \mathbf{e})\overline{w}_i^j \Delta e_m \qquad (6.27)$$

where $\eta_p, \eta_d > 0$ are the adaptation rates. By substituting 6.26 and 6.27 into (6.24) and (6.25), respectively, the results are written as

$$\left[\dfrac{\partial \mathbf{f}_j \mathbf{e}}{\partial \mathbf{P}_j}\right]^T \dot{\mathbf{P}}_j = \eta_p(\dot{\theta}_j - \mathbf{f}_j \mathbf{e}) \sum_{i=1}^{r} \left(\sum_{m=1}^{M} (w_i^j e_m)^2\right) \qquad (6.28)$$

$$\left[\dfrac{\partial \mathbf{f}_j \mathbf{e}}{\partial \mathbf{D}_j}\right]^T \dot{\mathbf{D}}_j = \eta_d(\dot{\theta}_j - \mathbf{f}_j \mathbf{e}) \sum_{i=1}^{r} \left(\sum_{m=1}^{M} (w_i^j \Delta e_m)^2\right) \qquad (6.29)$$

By substituting (6.28) and (6.29) into (6.23), the derivative of the Lyapunov candidate function is described as

$$\dot{V} = -(\dot{\theta}_j - \mathbf{f}_j \mathbf{e})^2 \begin{bmatrix} \eta_p \sum_{i=1}^{r} \left(\sum_{m=1}^{M} (w_i^j e_m)^2\right) \\ \eta_d \sum_{i=1}^{r} \left(\sum_{m=1}^{M} (w_i^j \Delta e_m)^2\right) \end{bmatrix} \qquad (6.30)$$

where $\dot{V}$ is negative definite. Thus the proposed updates (6.26) and (6.27) of the PD parameters ensures convergence in tracking errors.

6.2.3 Stability Analysis

The stability of the T-S fuzzy PD closed loop system can be verified in terms of Lyapunov stability. Let us consider the Lyapunov function candidate as

$$V = \dfrac{1}{2} \| \mathbf{e} \|^2 \qquad (6.31)$$

where $\mathbf{e} = \mathbf{s} - \mathbf{s}^*$ and $\mathbf{s}^*$ is constant. The derivative of the Lyapunov candidate function V is derived as

Adaptive Distributed T-S Fuzzy PD Controller

$$\begin{aligned}\dot{V} &= \mathbf{e}^T \dot{\mathbf{e}} \\ &= \mathbf{e}^T \mathbf{L_{s^*}} \mathbf{J}_e \dot{\hat{\boldsymbol{\theta}}} \\ &= \mathbf{e}^T \mathbf{L_{s^*}} \mathbf{J}_e \hat{\mathbf{f}}(\mathbf{s},\mathbf{w},\mathbf{P},\mathbf{D}\phi)\mathbf{e}\end{aligned} \qquad (6.32)$$

Assuming the universal approximation capability of the T-S fuzzy model, the T-S fuzzy function using the trained data set can be described as

$$\hat{\mathbf{f}}(\mathbf{s},\mathbf{w},\mathbf{P},\mathbf{D}\phi)\mathbf{e} = -\kappa \hat{\mathbf{J}}_e^\dagger \hat{\mathbf{L}}_{\mathbf{s}^*}^\dagger \mathbf{e} + \epsilon \qquad (6.33)$$

where $\epsilon = \dot{\boldsymbol{\theta}} - \dot{\hat{\boldsymbol{\theta}}}$ is the approximation error. Substituting (6.33) into (6.32), the derivative of the Lyapunov candidate function V is redefined as

$$\begin{aligned}\dot{V} &= \mathbf{e}^T \mathbf{A}(-\kappa \hat{\mathbf{A}}^\dagger \mathbf{e} + \epsilon) \\ &= -\kappa \mathbf{e}^T \mathbf{A} \hat{\mathbf{A}}^\dagger \mathbf{e} + \mathbf{e}^T \mathbf{A} \epsilon\end{aligned} \qquad (6.34)$$

where $\mathbf{A} = \mathbf{L_{s^*}} \mathbf{J}_e$ and $\hat{\mathbf{A}}^\dagger = \hat{\mathbf{J}}_e^\dagger \hat{\mathbf{L}}_{\mathbf{s}^*}^\dagger$.

Initially, if $\epsilon = 0$, there is no approximation error, then $\hat{\mathbf{A}} = \mathbf{A}$. The rank of matrix $\mathbf{A} \in \mathbb{R}^{m,n}$ cannot be greater than m nor n. The rank of $\mathbf{A}$ can be written as $\text{rank}(\mathbf{A}) = \min(m,n)$. In this case $m < n$ therefore $\mathbf{A}$ has full rank of m. If $\mathbf{A}$ is full rank then $\mathbf{A}^T \mathbf{A}$ is invertible. Thus, the left inverse of $\mathbf{A}$ can be applied which is defined as $\mathbf{A}^\dagger = (\mathbf{A}^T \mathbf{A})^{-1} \mathbf{A}^T$. For the case where no approximation error is assumed, it can be said that $(\mathbf{A}\hat{\mathbf{A}}^\dagger) = \mathbf{I}$, where $\mathbf{I}$ is an identity matrix. In this case, the derivative of the Lyapunov candidate function

$$\dot{V} = -\kappa \parallel \mathbf{e} \parallel^2 \text{ is negative definite} \qquad (6.35)$$

and the system is globally asymptotically stable.

If $\hat{\mathbf{A}} \neq \mathbf{A}$ and $\mathbf{A}$ is full rank then $(\mathbf{A}\hat{\mathbf{A}}^\dagger) = \mathbf{I}'$ and

$$\dot{V} = -\kappa \mathbf{e}^T \mathbf{I}' \mathbf{e} \qquad (6.36)$$

For a positive definite matrix $\mathbf{I}'$, a matrix definiteness can be described as

$$\lambda_{min}(\mathbf{I}') \parallel \mathbf{e} \parallel^2 \leq \mathbf{e}^T \mathbf{I}' \mathbf{e} \leq \lambda_{max}(\mathbf{I}') \parallel \mathbf{e} \parallel^2 \qquad (6.37)$$

where $\lambda_{min}(\cdot)$ and $\lambda_{max}(\cdot)$ denote the minimum and the maximum eigenvalues of a matrix, respectively. Thus the derivative of Lyapunov candidate function can be derived as

$$\dot{V} \leq -\kappa \lambda_{min}(\mathbf{I}') \parallel \mathbf{e} \parallel^2 \qquad (6.38)$$

The derivative of the Lyapunov candidate function will be negative definite if $\lambda_{min}(\mathbf{I}') > 0$ and $\mathbf{I}'$ is positive definite. Hence the system is locally stable. If $\epsilon \neq 0$ is considered then (6.34) can be described as

$$\begin{aligned}\dot{V} &\leq -\kappa \parallel \mathbf{e} \parallel (\lambda_{min}(\mathbf{I}') \parallel \mathbf{e} \parallel -\sigma_{max}(\mathbf{A})\epsilon_{max}) \\ &\leq \frac{-\kappa \parallel \mathbf{e} \parallel}{\lambda_{min}(\mathbf{I}')} \left(\parallel \mathbf{e} \parallel -\frac{\sigma_{max}(\mathbf{A})\epsilon_{max}}{\lambda_{min}(\mathbf{I}')} \right)\end{aligned} \qquad (6.39)$$

where $\|\epsilon\| \leq \epsilon_{max}$ and $\sigma_{max}(\cdot)$ denotes the largest singular value of a matrix. $\dot{V}$ is negative definite if $\lambda_{min}(\mathbf{I'}) > 0$ and if

$$\|\mathbf{e}\| > \frac{\sigma_{max}(\mathbf{A})\epsilon_{max}}{\lambda_{min}(\mathbf{I'})}$$

Hence the system is locally uniformly ultimately bounded (U.U.B) with the approximation error. It immediately follows that

$$\dot{V}(\mathbf{e}) < 0, \quad \forall \|\mathbf{e}\| > \delta \qquad (6.40)$$

where $\delta = \frac{\sigma_{max}(\mathbf{A})\epsilon_{max}}{\lambda_{min}(\mathbf{I'})}$. In other words, the derivative of the Lyapunov candidate function is negative outside of the compact set $B_\delta = \{\|\mathbf{e}\| \leq \delta\}$, all the solutions that start outside B_δ will enter this set within a finite time, and will remain inside the set for future times [199].

6.3 Experimental Results

The experimental setup (see Figure 6.3) consists of a 7 DOF robot manipulator, a firewire camera which is placed on the robot end-effector, and a PC. The distributed fuzzy PD controller based on the T-S fuzzy model learns the mapping between the velocity of the image feature $\mathbf{s}$ and the joint angle velocity $\dot{\boldsymbol{\theta}}$. The camera is interfaced through a firewire port to the PC and the images are processed using an OpenCV library. The camera provides color images at a resolution of 320×240 pixels. An image histogram back projection algorithm and image moments computation are used to identify the target object. Verification of the trajectories of an object and the robot end-effector are obtained

(a) Vicon system working space (b) Vicon markers on the target and the end-effector

FIGURE 6.3: Experimental setup.

Experimental Results 217

using a Vicon system. A ball is used as the target in an uncluttered environment in these experiments. Figure 6.3(b) shows that Vicon markers are positioned on the robot end-effector and on the target ball in order for each to be tracked by the Vicon system.

Experiments using MBVS were conducted to obtain data variables that are required for the learning process. For a defined working space, a number of trials were carried out using different initial positions of the robot end-effector such that the initial target would generate a wide range of data for joint velocities $\dot{\boldsymbol{\theta}}$, current image feature vector $\mathbf{s} = [s_x, s_y, s_a]^T = [s_1, s_2, s_3]^T$, image feature error vector $\mathbf{e} = [e_{s_x}, e_{s_y}, e_{s_a}]^T = [e_1, e_2, e_3]^T$ and the rate of image feature error vector $\Delta \mathbf{e} = [\Delta e_1, \Delta e_2, \Delta e_3]^T$. The experiment was comprised of the following steps:

1. Each of the fuzzifier inputs in a vector $\mathbf{s}$ have three Gaussian membership functions described by its mean and variance (c, σ). Initially, the Gaussian functions were distributed within a normalised range $[-1, 1]$ such that the overlapped area between two Gaussian functions was 25%, approximately. The consequent parameters $\mathbf{P}$ and $\mathbf{D}$ in each joint velocity model were initialized with random numbers within the range of $[-1, 1]$

2. The offline learning of the T-S fuzzy PD model was verified using previously collected data. As suggested in [200], the offline learning reduces the demand on online data generation. The offline training data was obtained by moving the target object by hand to get a sufficient data set of the joint velocities and the image features in the workspace.

3. The implementation of the T-S fuzzy PD model in real-time was considered to compare the performance between the adaptive learning and non-adaptive learning.

Seven T-S fuzzy PD models of joint velocity $\widehat{\dot{\theta}}_j$ have been obtained in the training process, and every joint velocity model contains 27×3 proportional gain parameters in a vector of $\mathbf{P}_j \in R^{81}$ and 3×3 Gaussian membership functions' parameters $(c_{n_m m}^j, \sigma_{n_m m}^j)$, where $n_m, m \in [1, 2, 3]$ and $j \in [1, 2, \cdots, 7]$.

Additionally there are 27×3 derivative gain parameters in a vector of $\mathbf{D}_j \in R^{81}$. The identified Gaussian membership functions and consequent parameters of the modeled joint velocities $\widehat{\dot{\boldsymbol{\theta}}} = [\widehat{\dot{\theta}}_1, \widehat{\dot{\theta}}_2 ..., \widehat{\dot{\theta}}_7]^T$ may vary since the algorithm depends on random initialized values of consequent parameters, a learning rate η, a stopping condition and the data set. Using the learning rate $\eta = 0.1$ and 200 iterations, the RMSE of $\widehat{\dot{\boldsymbol{\theta}}}$ is shown in Figure (6.4).

In this chapter we are primarily interested in the real-time implementation results rather than the presentation of identified T-S fuzzy PD parameters. However the identified consequent parameters $\widehat{\dot{\theta}}_1$ and $\widehat{\dot{\theta}}_2$ of the T-S fuzzy PD model are considered to give a more detailed explanation of the proposed schemes, as shown in Table 6.2. The consequent parameters identified for joint velocity $\widehat{\dot{\theta}}_1$ of the T-S fuzzy PD are:

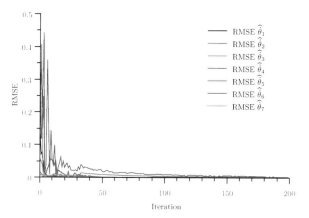

FIGURE 6.4: RMSE of $\widehat{\theta}_j$ Vicon markers on the target and the end-effector.

TABLE 6.2: The updated membership function of the T-S fuzzy PD controller of $\widehat{\theta}_1$ and $\widehat{\theta}_2$

	Gaussian parameters						
$\widehat{\theta}_j$	Membership functions	s_1		s_2		s_3	
		c	σ	c	σ	c	σ
$\widehat{\theta}_1$	μ^{1m}	0.00	1.39	0.00	1.95	0.00	1.49
	μ^{2m}	1.00	0.54	0.46	0.10	1.90	1.99
	μ^{3m}	1.00	0.20	0.00	2.00	1.00	0.20
$\widehat{\theta}_2$	μ^{1m}	-0.50	1.39	0.00	1.95	0.00	1.49
	μ^{2m}	0.00	0.46	0.56	1.90	0.10	1.73
	μ^{3m}	-0.27	1.42	1.00	1.87	1.00	0.20

R_1^1 :**if** s_1 is A_{11}^1 **and** s_2 is A_{12}^1 **and** s_3 is A_{13}^1
then $y_1^1 = 1.60974(e_1) - 0.0378469(e_2) + 1.70716(e_3)$
$+ 0.902445(\Delta e_1) - 0.0506251(\Delta e_2) + 1.78611(\Delta e_3)$

R_2^1 :**if** s_1 is A_{21}^1 **and** s_2 is A_{22}^1 **and** s_3 is A_{23}^1
then $y_2^1 = 2.55031(e_1) - 0.437355(e_2) + 3.26196(e_3)$
$+ 1.45803(\Delta e_1) - 0.43672(\Delta e_2) + 3.2049(\Delta e_3)$

$\vdots \quad \vdots$

R_{27}^1 :**if** s_1 is A_{31}^1 **and** s_2 is A_{32}^1 **and** s_3 is A_{33}^1
then $y_{27}^1 = 2.4745(e_1) - 0.487684(e_2) - 0.208851(e_3)$
$+ 1.41299(\Delta e_1) - 0.499181(\Delta e_2) - 0.223516(\Delta e_3)$

Experimental Results

The consequent parameters identified for joint velocity $\widehat{\dot{\theta}}_2$ of the T-S fuzzy PD are:

$$R_1^2 : \text{if } s_1 \text{ is } A_{11}^2 \text{ and } s_2 \text{ is } A_{12}^2 \text{ and } s_3 \text{ is } A_{13}^2$$
$$\text{then } y_1^2 = -1.338(e_1) + 2.49029(e_2) - 8.50078(e_3)$$
$$- 2.48748(\Delta e_1) - 5.46609(\Delta e_2) - 10.217(\Delta e_3)$$

$$R_2^2 : \text{if } s_1 \text{ is } A_{21}^2 \text{ and } s_2 \text{ is } A_{22}^2 \text{ and } s_3 \text{ is } A_{23}^2$$
$$\text{then } y_2^2 = -3.26182(e_1) + 12.1912(e_2) - 3.42069(e_3)$$
$$- 3.67392(\Delta e_1) + 1.22814(\Delta e_2) - 5.77275(\Delta e_3)$$

$$\vdots \qquad \vdots$$

$$R_{27}^2 : \text{if } s_1 \text{ is } A_{31}^1 \text{ and } s_2 \text{ is } A_{32}^1 \text{ and } s_3 \text{ is } A_{33}^1$$
$$\text{then } y_{27}^2 = 2.84481(e_1) + 12.743(e_2) + 2.50398(e_3)$$
$$+ 0.439647(\Delta e_1) + 2.34862(\Delta e_2) + 0.119995(\Delta e_3)$$

The identified parameters are used to verify the system using the same data as used in the training process. Given the identified T-S fuzzy PD parameters, the data set of the current image feature vector and the image feature error vector, the T-S fuzzy PD joint velocity models are tested. Comparisons between the MBVS joint velocity of $\dot{\theta}_1$ and $\dot{\theta}_2$ of the T-S fuzzy PD scheme are shown in Figure (6.5(a)) and Figure 6.5(b), respectively. It can be seen from Figure (6.5(a)) and Figure 6.5(b) that the T-S fuzzy PD schemes of $\widehat{\dot{\theta}}_1$ and $\widehat{\dot{\theta}}_2$ are very similar to the model-based joint velocities. The average error $\|\dot{\theta}_1 - \widehat{\dot{\theta}}_1\|$ and $\|\dot{\theta}_2 - \widehat{\dot{\theta}}_2\|$ of the T-S fuzzy PD during the offline learning process was computed to be $0.160°/s$ and $0.905°/s$, respectively.

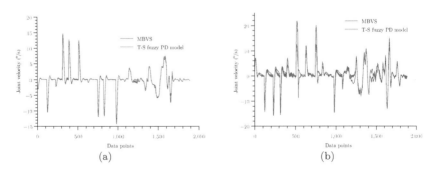

FIGURE 6.5: Input-output data verification of the learned T-S fuzzy parameters (a) Comparison between $\dot{\theta}_1$ and $\widehat{\dot{\theta}}_1$, (b) Comparison between $\dot{\theta}_2$ and $\widehat{\dot{\theta}}_2$.

6.3.1 Visual Servoing for a Static Target

We provide real-time implementation results for visual servoing using the model-based, adaptive and non-adaptive T-S fuzzy PD controllers for a static target. Ten experiments for each scheme have been carried out. In each experiment either the position of the target or the position of the end-effector was placed in an arbitrary position. Once those positions were decided each of the controller schemes was run; this procedure was needed in order to make a fair comparison of the performances between all of those schemes. In each experiment, the new updated T-S fuzzy parameters were used by the adaptive T-S fuzzy PD; in contrast, the non-adaptive T-S fuzzy PD used the same T-S fuzzy PD parameters which had been obtained during offline learning. In the experiments, the orientation of the robot end-effector has not been considered. Thus, given three image features, the MBVS scheme will not change the orientation of the robot end-effector.

Figures 6.6(a) - 6.6(c) show the image feature tracking percentage error between the desired features and the current features in the final experiment. These figures show that the adaptive T-S fuzzy PD can dynamically change the parameters to reduce the modeling error.

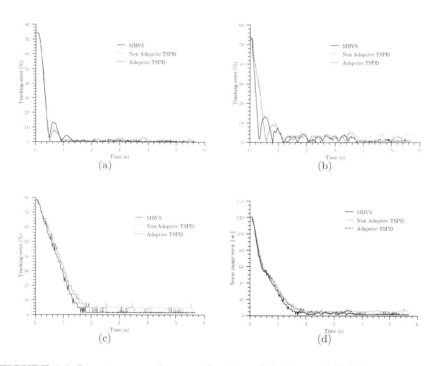

FIGURE 6.6: Input-output data verification of the learned T-S fuzzy parameters (a) $\dot{\theta}_1$, (b) $\dot{\theta}_1$, (c) area, (d) normalized error.

Experimental Results

The steady state was reached at $t \geq 2s$, the average percentage errors of image area e_3 were computed as 1.59%, 4.40% and 3.07% for MBVS, T-S fuzzy PD non-adaptive and adaptive, respectively. The image centroid error was relatively small between 2 and 10 pixels which would not significantly change the robot end-effector's desired position. Image noise caused error in the image space. Once the error of the image features is decreased, as an example, the joint velocities $\dot{\theta}_1$ and $\dot{\theta}_2$ go to zero as shown in Figure 6.7(a) and Figure 6.7(b), respectively, which means the desired position of the robot end-effector is reached. The controller will maintain the robot end-effector position in the desired position as long as the targeted object remains in the same position. For $2s \leq t \leq 5.6s$, the velocities of joint-1 were computed as: $\dot{\theta}_1 \leq |0.11|°/s$, $\widehat{\dot{\theta}}_1 \leq |0.40|°/s$, and for MBVS, T-S fuzzy PD non-adaptive and adaptive, respectively. Whereas, the velocities of joint-2 were computed as: $\dot{\theta}_2 \leq |0.70|°/s$, $\widehat{\dot{\theta}}_2 \leq |0.99|°/s$, and $\widehat{\dot{\theta}}_2 \leq |0.77|°/s$ for MBVS, T-S fuzzy PD non-adaptive and adaptive, respectively. Figure 6.8 depicts the initial and final robot end-effector position with respect to the ball position.

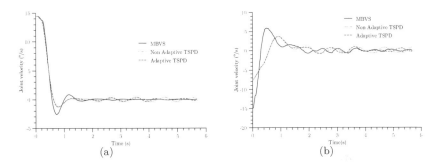

FIGURE 6.7: Comparison of the joint velocities between MBVS, non-adaptive and adaptive T-S fuzzy PD controller (a) $\dot{\theta}_1$, (b) $\dot{\theta}_2$.

(a) Initial robot end-effector position   (b) Final robot end-effector position

FIGURE 6.8: T-S fuzzy PD visual servoing for a static target.

6.3.2 Compensation of Model Uncertainties

An experiment for the adaptive T-S fuzzy PD controller introducing model uncertainties has been conducted. In offline learning, the T-S fuzzy PD parameters were learned using the centered camera configuration with respect to the end-effector frame. To demonstrate the capability of the adaptive T-S fuzzy PD controller in the presence of model uncertainty, the camera position was altered. The camera pose on the end-effector was displaced at $x_{cam} = o_x - 7cm$, $y_{cam} = o_y + 4cm$ and $\alpha_{cam} = 30°$ where (x_{cam}, y_{cam}) is the center coordinate of the optical axis of the camera, (o_x, o_y) is the origin of the end-effector frame and α_{cam} is the orientation of the camera optical axis with respect to the z-axis of the end-effector frame. Even though the position of the camera was not in an accurate position (center of the end-effector frame), convergence was achieved and the error **e** decreased as shown in Figure 6.9.

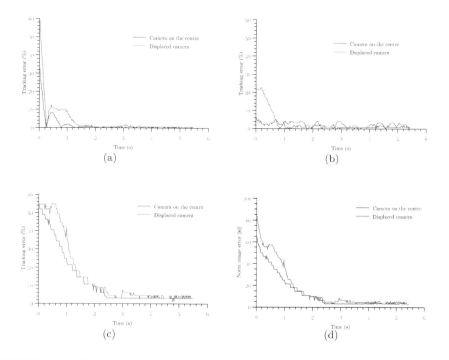

FIGURE 6.9: Compensation of model uncertainties: camera position has been altered (a) The centroid image coordinate error in x-axis, (b) The centroid image coordinate error in y-axis, (c) The image area error, (d) The norm of the error **e**.

Experimental Results

6.3.3 Visual Servoing for a Moving Target

An experiment for tracking a moving target for both schemes has also been conducted. The Vicon system has been used in this experiment to verify the trajectory of the target object and the robot end-effector. The Vicon system that is used in this experiment has nine infrared cameras which are able to detect the position of the marker in a 64 m^3 working area (8 m in length, 4 m in width, 2 m in height) with 1 mm precision. The markers were placed on the target object and the robot end-effector. The minimum number of markers is three in order to create a segment model in the Vicon system. The target object was moved by a demonstrator in a circular motion. It was difficult to maintain the same circular trajectory of the target in every experiment. However, Figure 6.10 shows that the movement of the target can be followed

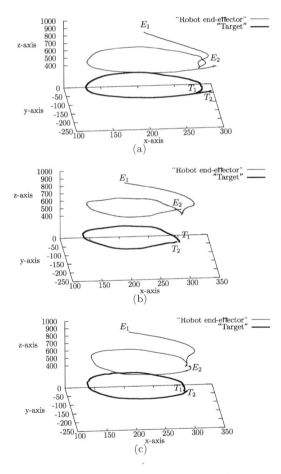

FIGURE 6.10: Target and robot end-effector trajectories of model-based, non-adaptive and adaptive T-S fuzzy PD controller (a) MBVS, (b) Non-Adaptive T-S Fuzzy PD, (c) Adaptive T-S Fuzzy PD.

by the robot end-effector which has an attached camera on top of the gripper. The desired features were measured from a distance of 25 cm along the z-axis between the camera and the target object. T_1 and E_1 denote the starting position of the target object and the robot end-effector, respectively. T_2 and E_2 denote the final positions. The corresponding performance of non-adaptive and adaptive distributed T-S fuzzy PD can be seen in Figure 6.10(b) and Figure 6.10(c) respectively.

At the end of the measurement, T_2 and E_2, the image features error of the adaptive and non-adaptive T-S Fuzzy PD are presented in Table 6.3.

A snapshot of the video files have been recorded and can be seen in Figure 6.11 to show the movement of the robot end-effector and the target object as

TABLE 6.3: Image features error: MBVS, adaptive, and non-adaptive ditributed T-S fuzzy PD controller

Scheme	Target error		
	Centroid		Area
	x	y	
MBVS	1	2	80
Adaptive	2	5	100
Non Adaptive	4	12	461

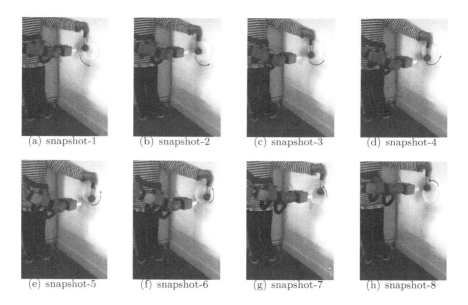

(a) snapshot-1 (b) snapshot-2 (c) snapshot-3 (d) snapshot-4
(e) snapshot-5 (f) snapshot-6 (g) snapshot-7 (h) snapshot-8

FIGURE 6.11: Snapshot pictures of T-S fuzzy PD visual servoing for tracking a circular trajectory of the object.

well as the last three joint configurations. Figure 6.11(a) shows the starting position of the robot end-effector before the object was moved in a circular trajectory using a white circle plate as a guide. One can see in the snapshots in Figure 6.11(a) - Figure 6.11(h) that the last three joint configurations were changed to follow the target during servoing while the system was maintaining the offset between the target object and the robot end-effector and also maintaining the center of the target object image in the center of the image plane.

6.4 Computational Complexity

MBVS for a redundant manipulator requires the computation of the pseudo-inverse Jacobian. Singular value decomposition is used to compute this. In [201], the computational cost of the pseudo-inverse has been evaluated and it has been determined that this involves a total of $3NM^2 + M^3 + N^2M$ floating point operations (flops). In the offline learning process, the parameters of the distributed T-S fuzzy system are learned. Given the input vectors of $\mathbf{s}$, $\mathbf{e}$ and $\Delta \mathbf{e}$ in the image space, the learned T-S fuzzy is used to compute the joint velocity $\hat{\boldsymbol{\theta}}$ in real-time. In this work, the total number of the Gaussian membership functions in each fuzzifier inputs are the same, therefore the total number of rules r_j for every joint is the same and is denoted as r.

The computation of the normalised weight vector $\overline{\mathbf{w}}$ for j-th T-S fuzzy network requires $rM + r$ flops. The computation of $\hat{\theta}_j$ requires $2r(M+1)$ flops approximately. Therefore, total computation of $\hat{\boldsymbol{\theta}}$ requires $N\{rM + r + 2r(M+1)\} = 3Nr(M+1)$ flops. The computational complexity for the forward computation of the T-S fuzzy network is in the order of $O(N)$. Similarly, the computational complexity of the online adaptation algorithm is computed in the order of $O(N)$. The total computation of the adaptive distributed fuzzy PD controller is linear with the number of DOF of a robot manipulator while the computation of the pseudo-inverse Jacobian is in the order of $O(N^2)$. The computation of the adaptive distributed fuzzy PD controller is significantly more efficient than MBVS for robot manipulators with greater DOF.

6.5 Summary

A simple yet elegant approach to a visual servoing control scheme has been presented in this chapter. The control action has been learned as a function of the error in the visual space using a T-S fuzzy framework. Thus the parametric space of the controller consists of locally valid PD gains in a distributed fashion. The initial values of these parameters are learned while mimicking the model-based controller in the first phase. An online adaptation scheme has been proposed that can fine-tune the controller parameters to compensate the uncertainities associated with the system model and environment. The conceptual novelty in this proposed scheme is that the manipulator can be controlled without having to compute its own inverse Jacobians. Thus the model reflects a more cognitive learning architecture just like a child learns to actuate his/her hands and legs without the need for understanding the complexities of the involved kinematics.

The proposed scheme has been validated through exhaustive experimentation on a 7 DOF robot manipulator. The robot has been actuated using the model-based controller given in Eq. (6.9). The controller input-output data have been used to learn the intial parameters of the distributed fuzzy PD controller which has been termed as a non-adaptive fuzzy controller in this chapter. These controller parameters have been fine tuned using the proposed adaptation scheme. It has been shown that the controller is effective in visual servoing for both static and moving targets. During experimentation, tracking errors are always less than five pixel counts in the x and y visual features.

The proposed controller works efficiently within the offline trained workspace. The online update will expand this workspace locally but will not be able to function globally. Given an initial joint configuration for which the T-S fuzzy network has been trained offline, the online update scheme ensures tracking in the local neighborhood. If a new initial joint configuration demanded by a specific application is far away from this initial joint configuration, then further offline training of the T-S fuzzy network will be necessary. The main purpose of the adaptive tuning is to ensure compensation for uncertainties in the model as well as in the environment. However a set of initial joint configurations that will span the entire workspace can be selected. A T-S fuzzy controller for each of these configurations can be trained offline. In that case, given a situation, a specific T-S fuzzy network is selected for online update.

The readers may refer [11, 77, 152] for methods of computing the Jacobian matrix include closed-form solutions and [202–205] for adaptive schemes. More recently, approaches to visual servoing have been developed that do not rely on the computation of both the robot Jacobian matrix and the image Jacobian matrix. In [206], the robot-image Jacobian is estimated using a nonlinear optimization algorithm [207]. Similarly, in [208] Broyden's method is generalized. However like other nonlinear optimisation algorithms [209], the approaches are sensitive to noise as well as the initial robot-image Jacobian matrix approximation.

There are also approaches to estimate the robot-image Jacobian using learning algorithms for example, self organising maps are used in [200, 210]. Two calibrated static cameras were used to learn the inverse kinematics relationship of a 7 DOF robot manipulator while redundancy resolution was also addressed. The obtained inverse robot-image Jacobian approximation eliminates the necessity for online pseudo-inverse computation. In [211], iterative learning was proposed to approximate the Jacobian given a set of sample points of the demonstrated trajectory. An eye-to-hand configuration was used so that the system could compute the difference between the demonstrated trajectory and the current position of the robot end-effector in every iteration. Fuzzy modeling techniques have become increasingly popular. System identification using inverse fuzzy modeling has been proposed in [212]. In [213], fuzzy clustering and inverse fuzzy models are derived; this method learned the inverse model of the system where the robot-image Jacobian was estimated and has also been applied in an eye-to-hand image based visual servoing system. The experimental results show convergence of a 6 DOF robot manipulator can be reached in a two dimensional trajectory (X-Z axes). The fuzzy model suggested by Takagi-Sugeno (1985) [192] can represent a general class of static or dynamic nonlinear systems [214]. In [212] the integration of learning algorithms such as neural-fuzzy model and genetic algorithms was proposed to solve the inverse dynamic model of a two-axis pneumatic manipulator system. As the robotic manipulator, together with the visual system, is a nonlinear system, it is advantageous to use the T-S fuzzy model in the system [192].

7
Kinematic Control using Single Network Adaptive Critic

7.1 Introduction

Consider the problem of a robot cutting out a circular sheet from a large metalic sheet. The objective is to complete the task in minimum time while minimizing the jerks in joint movements. Thus the control policy must optimize a global cost function. Consider another example. Any government tries to preserve its forest resources. There are many takers for forest resources. Humans require woods for making houses and furnitures. Jungle habitats depend on the density of trees. So new trees must be planted while old trees are allowed to be cut. If one cuts trees without any policy, then soon the forest will evaporate as it requires some incubation period for new trees to grow. Again we must learn the art of optimal control policy that defines the rate of cutting trees and the rate of planting new trees. Such optimal control policy can be derived using Pontryagin's maximum principle (a necessary condition also known as Pontryagin's minimum principle or simply Pontryagin's Principle) [2] or by solving the Hamilton–Jacobi–Bellman equation (a sufficient condition).

In general such dynamical systems are represented by n-dimensional vector differential equation which is usually nonlinear. For such nonlinear systems, there is no closed loop solution to an optimal control problem. Moreover, the iterative solution is obtained by offline computation.

In applications related to robotic systems, we require real-time solutions. The Approximate Dynamic Programming (ADP) has become very popular to find near-optimal solution. Most of the robotic systems can be presented as input-affine nonlinear systems. Thus among many variants of ADP, single network adaptive critic (SNAC) [215] based schemes are effective to design optimal control policy for robotic systems.

This chapter will introduce the concept of SNAC and its application to simple nonlinear systems, kinematic control, and visual kinematic control of robot manipulators.

7.1.1 Discrete-Time Optimal Control Problem

Consider the nonlinear dynamical system,

$$x(k+1) = f(x(k), u(k)) \qquad (7.1)$$

where $x(k) \in R^n$ is the state and $u(k) \in R^m$ is the input to the system. The control task is to stabilize the plant while minimizing the cost function,

$$J_c = \sum_{k=0}^{\infty} L(x(k), u(k)) \qquad (7.2)$$

where J_c is the cost function. The main interest of this thesis work is on the optimal control problem with boundary condition $x(k) = 0$ as $k \to \infty$.

The cost function J_c can be expressed as,

$$\begin{aligned} J_c(k) &= L(x(k), u(k)) + \sum_{\tilde{k}=k+1}^{\infty} L(x(\tilde{k}), u(\tilde{k})) \\ &= L(x(k), u(k)) + J_c(k+1) \end{aligned} \qquad (7.3)$$

where $L(x(k), u(k))$ is the utility function at instant k, $J_c(k)$ and $J_c(k+1)$ are the cost-to-go functions at the instants k and $k+1$ respectively. The above equation is known as Bellman's equation and it evaluates the value of the current policy $u(k)$ in terms of the value $J_c(k+1)$.

The optimal value is obtained using Bellman's equation as,

$$J_c^*(k) = \max_{u(k)} \left(L(x(k), u(k)) + J_c(k+1) \right). \qquad (7.4)$$

Bellman's optimality condition [216] states that "An optimal policy has the property that no matter what the previous decisions (i.e. controls) have been, the remaining decisions must constitute an optimal policy with regard to the state resulting from those previous decisions." It is expressed in terms of equation as,

$$J_c^*(k) = \max_{u(k)} \left(L(x(k), u(k)) + J_c^*(k+1) \right). \qquad (7.5)$$

Bellman's optimality condition is known as discrete-time HJB equation.

The optimal control problem is solved by defining the costate vector $\lambda(k)$ as,

$$\lambda(k) = \frac{\partial J_c(k)}{\partial x(k)} \qquad (7.6)$$

where,

$$\frac{\partial J_c(k)}{\partial x(k)} = \begin{bmatrix} \frac{\partial J_c(k)}{\partial x_1(k)} & \frac{\partial J_c(k)}{\partial x_2(k)} & \cdots & \frac{\partial J_c(k)}{\partial x_n(k)} \end{bmatrix} \qquad (7.7)$$

Introduction

Here, $J_c(k) \in \mathbb{R}$ and $\boldsymbol{x}(k) \in \mathbb{R}^n$. Consider $f(x): \mathbb{R}^n \Rightarrow \mathbb{R}^m$. Then,

$$\frac{\partial f}{\partial \boldsymbol{x}} = \begin{bmatrix} \frac{\partial f_1}{\partial x_1} & \frac{\partial f_1}{\partial x_2} & \cdots & \frac{\partial f_1}{\partial x_n} \\ \cdot & \cdot & & \cdot \\ \cdot & \cdot & & \cdot \\ \frac{\partial f_m}{\partial x_1} & \frac{\partial f_m}{\partial x_2} & \cdots & \frac{\partial f_m}{\partial x_n} \end{bmatrix} \tag{7.8}$$

It is clear that $\frac{\partial f}{\partial \boldsymbol{x}} \in \mathbb{R}^{n \times m}$. Using the same idea, $\boldsymbol{\lambda}(k) \in \mathbb{R}^{1 \times n}$ The optimal control law is derived from the necessary condition of optimality,

$$\frac{\partial J_c(k)}{\partial \boldsymbol{u}(k)} = \boldsymbol{0}. \tag{7.9}$$

$\frac{\partial J_c(k)}{\partial \boldsymbol{u}(k)}$ is evaluated using the Bellman's equation (7.3) as,

$$\frac{\partial J_c(k)}{\partial \boldsymbol{u}(k)} = \frac{\partial L}{\partial \boldsymbol{u}(k)} + \frac{\partial J_c(k+1)}{\partial \boldsymbol{u}(k)}$$

Optimal cost $J_c(k)$ at any instant k is a function of the state at that instant $\boldsymbol{x}(k)$. Hence, simplifying the above equation by applying the necessary condition of optimality (7.9) results in,

$$\frac{\partial L}{\partial \boldsymbol{u}(k)} + \frac{\partial J_c(k+1)}{\partial \boldsymbol{x}(k+1)} \frac{\partial \boldsymbol{x}(k+1)}{\partial \boldsymbol{u}(k)} = \boldsymbol{0}. \tag{7.10}$$

The above equation gives the optimal control policy in terms of $\boldsymbol{\lambda}(k+1)$. The costate vector $\boldsymbol{\lambda}(k)$ is evaluated as,

$$\begin{aligned} \frac{\partial J_c(k)}{\partial \boldsymbol{x}(k)} &= \frac{\partial L}{\partial \boldsymbol{x}(k)} + \frac{\partial J_c(k+1)}{\partial \boldsymbol{x}(k)} \\ &= \frac{\partial L}{\partial \boldsymbol{x}(k)} + \frac{\partial J_c(k+1)}{\partial \boldsymbol{x}(k+1)} \frac{\partial \boldsymbol{x}(k+1)}{\partial \boldsymbol{x}(k)} \end{aligned} \tag{7.11}$$

Along the optimal path, the costate vector is simplified to,

$$\boldsymbol{\lambda}(k) = \frac{\partial L}{\partial \boldsymbol{x}(k)} + \boldsymbol{\lambda}(k+1) \frac{\partial \boldsymbol{x}(k+1)}{\partial \boldsymbol{x}(k)} \tag{7.12}$$

The optimal control policy is obtained using equations (7.12) and (7.10). The costate vector $\boldsymbol{\lambda}(k)$ is computed in backward direction of time with the terminal condition $\boldsymbol{\lambda}(k = \infty) = \boldsymbol{0}$.

7.1.2 Adaptive Critic Based Control

The optimal control policy $\boldsymbol{u}^*(k)$ depends on $\boldsymbol{\lambda}(k+1)$ and, hence, it depends on future instances. The costate vector $\boldsymbol{\lambda}(k)$ is computed in backward direction of time using equation (7.10). The major issue in implementing the optimal

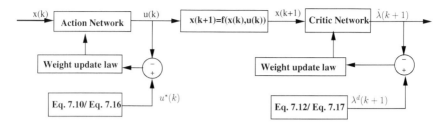

FIGURE 7.1: Adaptive critic-based optimal control scheme.

control is that the analytic expression is not available for many cases and it is to be computed backward in time.

Adaptive critic [217] has been proposed to learn the optimal control forward in time. Adaptive critic approximates the optimal controller in state-feedback form with two neural networks: (i) action network (ii) critic network. Action network approximates the control input, $u(k)$ as a function of the system state, $x(k)$. The performance of the control input generated by the action network is evaluated using the critic network. The critic network approximates either the value function $J_c(k)$ or the costate vector, $\lambda(k)$ as a function of the system state, $x(k)$. The value function is approximated using (7.10) while the costate vector is approximated with the critic network using (7.12). The control policy is learned with an action network using equation (7.10). $\lambda(k+1)$ required in equation (7.10) is obtained either modeling the value function using equation (7.3) or the costate vector using equation (7.12) with a critic network. Both action and critic networks are usually modeled by a MLN with single hidden layer, where the output layer remains linear. The block schematic of the proposed scheme is shown in Figure 7.1.

7.1.2.1 Training of Action and Critic Network

The near-optimal solution is obtained online through learning in critic based methodology. As the learning progresses, the control policy from AC methodology converges to the optimality. The training of action and critic network is explained with an example as shown below. Consider the example:

$$\begin{aligned} \dot{x}_1 &= x_2 \\ \dot{x}_2 &= \alpha(1 - x_1^2)x_2 - x_1 + (1 + x_1^2 + x_2^2)u \end{aligned} \quad (7.13)$$

This represents the dynamics of the Vanderpol oscillator system. Discretizing (7.14), we get

$$\begin{aligned} x_1(k+1) &= x_1(k) + Tx_2(k) \\ x_2(k+1) &= x_2(k) + T(\alpha(1 - x_1(k)^2)x_2(k) - x_1(k) + \\ &\quad (1 + x_1(k)^2 + x_2(k)^2)u(k)) \end{aligned} \quad (7.14)$$

Introduction

where, T is the sampling time. Now, the optimal control policy and the costate vector for this system may be found using (7.10) and (7.12) respectively. Here, $x(k) = [x_1(k) \ x_2(k)]^T$. Let the cost function to be minimized be $J_c = \sum_{k=0}^{\infty} L(x(k), u(k))$, where $L(k) = \frac{1}{2}(x(k)^T x(k) + u(k)^T u(k))$. Then, (7.10) may be invoked and re-written as,

$$\frac{\partial L}{\partial u(k)} + \frac{\partial J_c(k+1)}{\partial x(k+1)} \frac{\partial x(k+1)}{\partial u(k)} = 0$$

$$u_d^T(k) + \lambda(k+1)g = 0 \tag{7.15}$$

where, $g = [0 \ \ T(1 + x_1(k)^2 + x_2(k)^2)]^T$. Therefore, optimal control policy may be written as,

$$u_d(k) = -g^T \lambda(k+1)^T \tag{7.16}$$

Now, (7.12) may be re-written for this system to find out the desired costate vector which may be used to tune the critic weights.

$$\lambda_d(k) = x(k)^T + \lambda(k+1)f \tag{7.17}$$

where,

$$f = \begin{bmatrix} 1 & T \\ 2Tx_1(k)(u(k) - \alpha x_2(k)) & 1 + T(\alpha - \alpha x_1(k)^2 + 2u(k)x_2(k)) \end{bmatrix}.$$

Given below is the algorithm for training of action and critic network for the system (7.14).

1. Input $x(k)$ to the action network and get $u(k)$.

2. Compute $x(k+1)$ using the system dynamics (7.14) with the control input $u(k)$.

3. Compute $\lambda(k+1)$ from the critic network with the input $x(k+1)$.

4. Compute the desired input $u_d(k)$ using the optimal control law equation (7.16) with $x(k)$ and $\lambda(k+1)$.

5. Update the action network by minimizing $\| u_d(k) - u(k) \|$.

The critic network which represents the relationship between the costate $\lambda(k)$ and the state $x(k)$ is trained as follows:

1. Compute the costate vector $\lambda(k)$ and the input $u(k)$ with $x(k)$ using the critic and the action network respectively.

2. Obtain the system state at $(k+1)$th instant for $u(k)$ using equation (7.14).

3. Compute $\lambda(k+1)$ with $x(k+1)$ using the critic network.

4. Compute $\lambda_d(k)$ using $\lambda(k+1)$ and $x(k)$ in the costate equation (7.17).

5. Update the critic network by minimizing $\| \lambda_d(k) - \lambda(k) \|$.

Adaptive critic learns the optimal solution in the forward direction of time using the action and the critic networks. The action and the critic networks are updated alternatively. The critic network is updated assuming that the action network is giving an admissible control input. The action network is trained assuming that the output of the critic network is optimal. The critic network is used only during the training phase to learn the action network. It is not required in the real-time control process.

There exists a class of systems for which the analytic expression for the optimal controller is available in terms of the costate vector. In such cases, the action network is not necessary during both the training and the implementation stages. It is enough to learn the costate vector during the training phase. The costate vector estimated with the critic can be used to implement the optimal control policy in real-time. Such adaptive critic architecture is computationally efficient since the computation associated with the action network is not required. In addition, the control architecture would be robust due to the absence of inaccuracies in modeling the action network. Considering this simplicity, single network adaptive critic has been proposed for optimal control problems whose optimal control policy can be analytically represented in terms of the costate vectors. A brief introduction of SNAC is presented further in the context of optimal control of input affine systems with quadratic cost function, since the global optimal redundancy resolution problem is to be formulated in this framework.

7.1.3 Single Network Adaptive Critic (DT-SNAC)

Consider an input affine dynamical system,

$$\boldsymbol{x}(k+1) = \boldsymbol{f}(\boldsymbol{x}(k)) + \boldsymbol{g}(\boldsymbol{x}(k))\,\boldsymbol{u}(k) \tag{7.18}$$

with a quadratic cost function,

$$J_c = \frac{1}{2}\sum_{k=0}^{\infty}(\,\boldsymbol{x}^T(k)\,\mathbf{Q}\,\boldsymbol{x}(k) + \boldsymbol{u}^T(k)\,\mathbf{R}\,\boldsymbol{u}(k)\,) \tag{7.19}$$

where $\mathbf{Q} \in R^{n \times n}$ is a positive semi-definite matrix and $\mathbf{R} \in R^{m \times m}$ is a positive definite matrix which penalizes the states and inputs respectively.

The analytical expression for the optimal control input can be obtained using equation (7.10) as,

$$\boldsymbol{u}^*(k) = -\mathbf{R}^{-1}\boldsymbol{g}^T(\boldsymbol{x}(k))\,\boldsymbol{\lambda}^*(k+1) \tag{7.20}$$

where $\boldsymbol{\lambda}^*(k+1) = \frac{\partial J_c^*(k+1)}{\partial \boldsymbol{x}(k+1)}$ is the optimal costate vector of the system. The optimal control policy can be implemented in real-time, if either the optimal costate vector $\boldsymbol{\lambda}^*(k+1)$ or the estimate of the optimal costate vector $\hat{\boldsymbol{\lambda}}(k+1)$ is known. The closed loop control with adaptive critic is possible for the above optimal control problem, if the costate vector $\boldsymbol{\lambda}^*(k+1)$ is estimated with a

Introduction

critic network in contrast to $\boldsymbol{\lambda}^*(k)$. In such cases, there is no need for an action network since the closed form solution exists in terms of $\hat{\boldsymbol{\lambda}}(k+1)$. The methodology is computationally efficient and robust due to the absence of the action network.

In this context, a single network adaptive critic has been proposed in [215] for optimal control problem with a closed form solution to the control input $\boldsymbol{u}^*(k)$ in terms of the costate vector $\boldsymbol{\lambda}^*(k+1)$. In SNAC, the costate vector at $(k+1)$th instant is modeled as $\hat{\boldsymbol{\lambda}}(k+1) = \mathbf{f}_N(\mathbf{x}(k))$.

In case of LTI systems, a critic network of architecture $\hat{\boldsymbol{\lambda}}(k+1) = \mathbf{W}\boldsymbol{x}(k)$ was considered in [215]. The optimal value of $\mathbf{W}$ is known for LTI system as given by,

$$\mathbf{W} = (\mathbf{I} + \boldsymbol{P}\mathbf{B}\mathbf{R}^{-1}\mathbf{B}^T)^{-1}\boldsymbol{P}\mathbf{A} \tag{7.21}$$

where $\boldsymbol{P}$ is the solution of discrete-time algebraic Riccati equation (DARE). It has been shown that the critic network converges to the solution of the ARE, if the update converges. For a nonlinear system, this critic network can be modeled as feedforward network or a T-S fuzzy model-based architecture. There are other possible neural architectures that can be used to model the critic network.

7.1.4 Choice of Critic Network Model

7.1.4.1 Costate Vector Modeling with MLN Critic Network

Here, the costate vector is estimated using a MLN from the present states as given by, $\boldsymbol{\lambda}(k+1) = f(W, \boldsymbol{x}(k))$, where W represents the weights of the critic network. Usually, MLN has a single hidden layer with a nonlinear activation function and the output of the MLN is taken to be linear. The steps for training the critic network which captures the relationship between $\boldsymbol{x}(k)$ and $\boldsymbol{\lambda}(k+1)$ using the architecture shown in Figure 7.2 are as follows:

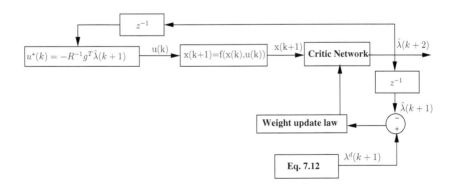

FIGURE 7.2: Control scheme with discrete-time single network adaptive critic (DT-SNAC).

1. Generate state $\boldsymbol{x}(k)$ in the domain of interest.
2. For each element $\boldsymbol{x}(k)$, follow the steps below:
    (a) Input $\boldsymbol{x}(k)$ to the critic network to obtain $\boldsymbol{\lambda}(k+1) = \boldsymbol{\lambda}^a(k+1)$.
    (b) Calculate the optimal control $\boldsymbol{u}(k)$ using (7.20).
    (c) Get the next state $\boldsymbol{x}(k+1)$ from (7.18) using $\boldsymbol{x}(k)$ and $\boldsymbol{u}(k)$.
    (d) Input $\boldsymbol{x}(k+1)$ to the critic network to obtain $\boldsymbol{\lambda}(k+2)$.
    (e) Using $\boldsymbol{x}(k+1)$ and $\boldsymbol{\lambda}(k+2)$, calculate $\boldsymbol{\lambda}^d(k+1)$ from the costate equation.
3. Train the critic network for all states $\boldsymbol{x}(k)$ in the domain of operation; the output being corresponding $\boldsymbol{\lambda}^d(k+1)$.
4. Check the convergence of the critic network. If the convergence is achieved, revert to step 2 with $n = n+1$. Otherwise repeat steps 2-3.
5. Continue steps 2-4 until the convergence is achieved.

7.1.4.2 Costate Vector Modeling with T-S Fuzzy Model-Based Critic Network

T-S fuzzy model is used to model the nonlinear costate vector in equation (7.12) as a fuzzy cluster of the costate vector of the local linear models. With such architecture, the individual zones of the T-S fuzzy model correspond to the optimal weights of the local linear model within its fuzzy boundary.

The ith rule of the T-S fuzzy model-based critic network for nonlinear system with the current state vector $\boldsymbol{x}(k)$ is defined as,

$$\text{IF } x_1(k) \text{ is } F_1^i \text{ AND } \cdots \text{ AND } x_n(k) \text{ is } F_n^i \text{ THEN}$$

$$\hat{\boldsymbol{\lambda}}_i(k+1) = \mathbf{W}_i \boldsymbol{x}(k)$$

where F_j^i, $j = 1, 2, \cdots, n$, is the jth fuzzy set of the ith rule. $\mathbf{W}_i \in R^{n \times n}$ is the linear map associated with the ith fuzzy zone to model the costate vector. Fuzzy zone F_j^i is defined using Gaussian function. It is associated with mean c_j^i and the standard deviation σ_j^i to define the fuzzy space. The fuzzy membership value associated with F_j^i is computed as,

$$\mu_i^j(x_j) = e^{\left(\frac{-(x_j - c_j^i)^2}{2(\sigma_j^i)^2}\right)}. \tag{7.22}$$

Let,

$$\mu_i(\boldsymbol{x}) = \prod_{j=1}^{n} \mu_i^j(x_j) \tag{7.23}$$

where $i = 1, 2, \cdots, N_n$ and $\mu_i(\boldsymbol{x})$ is the membership value of the ith fuzzy zone. The ith fuzzy rule is defined with fuzzy center $\boldsymbol{c}_i = [c_1^i \ c_2^i \ \cdots \ c_n^i]^T$, and standard deviation $\boldsymbol{\sigma}_i = [\sigma_1^i \ \sigma_2^i \ \cdots \ \sigma_n^i]^T$.

Introduction

The standard deviation of the Gaussian function is chosen such that at least one fuzzy zone is active at every operating point. The maximum overlap between two fuzzy zones at the center of the fuzzy set is assumed and then the standard deviation of the Gaussian function is computed. Let d_j^i be the distance to the center of the adjacent fuzzy set from center of the fuzzy set F_j^i, then the standard deviation is computed as,

$$\sigma_j^i = \sqrt{\frac{(d_j^i)^2}{2\log(\mu^{cen})}} \qquad (7.24)$$

where μ^{cen} is the overlap of the closest neighbor at the fuzzy center c_j^i. The above method of computing the standard deviation of the Gaussian function to estimate the spread of the fuzzy sets is known as nearest neighbor heuristic.

Given the current state vector $x(k)$, the fuzzy model around the operating point is constructed as the weighted average of the local models and has the form,

$$\hat{\boldsymbol{\lambda}}(k+1) = \frac{\sum_{i=1}^{N_n} \mu_i \mathbf{W}_i x(k)}{\sum_{i=1}^{N_n} \mu_i}. \qquad (7.25)$$

In case of a linear system, only one fuzzy zone is considered to represent the entire workspace, and then the T-S fuzzy model gets simplified to the critic network suggested in [215]. It should be noted that the costate vector in each fuzzy zone is learned with a network similar to that of the costate vector of a linear system. It is shown in [215] that the weights would converge to optimum value for linear systems, if the iteration converges. Hence, with the proposed T-S fuzzy model-based critic network, it is expected that the network would converge to the optimal value corresponding to the local linear model of each fuzzy zone.

The control scheme is shown in Figure 7.3. In general, any network architecture can be chosen to model the co-state vector. The critic network approximates the costate vector using the current state $x(k)$, and the output of the critic network is used to compute the input $u(k)$ using equation (7.20). T-S fuzzy model is used in this thesis work, to decompose the global nonlinear costate vector as a fuzzy cluster of the costate vector of the local linear systems. Such a model gives a meaningful insight about the costate vector in terms of the local linear model.

The critic network is updated using equation (7.12) with $x(k+1)$ and $\lambda(k+2)$ as explained below. The network is to be trained such that the weights converge to the optimal value in each fuzzy zone. The critic network is learned within a fuzzy zone initially, and then the zone of learning is expanded gradually toward the entire universe by including the neighbor zones. In this work, the network is learned initially in the fuzzy zone defined around the

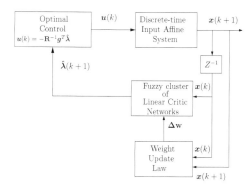

FIGURE 7.3: Control scheme with discrete-time single network adaptive critic(DT-SNAC) using T-S Model.

origin of the state space. To achieve uniform network convergence over the entire workspace, the learning zones $S_i = \{x(k) : \| x(k) \| < C_i\}$ are defined, where C_i is a positive constant and $i = 1, 2, \ldots, I$. C_i is chosen such that $C_i < C_{i+1}$. Initially C_1 is chosen a small value such that the network learns the optimal weights corresponding to the fuzzy zone around the origin. Then the region of learning is gradually increased. With such training, the network weights would vary smoothly from one zone of operation to another zone of operation.

The critic network is learned as follows:

1. Generate N_D random initial operating points for S_i as explained above for each learning stages. Initialize $i = 1$ and $k = 0$. Repeat the following steps for each member of $x(k)$ in S_i.

2. Compute $\hat{\lambda}(k+1)$ with $x(k)$ using the critic network (7.25).

3. Compute the control input $u(k)$ using equation (7.20) with $\hat{\lambda}(k+1)$.

4. Give the input $u(k)$ and obtain the next state $x(k+1)$.

5. Compute $\hat{\lambda}(k+2)$ with $x(k+1)$ using the T-S fuzzy model-based critic network (7.25).

6. Compute $\lambda_d(k+1)$ using $x(k+1)$ and $\hat{\lambda}(k+2)$ in equation (7.12).

7. Update the weights of the critic network by minimizing $\| \lambda_d(k+1) - \hat{\lambda}(k+1) \|$.

8. Increment k and repeat from step (2) for $N_{\max}$ instants.

9. Repeat from step (2) for N_D random points in S_i with $k = 0$.

Introduction

10. Check for the convergence of the weights of the critic network. If convergence is achieved, go to step 1) with $i = i + 1$. Otherwise, repeat steps (2) – (8) for all the members of S_i.

11. Repeat steps (1) – (9) till $i = I$.

$N_{\max}$ represents the number of evolutions in time from the initial state $\boldsymbol{x}(k)$ defined in S_i. If the system is unstable with the initial critic network, then the system is not evolved to instability, and $N_{\max}$ is initialized as $N_{\max} = 1$. The generation of training data with learning zones S_i is known as telescopic method [215].

The weights of the sub-critic networks in the T-S fuzzy model represent the costate vector corresponding to the local zone defined by fuzzification. The weights of the critic network vary smoothly from the optimal value at the origin and, hence, the weights of the critic network are initialized with the optimal value at the origin. The optimal weights corresponding to the origin is computed for the linearized model at the origin using discrete-time algebraic Riccati equation (DARE). It will be shown in experimental results, that the weights gradually change from the optimal values at the origin, as the system states move away from the origin.

Example 7.1. Consider the first order nonlinear system dynamics given by,

$$\dot{x} = -x^3 + u \tag{7.26}$$

where x is the system state and u is the input to the system. The objective is to stabilize the system such that the input would minimize the cost function,

$$J_c = \frac{1}{2}\int_0^\infty (x^2 + u^2)dt. \tag{7.27}$$

The analytic expression of the optimal controller is known for the system and is given by,

$$u^* = x^3 - \sqrt{x^6 + x^2}. \tag{7.28}$$

The main objective of this simulation is to show that the T-S fuzzy model-based SNAC approximates the costate vector effectively for discrete-time systems, and converges to global optimal solution. The discrete-time SNAC proposed in 7.1.3 can be applied to this optimal control problem by discretizing the dynamics and the cost function.

Solution 7.1. The discrete-time representation of the dynamics is given by,

$$x(k+1) = x(k) + \Delta T(-x^3(k) + u(k)) \tag{7.29}$$

where $x(k)$ is the system state, and $u(k)$ is the input to the system at the instant k, and ΔT is the sampling time. The objective is to stabilize the system such that the input would minimize the cost function,

$$J_c = \frac{1}{2}\sum_0^\infty (x^2 + u^2)\Delta T. \tag{7.30}$$

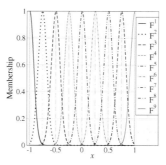

FIGURE 7.4: T-S fuzzy model with nine equally spaced fuzzy zones in the workspace $(-1, 1)$.

The operating zone is considered as $(-1, 1)$, and the costate vector is approximated with nine equally spaced fuzzy zones. The standard deviation of the Gaussian function is chosen using the first nearest neighbor heuristic with a maximum overlap of 0.01 between two adjacent fuzzy sets. The fuzzy zones are shown in Figure 7.4. The mean value of the fuzzy membership functions are $c = [-1\ -0.75\ -0.5\ -0.25\ 0\ 0.25\ 0.5\ 0.75\ 1.0]^T$ and the standard deviation corresponding to the overlap 0.01 is computed using equation (7.24) as 0.0824. The system is simulated with a sampling time of 0.1s for a duration of 10s and, hence, $N_{\max} = 100$. The critic network is trained with 5,000 random points in the operating zone, and the evolution of weights at different fuzzy zone during training is shown in Figure 7.5(a). It is clear from the figure that the weights are converging and are varying smoothly from the costate vector at the origin. The controller performance is tested from different initial states and compared with the optimal control policy. The performance comparison of the discrete-time SNAC with the optimal control policy is shown in Figure 7.6, which shows that the critic performs closer to the optimal cost. The corresponding control cost is tabulated in Table 7.1. It is evident that, the cost incurred with the T-S fuzzy model-based SNAC is closer to that of the optimal control policy. The cost incurred from various initial state x_0 in the operating zone is shown in Figure 7.5(b). The figure clearly shows that the fuzzy clustering with the linear costate model approximates the global costate effectively.

TABLE 7.1: First order system controlled with DT-SNAC and optimal control policy: Control cost at simulated points

x_0	-0.4248	0.464	-0.8348	0.5690	-0.5696
Critic	0.1048	0.1222	0.3147	0.1730	0.1733
Optimal	0.1046	0.1222	0.3147	0.1731	0.1734

Adaptive Critic Based Optimal Controller Design 241

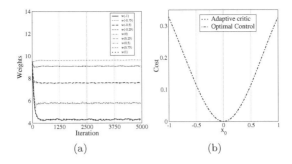

FIGURE 7.5: First order system with DT-SNAC: (a) Evolution of weights of the critic network during the training phase, (b) Control cost from different initial states for DT-SNAC and optimal control policy.

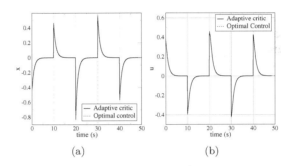

FIGURE 7.6: First order system controlled with DT-SNAC and optimal control policy: Controller performance: (a) State, (b) Input.

7.2 Adaptive Critic Based Optimal Controller Design for Continuous-time Systems

The real-time implementation of the optimal control policy for a discrete time nonlinear system is constrained by computational complexity, absence of closed form solution and offline numerical computation. These challenges are equally valid for continuous-time systems also and, hence, AC based methodologies have been proposed for the optimal control of continuous time nonlinear systems [218]. Baird [219] proposed AC based controller for continuous time systems using advantage updating and discussed the effect of sampling time on training. Similar to the discrete-time systems, AC methodologies use dual network architecture to learn the optimal solution of continuous-time systems. SNAC has been proposed for continuous-time input affine systems in [220, 221]. The convergence to optimality is shown for input affine systems with initial stabilizing control policy in [222, 223]. The convergence to optimality for a general nonlinear system with constrained input is shown in [224].

This chapter aims to develop adaptive critic based control strategy for visually controlled manipulator, whose nonlinear dynamics occur in input affine form. Such systems can be optimally controlled using SNAC, with a critic network modeling either the value function or the costate vector of the optimal control problem. The major intent of this thesis work is to model the value function of the optimal control policy with a critic network. Hence, further discussion primarily focuses on SNAC based optimal control methodology for input affine systems, where the critic network approximates the value function.

7.2.1 Continuous-time Single Network Adaptive Critic (CT-SNAC)

The dynamics of a continuous-time nonlinear input affine system is represented as,

$$\dot{x} = f(x) + g(x)u. \tag{7.31}$$

The task is to find a control input u^* which stabilizes the system, while minimizing the quadratic cost function,

$$\begin{aligned} J_c &= \int_0^\infty \frac{1}{2}(x^T Q x + u^T R u) dt \\ &= \int_0^\infty \psi(x, u) dt \end{aligned} \tag{7.32}$$

where Q is a positive semi-definite matrix, R is a positive-definite matrix, and $\psi(x, u)$ is the utility function of the continuous-time optimal control problem. The optimal control problem with boundary conditions $t_f = \infty$ and $x_f = 0$ is analyzed in this thesis work. The optimal control policy is derived by defining the Hamiltonian of the control problem as,

$$H(x, \lambda, u) = \psi(x, u) + \lambda \dot{x} \tag{7.33}$$

where $\lambda = \frac{\partial J_c}{\partial x} \in \mathbb{R}^{1 \times n}$ is the costate vector of the system.

The optimal control law satisfies the necessary condition of extremum given by,

$$\frac{\partial H}{\partial u} = \mathbf{0}. \tag{7.34}$$

The above condition gives the optimal control policy as a function of x and λ. The closed form solution to the above condition does not exist for general nonlinear systems. But, the closed form solution can be obtained as a function of λ^*, in case of input affine systems. The necessary condition of optimality (7.34) reduces to,

$$\frac{\partial \psi}{\partial u} + \lambda^* \frac{\partial}{\partial u} \left(f(x) + g(x) u \right) = \mathbf{0}. \tag{7.35}$$

Adaptive Critic Based Optimal Controller Design

The above condition gives the optimal control policy,

$$u^* = -\mathbf{R}^{-1}g^T(x)\boldsymbol{\lambda}^{*T}. \qquad (7.36)$$

The optimal cost satisfies the HJB equation,

$$\frac{\partial J_c^*}{\partial t} + \min_u H(x, \boldsymbol{\lambda}^*, u) = 0. \qquad (7.37)$$

The costate vector $\boldsymbol{\lambda}^*$ is computed using HJB equation for the optimal control policy u^*. The HJB equation gives the solution to the optimal control problem for any class of systems. However, the analytical solution to the HJB equation is difficult to obtain in most of the cases.

Adaptive critic methodology solves the above optimal control problem with action and critic networks. Action network solves equation (7.34) and approximates the relationship between the input u and the system state x. Critic network solves HJB equation (7.37) and estimates either the value function J_c or the costate vector $\boldsymbol{\lambda}$ as a function of system state x.

This chapter explains the SNAC based approaches for continuous-time system (CT-SNAC) by approximating the value function with a critic network $V(x, \mathbf{w})$, where $\mathbf{w}$ is the parameter vector of the critic network. After learning the optimal value function J^* with the critic network, the costate vector $\frac{\partial J_c^*}{\partial x}$, is calculated from the network as $\boldsymbol{\lambda} = \frac{\partial J_c}{\partial x}$, to compute the control input using equation (7.36).

Action network is not necessary for input affine systems since the analytic expression for the input is available in terms of the costate vector. Any parameterized model $J_c(\mathbf{w}, x)$ can be used to learn the optimal cost, provided that $J_c(\mathbf{w}, 0) = 0$ and $\frac{\partial J_c}{\partial x}$ exists.

7.2.2 Critic Network: Weight Update Law

The critic network has to be updated such that it satisfies the HJB equation (7.37). HJB requires the costate vector $\boldsymbol{\lambda}$, and the system dynamics (7.31) to compute the value function V. Exact knowledge of the system dynamics is required to update the critic network with equation (7.37). If the dynamics is unknown, the critic network can be updated using the derivative of the system states. This thesis work proposes an alternative approach which does not require either the exact knowledge of the model or the derivatives of the state to update the critic network. The update methodology for the critic network is derived as follows:

The critic network must satisfy the HJB equation since it approximates the value function. It is expressed in terms of equation as,

$$\frac{\partial V(x)}{\partial t} + \min_u H(x, \boldsymbol{\lambda}, u) = 0. \qquad (7.38)$$

The above equation is rewritten by substituting equation (7.33) for input affine system as,

$$\frac{\partial V(\boldsymbol{x})}{\partial t} + \psi(\boldsymbol{x}, \boldsymbol{u}) + \boldsymbol{\lambda}(\boldsymbol{f}(\boldsymbol{x}) + \boldsymbol{g}(\boldsymbol{x})\boldsymbol{u}) = 0$$

$$\frac{\partial V(\boldsymbol{x})}{\partial t} + \psi(\boldsymbol{x}, \boldsymbol{u}) + \left(\frac{\partial V}{\partial \boldsymbol{x}}\right)(\boldsymbol{f}(\boldsymbol{x}) + \boldsymbol{g}(\boldsymbol{x})\boldsymbol{u}) = 0. \quad (7.39)$$

The aforementioned equation simplifies to,

$$\dot{V}(\boldsymbol{x}) = -\psi(\boldsymbol{x}, \boldsymbol{u}) \quad (7.40)$$

where

$$\begin{aligned}
\dot{V}(\boldsymbol{x}) &= \frac{\partial V(\boldsymbol{x})}{\partial t} + \left(\frac{\partial V}{\partial \boldsymbol{x}}\right)\dot{\boldsymbol{x}} \\
&= \frac{\partial V(\boldsymbol{x})}{\partial t} + \left(\frac{\partial V}{\partial \boldsymbol{x}}\right)(\boldsymbol{f}(\boldsymbol{x}) + \boldsymbol{g}(\boldsymbol{x})\boldsymbol{u}).
\end{aligned} \quad (7.41)$$

The above equation gives the expression for change in the value function as the system moves along the optimal path. $\frac{\partial V(\boldsymbol{x})}{\partial t}$ represents the explicit time dependence of the value function on time, which occurs due to the evolution of weights during the training phase. Hence, a weight update law has to be proposed which ensures that the critic network satisfies equation (7.40).

In [220] a continuous-time weight update has been proposed with the minimum norm solution to equation (7.39) which requires either the dynamics of the system or the derivative of the system state. The major drawback is that the convergence to optimality is not guaranteed with the minimum norm solution. A discrete-time weight update is proposed in this work by discretizing equation (7.40) and then the weights are updated with gradient descent rule. Hence, further discussions will be continued in discrete-steps of sampling time ΔT. The proposed approach is motivated by the relationship,

$$\begin{aligned}
J(\boldsymbol{x}(t_0), \boldsymbol{u}(t_0)) &= \int_{t_0}^{\infty} \psi(\boldsymbol{x}(\tau), \boldsymbol{u}(\tau))d\tau \\
&= \int_{t_0}^{t_1} \psi(\boldsymbol{x}(\tau), \boldsymbol{u}(\tau))d\tau \\
&\quad + \int_{t_1}^{\infty} \psi(\boldsymbol{x}(\tau), \boldsymbol{u}(\tau))d\tau \\
&= \int_{t_0}^{t_1} \psi(\boldsymbol{x}(\tau), \boldsymbol{u}(\tau))d\tau + J(\boldsymbol{x}(t_1), \boldsymbol{u}(t_1)) \\
&= \psi(\boldsymbol{x}(t_0), \boldsymbol{u}(t_0))\Delta T + J(\boldsymbol{x}(t_1), \boldsymbol{u}(t_1)) \quad (7.42)
\end{aligned}$$

where $\Delta T = t_1 - t_0$. Such discrete-time approximations are always valid with high sampling rate. The discrete-time form of equation (7.40) is written as,

$$\Delta V(\boldsymbol{x}(k)) = -\psi(\boldsymbol{x}(k), \boldsymbol{u}(k))\Delta T. \quad (7.43)$$

Adaptive Critic Based Optimal Controller Design

Following equation (7.42), the equation (7.40) is expanded as,

$$V(\boldsymbol{x}(k)) = \psi(\boldsymbol{x}(k), \boldsymbol{u}(k))\Delta T + V(\boldsymbol{x}(k+1)) \qquad (7.44)$$

where $\Delta V(\boldsymbol{x}(k)) = V(\boldsymbol{x}(k+1)) - V(\boldsymbol{x}(k))$, k and $k+1$ are sample instants. The above equation is analogous to the Bellman's equation (7.3) of the discrete-time system. Value function has been updated using Bellman's equation for discrete-time systems, and the convergence has been proved in [225]. Hence a weight update with equation (7.43) is expected to converge to the optimal values, if the Euler approximation is valid.

With such analogy, the weight update law is further derived as follows. At any instant the weights of the critic network must satisfy equation (7.44). The weights are updated such that the above equation is satisfied. The optimal value function $V^d(k)$ is predicted with the weight vector $\mathbf{w}(k)$ as,

$$V^d(\mathbf{w}(k), \boldsymbol{x}(k)) = \psi(\boldsymbol{x}(k), \boldsymbol{u}(k))\Delta T + V(\mathbf{w}(k), \boldsymbol{x}(k+1)) \qquad (7.45)$$

where $V^d(\mathbf{w}(k), \boldsymbol{x}(k))$ is the predicted optimal cost. The weights are updated using gradient descent law, to minimize $\| V^d(\mathbf{w}(k), \boldsymbol{x}(k)) - V(\mathbf{w}(k), \boldsymbol{x}(k)) \|$, where $V(\mathbf{w}(k), \boldsymbol{x}(k))$ is the actual value estimated from the critic network. Thus, the CT-SNAC can be updated to optimality without the complete knowledge of the system with the proposed Euler approximation based weight update scheme. This is one of the key contributions achieved. The accurate model of the system is not available in real-time and, hence, the proposed approach can be executed to tune the controller parameters in real-time.

This approach is similar to the approach proposed in [219]. But the advantage of the action is computed at each stage in [219] and then the critic network is updated. In our approach there is no need for separate computation of the advantage and the critic network is updated directly by predicting the desired value function (7.45).

The control scheme with SNAC for continuous-time system is shown in Figure 7.7.

7.2.3 Choice of Critic Network

7.2.3.1 Critic Network using MLN

Just as in DT-SNAC, here also we can use a MLN to model the value function or the costate vector. Usually we take MLN with single hidden layer with non-linear activation function. The ouput is usually linear. Considering the MLN to estimate the value function, the steps for training are:

1. Generate state $\boldsymbol{x}(k)$ in the domain of interest.

2. For each element $\boldsymbol{x}(k)$, follow the steps below:

    (a) Input $\boldsymbol{x}(k)$ to the critic network to obtain $V(k+1)$. Compute the costate vector $\boldsymbol{\lambda}(k+1)$ from $V(k+1)$.

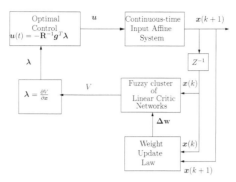

FIGURE 7.7: Control scheme with continuous-time single network adaptive critic (CT-SNAC).

    (b) Calculate the optimal control $u(k)$ using (7.20).

    (c) Get the next state $x(k+1)$ by giving the input $u(k)$.

    (d) Input $x(k+1)$ to the critic network to obtain $V(k+2)$ and compute $\lambda(k+2)$.

    (e) Using $x(k+1), u(k+1)$, compute $\psi(k+1)$. Using $\psi(k+1)$ and $V(k+2)$, calculate $V^d(k+1)$ using (7.45).

3. Train the critic network for all states $x(k)$ in the domain of operation.

4. Check the convergence of the critic network. If the convergence is achieved, revert to step 2 with $n = n+1$. Otherwise repeat steps 2-3.

5. Continue steps 2-4 until the convergence is achieved.

7.2.3.2   T-S Fuzzy Model-Based Critic Network with Cluster of Local Quadratic Cost Functions

In this section, T-S fuzzy model-based critic network is proposed to model the value function. The sub-critic networks in each fuzzy zone, model the value function with a quadratic cost function. The optimal cost of the LTI system with quadratic cost function (7.32), is given by,

$$J_c^* = \frac{1}{2} x^T P x \tag{7.46}$$

where P is a symmetric positive definite matrix, computed from ARE,

$$A^T P + PA - PBR^{-1}B^T P + Q = 0. \tag{7.47}$$

Nonlinear system behaves like a linear system in a small operating zone. It is intuitive and reasonable that a quadratic cost is a valid value function in a small zone of operation. The nonlinear dynamics can be represented with fuzzy cluster of local linear models in global operating zone and, hence, the global value can be better approximated with cluster of quadratic cost functions by fuzzifying the system states. With this motivation, T-S fuzzy model with a quadratic cost function in every fuzzy zone is chosen as the critic network to approximate the value function.

The ith zone of the T-S fuzzy model-based critic network to approximate the value function of the continuous-time nonlinear dynamical system is defined as,

$$\text{IF } x_1(t) \text{ is } F_1^i \text{ AND } \cdots \text{ AND } x_n(t) \text{ is } F_n^i \text{ THEN}$$

$$V_i(\boldsymbol{x}) = \frac{1}{2} \boldsymbol{x}^T \mathbf{W}_i^P \boldsymbol{x}$$

where F_j^i, $j = 1, 2, \cdots, n$, is the jth fuzzy set of the ith zone. $\mathbf{W}_i^P$ is a symmetric matrix which represents the linear map associated with the ith fuzzy zone to model the local value function. The fuzzy membership value μ_i associated with the ith zone is defined as,

$$\mu_i = \prod_{j=1}^n \mu_i^j(x_j) \tag{7.48}$$

where $\mu_i^j(x_j)$ is the membership function of the fuzzy set F_j^i, $i = 1, 2, \cdots, N_n$ and $V_i(\boldsymbol{x})$ represents the value corresponding to the operating zone defined by the ith fuzzy zone. Gaussian membership is used to define the fuzzy space associated with F_j^i similar to the discrete-time systems. $\mu_j^i(x_j)$ and $\mu_i(\boldsymbol{x})$ are computed as discussed in section 7.1.4.2.

The fuzzy model around the operating point $\boldsymbol{x}(t)$ is constructed as the weighted average of the local models as,

$$V(\boldsymbol{x}) = \frac{1}{2} \frac{\sum_{i=1}^{N_n} \mu_i \boldsymbol{x}^T \mathbf{W}_i^P \boldsymbol{x}}{\sum_{i=1}^{N_n} \mu_i} \tag{7.49}$$

where $V(\boldsymbol{x})$ is the control cost to stabilize from the state $\boldsymbol{x}(t)$.

The T-S fuzzy model approximates the value function with a fuzzy cluster of the value function of the linear models. Hence, T-S fuzzy model explains the relationship between the value function of the local linear models and that of the global nonlinear model. It is expected that such an architecture would converge to optimality of the local models in each fuzzy zone.

The network should be trained such that the weights converge to the optimal values in each fuzzy zone. The network is trained by defining the training zones of increasing size, with the telescopic method discussed for discrete-time system in section 7.1.4.2.

The critic network is learned as follows:

1. Generate N_D random initial operating points for S_i where $i = 1, 2, \ldots, I$, as explained in section 7.1.4.2 for each zone of operation. Initialize $i = 1$ and $k = 0$. Repeat the following steps for each member of $\boldsymbol{x}(k)$ in S_i.

2. Compute the control input $\boldsymbol{u}(t) = \boldsymbol{u}(k)$ using equation (7.36) with the costate vector estimated from the T-S fuzzy model-based critic network (7.49).

3. Give the input $\boldsymbol{u}(k)$ and obtain the next state $\boldsymbol{x}(k+1)$.

4. Compute $\psi(\boldsymbol{x}(k), \boldsymbol{u}(k))$, $V(\boldsymbol{x}(k+1))$ and $V(\boldsymbol{x}(k))$, with current instant weight vector $\mathbf{w}(k)$.

5. Compute $V^d(\mathbf{w}(k), \boldsymbol{x}(k))$ from $V(\mathbf{w}(k), \boldsymbol{x}(k+1))$ and $\psi(\boldsymbol{x}(k), \boldsymbol{u}(k))$ using equation (7.45).

6. Update the weights to minimize, $\| V^d(\mathbf{w}(k), \boldsymbol{x}(k)) - V(\mathbf{w}(k), \boldsymbol{x}(k)) \|$.

7. Increment k and repeat from step (2) for $N_{\max}$ instants.

8. Repeat from step (2) for N_D random points in S_i, with $k = 0$.

9. Check for convergence of the weights of the critic network. If convergence is achieved, go to step (1) with $i = i + 1$. Otherwise, repeat steps (2) − (8) for all the members of S_i.

10. Repeat steps (1) − (9) till $i = I$.

Since the T-S fuzzy model represents the value function, it must be a positive definite function. Hence, the weight is to be properly initialized such that the T-S fuzzy model is positive definite from the beginning of the training. In this approach, the weights in each fuzzy zone are initialized to the weights corresponding to the optimal weights of the linearized model around the origin, which can be easily obtained from ARE. It will be shown further in the experiments that the weights gradually change from the optimal values at the origin as the system states move away from the origin. If the system is not stable initially, then $N_{\max}$ is chosen as 1.

7.2.4 CT-SNAC

This section discusses the simulation results obtained with the continuous-time single network adaptive critic (CT-SNAC) proposed in 7.2. The proposed continuous-time adaptive critic is tested on four systems. At first, a second order LTI system is selected to show that the critic network converges to the optimal cost of the LTI system with the proposed learning scheme. Then the first order nonlinear system considered in the section 7.2.1 is considered, and it is shown through the simulation that the weights of each zone finally settle to

Adaptive Critic Based Optimal Controller Design

a value, which closely approximates the optimal cost of the nonlinear system. Finally, the control scheme is tested on benchmark systems-Vanderpol oscillator and single link manipulator. The controller performance is compared with linear quadratic regulator (LQR) obtained for the linearized model around the origin, since the optimal cost is not known for the aforementioned benchmark systems.

Example 7.2. Consider an LTI system [220] with dynamics,

$$\begin{bmatrix} \dot{x}_1 \\ \dot{x}_2 \end{bmatrix} = \begin{bmatrix} 0 & 1 \\ 0.4 & 0.1 \end{bmatrix} \begin{bmatrix} x_1 \\ x_2 \end{bmatrix} + \begin{bmatrix} 0 \\ 1 \end{bmatrix} u. \qquad (7.50)$$

The task is to find the control law u, which minimizes the cost (7.32), where

$$\mathbf{Q} = \begin{bmatrix} 1 & 0 \\ 0 & 1 \end{bmatrix} \text{ and } R = 1.$$

The optimal cost (7.46) of the system is obtained from ARE (7.47) as,

$$\mathbf{P} = \begin{bmatrix} 2.10456 & 1.4722 \\ 1.4722 & 2.09112 \end{bmatrix}. \qquad (7.51)$$

The critic network with only one fuzzy zone is considered since the system dynamics is linear. The critic network is expressed as,

$$V = \frac{1}{2}(w_1 x_1^2 + w_2 x_2^2 + 2w_3 x_1 x_2). \qquad (7.52)$$

Solution 7.2. The critic network is trained from 200 random initial states with sampling instant 0.01s and $N_{\max} = 200$. The weight evolution during the training phase is shown in Figure 7.8(a). The critic network has converged to $w_1 = 2.0889$, $w_2 = 2.051$ and $w_3 = 1.449$, which clearly demonstrates that the weights of the critic network converge to the optimum values of the LTI system with the proposed weight update.

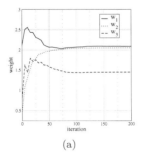
(a)

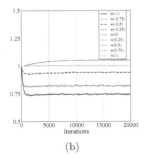
(b)

FIGURE 7.8: Evolution of the weights of the critic network during training phase: (a) LTI system controlled with CT-SNAC, (b) First order system with CT-SNAC.

Example 7.3. The first order nonlinear system discussed in Section 7.2.1 is considered to analyze the performance of the proposed SNAC based control methodology for continuous-time systems. The cost function is chosen same as the discrete-time case so that the results can be compared. DT-SNAC was designed for the discretized system in Section 7.5. In this simulation the actual continuous-time dynamics is considered while learning the optimal control policy with SNAC based methodology. The main objective of this simulation is to show that the nonlinear optimal cost can be modeled better with the fuzzy cluster of the cost function of the local linear model.

Solution 7.3. The operating zone is considered as $(-1, 1)$ similar to Section 7.5, and the value function is approximated with nine equally spaced fuzzy zones over the workspace. The fuzzy zones are chosen as same as that of discrete-time system, as discussed in Section 7.5. The fuzzy zones shown in Figure 7.4 are valid for this experiment too. The system is simulated with a sampling time of 0.1s for a duration of 10s. The weight of the critic network in every fuzzy zone is initialized to the weight corresponding to the fuzzy zone around the origin. The critic network is trained with 20,000 random points in the operating zone. The evolution of the weights in different zones during the training phase is shown in Figure 7.8(b) which clearly shows that the weights of the critic network converge with the proposed update scheme.

The final value of the weights in different operating zone are tabulated in Table 7.2. It is clear from the table that the weights in individual zone smoothly vary from the optimal weights corresponding to the origin. The learned weight around the fuzzy zone $x_1 = 0$ is 1.0009, which is closer to the optimal cost $P = 1$ of the linearized system. This corroborates the claim that the fuzzy clustering with local linear system modeling gives meaningful insight of relation between the local and the global optimal costs. After initial training, the system is controlled from different initial states and the controller performance is compared with the optimal control policy. The comparative performance of the proposed CT-SNAC based controller and the optimal controller is shown in Figure 7.9. The corresponding control cost is tabulated in Table 7.3. It is evident that, the fuzzy cluster of linear costs approximate the nonlinear optimal value function effectively and performs closer to the optimal controller.

To evaluate further, the optimal cost from various initial operating points x_0 in the operation zone is shown in Figure 7.10. The figure clearly shows that the nonlinear optimal cost is effectively approximated using T-S fuzzy model-based critic network. Comparing Figure 7.10 and Table 7.2, it is easy to infer that the weight in individual fuzzy zones approaches the cost of the

TABLE 7.2: First order system with CT-SNAC: Estimated optimal weights

Zone	-1	-0.75	-0.5	-0.25	0	0.25	0.5	0.75	1
Weights	0.745	0.818	0.941	1.052	1.00	1.051	0.938	0.821	0.747

Adaptive Critic Based Optimal Controller Design

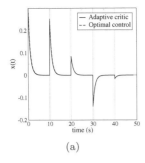 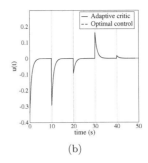

FIGURE 7.9: First order system controlled with CT-SNAC and optimal control policy: Controller performance (a) State, (b) Input.

TABLE 7.3: First order system controlled with CT-SNAC and optimal controller: Control cost at simulated points

x_0	0.2915	0.2221	0.0762	-0.1237	-0.0149
Critic cost	0.0563	0.0415	0.0044	0.01248	0.00013
Optimal cost	0.0540	0.0400	0.0046	0.01244	0.00014

corresponding operating zone. Figure 7.6 and and Figure 7.10 show that the DT-SNAC has approached optimality better than the CT-SNAC. In addition, this performance is observed after training with 20,000 random points in CT-SNAC, while DT-SNAC has used 5,000 initial points only. It may be due to the choice of sampling time to collect the data. If the sampling time is too small, then the data may not carry enough information about the system dynamics and, hence, the learning is slow. The effect of the sampling time on the network convergence has been analyzed in [219] and it has been demonstrated that a high sampling rate may reduce the rate of the convergence.

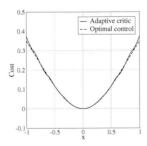

FIGURE 7.10: First order system with CT-SNAC: Control cost from various initial states in the workspace.

Example 7.4 (Vanderpol oscillator). The objective of this simulation is to show that an improved cost can be obtained using the fuzzy cluster of the linear models to approximate the value function of the nonlinear system. The performance of the T-S fuzzy model-based critic network is compared to the conventional LQR controller obtained for the linearized model around the origin. The optimal controller obtained from LQR is locally valid and may be unstable for wide range of operation. With T-S fuzzy model-based critic network, the controller guarantees the stability and optimality for a wide range of operation.

The Vanderpol oscillator system is a benchmark system with unstable equilibrium point at the origin and exhibits limit cycle too. The dynamics of the system is given by,

$$\begin{aligned} \dot{x}_1 &= x_2 \\ \dot{x}_2 &= \alpha(1-x_1^2)x_2 - x_1 + (1+x_1^2+x_2^2)u \end{aligned} \quad (7.53)$$

where $\alpha = 0.5$ is considered in the simulation. The control task is to compute the input u to asymptotically stabilize the system around the origin while minimizing the quadratic cost function (7.32), where $R = 1$ and $\mathbf{Q}$ is chosen as Identity matrix.

Solution 7.4. The CT-SNAC proposed in section 7.2.1 is employed, and the optimal cost is approximated by training a T-S fuzzy model-based critic network with nine equally spaced fuzzy zones for both the states in the operating zone $[-1, 1]$. The standard deviation of the Gaussian membership function zone is chosen by considering the maximum overlap between the adjacent zones as 0.01. The fuzzy sets defined along the state x_1 and x_2 are shown in Figure 7.11(a) and Figure 7.11(b) respectively. The mean and the standard deviation of the Gaussian function are same as chosen in Section 7.5. The fuzzy sets defined on x_1 and x_2 create 81 fuzzy zones within the workspace. The system is simulated with a sampling time $\Delta T = 5$ms and it is evolved from

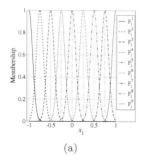
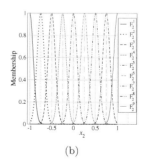

FIGURE 7.11: Fuzzy zones for Vanderpol oscillator system (a) x_1, (b) x_2.

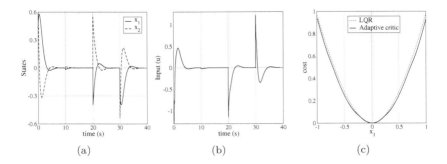

FIGURE 7.12: Vanderpol oscillator controlled with CT-SNAC: Controller performance (a) State, (b) Input, (c) Control cost from different initial states.

the initial state for $N_{\max} = 500$. The critic network is trained with 50,000 random points in the operating zone, and then the controller performance is analyzed. The closed loop control performance from different initial operating points is shown in Figure 7.12. It is easy to infer that the T-S fuzzy model-based SNAC stabilizes the system in the considered operating zone. The cost incurred from various initial states are computed and analyzed to evaluate further. The cost incurred with LQR and T-S fuzzy based SNAC is compared in Figure 7.12(c). The figure shows the cost incurred from the initial state $x = [x_1, 0]$ in the chosen universe of operation. These initial states are chosen in particular, because LQR is designed for origin, i.e., $x = [0, 0]$. Hence, the figure clearly shows the improvement in the controller performance obtained using the fuzzy cluster of the critic networks corresponding the linear model, as the system state x_1 varies around the origin. The optimal cost modeling with T-S fuzzy model-based critic network results in a solution better than the LQR based control, and the observed improvement is more as the zone of operation increases. The improvement is more as the initial state moves away from the origin since the effect of nonlinearity in the dynamics increases for the initial state far away from the origin. The control cost near the origin is closer to the optimal cost of the linearized model around $x = [0, 0]$, which shows that the T-S fuzzy model-based critic network converges to the value function of the local linear models within each fuzzy zone.

Example 7.5 (Single link manipulator). The dynamics of the single link manipulator is given by,

$$\begin{aligned} \dot{x}_1 &= x_2 \\ \dot{x}_2 &= -10\sin(x_1) + u. \end{aligned} \tag{7.54}$$

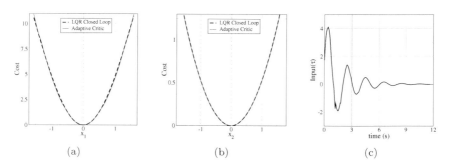

FIGURE 7.13: Single link manipulator with CT-SNAC: (a) Control cost: x_1, (b) Control cost: x_2, (c) Control input.

The task is to find the control law u, which minimizes the cost (7.32), with the following parameters:

$$\mathbf{Q} = \begin{bmatrix} 1 & 0 \\ 0 & 1 \end{bmatrix} \text{ and } R = 1.$$

Solution 7.5. The region of operation is chosen as $[-\pi/2, \pi/2]$ for both x_1 and x_2. The optimal cost is learned with a T-S fuzzy model with nine equally spaced fuzzy zones for both the states, similar to the critic network chosen for Vanderpol system in the previous section. Gaussian function is chosen similar to the previous simulation, to define the fuzzy membership functions. The critic is learned initially with 50,000 random points in the operating universe with a sampling time 0.01s, and the performance is compared with LQR gains. The closed loop control performance with the critic is shown in Figure 7.13(c), which shows that the critic network effectively stabilizes the system. The cost incurred with the LQR and the T-S fuzzy model-based critic network is compared in Figure 7.13. Figure 7.13(a) shows the control cost incurred from initial states $\boldsymbol{x} = [x_1, 0.0]$, and Figure 7.13(b) shows the control cost incurred from initial states $\boldsymbol{x} = [0.0, x_2]$. It is easy to note from the figures that the SNAC minimizes the performance index better than the LQR in most of the region, but the improvement is not as significant as in Vanderpol oscillator system. This performance can be improved by updating the critic network further.

Example 7.6 (3 link manipulator). The dynamics of the 3 link manipulator is given by,

$$\begin{aligned} x_1 &= R\cos\theta_1 \\ x_2 &= R\sin\theta_1 \\ x_3 &= l_2\sin\theta_2 + l_3\sin\theta_3 + l_1 \end{aligned} \tag{7.55}$$

where $\boldsymbol{x}(k) = [x_1(k) \; x_2(k) \; x_3(k)]^T$ and $\boldsymbol{\theta}(k) = [\theta_1(k) \; \theta_2(k) \; \theta_3(k)]^T$. Also, $R = l_2 cos\theta_2 + l_3 cos\theta_3 + t$. The respective link lengths are l_1, l_2, l_3 and t is the length of the rigid portion of the wrist. Here, we take the parameters from CRS PLUS manipulator, $l_1 = l_2 = l_3 = 254mm$ and $t = 50mm$. It is assumed that the wrist of the manipulator is rigid, assured by locking the joint for our purpose, and always holds the end-effector parallel to the work table (the XY plane).

The task is to find the control law $\boldsymbol{u}(k) = \Delta\boldsymbol{\theta}(k)$, which minimizes the cost (7.32) and the critic is to be modeled using a MLN. The parameters to be used are:

$$\mathbf{Q} = \begin{bmatrix} 1 & 0 & 0 \\ 0 & 1 & 0 \\ 0 & 0 & 1 \end{bmatrix} R = Q.$$

Solution 7.6. The workspace is chosen as $200 \times 200 \times 200 \; mm^3$, with $x_1 \in [150 \; 350]mm$, $x_2 \in [150 \; 350]mm$ and $x_3 \in [150 \; 350]mm$. The critic is a MLN with a single hidden layer. The hidden layer has a sigmoidal activation function and the output layer is linear. The critic is trained initially with 500 random points in the operating universe. Each point is iterated twenty times with a sampling time 0.01s. Then, it is tested for a fixed desired point and a time-varying circular reference trajectory. The results of the simulation for testing are given in the Figure 7.16 and 7.14. Figure 7.14 shows the end-effector position, end-effector position error and joint angle throughout time whereas Figure 7.15 shows the same for an initial period of 0.5 seconds. Figure 7.16 shows the end-effector position, end-effector position error, joint angles and joint angle control input throughout time respectively whereas

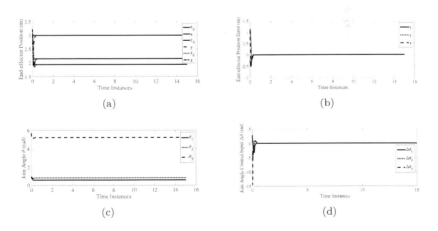

FIGURE 7.14: Simulation: End-effector motion while reaching a fixed desired point: (a) End-effector position (m), (b) End-effector position error (m), (c) Joint angles (rad), (d) Joint angle control input (rad).

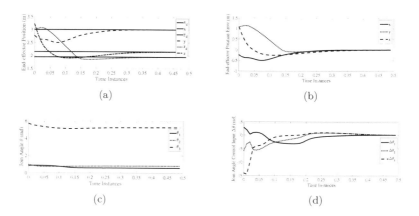

FIGURE 7.15: Simulation: End-effector motion while reaching a fixed desired point: (a) End-effector position (m), (b) End-effector position error (m), (c) Joint angles (rad), (d) Joint angle control input (rad).

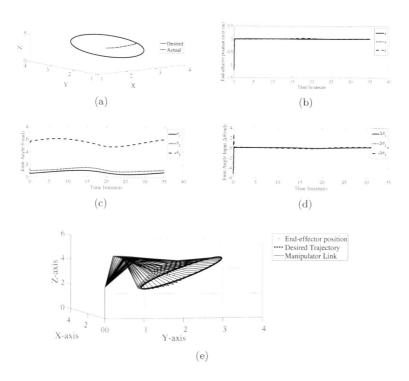

FIGURE 7.16: End-effector tracking a time-varying reference trajectory (a) Desired and actual end-effector positions, (b) End-effector position error (m), (c) Joint angle (rad), (d) Joint angle control input (rad), (e) Manipulator link position.

Discrete-time input affine system representation of forward kinematics 257

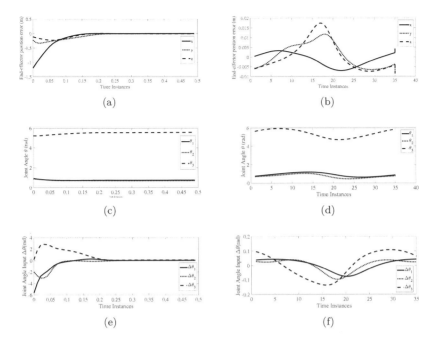

FIGURE 7.17: End-effector tracking a time-varying reference trajectory (a)(b) End-effector position error (m), (c)(d) Joint angle (rad), (e)(f) Joint angle control input (rad).

Figure 7.17 shows the end-effector error, joint angles and joint angle control input once the end-effector starts tracking the time-varying reference trajectory. As for tracking a time-varying reference trajectory, Figure 7.16(e) shows the manipulator link position once the manipulator starts tracking the time-varying reference trajectory.

7.3 Discrete-Time Input Affine System Representation of Forward Kinematics

Global optimal redundancy resolution can be achieved through adaptive critic formalism only if the redundancy resolution is formulated as an optimal control problem. An optimal control problem involves an integral cost function which defines the performance measure, and a dynamical system which explains the evolution of the state x with time due to the control input u. Hence, the redundancy resolution during closed loop positioning task has to be expressed as a dynamical system with an integral cost function to compute the optimal solution with adaptive critic methodologies. In this thesis work, the forward kinematic relationship of the redundant manipulator is formulated as a discrete-time input affine system, and the environmental constraints are modeled as an integral cost function to solve the redundancy resolution as an optimal control problem.

Consider the velocity level forward kinematic relationship from the joint angle space to the end-effector position in the Cartesian space,

$$\dot{\mathbf{x}} = \mathbf{J}\dot{\boldsymbol{\theta}}. \tag{7.56}$$

The dynamical system representation of the positioning task is obtained by considering the forward differential kinematics,

$$\boldsymbol{\Delta}\mathbf{x} = \mathbf{J}\boldsymbol{\Delta}\boldsymbol{\theta} \tag{7.57}$$

where $\boldsymbol{\Delta}\mathbf{x} = \begin{bmatrix} \Delta x & \Delta y & \Delta z \end{bmatrix}^T$,

and $\boldsymbol{\Delta}\boldsymbol{\theta} = \begin{bmatrix} \Delta\theta_1 & \Delta\theta_2 & \Delta\theta_3 & \Delta\theta_4 & \Delta\theta_5 & \Delta\theta_6 & \Delta\theta_7 \end{bmatrix}^T$. The forward differential kinematics can be expressed as a set of discrete motion of the end-effector at different instants as follows:

$$\begin{aligned} \boldsymbol{\Delta}\mathbf{x} &= \mathbf{J}\boldsymbol{\Delta}\boldsymbol{\theta} \\ \mathbf{x}(k+1) - \mathbf{x}(k) &= \mathbf{J}\boldsymbol{\Delta}\boldsymbol{\theta}(k) \end{aligned} \tag{7.58}$$

where $\mathbf{x}(k+1)$, $\mathbf{x}(k)$ are the end-effector position at $(k+1)$th and kth instants respectively and $\boldsymbol{\Delta}\boldsymbol{\theta}(k)$ is the change in the joint angle at the kth instant. In the above equation, the instant k represents the number of steps taken by the manipulator to reach the current position, and it is not specifying a fixed time duration. The discrete motion can be expressed as a dynamical system as,

$$\mathbf{x}(k+1) = \mathbf{x}(k) + \mathbf{J}\boldsymbol{\Delta}\boldsymbol{\theta}(k). \tag{7.59}$$

The above equation represents the positioning task as a discrete-time input affine system. The closed loop error dynamics which move the end-effector from the current position $\mathbf{x}$ to the desired position $\mathbf{x}_d$ is derived as,

$$\begin{aligned} \mathbf{e}(k+1) &= \mathbf{e}(k) - \mathbf{J}\boldsymbol{\Delta}\boldsymbol{\theta}(k) \\ &= \mathbf{A}\mathbf{e}(k) + \mathbf{B}\boldsymbol{u}(k) \end{aligned} \tag{7.60}$$

where $\mathbf{e}(k) = \mathbf{x}_d(k) - \mathbf{x}_c(k)$, $\mathbf{A} = \mathbf{I}$, $\mathbf{B} = -\mathbf{J}$ and $\boldsymbol{u}(k) = \boldsymbol{\Delta}\boldsymbol{\theta}(k)$. It is assumed in the dynamical system formulation, that $\mathbf{x}_d(k+1) = \mathbf{x}_d(k)$, i.e. the position of the object is fixed and the manipulator reaches the object at the desired position in multi-step movement. The number of steps taken to reach the desired position depends on the required accuracy, training, and also the distance between the initial and final positions of the end-effector.

The error dynamics (7.60) represents the closed loop kinematic control of the manipulator as a discrete-time input affine system. In discrete-time dynamical system formulation of the closed loop positioning task, the system states and the inputs are taken as $x(k) = e(k)$, and $u(k) = \Delta\theta(k)$ respectively. The above formulation enables to address the redundancy resolution problem as an optimal control problem, if the additional cost is represented as an integral cost function. Existing global optimal redundancy resolution schemes [44, 45] focus on a particular Cartesian space trajectory only. In contrast, the end-effector positioning task over a workspace is targeted in this work. Model-based global optimal redundancy resolution schemes [46, 47] define the state space in the joint angle space by using the generalized inverse kinematic relationship (1.5), and ϕ is considered as the input. In contrast, the state space is defined on the task space x, and the change in the joint angle is considered as the input, in the proposed approach. Such formalism fits better to the real-world implementation, since the end-effector is moved in the Cartesian space by controlling the joint angle input. It will be shown that such a formulation really simplifies the controller design and allows to improve the performance in real-time.

7.4 Modeling the Primary and Additional Tasks as an Integral Cost Function

The primary positioning task and the additional tasks introduced by the environment act as constraints on the motion of the manipulator. The cost function has to be appropriately chosen such that both the primary and the additional tasks are achieved. In this thesis work, the primary and the additional tasks are represented as a quadratic cost function,

$$J_c = \frac{1}{2}\sum_{k=0}^{\infty}(\mathbf{e}^T(k)\mathbf{Q}\mathbf{e}(k) + \Delta\boldsymbol{\theta}^T(k)\mathbf{R}\Delta\boldsymbol{\theta}(k)). \tag{7.61}$$

It has been demonstrated that an input affine system stabilized while minimizing a quadratic cost function can be optimally controlled with SNAC, and an action network is not required to learn the optimal control policy. Hence, the redundancy can be resolved optimally using SNAC without an action network, if the primary positioning task and the additional task is specified in the form of a quadratic cost. Such formalism simplifies the critic implementation and it is computationally efficient.

The state weight matrix $\mathbf{Q}$ specifies the primary task. An infinite-time optimal control problem results in zero terminal condition, which indicates the successful completion of the primary positioning task. In the current work, the state weight matrix **Q** is chosen as an identity matrix to ensure uniform convergence toward the desired position in every coordinate direction. The input weight matrix **R** has to be chosen to penalize the individual joints based on the additional task.

7.4.1 Quadratic Cost Minimization (Global Minimum Norm Motion)

Quadratic cost measure has been used to design the controller which minimizes the variation of the system state and input. **R** is chosen to control the variation in the input. In this study, a diagonal matrix with constant entry R_{gain} for all the joints is chosen, which uniformly weighs the motion of all the joints. The speed of convergence to the desired position can be controlled by varying R_{gain}. Hence, a simple quadratic cost minimization results in the global minimization of the Euclidean norm of the joint angular velocity.

7.4.2 Joint Limit Avoidance

The joint limit avoidance is one of the key additional tasks expected from the redundant manipulator due to its physical limitation. In this chapter, the performance criterion proposed by Zghal et al. [226] is used to constrain the joints within the kinematic limits. The performance criterion is expressed as,

$$H(\boldsymbol{\theta}) = \sum_{i=1}^{m} \frac{1}{4} \frac{(\theta_{i_{max}} - \theta_{i_{min}})^2}{(\theta_{i_{max}} - \theta_i)(\theta_i - \theta_{i_{min}})} \tag{7.62}$$

The input weight matrix R of the quadratic cost function is defined with the above performance criterion as,

$$\mathbf{R}(\boldsymbol{\theta}) = \begin{bmatrix} R_1 & 0 & 0 & \cdots & 0 \\ 0 & R_2 & 0 & \cdots & 0 \\ \cdots & \cdots & \cdots & \ddots & \cdots \\ 0 & 0 & 0 & \cdots & R_7 \end{bmatrix}. \tag{7.63}$$

$\mathbf{R}(\boldsymbol{\theta})$ is a diagonal matrix with ith diagonal entry defined as,

$$R_i(\boldsymbol{\theta}) = \begin{cases} 1 + \left| \frac{\partial H(\boldsymbol{\theta})}{\partial \theta_i} \right| & \text{if } \Delta \left| \frac{\partial H(\boldsymbol{\theta})}{\partial \theta_i} \right| \geq 0 \\ 1 & \text{if } \Delta \left| \frac{\partial H(\boldsymbol{\theta})}{\partial \theta_i} \right| < 0. \end{cases} \tag{7.64}$$

The input weight $R_i(\boldsymbol{\theta})$ penalizes the individual joints based on its position relative to its limits. The joint angle space motion is penalized less if the current joint angle configuration is in the middle of the physical limits and the penalization increases as the link moves toward its limit. Individual joints are penalized only for its relative position with respect to its own limits, i.e., ith joint is not penalized for the relative position of the jth joint with its limit. Since $\mathbf{R}(\boldsymbol{\theta})$ is defined as a diagonal matrix its inverse can be computed with the reciprocal of the corresponding elements which makes the chosen performance criterion computationally efficient.

7.5 Single Network Adaptive Critic Based Optimal Redundancy Resolution

The discrete-time input affine system formulation of the closed loop positioning dynamics discussed in equation (7.60) is suitable for single network adaptive critic based control since the primary and the additional tasks are expressed as a quadratic cost function. The optimal redundancy resolution scheme with single network adaptive critic is shown in Figure 7.18. With SNAC based redundancy resolution scheme, the joint angle input $\boldsymbol{\Delta\theta}$ is computed as

$$\boldsymbol{\Delta\theta}(k) = \mathbf{R}^{-1}\mathbf{J}^T\hat{\boldsymbol{\lambda}}(k+1) \qquad (7.65)$$

where $\hat{\boldsymbol{\lambda}}(k+1)$ is obtained from the critic network, which optimizes the given cost. The advantage of such an approach is that the optimal closed loop positioning of the redundant manipulator is achieved in real-time without the computation of pseudo-inverse of the forward Jacobian. Hence, a computationally efficient global optimal control strategy is obtained with critic based approach. This claim will be further confirmed with computational complexity analysis.

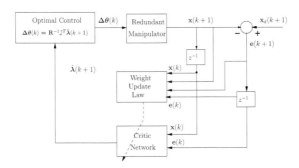

FIGURE 7.18: Schematic diagram: Optimal redundancy resolution scheme with SNAC.

7.5.1 T-S Fuzzy Model-Based Critic Network for Closed Loop Positioning Task

The T-S fuzzy model-based critic network proposed in Section 7.1.4.2 is used to learn the costate vector. The critic network is designed for the stabilization of the states, and in general the states are fuzzified to learn the costate vector. Hence, the critic network should be fuzzified with end-effector position error $\mathbf{e}(k)$ for closed loop positioning task. But, it is always desired to achieve a global positioning where the end-effector can be moved from any initial position to arbitrary desired position over the entire workspace. The global positioning depends on both the current and the desired positions of the end-effector. The performance of the critic network will be poor if the error is considered as the input to the fuzzifier, since the error changes based on both the current and desired positions. The current position of the end-effector specifies the kinematic state of the manipulator better than the positioning error since the forward Jacobian changes as the end-effector moves along the trajectory.

To achieve a better global positioning the current position of the end-effector $\mathbf{x}$ is given as the input to the fuzzifier, instead of the error $\mathbf{e}$. As the manipulator changes its joint angle configuration along the trajectory, different fuzzy zones of the critic get activated and the corresponding weights are trained accordingly. The current position of the end-effector has been suggested in [170] and [10] for better dynamic performance.

With above modification, the ith rule of the T-S fuzzy model-based critic network for manipulator end-effector positioning task is given by,

IF $x(k)$ is F_1^i AND $y(k)$ is F_2^i AND $z(k)$ is F_3^i THEN,

$$\hat{\lambda}_i(k+1) = \mathbf{W}_i \mathbf{e}(k)$$

where $\mathbf{W}_i \in R^{3 \times 3}$ is the linear map to learn the costate vector from the Cartesian space, in the ith fuzzy zone. The fuzzy space is defined along individual coordinate direction of the Cartesian space as,

$$\begin{aligned}
\mu_x^i(x) &= e^{\left(\frac{-(x-c_x^i)^2}{2(\sigma_x^i)^2}\right)} \\
\mu_y^i(y) &= e^{\left(\frac{-(y-c_y^i)^2}{2(\sigma_y^i)^2}\right)} \\
\mu_z^i(z) &= e^{\left(\frac{-(z-c_z^i)^2}{2(\sigma_z^i)^2}\right)}.
\end{aligned} \quad (7.66)$$

The fuzzy membership value of the ith rule is computed using the product rule,

$$\mu_i(\mathbf{x}) = \mu_x^i \, \mu_y^i \, \mu_z^i. \quad (7.67)$$

7.5.2 Training Algorithm

The critic network is trained similar to the DT-SNAC training algorithm discussed in Section 7.2.2. To achieve the smooth network convergence, the network is learned from a selected fuzzy zone $\mathbf{x}_{f0}$, to the entire workspace such that the weights would converge to optimal values in each zone. The successive learning zones are defined as $S_i = \{\mathbf{x}_d(k) : \| \mathbf{x}_d(k) - \mathbf{x}_{f0} \| < C_i, i = 1, 2, \ldots, I\}$, where C_i is a positive constant, and $C_i < C_{i+1}$. Initially C_1 is chosen a small value so that the network would learn the optimal weights corresponding to the selected zone. Then, the operating zone is gradually increased.

The critic network is learned as follows:

1. Generate N_D random desired positions for each S_i. Initialize $i = 1$.

2. Set $k = 0$ and $n = 0$. Choose a random initial point $\mathbf{x}(k)$.

3. Compute the initial state $\mathbf{e}(k) = \mathbf{x}_d(n) - \mathbf{x}(k)$. Give error $\mathbf{e}(k)$ as the input to compute the costate vector and $\mathbf{x}(k)$ as the input to the fuzzifier to compute $\hat{\boldsymbol{\lambda}}(k+1)$.

4. Compute the input $\boldsymbol{\Delta\theta}(k)$, from equation (7.65), using $\hat{\boldsymbol{\lambda}}(k+1)$.

5. Give the input to the dynamical system representation equation (7.60) of the manipulator, and compute the next instant end-effector position $\mathbf{x}(k+1)$ and the positioning error $\mathbf{e}(k+1)$.

6. Compute $\hat{\boldsymbol{\lambda}}(k+2)$ from the critic network using $\mathbf{e}(k+1)$.

7. Compute $\boldsymbol{\lambda}_d(k+1)$ using $\hat{\boldsymbol{\lambda}}(k+2)$ and $\mathbf{e}(k+1)$ in the costate vector equation (7.12). Consider $\boldsymbol{\lambda}_d(k+1)$ as the desired costate vector and update the network weights to minimize $\| \boldsymbol{\lambda}_d(k+1) - \hat{\boldsymbol{\lambda}}(k+1) \|$.

8. Check if $\| \mathbf{e}(k+1) \| < \epsilon_e$. If $\| \mathbf{e}(k+1) \| > \epsilon_e$ then repeat from step (3) with $k = k+1$ till $N_{\max}$ iterative steps.

9. If the desired position is reached with chosen accuracy, set $n = n+1$ and repeat from step (2) for N_D random points.

10. Check $\| \boldsymbol{\lambda}_d(k+1) - \hat{\boldsymbol{\lambda}}(k+1) \| < \epsilon_\lambda$ for N_D points in S_i. If no, repeat steps $(2) - (9)$. Otherwise, set i=i+1 and repeat steps $2 - 9$ till $i = I$.

The optimal weights in each zone will vary smoothly from the optimal value of the fuzzy zone $\mathbf{x}_{f0}$, since the system dynamics gradually deviate from the linear behavior as the zone of operation increases. Considering this fact, the weights of the critic network are always initialized with optimal value corresponding to the fuzzy zone $\mathbf{x}_{f0}$ which can be computed using ARE with the linearized model. The Cartesian position $\mathbf{x}_{f0}$ is reached with the joint angle configuration $\boldsymbol{\theta}_0$, and the computation of the input requires the computation of the Jacobian $\mathbf{J}$. Hence, the network training is initiated from the joint angle configuration $\boldsymbol{\theta}_0$ which results in $\mathbf{x}(0) = \mathbf{x}_{f0}$, if $n = 0$ and $\mathbf{x}(0) = \mathbf{x}_d^{n-1}$ if $n > 0$, where $\mathbf{x}_d^{n-1}$ is the desired position of the $(n-1)$th point in S_i. If the end-effector moves outside the workspace during the training phase then the training is initialized from $\mathbf{x}_{f0}$. The evolution of state with multi-step movement i.e., $N_{\max} > 1$ is necessary in case of SNAC based control of the redundant manipulator so that the Jacobian $\mathbf{J}$ can be estimated with $\boldsymbol{\theta}(k)$.

The network has to be trained by penalizing the joint angle space motion based on the additional task requirement. Hence, a smooth joint angle space motion is necessary so that the links are penalized from the initial position. To achieve such a smooth motion the input to the critic network is given as $K\mathbf{e}(k)$, where K is the feedback gain which is typically chosen as $K \leq 1$. The smaller gain value results in a slower motion and the movement will not be jerky due to the large value of initial error. A small value of K reduces the error presented to the critic network. A smaller error indicates that the desired position is closer to the current position. In such a case, the manipulator moves slowly, and more iterations are taken to reach the desired position. With a slower motion, the Jacobian will be evaluated at more number of positions resulting in an effective redundancy resolution.

7.6 Computational Complexity

The optimal costate vector is learned during the training phase with the kinematic model given in equation (2.1), and the trained network is used to compute the input in real-time. It will be shown that the computational requirement of the T-S fuzzy model-based critic network is low since the nonlinear costate vector is approximated with clusters of local linear models.

The local costate vector is represented by $n \times n$, where $n = 3$ is the dimension of the task space. The computation of the local costate vector requires n^2 order flops, where an individual flop represents an addition and a multiplication operation. If N_r rules are firing at each operating point then the local costate vectors are computed with $N_r n^2$ order flops and further $N_r n$ order flops are required for computing the overall costate vector. It is clearly evident that the computation of the costate vector is independent of the DOF of the manipulator and in general, $N_r \ll N_n$.

The computation of the input from the costate vector requires nm order flops for computing $\mathbf{J}^T \hat{\boldsymbol{\lambda}}$ and m order flops for $\mathbf{R}^{-1} \mathbf{J}^T \hat{\boldsymbol{\lambda}}$, since $\mathbf{R}$ is a diagonal matrix. The proposed adaptive critic based redundancy resolution scheme requires a total of $N_r(n^2 + n) + m(n + 1)$ order flops for computing the input which is linear with the DOFs of the manipulator.

The pseudo-inverse based redundancy resolution scheme [18] involves the computation of the minimum norm motion for the primary task and the self-motion for accomplishing the additional task. The redundancy resolution requires computation of Moore-Penrose pseudo-inverse. The computation of Moore-Penrose pseudo-inverse [227] involves singular value decomposition (SVD) and matrix multiplications for inverse computation. The SVD computation is of order $O(mn^2)$ and requires $2mn^2 + n^3$ flops approximately. The multiplications involve $mn^2 + m^2n$ order flops and, hence, the pseudo-inverse computation requires a total of $3mn^2 + n^3 + nm^2$ order flops. The minimum norm input computation involves nm order flops. The self motion is computed with $m(m-1)n/2 + m^2$ order flops where $m(m-1)n/2$ is required for computing $(\mathbf{I} - \mathbf{J}^+\mathbf{J})$ and m^2 is for $(\mathbf{I} - \mathbf{J}^+\mathbf{J})\phi$ respectively. The total computational cost of the pseudo-inverse based technique is $(3mn^2 + n^3 + 3/2m^2n - mn/2)$ order flops which is more than the adaptive critic based method.

The computational requirement increases in the order $O(m^2)$ for the pseudo-inverse based technique, while it is linear with DOF for the critic based approach which makes it a better approach for real-time implementation. In addition to low computational requirements, critic based methodology guarantees a global optimal solution while optimality is not ensured with the pseudo-inverse based technique. In addition, the analysis shows that the computational load increases in cubic order of the dimension of the workspace, in the case of pseudo-inverse based methods while it is of quadratic order in the critic based approach. Hence, the proposed SNAC based approach is computationally cost effective than the pseudo-inverse based method, even if the dimension of the task space increases. This is an another key finding in this thesis work.

7.7 Simulation Results

The forward kinematic model discussed in Chapter 2 is used to train the network. The critic network is trained within a cubic volume of workspace with diagonal vertices $(0.2, -0.25, 0.0)$ and $(0.7, 0.25, 0.3)$. The workspace is fuzzified with five equally spaced fuzzy zones in each coordinate direction. The standard deviation of the fuzzy membership function is selected such that the effect of two fuzzy zones will be predominant at each operating point, and the effect of neighboring fuzzy zones at the fuzzy center will be less than 5%. The fuzzy sets defined on the coordinate directions x, y, z are shown in Figure 7.19(a) and Figure 7.19(b) and Figure 7.19(c) respectively. The mean value of the Gaussian functions which defines the fuzzy sets along x-axis are chosen as $(0.2, 0.325, 0.45, 0.575, 0.7)$ and the standard deviation corresponding to the 5% overlap is computed as to be 0.0511. The fuzzy sets are defined along the y coordinate direction with Gaussian functions of mean $(-0.25, -0.125, 0.0, 0.125, 0.25)$ and standard deviation 0.0511. The fuzzy centers are at $(0.0, 0.075, 0.15, 0.225, 0.30)$ along the z coordinate direction, and the standard deviation is computed using equation (7.24) as to

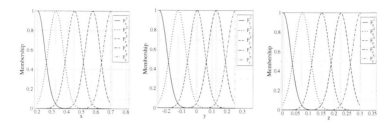

FIGURE 7.19: Fuzzy sets for positioning task: (a) x-axis, (b) y-axis, (c) z-axis.

be 0.0306. The critic network totally comprises 125 fuzzy zones with a linear weight matrix $\mathbf{W}_i$ to model the local costate vector in each fuzzy zone. The number of fuzzy zones effective at each operating point for the critic network is $N_r = 2^3 = 8$. The critic network is trained with 5,000,000 random points with a learning rate of 0.01, $\epsilon_e = 0.5mm$ and $N_{\max} = 50$. The seventh link does not contribute to any position change and $\Delta\theta_7$ remains zero throughout the simulations. Hence, the joint angle input to the seventh link is not shown while discussing the results obtained in the following experiments.

7.7.1 Global Minimum Norm Motion

In this simulation, the state weight matrix $\mathbf{Q}$ and the input weight matrix $\mathbf{R}$ are chosen as identity matrices as discussed in 7.4.1. R_{gain} is chosen as 1 and the feedback gain K is chosen as 0.5 in the experiment.

After critic training, the network is analyzed in two stages. In first stage, the desired position of the end-effector is fixed as $\mathbf{x}_d = (0.4, 0.1, 0.2)$m and the robot manipulator is started from random initial positions. The position of the end-effector at successive instants and the corresponding errors are shown in Figure 7.20. The initial random points are explicitly shown with $\star$ in the figures. It is observed in simulation that the end-effector reaches the desired position with an accuracy of 0.1mm in 20 iterative steps indicating that the network is trained adequately. An instant in the figures corresponds to one iterative step taken by the manipulator to move from the current position toward the desired position.

The corresponding joint angle movement is shown in Figure 7.21. The initial joint angle corresponding to the random initial positions are represented as $\star$, and show that the manipulator starts from random initial joint angle and reaches the desired position with smooth joint movements. The figure also shows that the final joint angle depends on the initial joint configuration of the manipulator.

Since global positioning is desired from random initial position to arbitrary desired position, the controller performance is further analyzed for various desired positions over the entire workspace. The successive position of

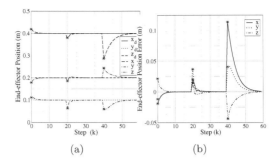

FIGURE 7.20: Simulation: Closed loop positioning at the desired position $(0.4, 0.1, 0.2)$m with global minimum norm motion: (a) Position, (b) Error.

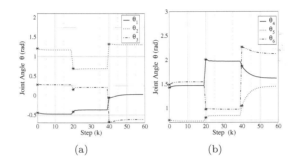

FIGURE 7.21: Simulation: Joint angle θ during positioning at the desired position $(0.4, 0.1, 0.2)$m with global minimum norm motion: (a) $(\theta_1, \theta_2, \theta_3)$, (b) $(\theta_4, \theta_5, \theta_6)$.

the end-effector and the corresponding errors are shown in Figure 7.22. The instant at which the desired position changes is shown with $\star$. The end-effector reaches the various desired positions with an average accuracy of 0.1mm in 20 iterative steps and stays within an error of 1mm after ten iterative steps. The positioning accuracy varies over the workspace since the uniform convergence is possible only with exhaustive training. As discussed in section 7.1.3, the adaptive critic based approaches guarantee the performance improvement with time, and the controller approaches the optimal value with continuous weight update. Hence, a better positioning accuracy and optimality is feasible with further weight update which makes the critic based approach superior than the existing sub-optimal approaches.

Figure 7.22(a) shows that the closed loop positioning exhibits an oscillatory behavior for the chosen $\mathbf{Q}$ and $\mathbf{R}$. It is well-known that a non-oscillatory performance can be achieved in optimal control based strategies, by choosing appropriate $\mathbf{Q}$ and $\mathbf{R}$ which is an iterative process. The corresponding joint

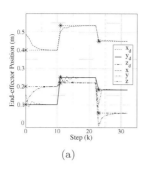 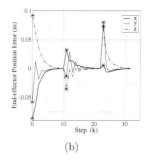

(a)   (b)

FIGURE 7.22: Simulation: Closed loop positioning at the arbitrary desired position from random initial position, with global minimum norm motion: (a) Position, (b) Error.

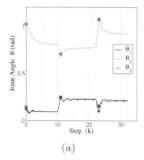 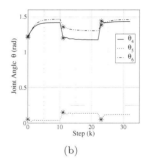

(a)   (b)

FIGURE 7.23: Simulation: Joint angle $\boldsymbol{\theta}$ during positioning at the arbitrary desired position from random initial position, with global minimum norm motion: (a) $(\theta_1, \theta_2, \theta_3)$, (b) $(\theta_4, \theta_5, \theta_6)$.

angle movement for random positioning task is shown in Figure 7.23. It is clear from the $\star$ mark in the figures that initially there is a large change in the joint angle due to huge error and the movement slows down with decreasing error. The performance of the controller is tested with a closed elliptical trajectory. A complex closed trajectory is particularly chosen to test the performance of the controller over the entire workspace. The critic based methodology is devised for closed loop positioning task with the assumption that $\mathbf{x}_d(k+1) = \mathbf{x}_d(k)$, i.e., the manipulator is moving to reach a stationary object. The above assumption is valid at each sampling instant while tracking a continuous trajectory such as ellipse, where the desired position has to be reached at each time instant, as explained in the following steps. Let the position of the end-effector at t_kth time instant is $\mathbf{x}(t_k)$ owing to a joint angle configuration $\boldsymbol{\theta}(t_k)$ and the desired position at the $(t_k + 1)$th time instant is $\mathbf{x}_d(t_k + 1)$.

Simulation Results

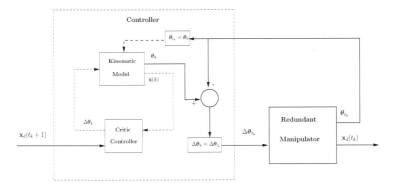

FIGURE 7.24: Schematic diagram: Controlling the manipulator to track the ellipse.

- The desired position to estimate the control law is fixed as $\mathbf{x}_{desired} = \mathbf{x}_d(t_k + 1)$. The initial position and joint angle are taken as $\mathbf{x}(0) = \mathbf{x}(t_k)$ and $\boldsymbol{\theta}(0) = \boldsymbol{\theta}(t_k)$ respectively.

- The adaptive critic is iterated with the kinematic model (4.1) until the end-effector reaches the position $\hat{\mathbf{x}}(k)$ with the joint angle $\boldsymbol{\theta}(k)$ such that the positioning error $\|\mathbf{x}_{desired} - \hat{\mathbf{x}}(k)\| < \epsilon_{max}$ where ϵ_{max} is the desired positioning error tolerance and k is the number of steps as discussed in section 7.3. The end-effector position $\hat{\mathbf{x}}(k)$ is predicted using the forward kinematic model with $\boldsymbol{\theta}(k)$.

- The control input is calculated as $\Delta\boldsymbol{\theta}(t_k) = \boldsymbol{\theta}(k) - \boldsymbol{\theta}(0)$ and is applied to the manipulator.

- The end-effector moves to the position $\mathbf{x}(t_k + 1)$ with the given input $\Delta\boldsymbol{\theta}(t_k)$. In real-time, $\mathbf{x}(t_k+1)$ differs from $\hat{\mathbf{x}}(k)$ due to model inaccuracies.

- The critic is then presented with the next desired position to move along the trajectory.

The schematic diagram of controlling the manipulator to follow the elliptical trajectory is shown in Figure 7.24. The kinematic model is used to compute the optimal joint angle input iteratively, and then the manipulator is presented with the computed input in both simulation and real-time. To analyze the controller performance, the number of steps taken by the critic at each instant to reach within the chosen accuracy is computed in the following simulations. The desired elliptical trajectory is taken as,

$$\begin{aligned} x &= 0.45 + 0.15\cos(0.05t_k) \\ y &= 0.15\sin(0.05t_k) \\ z &= 0.15. \end{aligned} \quad (7.68)$$

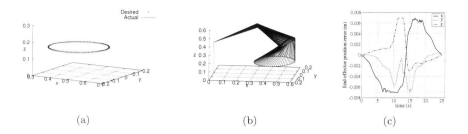

FIGURE 7.25: Simulation: End-effector motion while tracking the elliptical trajectory with global minimum norm motion: (a) Trajectory (m), (b) Manipulator configuration (m), (c) Position error.

where t_k is the sampling instant and the r.m.s. positioning error tolerance, $\epsilon_{\max}$ is chosen as 1.0cm to iterate the critic based optimal control law before applying to the robot manipulator. The experiment is performed in real-time with a sampling time of 200ms. Hence, the simulation results are shown with the same sampling time for easier comparison with the real-time results. The end-effector trajectory and its corresponding manipulator configuration is shown in Figure 7.25. It is clear from the figure that the closed loop system tracks the trajectory with smooth joint angle space motion. The positioning error shown in Figure 7.25(c) indicates that the manipulator tracks the trajectory with an accuracy of 7mm. The corresponding joint angle input and the joint angle configurations are shown in Figure 7.26 and Figure 7.27 respectively to demonstrate the smoothness of the learned solution. It is observed that the manipulator tracks the ellipse after 1.03 iterative steps on average. The number of iterations can be reduced and also the speed of the tracking can be increased by increasing the feedback gain K but it may result in an oscillatory behavior.

The computation time requirement for the critic based approach is further analyzed during simulation. The computation time for drawing the elliptical trajectory for 100 cycles is measured. The complete cycle of the elliptical trajectory (7.68) is represented by 126 points and, hence, the end-effector moves in between 12,600 operating points over the experiment. The simulation is run on the personal computer discussed in section 2.1. The simulations are performed at different time of the day with various CPU load conditions. The average and standard deviation of the computation time at each operating point over various runs is presented in Table 7.4. The computation time is compared with the minimum norm motion from the pseudo-inverse based technique and it is clear from the table that the proposed approach takes approximately 75% time that is required with pseudo-inverse based control for 7 DOF robot manipulator. The critic based approach takes approximately $33\mu s$ at each operating point to compute the input which is negligible compared

Simulation Results

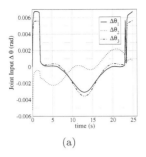 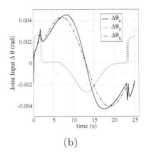

FIGURE 7.26: Simulation: Joint angle input $\Delta\boldsymbol{\theta}$ while tracking the ellipse with global minimum norm motion: (a) $(\Delta\theta_1, \Delta\theta_2, \Delta\theta_3)$, (b) $(\Delta\theta_4, \Delta\theta_5, \Delta\theta_6)$.

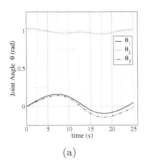 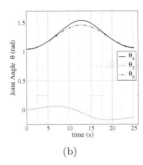

FIGURE 7.27: Simulation: Joint angle $\boldsymbol{\theta}$ while tracking the elliptical trajectory with global minimum norm motion: (a) $(\theta_1, \theta_2, \theta_3)$, (b) $(\theta_4, \theta_5, \theta_6)$.

TABLE 7.4: Computation time requirement for global minimum norm motion

critic (μs)		Pseudo-inverse (μs)	
Average	Std. Deviation	Average	Std. Deviation
32.54	16.25	41.92	20.27
32.14	16.55	40.91	21.06
32.14	15.57	40.04	20.20
31.35	17.11	40.91	19.81
32.54	15.25	40.91	21.06

to the sampling time 200ms of the redundant manipulator. This confirms the claim that the adaptive critic based method results in an optimal redundancy resolution which can be implemented in real-time. As discussed earlier, the pseudo-inverse based technique results in local optimal solution and the optimality over the entire trajectory is not guaranteed. Hence, the proposed SNAC based redundancy resolution scheme results in a computationally efficient control strategy to obtain global optimal solution.

7.7.2 Joint Limit Avoidance

The control task considered is to compute the input $\boldsymbol{u}(k) = \boldsymbol{\Delta\theta}(k)$ to position the end-effector, which minimizes the cost function,

$$J_c = \frac{1}{2} \sum_{k=0}^{\infty} (\; \mathbf{e}^T(k)\mathbf{Q}\mathbf{e}(k) + \boldsymbol{\Delta\theta}^T(k)\mathbf{R}(\boldsymbol{\theta})\boldsymbol{\Delta\theta}(k) \;) \tag{7.69}$$

where $\mathbf{R}(\boldsymbol{\theta})$ is computed as discussed in section 7.4.2. In this simulation, the kinematic constraint is applied to the fourth joint with limit given as,

$$-1.25 \;<\; \theta_4 < 1.25 \text{ rad.} \tag{7.70}$$

The above operating range is chosen based on the observations from the global minimum norm motion. The joint angle of the fourth link varies in the range of $(1.04, 1.53)$ radian, while tracking the elliptical trajectory chosen in (7.68). Global minimum norm motion violates the above kinematic limit. The critic network is trained with feedback gain $K = 0.1$ to accurately learn the cost penalization.

After initial training, the network is analyzed in two stages as discussed in Section 7.7.1. Initially, the desired position of the manipulator is chosen as same as that of the global minimum norm motion, and then the manipulator is started from different initial positions. The end-effector position at successive instants and the corresponding positioning error are plotted in Figure 7.28. The joint angle while moving toward the desired position are shown in Figure 7.29. Then the controller is checked for arbitrary positioning from various initial position over the entire workspace. The corresponding positioning results are shown in Figure 7.30 with joint angles in Figure 7.31. The initial random point is explicitly shown with $\star$ in the figures. The results are similar to that of quadratic cost minimization with two major differences. It is observed that the desired position is reached with an accuracy of 0.1mm in fifty iterative steps in contrast to twenty steps. More iterative steps are taken since the feedback gain is chosen as $K = 0.1$ so that the joints are penalized properly. In addition, the joint angle of the fourth link is within the chosen limits. The performance of the controller is tested with the closed elliptical trajectory (7.68). The same trajectory is chosen so that the performance can be compared with the results obtained for the global minimum norm motion. The end-effector trajectory and the corresponding manipulator configuration are shown in Figure 7.33. The tracking error shown in Figure 7.33(c) indicates that the end-effector tracks the trajectory smoothly within an accuracy of 7.4mm with the proposed adaptive critic based kinematic limit avoidance scheme. It is observed in experiments that the critic required 1.508 iterative steps in each instant on average along the trajectory. The average number of

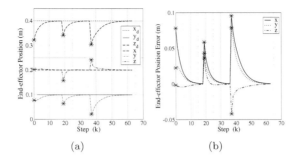

FIGURE 7.28: Simulation: Closed loop positioning at the desired position $(0.4, 0.1, 0.2)$m with kinematic limit avoidance: (a) Position, (b) Error.

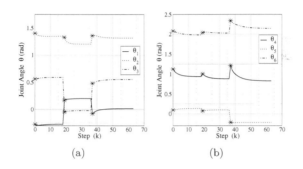

FIGURE 7.29: Simulation: Joint angle $\boldsymbol{\theta}$ during positioning at the desired position $(0.4, 0.1, 0.2)$m with kinematic limit avoidance: (a) $(\theta_1, \theta_2, \theta_3)$, (b) $(\theta_4, \theta_5, \theta_6)$.

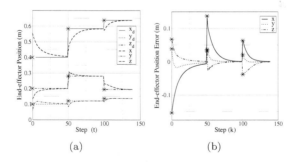

FIGURE 7.30: Simulation: Closed loop positioning at the arbitrary desired position from random initial position with kinematic limit avoidance: (a) Position, (b) Error.

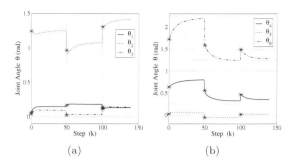

FIGURE 7.31: Simulation: Joint angle θ during positioning at arbitrary desired point from random initial point with kinematic limit avoidance: (a) $(\theta_1, \theta_2, \theta_3)$, (b) $(\theta_4, \theta_5, \theta_6)$.

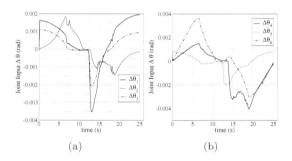

FIGURE 7.32: Simulation: Joint angle input $\Delta\theta$ while tracking the elliptical trajectory with kinematic limit avoidance : (a) $(\Delta\theta_1, \Delta\theta_2, \Delta\theta_3)$, (b) $(\Delta\theta_4, \Delta\theta_5, \Delta\theta_6)$.

iterative steps is higher than the global minimum norm motion since $K = 0.1$. The joint angle input and the joint angle trajectory are shown in Figure 7.32 and Figure 7.34 respectively. It is clear from the figures that the joint angle of the fourth link is within the kinematic limit. The joint angle of the fourth link is separately shown in Figure 7.34(c) for better understanding of the effect of kinematic limit avoidance scheme on the fourth link. It is clear from the figure that the manipulator avoids the kinematic limit while tracking the elliptical trajectory, with the proposed SNAC based redundancy resolution scheme. The computation time is then compared with the pseudo-inverse based joint limit avoidance scheme. The manipulator is simulated to draw the elliptical trajectory defined by equation (7.68) for 100 cycles, similar to the analysis performed for global minimum norm motion. The average and standard deviation of the computation time taken at each operating point over various runs is tabulated in Table 7.5. It is clear from the table that the proposed approach

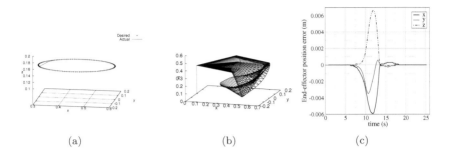

FIGURE 7.33: Simulation: End-effector motion while tracking the elliptical trajectory with kinematic limit avoidance: (a) Trajectory (m), (b) Manipulator configuration (m), (c) Position error.

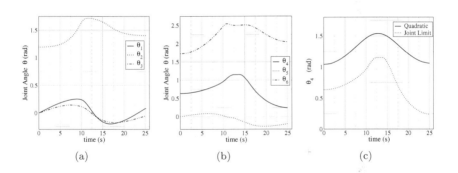

FIGURE 7.34: Simulation: Joint angle θ while tracking the elliptical trajectory with kinematic limit avoidance: (a) $(\theta_1, \theta_2, \theta_3)$, (b) $(\theta_4, \theta_5, \theta_6)$, (c) Joint angle of the fourth link with and without kinematic limit avoidance.

TABLE 7.5: Computation time requirement for joint limit avoidance

critic (μs)		Pseudo-inverse (μs)	
Average	Std. Deviation	Average	Std. Deviation
33.33	6.57	44.44	5.73
34.13	5.89	44.29	5.88
33.33	6.57	44.13	6.78
32.94	5.93	44.29	6.66
32.94	7.8	44.76	6.27

takes $33\mu s$ to compute input for kinematic limit avoidance while the pseudo-inverse based technique requires $44\mu s$. The computation time with SNAC is 25% less than the pseudo-inverse based method, which matches with the result of global minimum norm simulations presented in Section 7.7.1. These observations corroborate the claim that critic based redundancy resolution is computationally efficient than pseudo-inverse based methods, while ensuring global optimal solution.

7.8 Experimental Results

The real-time experiment is performed on the PowerCube manipulator discussed in Chapter 2. The real-time experiments have been performed for the elliptical trajectory (7.68). The tip of the end-effector and the manipulator configuration are observed with stereo-vision while tracking the ellipse. As mentioned in the simulation, the critic is iterated with the model at each sampling instant and then the final control law is given to the manipulator. The sampling interval is chosen as 200ms based on the computational requirements for image acquisition and processing, as well as the manipulator speed. A smaller sampling interval would result in a jerky motion of the manipulator without reaching the commanded position.

7.8.1 Global Minimum Norm Motion

The network trained in Section 7.7.1 is used in the experiment and the forward kinematic model is used to compute the output during iteration. The end-effector of the robot manipulator is observed using stereo vision over the entire trajectory.

The end-effector position while tracking the trajectory and the tracking error in real-time experiment is shown in Figure 7.35. It is clear from the figure that the manipulator tracks the trajectory with an error of 1.4cm which is more than that of the simulation results. The increase in the error is observed due to model inaccuracies. This performance can be further improved by updating the critic in real-time. The corresponding input and the joint angle configurations are shown in Figure 7.36 and Figure 7.37 respectively which clearly show that the real-time performance is closer to the simulation results in spite of model inaccuracies. The end-effector trajectory in the vision space is shown in Figure 7.38. The trajectory is noisy due to the image processing inaccuracy associated with the identification of the end-effector using the centroid of the red tape.

Experimental Results

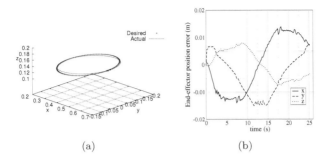

FIGURE 7.35: End-effector motion in real-time experiment with global minimum norm motion: (a) Trajectory (m), (b) Position error.

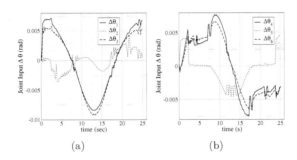

FIGURE 7.36: Joint angle Input $\Delta\boldsymbol{\theta}$ in real-time experiment with global minimum norm motion: (a) $(\Delta\theta_1, \Delta\theta_2, \Delta\theta_3)$, (b) $(\Delta\theta_4, \Delta\theta_5, \Delta\theta_6)$.

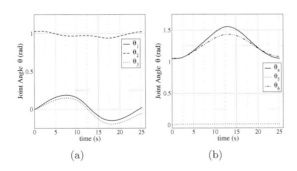

FIGURE 7.37: Joint angle $\boldsymbol{\theta}$ in real-time experiment with global minimum norm motion: (a) $(\theta_1, \theta_2, \theta_3)$, (b) $(\theta_4, \theta_5, \theta_6)$.

278   *Kinematic Control using Single Network Adaptive Critic*

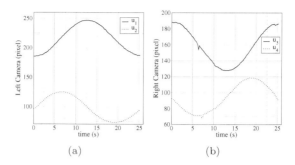

FIGURE 7.38: Trajectory of the end-effector in the vision space during global minimum norm motion: (a) Left camera x-y coordinates (u_1, u_2), (b) Right camera x-y coordinates (u_3, u_4).

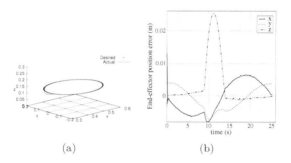

FIGURE 7.39: End-effector motion in real-time experiment with kinematic limit avoidance: (a) Trajectory (m), (b) Error.

7.8.2 Joint Limit Avoidance

The elliptical trajectory is tracked considering the kinematic limits with the network obtained in Section 7.7.2. The kinematic limit of the fourth joint due to engineering design is $[-90^0, 90^0]$ as tabulated in Table 2.2. The additional constraint (7.70) is considered and the critic network is learned. The learned network is used in real-time experiment. The end-effector trajectory and its corresponding manipulator configuration are shown in Figure 7.39. It is observed that the end-effector tracks the trajectory with an accuracy of 2.55cm in a small operating zone which is more than the theoretical accuracy of 7mm indicating that the network has to be updated more in those operating zones. The corresponding input and joint angle configurations are shown in Figure 7.40 and Figure 7.41 respectively. It is clear from the figures that the joint angle of the fourth link is within the kinematic limit as similar

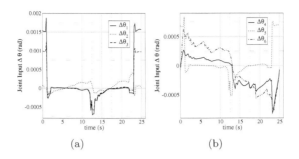

FIGURE 7.40: Real-time experiment: Joint angle input $\Delta\boldsymbol{\theta}$ while tracking an elliptical trajectory with kinematic limit avoidance: (a) $(\Delta\theta_1, \Delta\theta_2, \Delta\theta_3)$, (b) $(\Delta\theta_4, \Delta\theta_5, \Delta\theta_6)$.

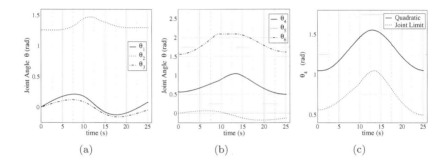

FIGURE 7.41: Real-time experiment: Joint angle $\boldsymbol{\theta}$ while tracking an elliptical trajectory with kinematic limit avoidance: (a) $(\theta_1, \theta_2, \theta_3)$, (b) $(\theta_4, \theta_5, \theta_6)$, (c) Joint angle of the fourth link with and without kinematic limit avoidance.

to the simulation results. The fourth joint angle is separately shown in Figure 7.41(c) for better comparison of simulation and experimental results. The experimental result is similar to the simulation and the manipulator avoids the kinematic limit with single network adaptive critic based approach effectively. The trajectory of the end-effector as seen through the vision space is shown in Figure 7.42. It is easy to note that the trajectory is similar to that of the quadratic cost minimization shown in Figure 7.38 due to accurate tracking.

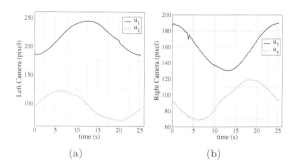

FIGURE 7.42: Trajectory of the end-effector in the vision space during kinematic limit avoidance: (a) Left camera x-y coordinates (u_1, u_2), (b) Right camera x-y coordinates (u_3, u_4).

7.9 Conclusion

In this chapter, we first introduced the concept of Discrete Time Adaptive Critic Networks. The concept of Discrete Time Adaptive Critic is first explained using an example given in (7.14). Later, we moved on to the introduction of Single Network Adaptive Critic (SNAC). It was shown that SNAC approach may be used to find out the optimal control law of discrete time systems as well as continuous time systems. The costate vector is modeled by using a critic network which may be a MLN or a T-S Fuzzy based model. Example 7.1 demonstrates the application of a T-S Fuzzy Model-based DT-SNAC to a a first order system to evaluate the optimal control policy. Examples 7.2, 7.3 and 7.4 discussed the simulation results obtained with CT-SNAC. Example 7.5 and 7.6 demonstrated the Single network Adaptive Critic-based kinematic control of single link and 3-link Robot Manipulator respectively. Here, the critic that models the costate vector is a MLN. It is seen that it achieves the tasks of reaching a fixed desired end-effector position as well as tracking a desired time-varying reference trajectory with optimal control input to the joints which eventually minimizes the selected global cost function.

Later, we discussed a SNAC based redundacy solution schemes for redundant manipulators. For demonstration, the optimal redundancy resolution is implemented in real-time in single network adaptive critic framework. The computational cost of the proposed SNAC based scheme is analyzed and compared with the pseudo-inverse based redundancy resolution. It is shown that the computational requirement of the critic based method is linear with the degree of redundancy, while the pseudo-inverse based technique is in quadratic order. SNAC is trained with the available kinematic model and tested in both

Conclusion

simulation and real-time. Simulations show that the arbitrary accuracy can be achieved with multi-step movement and an accuracy of ±1cm is achieved with two step movements for smooth trajectories.

The optimal controllers for a general class of nonlinear system is not known and demand high computational cost. This chapter introduces optimal controller design for input affine nonlinear systems in adaptive critic framework. The T-S fuzzy model critic network has been used for both the continuous-time (CT-SNAC) and discrete-time (DT-SNAC) input affine systems. The optimal controller for a nonlinear system has been obtained as a fuzzy cluster of the optimal controllers of the local linear models, establishing the relationship between the local linear dynamics and the global nonlinear dynamics. In case of discrete-time system, SNAC proposed in [215] has been used to learn the optimal costate vector as a fuzzy cluster of local linear costate vectors. A Euler approximation based weight update is proposed for continuous-time systems controlled with SNAC. The approach approximates the value function with a critic network, by discretizing continuous-time cost function in a form similar to the Bellman's equation of the discrete-time system. The optimal kinematic control using this SNAC architecture has been introduced for a 3-link robot manipulator. This architecture is further extended for optimal kinematic control of a 7 DOF robot manipulator where optimal redundancy resolution becomes a key factor.

The global optimal redundancy resolution is achieved by formulating an optimal control problem for the positioning task. The proposed scheme focuses on resolving the redundancy while the end-effector moves in discrete steps toward the desired position. The proposed scheme formulates the end-effector closed loop positioning task as a discrete-time nonlinear input affine system, by defining the state space in the task space. The primary positioning task and the desired additional task is modeled as an integral quadratic cost function. With such simplicity, the global optimal solution is achieved over the workspace by minimizing an integral cost function. The optimal redundancy resolution is implemented in real-time in single network adaptive critic framework. The computational cost of the proposed SNAC based scheme is analyzed and compared with the pseudo-inverse based redundancy resolution. It is shown that the computational requirement of the critic based method is linear with the degree of redundancy, while the pseudo-inverse based technique is in quadratic order. SNAC is trained with the available kinematic model and tested in both simulation and real-time. Simulations show that the arbitrary accuracy can be achieved with multi-step movement and an accuracy of ±1cm is achieved with two step movements for smooth trajectories.

8

Dynamic Control using Single Network Adaptive Critic

8.1 Introduction

The intention behind designing a controller is to generate stable control actions associated to a linear or nonlinear dynamical system. These controllers work in the *inner loop* of the control hierarchy of nonlinear systems such as robotic manipulators. It is often desired that the control action to be the *best* or *optimal* in terms of effectiveness. This is achieved by minimizing some performance index while maintaining the stability of the closed loop system. Performance index is a numerical value, associated with the quality of the control action. The idea is that the desired quality of the system's performance improves as the performance index minimizes. Our task as a control engineer is to design an admissible controller that minimizes the performance index while fulfilling the system objective. In this chapter we will discuss about the design of such a controller applicable to control robotic manipulators.

Let us review few definitions which are in use in the following sections.

Definition 8.1. *A real $n \times n$ symmetric matrix $\mathbf{Q}$ is called positive $(\mathbf{Q} > 0)$ if and only if $\mathbf{x}^T \mathbf{Q} \mathbf{x} > 0$, $\forall\ \mathbf{x} \in \Re^n$, where $\mathbf{x}$ is a real vector.*

Definition 8.2. *A real $n \times n$ symmetric matrix $\mathbf{Q}$ is called positive semi-definite $(\mathbf{Q} \geq 0)$ if and only if $\mathbf{x}^T \mathbf{Q} \mathbf{x} \geq 0$, $\forall\ \mathbf{x} \in \Re^n$, where $\mathbf{x}$ is a real vector of non-zero numbers.*

Definition 8.3. *A real $n \times n$ matrix $\bar{\mathbf{Q}}$ is called negative $(\bar{\mathbf{Q}} < 0)$ if $-\bar{\mathbf{Q}}$ is positive.*

Definition 8.4. *(Zero-State Observability [228]) System (8.40) with measured output $\mathbf{y} = \mathbf{h}(\mathbf{x})$ is zero-state observable if $\mathbf{y}(t) \equiv 0\ \forall t \geq 0$ implies that $\mathbf{x}(t) \equiv 0\ \forall t \geq 0$.*

Definition 8.5. *The fuzzy system (8.52) would be called* asymptotically stable *if the equilibrium state $\mathbf{x}_e$ of the system is Lyapunov stable and there exists $\bar{\delta} > 0$ such that for an initial state $\mathbf{x}(0)$, if $\|\mathbf{x}(0) - \mathbf{x}_e\| < \bar{\delta}$, then $\lim_{n \to \infty} \|\mathbf{x}(0) - \mathbf{x}_e\| = 0$.*

Definition 8.6. *A control action* $\mathbf{u}_i(\mathbf{x}(t))$ *is said to be an admissible policy for ith iteration where,* $i = 1, 2, 3, ...\infty$ *if it is continuous on* $\bar{\mathbb{Z}}$, *stabilizes system* (8.40) *on* $\bar{\mathbb{Z}}$ *where* $\forall \mathbf{x}_0$ *on* $\bar{\mathbb{Z}}$, $\mathbf{u}_i(\mathbf{x}(t) = 0) = 0$ *and* J_0 *is finite.*

8.2 Optimal Control Problem of Continuous Time Nonlinear System

Selection of the performance index is very important for optimal performance of the controller. The design of the performance metric can be considered as a part of the system modelling. Here, we shall discuss some of the performance indices in the following.

Consider an autonomous continuous nonlinear system is represented by the following equation

$$\dot{\mathbf{x}}(t) = \mathbf{f}(\mathbf{x}(t), \mathbf{u}(t)) \quad with \quad \mathbf{x}(t_0) = \mathbf{x}_0 \tag{8.1}$$

$$\mathbf{y}(t) = \mathbf{C}\mathbf{x}(t) \tag{8.2}$$

where, $\mathbf{x} \in \Re^n$ and $\mathbf{u} \in \Re^m$ are the state and the control input vector respectively. As an optimal control problem, we intend to design the control action $\mathbf{u}(t)$ such that the following performance index is minimized.

$$J(\mathbf{x}) = \int_{t_0}^{T} L(\mathbf{x}) d\tau \tag{8.3}$$

where $\phi(\mathbf{x}(T))$ is the terminal cost and $L(\mathbf{x})$ is the utility / instantaneous cost[1]. Minimizing (8.3) ensures that the system in (8.1) performs optimally in the interval $[t_0, T]$. Here the utility cost defines what optimal feature we want in our system. If we desire the system state to be *small*, an appropriate utility cost could be

$$L_1(\mathbf{x}) = \mathbf{x}^T \mathbf{x} \tag{8.4}$$

This is intuitive from the above expression that ensuring minimization the performance index (8.3) would keep the states *small*. By selecting

$$L_2(\mathbf{x}) = \mathbf{x}^T \mathbf{Q} \mathbf{x} \tag{8.5}$$

(where, $\mathbf{Q} = \mathbf{C}^T \mathbf{C}$) the output state can be kept small. If one wish to keep the magnitude of the control action *small*, the utility function can be of the following form

$$L_3(\mathbf{x}) = \mathbf{u}^T \mathbf{R} \mathbf{u} \tag{8.6}$$

---
[1] The utility function can be $L(\mathbf{x}, \mathbf{u})$.

Here, **R** is a symmetric positive definite matrix, to weight the components of control action **u**. This should be noted here that performance indices made by (8.4) (or (8.5)) and (8.6) are conflicting objectives and cannot be achieved simultaneously. To keep the states *small* in the duration $[t, T]$, the control action has to be *large*. If the control action is kept *small* in the duration $[t, T]$, the state becomes large. Hence, a trade off has to be made between these objectives and can be obtained by taking convex combination to embed in a single performance index. The more generic form of the utility cost function can be given as follows.

$$L(\mathbf{x}, \mathbf{u}) = \frac{1}{2} \left[\mathbf{x}^T \mathbf{Q} \mathbf{x} + \mathbf{u}^T \mathbf{R} \mathbf{u} \right] \qquad (8.7)$$

Here **Q** and **R** can selected to make the trade off between the conflicting objectives. The introduction of $\frac{1}{2}$ is to simplify to subsequent algebraic manipulations.

Certain control applications such as position control of robotic arm require the final state $\mathbf{x}(T)$ to be as close to **0** as possible. It suggests the control action also needs to minimize terminal cost $\Phi(\mathbf{x}(T))$. This can be achieved by adding the term $\Phi(\mathbf{x}(T))$ in the performance index as follows.

$$J(\mathbf{x}) = \Phi(\mathbf{x}(T)) + \int_{t_0}^{T} L(\mathbf{x}) d\tau \qquad (8.8)$$

A suitable choice for $\Phi(\mathbf{x}(T))$ is $\frac{1}{2}\mathbf{x}^T \mathbf{F} \mathbf{x}$ which helps to keep the terminal state as close to **0**. Here, **F** is a symmetric positive definite matrix. The problem of minimizing (8.8) subject to (8.1) is termed as *linear quadratic regulator* (LQR) problem.

8.2.1 Linear Quadratic Regulator

Consider a linear time invariant system

$$\dot{\mathbf{x}} = \mathbf{A}\mathbf{x} + \mathbf{B}\mathbf{u}, \quad \mathbf{x}(t_0) = \mathbf{0} \qquad (8.9)$$

The linear quadratic regulator (LQR) problem of the system (8.9) is about designing a control law

$$\mathbf{u} = -\mathbf{K}\mathbf{x} \qquad (8.10)$$

such that the system (8.9) is asymptotically stable in the closed loop while the performance index

$$J(\mathbf{x}) = \int_{t_0}^{\infty} \left[\mathbf{x}^T \mathbf{Q} \mathbf{x} + \mathbf{u}^T \mathbf{R} \mathbf{u} \right] d\tau \qquad (8.11)$$

is minimized. The optimal control law is represented as $\mathbf{u}^*$. The asymptotic stability is ensured by finding a Lyapunov function $V = \mathbf{x}^T \mathbf{P} \mathbf{x}$ where **P** is a positive definite matrix, such that, $\frac{dV}{dt}$ is negative definite on closed loop trajectory. It can be noted here that the significance of the terminal cost $\Phi(\mathbf{x}(T))$ is lost as $T \to \infty$.

The feedback gain $\mathbf{K}$ in (8.10) is designed such that

$$\min_{\mathbf{u}} \left(\frac{dV}{dt} + \mathbf{x}^T \mathbf{R} \mathbf{x} + \mathbf{x}^T \mathbf{Q} \mathbf{u} \right) = 0 \qquad (8.12)$$

subject to existence of a Lyapunov function $V = \mathbf{x}^T \mathbf{P} \mathbf{x}$. The application of such control action $\mathbf{u}^*$ results to the optimal cost

$$J(\mathbf{u}^*) = \mathbf{x}_0^T \mathbf{P} \mathbf{x}_0 \qquad (8.13)$$

where $\mathbf{x}_0$ is the initial state. The appropriate $\mathbf{P}$ is found by solving

$$\frac{\partial}{\partial \mathbf{u}} \left[\frac{dV}{dt} + \mathbf{x}^T \mathbf{R} \mathbf{x} + \mathbf{x}^T \mathbf{Q} \mathbf{u} \right] \bigg|_{\mathbf{u}=\mathbf{u}^*} = 0 \qquad (8.14)$$

Replacing $\frac{dV}{dt}$, the above can be written as

$$\frac{\partial}{\partial \mathbf{u}} \left[2\mathbf{x}^T \mathbf{P} \mathbf{A} \mathbf{x} + 2\mathbf{x}^T \mathbf{P} \mathbf{B} \mathbf{u} + \mathbf{x}^T \mathbf{R} \mathbf{x} + \mathbf{x}^T \mathbf{Q} \mathbf{u} \right] \bigg|_{\mathbf{u}=\mathbf{u}^*} = 0$$

After differentiation w.r.t. $\mathbf{u}$, the above becomes,

$$2 \left[\mathbf{x}^T \mathbf{P} \mathbf{B} + \mathbf{u}^T \mathbf{R} \right] \big|_{\mathbf{u}=\mathbf{u}^*} = 0 \qquad (8.15)$$

The optimal control law $\mathbf{u}^*$ can be derived from (8.15) as

$$\mathbf{u}^* = -\mathbf{R}^{-1} \mathbf{B}^T \mathbf{P} \mathbf{x} \qquad (8.16)$$

By relating the above equation with (8.10), it can be noted that the optimal gain,

$$\mathbf{K} = \mathbf{R}^{-1} \mathbf{B}^T \mathbf{P} \qquad (8.17)$$

It can be easily investigated that the second order sufficiency condition for the optimal control problem is also true by differentiating the left hand side of the equation (8.15). The derivation of the control law $\mathbf{u}^*$ was under the assumption that there exists an appropriate symmetric positive definite matrix $\mathbf{P}$ which governs the Lyapunov function of the system. Now we shall see how an "appropriate" $\mathbf{P}$ can be found.

Under the influence of the optimal control law $\mathbf{u}^*$, the closed loop system (8.9) can be represented as

$$\dot{\mathbf{x}} = \left(\mathbf{A} - \mathbf{B} \mathbf{R}^{-1} \mathbf{B}^T \mathbf{P} \right) \mathbf{x} \qquad (8.18)$$

and also the optimal controller satisfies the fact

$$\frac{dV}{dt} \bigg|^{\mathbf{u}=\mathbf{u}^*} + \mathbf{x}^T \mathbf{R} \mathbf{x} + \mathbf{u}^{*T} \mathbf{Q} \mathbf{u}^* = 0$$

Optimal Control Problem of Continuous Time Nonlinear System

By expanding $\frac{dV}{dt}\big|^{\mathbf{u}=\mathbf{u}^*}$ and $\mathbf{u}^*$, the above equation can be further written as

$$\mathbf{x}\left(\mathbf{A}^T\mathbf{P} + \mathbf{P}\mathbf{A} + \mathbf{Q} - \mathbf{P}\mathbf{B}\mathbf{R}^{-1}\mathbf{B}^T\mathbf{P}\right)\mathbf{x} = 0 \tag{8.19}$$

Since, (8.19) holds for any $\mathbf{x}$, the following has to be true

$$\mathbf{A}^T\mathbf{P} + \mathbf{P}\mathbf{A} + \mathbf{Q} - \mathbf{P}\mathbf{B}\mathbf{R}^{-1}\mathbf{B}^T\mathbf{P} = 0 \tag{8.20}$$

The equation (8.20) is known as *Algebraic Riccati Equation* (ARE). Therefore, the optimal control law $\mathbf{u}^*$ is found by solving the ARE.

8.2.2 Hamilton-Jacobi-Bellman Equation

In this subsection we shall discuss the analytic solution to the optimal control problem for a general nonlinear dynamical system. The solution can be achieved by solving the *Hamilton-Jacobi-Bellman* equation.

Consider a nonlinear dynamical system

$$\dot{\mathbf{x}}(t) = \mathbf{f}(t, \mathbf{x}(t), \mathbf{u}(t)) \quad with \quad \mathbf{x}(t_0) = \mathbf{x}_0 \tag{8.21}$$

and the performance index to be minimized

$$J(t_0, \mathbf{x}(t_0), \mathbf{u}(t)) = \phi(t_f, \mathbf{x}_f) + \int_{t_0}^{t_f} L(\tau, \mathbf{x}(\tau), \mathbf{u}(\tau)) \, d\tau \tag{8.22}$$

Defining the performance index for any time t

$$J(t, \mathbf{x}(t), \mathbf{u}(\tau)) = \phi(t_f, \mathbf{x}_f) + \int_{t}^{t_f} L(\tau, \mathbf{x}(\tau), \mathbf{u}(\tau)) \, d\tau \tag{8.23}$$

where $t \leq \tau \leq t_f$ and the optimal cost (performance index)

$$J^*(t, \mathbf{x}(t)) = \min_{\mathbf{u}} \, J(t, \mathbf{x}(t), \mathbf{u}(\tau)) \tag{8.24}$$

Therefore, the optimal cost can be rewritten as

$$J(t, \mathbf{x}(t), \mathbf{u}(\tau)) = \phi(t_f, \mathbf{x}_f) + \int_{t}^{t_f} L(\tau, \mathbf{x}(\tau), \mathbf{u}(\tau)) \, d\tau \tag{8.25}$$

where $t \leq \tau \leq t_f$ and the optimal cost (performance index)

$$J^*(t, \mathbf{x}(t)) = \min_{\mathbf{u}} \left[\int_{t}^{t+\Delta t} L(\tau, \mathbf{x}(\tau), \mathbf{u}(\tau)) \, d\tau \right. \\ \left. + \int_{t+\Delta t}^{t_f} L(\tau, \mathbf{x}(\tau), \mathbf{u}(\tau)) \, d\tau + \phi(t_f, \mathbf{x}_f) \right] \tag{8.26}$$

Using the Principle of optimality,

$$J^*(t, \mathbf{x}(t)) = \min_{\mathbf{u}} \left[\int_t^{t+\Delta t} L(\tau, \mathbf{x}(\tau), \mathbf{u}(\tau)) \, d\tau + J^*(t + \Delta t, \mathbf{x}(t + \Delta t)) \right] \tag{8.27}$$

By expanding $J^*(t + \Delta t, \mathbf{x}(t + \Delta t))$ about the point $(t, \mathbf{x}(t))$ and considering the fact that the optimal cost J^* is independent of $\mathbf{u}$

$$0 = \min_{\mathbf{u}} \left[\int_t^{t+\Delta t} L(\tau, \mathbf{x}(\tau), \mathbf{u}(\tau)) \, d\tau + \frac{\partial J^*}{\partial t} \Delta t + \frac{\partial J^*}{\partial \mathbf{x}} \dot{\mathbf{x}} \Delta t + H.O.T \right] \tag{8.28}$$

where $\dot{\mathbf{x}}$ is approximately $(\mathbf{x}(t + \Delta t) - \mathbf{x}(t))/\Delta t$ and H.O.T refers to higher order terms. As $\Delta t \to 0$,

$$0 = \frac{\partial J^*}{\partial t} + \min_{\mathbf{u}} \left[L(\mathbf{x}, \mathbf{u}) + \frac{\partial J^*}{\partial \mathbf{x}} \mathbf{f}(\mathbf{x}, \mathbf{u}) \right] \tag{8.29}$$

subject to the boundary condition

$$J^*(t_f, \mathbf{x}(t_f)) = \phi(t_f, \mathbf{x}_f) \tag{8.30}$$

In the above equation, $H = L + \frac{\partial J^*}{\partial \mathbf{x}} \mathbf{f}$ is called the Hamiltonian function. To solve the optimal control problem, we need to solve the above equation for a given system. In the optimal control literature, (8.29) is famous as *Hamilton-Jacobi-Bellman* (HJB) equation. When the cost function is evaluated in infinite time ($t_f = \infty$) for a time invariant system dynamics, then the terminal cost looses its significance and the performance index becomes

$$J(t_0, \mathbf{x}(t_0), \mathbf{u}(t)) = \int_{t_0}^{\infty} L(\tau, \mathbf{x}(\tau), \mathbf{u}(\tau)) \, d\tau \tag{8.31}$$

The HJB equation for the infinite time cost function becomes

$$\min_{\mathbf{u}} \left(L(\mathbf{x}, \mathbf{u}) + \frac{\partial J^*}{\partial \mathbf{x}} \mathbf{f}(\mathbf{x}, \mathbf{u}) \right) = 0 \tag{8.32}$$

8.2.3 Optimal Control Law for Input Affine System

Solution of (8.29) defines the optimal control policy that incurs the optimal cost. The problem can be simplified if we are able to represent the system dynamics in input affine form i.e. $\dot{\mathbf{x}} = \mathbf{f}(\mathbf{x}) + \mathbf{g}(\mathbf{x})\mathbf{u}$ and the optimal control problem is to minimize the performance index

$$J(t_0, \mathbf{x}, \mathbf{u}) = \int_{t_0}^{\infty} L(\tau, \mathbf{x}, \mathbf{u}) \, d\tau \tag{8.33}$$

where $L = \mathbf{x}^T \mathbf{Q} \mathbf{x} + \mathbf{u}^T \mathbf{R} \mathbf{u}$, subject to the constraint

$$\dot{\mathbf{x}}(t) = \mathbf{f}(\mathbf{x}(t)) + \mathbf{g}(\mathbf{x}(t))\mathbf{u}(t) \tag{8.34}$$

Optimal Control Problem of Continuous Time Nonlinear System   289

Then the optimal control policy $\mathbf{u}^*$ can be written as

$$\mathbf{u}^* = \arg\min_{\mathbf{u}} \left[\frac{1}{2} \left(\mathbf{x}^T \mathbf{Q} \mathbf{x} + \mathbf{u}^T \mathbf{R} \mathbf{u} \right) + \frac{\partial J^*}{\partial \mathbf{x}} \left(\mathbf{f}(\mathbf{x}) + \mathbf{g}(\mathbf{x})\mathbf{u} \right) \right] \quad (8.35)$$

Differentiating the RHS of (8.35) w.r.t $\mathbf{u}$ and setting it to zero, we can write

$$\mathbf{u}^* \mathbf{R} + \frac{\partial J^*}{\partial \mathbf{x}} \mathbf{g}(\mathbf{x}) = 0$$

$$\mathbf{u}^* = -\mathbf{R}^{-1} \mathbf{g}^T(\mathbf{x}) \left(\frac{\partial J^*}{\partial \mathbf{x}} \right)^T \quad (8.36)$$

The optimal cost J^* can be found by replacing the $\mathbf{u}^*$ in the HJB equation. We can see that the solution to the optimal control problem depends upon the optimal cost of the system as the optimal control law comes from HJB equation of the system [229]. Analytical computation of the optimal cost J^* (or the costate vector $\frac{\partial J^*}{\partial \mathbf{x}}$) requires valid mathematical models of the system dynamics $\mathbf{f}(\mathbf{x})$ and the input matrix $\mathbf{g}(\mathbf{x})$. Even for known nonlinear systems, the solution to the optimal control problem is non-trivial. The solution is obtained offline and is computed backward in time for finite time optimal control problems. If one wants to implement optimal policy in real-time, then the approximate solutions to the HJB equation using reinforcement learning have been proven to be efficient. Approximate dynamic programming (ADP) [191] on the other hand provides solution forward in time. ADP employs adaptive critic (AC) based methodologies [230] where the critic is trained to estimate the optimal cost J^*. Next, we shall discuss about the adaptive critic techniques.

8.2.4 Adaptive Critic Concept

The optimal control policy u^* which is obtained by solving the HJB equation, depends on the future cost. It also requires a known plant model in continuous or discrete time. Traditional solutions considered entire episodes of state evolution and the optimal control law was computed going backward in time. However, the major issue in implementing the optimal control is that the analytic expression of plant is not available in many cases. Systems with input affine form alleviate this problem as we can obtain an analytic solution for the optimal control law but one needs to know the future cost at any given instant. Adaptive critic [217] has been proposed to learn the optimal control forward in time. This technique utilizes an approximation of the future cost / value function for computing the optimal control law. Various architectures in adaptive critic methodology have been proposed by Werbos in early 1990s. An AC architecture generally employs two networks [231]: one to approximate the optimal cost (called *critic*) and the other to approximate the optimal policy (called *action*). The performance of the control input generated by the action network is evaluated using the critic network. The critic network approximates either the value function $V(\mathbf{x}(t))$ or the costate vector,

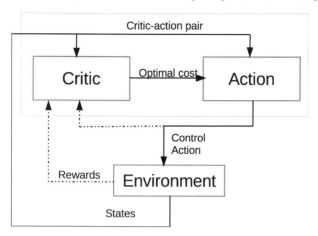

FIGURE 8.1: Adaptive dynamic programming (ADP) architecture using critic-action network. The action network provides optimal control action based on the current estimate of the optimal cost by the critic network, and the current state. The critic network corrects itself based on the reward received from the environment while evaluating the performance of the action network. The action network is also simultaneously trained based on the performance evaluation made by the critic.

$\frac{\partial J}{\partial \mathbf{x}}$ as a function of states of the system. The two networks are simultaneously trained until convergence to achieve the near optimal solution [232]. Figure 8.1 presents a simple ADP architecture using critic and action networks. AC methodologies include Heuristic Dynamic Programming (HDP), Dual Heuristic Programming (DHP), and Global Dual Heuristic Programming (GDHP). Each of the techniques has their action dependent and independent versions.

Critic architectures can be categorized based on the input to the system, output of the system and requirement of number of networks. In action independent architectures, only the states of the system is provided as input. Whereas action dependent architectures are provided with control signal as input along with the states of the system. An HDP architecture, the output of the critic network is the approximated value function and the gradient of the value function is available as output of a DHP architecture. The GDHP architecture provides both the value function and its gradient as output. When the system model is in input affine form, only one network is sufficient to implement the critic architecture. This kind of architecture is called single network adaptive critic (SNAC) [233] which requires the critic network to approximate the value function. The optimal control law can be obtained by solving the HJB equation. The SNAC architecture discards the use of action network as it is sufficient to know the optimal cost when the system is in input affine form. Figure 8.2 presents a basic SNAC architecture to learn a near optimal controller for a robot.

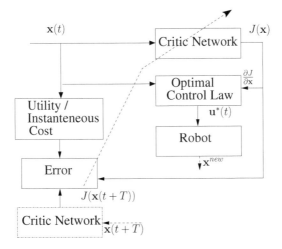

FIGURE 8.2: SNAC architecture for robot control employs only the critic network. The optimal control law is computed from the HJB equation using the costate derived from the optimal cost $J(\mathbf{X}_t)$ at $\mathbf{X}_t$, predicted by the critic. The critic network is updated to minimize the difference between target cost and the predicted cost.

8.3 Policy Iteration and SNAC for Unknown Continuous Time Nonlinear Systems

Policy iteration (PI) scheme is very powerful for implementing real-time learning in designing optimal controller. We next show how to exploit PI to learn a SNAC based optimal controller for an unknown nonlinear continuous time system.

A single network adaptive critic can be employed to control a continuous time nonlinear input affine system similar to the discrete-time case, since the analytic expression for the optimal control is known. The critic network should approximate either the value function J or the costate vector $\frac{\partial J}{\partial \mathbf{x}}$. to improve the control policy. In this approach the value function J is approximated with a critic network as $V(\mathbf{x})$, and the costate vector is computed as $\frac{\partial V}{\partial \mathbf{x}}$ from the critic network, to compute the control input in (8.36). It can be noted that SNAC needs the information of the system input matrix to compute the optimlal control signal. Later in this section we show how this information can be extracted from an unknown system.

8.3.1 Policy Iteration Scheme

Consider an initial stabilizing controller $\boldsymbol{u}_0$, and $V_0(\boldsymbol{x}_t)$ is the positive definite function identified with the critic network which satisfies the HJB equation (8.29), then $V_0(x)$ is the value function associated with control policy u_0 [222]. Let us define a new policy

$$\mathbf{u}_1 = \min_{\mathbf{u}} \quad (L(\mathbf{x}_t, \mathbf{u}_0) + V_0(\mathbf{x}_{t+1})) \quad (8.37)$$

The justification is presented in [234], where it is shown that the new policy $\mathbf{u}_1$ is *improved* as it results $V_1(\mathbf{x}_t) \leq V_0(\mathbf{x}_t)$. The above scheme gives an improved policy. This suggests an iterative scheme known as *policy iteration* which discovers the optimal controller. The scheme is given as follows:

Initialization step: Select an admissible / stabilizing control policy $\mathbf{u}_i$.

Evaluation step: Evaluate the current value function associated to the policy $\mathbf{u}_i$ as follows:

$$V_i(\mathbf{x}(t_0)) = \int_{t_0}^{t_0+T} (\mathbf{x}^T \mathbf{Q} \mathbf{x} + \mathbf{u} - i^T \mathbf{R} \mathbf{u}_i) \, dt + V_i(\mathbf{x}(t_0 + T)), \quad (8.38)$$

where T is the sampling time to compute the cost function. It should be noted that T is different from the system sampling time ΔT, and need not to be fixed. The value of T is chosen such that the reinforcement signal $\int_{t_0}^{t_0+T}(\mathbf{x}^T \mathbf{Q} \mathbf{x} + \mathbf{u}^T \mathbf{R} \mathbf{u}) \, dt$ carries enough information about the system dynamics. The above equation is analogous to the fixed point equation used in the reinforcement learning for the discrete-time system [234].

Policy improvement step: Determine the improved policy based on the HJB equation as follows:

$$\mathbf{u}_{i+1}(\mathbf{x}) = -\frac{1}{2} \mathbf{R}^{-1} \mathbf{g}^T(\mathbf{x}) \frac{\partial V_i}{\partial \mathbf{x}}, \quad (8.39)$$

$i = 0, 1, 2, \ldots, \infty$. This scheme is independent of $\mathbf{f}(\mathbf{x})$ in contrast to HJB equation. The aforementioned policy iteration scheme for nonlinear input affine system does not require the knowledge of $\mathbf{f}(\mathbf{x})$ during the training phase but the input matrix $\mathbf{g}(\mathbf{x})$ is needed to compute the input.

8.3.2 Optimal Control Problem of an Unknown Dynamic

Let us consider the following nonlinear system

$$\dot{\mathbf{x}} = \mathbf{f}(\mathbf{x}(t), \mathbf{u}(t)) \quad (8.40)$$

where, $\mathbf{x} \in \Re^n$ is measurable state vector of the system and $\mathbf{u} \in \Re^m$ is the control input to the system. Assume that $\mathbf{f}(\mathbf{x}(t), \mathbf{u}(t)) = 0$ when $\mathbf{x}(t) = 0$ and $\mathbf{u}(t) = 0$ and $\mathbf{f}(\mathbf{x}(t), \mathbf{u}(t))$ is Lipschitz continuous on a set $\hat{\mathbb{S}} \in \Re^n$, containing the origin.

Assumption 8.1. *System* (8.40) *has equilibrium state* $\mathbf{x_e}(\mathbf{t})$ *on a set* $\hat{\mathbb{S}} \in \Re^n$, *containing the origin, under the influence of control action* $\mathbf{u}(t) = 0$ *and is controllable in a sense that there exist a continuous control law on set* $\hat{\mathbb{S}}$ *that can stabilize the system asymptotically.*

Policy Iteration and SNAC

We formulate the optimal control problem as the following. Given a continuous time nonlinear system (8.40), design a control architecture $\mathbf{u} = \bar{\mathbf{g}}(\mathbf{x}(t))$ that stabilizes the system while minimizing the infinite-horizon cost (8.41) without the knowledge of system dynamics.

$$\begin{aligned} J(\mathbf{x}(t_0), \mathbf{u}(t_0)) &= \int_{t_0}^{\infty} (\mathbf{x}^T \mathbf{Q} \mathbf{x} + \mathbf{u}^T \mathbf{R} \mathbf{u}) \, \mathrm{d}t \\ &= \int_{t_0}^{\infty} \varphi(\mathbf{x}, \mathbf{u}) \, \mathrm{d}t, \end{aligned} \quad (8.41)$$

where, $\mathbf{Q} \in \Re^{n \times n}$, $\mathbf{Q} \geq 0$ and $\mathbf{R} \in \Re^{m \times m}$, $\mathbf{R} > 0$. Here, $\varphi(\mathbf{x}, \mathbf{u})$ is called the utility function that indicates a measure of the performance of the controller.

Assumption 8.2. *The performance index (8.41) satisfies zero-state observability [235].*

As we have seen earlier the Hamiltonian of system (8.40) related to infinite horizon cost (8.41) is given by

$$H(\mathbf{x}, J, \mathbf{u}) = \varphi(\mathbf{x}, \mathbf{u}) + \frac{\partial J}{\partial \mathbf{x}}^T \dot{\mathbf{x}} \quad (8.42)$$

An control policy that minimizes cost (8.41) can be found using the stationary condition of optimization yet the control action is a function of input matrix of the system. The optimal control law $\mathbf{u}^*$ satisfies the Hamilton-Jacobi-Bellman (HJB) equation

$$\frac{\partial J^*}{\partial t} + \min_{u} H(\mathbf{x}, J^*, \mathbf{u}) = 0. \quad (8.43)$$

Solution to the HJB equation depends on optimal cost J^* of the system. Single network adaptive critic can be used to estimate the optimal cost and is given by

$$J^*(\mathbf{x}(t)) \Leftarrow \Xi(\mathbf{x}(t)), \quad (8.44)$$

where Ξ is a fuzzy function that maps the states of the system to the optimal cost.

Assumption 8.3. *Unknown dynamical system (8.40) can be estimated using Takagi-Sugeno-Kang (TSK) fuzzy model with arbitrary precision.*

In the policy iteration framework the control policy in each iteration is dependent on system input matrix. TSK fuzzy representation of the system gives the flexibility to represent the system in input affine form. The TSK fuzzy representation of system (8.40) can be given by

$$\dot{\mathbf{x}} = \mathcal{A}(\mathbf{x})\mathbf{x} + \mathcal{B}(\mathbf{x})\mathbf{u}, \quad (8.45)$$

where $\mathcal{A}$ and $\mathcal{B}$ are fuzzy approximation of system and input matrix respectively.

Policy iteration can result a fuzzy network to approximate the optimal cost in subsequent updates, given the network is initialized by stable weights which in turn make first policy admissible. One way of searching initial weights could be learning an initial value function by a TSK fuzzy network if an initial controller exists but the method fails for highly nonlinear systems. Instead of learning an initial value function, the first policy can be made admissible if the weight matrices follow certain properties in relation to the system. These features of weight matrices also help to keep subsequent policies admissible during network updates.

It is experienced that approximating the initial value function for a highly nonlinear system is very difficult since, there exist multiple solution to the critic network parameters. During the search process, the parameters of the network change in a certain direction where it only minimize the training error and hence it do not care about the control policy to be admissible. The given approach selects those parameters/weights that lead to a stable control policy for system (8.45).

Since the dynamic model of the system is completely unknown, a TSK fuzzy approximation of the original dynamics as in (8.45) is considered as the design model to formulate the optimal control problem. The Hamiltonian of the above optimal control problem can be rewritten as

$$H(\mathbf{x}, \boldsymbol{\lambda}, \mathbf{u}) = \varphi(\mathbf{x}, \mathbf{u}) + \boldsymbol{\lambda}^T (\mathcal{A}\mathbf{x} + \mathcal{B}\mathbf{u}), \tag{8.46}$$

where $\boldsymbol{\lambda} = \frac{\partial J}{\partial \mathbf{x}}$ is the costate vector of the system. The control law $\mathbf{u}$ is optimal when $\mathbf{u} = \mathbf{u}^*$ where $\mathbf{u}^*$ needs to satisfy the necessary condition,

$$\frac{\partial H(\mathbf{x}, \boldsymbol{\lambda}^*, \mathbf{u})}{\partial \mathbf{u}} = \frac{\partial \varphi}{\partial \mathbf{u}} + \boldsymbol{\lambda}^{*T} \frac{\partial}{\partial \mathbf{u}} (\mathcal{A}\mathbf{x} + \mathcal{B}\mathbf{u}) = 0. \tag{8.47}$$

The solution of above expression gives the definition of the optimal control law and is stated as

$$\mathbf{u}^* = -\frac{1}{2} \mathbf{R}^{-1} \mathcal{B}^T \boldsymbol{\lambda}^*, \tag{8.48}$$

where costate vector $\boldsymbol{\lambda}^* = \frac{\partial J^*}{\partial \mathbf{x}}$ is along the optimal trajectory in the state space. It's well known that the optimal cost satisfies the HJB equation (8.43). Since $\boldsymbol{\lambda}^*$ being a function of the network (8.44), (8.48) can be rewritten as

$$\mathbf{u}^* = -\frac{1}{2} \mathbf{R}^{-1} \mathcal{B}^T \mho^*(\mathbf{P}^*, \mathbf{x}(t)), \tag{8.49}$$

where $\mho^*$ is the functional representation of $\boldsymbol{\lambda}^*$. $\mathbf{P}^* \in \bar{\mathcal{P}}$ represents the optimal network weights which has a significant role in keeping $\mathbf{u}^*$ admissible. Thus, the search process for $\mathbf{P}^*$ needs to discover those regions in $\bar{\mathcal{P}}$ where the solutions not only provide minimum network training error but also maintain $\mathbf{u}^*$'s admissibility.

8.3.3 Model Representation and Learning Scheme

A TSK fuzzy model represents a nonlinear complex function through a set of rules with linear consequent part which gives the insight of human reasoning in function approximation. TSK fuzzy modeling approach tries to decompose the input space into subspaces and then approximate the system in each subspace by a simple linear regression model [236]. This feature of TSK fuzzy model gives an efficient way to deal with nonlinear complexities.

8.3.3.1 TSK Fuzzy Representation of Nonlinear Dynamics

The nonlinear dynamics of the robotic manipulator (8.40) is represented in $n - dimensional$ state space model where states of the system are controlled by m dimensional input. The model is identified as fuzzy combinations of local linear models from input-output data. In general, fuzzy rules in the rule base that represents the model, contain all the possible combinations of fuzzy zones in the states space. This approach is highly computationally expensive for nonlinear MIMO systems. We incorporate a rule base that does not use all possible combinations of fuzzy zones in the states space. Instead, effective and efficient rules are searched that better represent the system. Thus, the following TSK fuzzy system of L rules represents the system (8.40).

lth rule:
IF $x_1(t)$ is Γ_1^l AND $\cdots$ AND $x_n(t)$ is Γ_n^l THEN

$$\dot{\mathbf{x}}^{(l)} = \mathbf{A}^{(l)}\mathbf{x} + \mathbf{B}^{(l)}\mathbf{u} \tag{8.50}$$

where $\mathbf{A}^{(l)} \in \Re^{n \times n}$, $\mathbf{B}^{(l)} \in \Re^{n \times m}$ and overall model is given by

$$\dot{\mathbf{x}} = \frac{\sum_{l=1}^{L} \omega^l \mathbf{A}^l}{\sum_{l=1}^{L} \omega^l} \mathbf{x} + \frac{\sum_{l=1}^{L} \omega^l \mathbf{B}^l}{\sum_{l=1}^{L} \omega^l} \mathbf{u} \tag{8.51}$$

$$= \mathcal{A}\mathbf{x} + \mathcal{B}\mathbf{u} \tag{8.52}$$

where $\Gamma_j^l = exp(-\frac{0.5(x_j - c_j^l)^2}{s_j^{l\,2}})$, c_j^l and $s_j^{l\,2}$ are the mean and variance of lth fuzzy zone of state x_j, $j = 1, 2, .., n$, is the jth fuzzy set of the lth rule. The rule membership degree ω^l, associated with the lth rule is defined as

$$\omega^l = \prod_{j=1}^{n} \mu_j^l(x_j), \tag{8.53}$$

where $\mu_j^l(x_j)$ is the membership function related to the fuzzy set Γ_j^l, $l = 1, 2, \cdots, L$.

8.3.3.2 Learning Scheme for the TSK Fuzzy Model

We invoke a TSK fuzzy logic based parametric system identification technique [237] that efficiently approximate the dynamic model and simultaneously keeps the rule base small. The technique uses genetic algorithm to find suitable fuzzy zones and least square method to find consequent parameter for a good representation of the unknown dynamical model. The rule base of the fuzzy network has L rules where $L = F_z$ and F_z is the number of fuzzy zones by which each state of the system is divided in the state space.

The parameters of the fuzzy model are learned in two steps from the input-output data set. First, GA finds out suitable mean and variance of the Gaussian membership functions which are used as fuzzy sets. And in the next step, least square technique is used to find consequent parameters given those means and variances and the data set. Thus, (8.52) can be written as

$$\dot{\mathbf{x}}^T = \begin{bmatrix} \sigma_1 \mathbf{x}^T & \sigma_2 \mathbf{x}^T & \cdots & \sigma_L \mathbf{x}^T \end{bmatrix} \begin{bmatrix} \mathbf{A}^{(1)T} \\ \mathbf{B}^{(1)T} \\ \mathbf{A}^{(2)T} \\ \mathbf{B}^{(2)T} \\ \vdots \\ \mathbf{A}^{(L)T} \\ \mathbf{B}^{(L)T} \end{bmatrix} \qquad (8.54)$$

where, $\sigma_l = \frac{\omega^l}{\Sigma_{l=1}^L \omega^l}$ is the normalized rule membership. Equation (8.54) is solved for $\mathbb{N}$ number of data points in the data set. The search continues until the TSK model satisfactorily represents the dynamic model.

8.3.4 Critic Design and Policy Update

Obtaining an analytical solution of the optimal control problem of such a system (8.40) is a difficult task since the solution depends upon J^*. If the complete knowledge of the system is available, one can arrive at the solution by going backward in time. Another approach is adaptive critic that obtains an approximate solution forward in time but still the system information is needed while the critic network is being trained. Instead of searching for an analytical solution, we opt for an near optimal solution by using a single critic network. The critic learns the optimal cost $J^*(\mathbf{x}, \mathbf{u})$ while maintaining the stability of the closed loop system. The critic is represented by a Takagi-Sugeno-Kang fuzzy model with a small rule base.

8.3.4.1 Construction of Initial Critic Network using Lyapunov Based LMI

Initial critic network is constructed in such a way that it results in a stable control law (8.49). The same fuzzy zones, those are found during the dynamic model identification, are chosen for constructing the rule base for the critic. As we learn the optimal controller by updating the critic weights, there has to be a set of starting weights, assigned to the network, for which the system (8.52) is stable in closed loop.

Policy Iteration and SNAC

Let us consider a fuzzy state feedback controller [238] that can stabilize the system (8.52)

$$\mathbf{u}_s = \mathbf{K}\mathbf{x} \qquad (8.55)$$

Control law (8.55) gives the close loop system dynamics as following

$$\dot{\mathbf{x}} = \sum_{l=1}^{L} \sigma_l \left(\mathbf{A}^{(l)} + \mathbf{B}^{(l)}\mathbf{K} \right) \mathbf{x} \qquad (8.56)$$

Here σ_l is the membership grade of *lth* rule that follows $\sum_{l=1}^{L} \sigma_l = 1$ and is defined by

$$\sigma_l = \frac{\omega^l}{\left(\sum_{l=1}^{L} \omega^l \right)} \qquad (8.57)$$

It should be noted that in this formulation, the total number of rules in the fuzzy rule base are same as the number of fuzzy zones for a state in the state space. This approach makes the rule base compact.

8.3.4.2 Lyapunov Function

From the theory of physics we know

$$Work = [Force] \cdot [Distance]$$

Here, $(\cdot)$ indicates the dot product between two quantities. In the literature [229] of stability of nonlinear systems, the Lyapunov function is described as energy like function. In the same way, we select an energy like fuzzy function (8.58) [239]

$$V(\mathbf{x(t)}) = 2 \int_0^{\mathbf{x}_0} \mathbf{\Omega(s)} \cdot \mathbf{ds} \qquad (8.58)$$

which is essentially a line integral function over the trajectory of system state evolution where $\mathbf{x}_0$ is a generic state. Here, $\mathbf{s} \in \Re^n$ and $\mathbf{\Omega(s)} = [\Omega_1(s) \; \Omega_2(s) \; ... \; \Omega_n(s)]^T$. However, $V(\mathbf{x(t)})$ to be considered as a Lyapunov function, it should be independent of state trajectory [239], i.e.,

$$\frac{\partial \Omega_i(\mathbf{x})}{\partial \mathbf{x_j}} = \frac{\partial \Omega_j(\mathbf{x})}{\partial \mathbf{x_i}} \qquad (8.59)$$

for $i, j=1,2,...,n$; Let us choose a fuzzy function $\mathbf{\Omega(s)}$ such that
  lth rule:
  IF $x_1(t)$ is Γ_1^l AND $\cdots$ AND $x_n(t)$ is Γ_n^l THEN

$$\mathbf{\Omega}^{(l)}(\mathbf{x}) = \mathbf{P}^{(l)}\mathbf{x} \qquad (8.60)$$

For $l=1,2,...,L$. Here, $\mathbf{P}^{(l)} \in \Re^{n \times n}$ is symmetric positive definite matrix and fuzzy aggregation of (8.60) is given as

$$\Omega(\mathbf{x}) = \mathbb{P}(\mathbf{x})\mathbf{x} = \frac{\Sigma_{l=1}^{L} w^{(l)} \mathbf{P}^{(l)}}{\Sigma_{l=1}^{L} w^{(l)}} \cdot \mathbf{x} \quad (8.61)$$

To define the structure of the Lyapunov function, we invoke the following theorem [239].

Theorem 8.1. $V(\mathbf{x})$ *is a Lyapunov function candidate if there exists a* $\mathbf{P}^{(l)} = \hat{\mathbf{P}} + \boldsymbol{\delta}^{(l)} > 0$ *such that*

$$\hat{\mathbf{P}} = \begin{bmatrix} 0 & p_{12} & \cdots & p_{1n} \\ p_{12} & 0 & \cdots & p_{2n} \\ \vdots & \vdots & \ddots & \vdots \\ p_{1n} & p_{2n} & \cdots & 0 \end{bmatrix} \quad (8.62)$$

and

$$\boldsymbol{\delta}^{(l)} = \begin{bmatrix} d_{11}^{(l)} & 0 & \cdots & 0 \\ 0 & d_{22}^{(l)} & \cdots & 0 \\ \vdots & \vdots & \ddots & \vdots \\ 0 & 0 & \cdots & d_{nn}^{(l)} \end{bmatrix} \quad (8.63)$$

The off diagonal terms of $\mathbf{P}^{(l)}$ *matrices will remain same for all the rules. The diagonal elements construct different* $\mathbf{P}$ *for each rule. It can be shown that (8.58) is a generic form of conventional Lyapunov candidate function* $V(\mathbf{x}) = \mathbf{x}^T \mathbf{P} \mathbf{x}$.

8.3.4.3 Conditions for Stabilization

We search for the $\mathbf{P}^{(l)}$ matrices [240] such that (8.58) is a Lyapunov function of the system (8.56).

Lemma 8.1. *The system described in (8.45) is asymptotically stable with the controller (8.55) if there exists* $\hat{\mathbf{P}}$, $\boldsymbol{\delta}^{(l)}$, *and* $\mathbf{K}$ *such that*

$$\mathbf{P}^{(l)} = \hat{\mathbf{P}} + \boldsymbol{\delta}^{(l)} > 0 \quad (8.64)$$

$$\mathbf{P}^{(l)} \mathbf{A}^{(l)} + \mathbf{P}^{(l)} \mathbf{B}^{(l)} \mathbf{K} + \mathbf{A}^{(l)^T} \mathbf{P}^{(l)} + \mathbf{K}^T \mathbf{B}^{(l)^T} \mathbf{P}^{(l)} < 0 \quad (8.65)$$

$$\mathbf{P}^{(i)} \mathbf{A}^{(j)} + \mathbf{P}^{(i)} \mathbf{B}^{(j)} \mathbf{K} + \mathbf{A}^{(j)^T} \mathbf{P}^{(i)} + \mathbf{K}^T \mathbf{B}^{(j)^T} \mathbf{P}^{(i)} +$$
$$\mathbf{P}^{(j)} \mathbf{A}^{(i)} + \mathbf{P}^{(j)} \mathbf{B}^{(i)} \mathbf{K} + \mathbf{A}^{(i)^T} \mathbf{P}^{(j)} + \mathbf{K}^T \mathbf{B}^{(i)^T} \mathbf{P}^{(j)} < 0 \quad (8.66)$$

where, $l = 1, 2, ..., L$ $i = 1, 2, ..., L-1$ and $j = i+1, ..., L$

Proof. Time derivative of Lyapunov function of the close loop system (8.56) can be given by

$$\dot{V}(\mathbf{x}(t)) = \mathbf{x}^T \mathbb{P}\dot{\mathbf{x}} + \dot{\mathbf{x}}^T \mathbb{P}\mathbf{x}$$

$$= \mathbf{x}^T \left(\sum_{i=1}^{L} \sigma_i \mathbf{P}^{(i)} \left(\sum_{j=1}^{L} \sigma_j \left(\mathbf{A}^{(j)} + \mathbf{B}^{(j)} \mathbf{K} \right) \right) \right.$$

$$\left. + \left(\sum_{i=1}^{L} \sigma_i \left(\mathbf{A}^{(i)T} + \mathbf{K}^T \mathbf{B}^{(i)T} \right) \right) \sum_{j=1}^{L} \sigma_j \mathbf{P}^{(j)} \right) \mathbf{x}$$

$$= \mathbf{x}^T \left(\sum_{i=1}^{L} \sum_{j=1}^{L} \sigma_i \sigma_j \left(\left(\mathbf{P}^{(i)} \mathbf{A}^{(j)} + \mathbf{P}^{(i)} \mathbf{B}^{(j)} \mathbf{K} \right) \right. \right.$$

$$\left. \left. + \left(\mathbf{A}^{(j)T} \mathbf{P}^{(i)} + \mathbf{K}^T \mathbf{B}^{(j)T} \mathbf{P}^{(i)} \right) \right) \right) \mathbf{x} \quad (8.67)$$

during the system identification, the algorithm ensures that $\sigma_i \sigma_j \geq 0 \ \forall \ \mathbf{x}$ and at least one rule is fired for any $\mathbf{x} \in \mathbb{D}$. Therefore, $\dot{V}(\mathbf{x})$ is negative if $\left(\mathbf{P}^{(i)} \mathbf{A}^{(j)} + \mathbf{P}^{(i)} \mathbf{B}^{(j)} \mathbf{K} \right) + \left(\mathbf{A}^{(j)T} \mathbf{P}^{(i)} + \mathbf{K}^T \mathbf{B}^{(j)T} \mathbf{P}^{(i)} \right) < 0$ for $\mathbf{x} \neq 0$ where all $i \in L$, $j \in L$. Equation (8.67) is rearranged in equation (8.64) – (8.66) □

Lemma 8.2. *Consider $V_0(\mathbf{x})$ as the Lyapunov function computed as in (8.58) of system (8.45) with stabilizing control input $\mathbf{u}_0$. Then, $\mathbf{u}_1$ in (8.68) also stabilizes the system (8.45)*

$$\mathbf{u}_1 = -\frac{1}{2} \mathbf{R}^{-1} \mathcal{B}^T \frac{\partial V_0}{\partial x} \quad (8.68)$$

if the following conditions are satisfied.

$$\mathbf{P}^{(l)} = \hat{\mathbf{P}} + \delta^{(l)} > 0, \quad \mathbf{R} > 0 \quad (8.69)$$

$$\mathbf{P}^{(l)} \mathbf{A}^{(l)} - \mathbf{P}^{(l)} \left(\mathbf{B}^{(l)} \mathbf{R}^{-1} \mathbf{B}^{(v)T} \mathbf{P}^{(k)} \right) + \mathbf{A}^{(l)T} \mathbf{P}^{(l)} -$$

$$\left(\mathbf{B}^{(l)} \mathbf{R}^{-1} \mathbf{B}^{(v)T} \mathbf{P}^{(k)} \right)^T \mathbf{P}^{(l)} < 0 \quad (8.70)$$

$$\mathbf{P}^{(i)} \mathbf{A}^{(j)} - \mathbf{P}^{(i)} \left(\mathbf{B}^{(j)} \mathbf{R}^{-1} \mathbf{B}^{(v)T} \mathbf{P}^{(k)} \right) + \mathbf{A}^{(j)T} \mathbf{P}^{(i)} -$$

$$\left(\mathbf{B}^{(j)} \mathbf{R}^{-1} \mathbf{B}^{(v)T} \mathbf{P}^{(k)} \right)^T \mathbf{P}^{(i)} < 0 \quad (8.71)$$

where $l, i, j, k, v = 1, 2, ..., L$. Here i, j, k and v iterates through all possible combinations of rules that can be fired simultaneously and $i \neq j \neq k \neq v$.

Proof. Let us replace the fixed gain matrix $\mathbf{K}$ by $-\mathbf{R}^{-1}\sum_{v=1}^{L}\sum_{k=1}^{L}\sigma_v\sigma_k\mathbf{B}^{(v)T}\mathbf{P}^{(k)}$ in *Lemma* 8.1. The close loop system (8.56) can be written as

$$\dot{\mathbf{x}} = \sum_{l=1}^{L}\sigma_l\left(\mathbf{A}^{(l)} - \mathbf{B}^{(l)}\sum_{v=1}^{L}\sum_{k=1}^{L}\sigma_v\sigma_k\mathbf{B}^{(v)T}\mathbf{P}^{(k)}\right)\mathbf{x} \qquad (8.72)$$

Derivative of the Lyapunov function of the close loop system (8.72) can be written by

$$\dot{V}(\mathbf{x}(t)) = \mathbf{x}^T\left(\sum_{i=1}^{L}\sigma_i\mathbf{P}^{(i)}\left(\sum_{j=1}^{L}\sigma_j\left(\mathbf{A}^{(j)} - \mathbf{B}^{(j)}\sum_{v=1}^{L}\sum_{k=1}^{L}\sigma_v\sigma_k\right.\right.\right.$$

$$\left.\left.\mathbf{R}^{-1}\mathbf{B}^{(v)T}\mathbf{P}^{(k)}\right)\right) + \left(\sum_{i=1}^{L}\sigma_i\left(\mathbf{A}^{(i)T} - \left(\mathbf{B}^{(i)}\sum_{v=1}^{L}\sum_{k=1}^{L}\sigma_v\sigma_k\right.\right.\right.$$

$$\left.\left.\left.\mathbf{R}^{-1}\mathbf{B}^{(v)T}\mathbf{P}^{(k)}\right)^T\right)\right)\sum_{j=1}^{L}\sigma_j\mathbf{P}^{(j)}\right)\mathbf{x}$$

$$= \mathbf{x}^T\left(\sum_{i=1}^{L}\sum_{j=1}^{L}\sum_{v=1}^{L}\sum_{k=1}^{L}\sigma_i\sigma_j\sigma_v\sigma_k\left(\mathbf{P}^{(i)}\mathbf{A}^{(j)} + \mathbf{A}^{(j)T}\mathbf{P}^{(i)} - \right.\right.$$

$$\left.\left.\mathbf{P}^{(i)}\left(\mathbf{B}^{(j)}\mathbf{R}^{-1}\mathbf{B}^{(v)T}\mathbf{P}^{(k)}\right) - \left(\mathbf{B}^{(j)}\mathbf{R}^{-1}\mathbf{B}^{(v)T}\mathbf{P}^{(k)}\right)^T\mathbf{P}^{(i)}\right)\right)\mathbf{x} \qquad (8.73)$$

Since $\sigma_i\sigma_j\sigma_v\sigma_k \geq 0\ \forall\ \mathbf{x}$ and at least one rule is fired for any $\mathbf{x} \in \mathbb{D}$, $\dot{V}(\mathbf{x})$ is negative provided $\left(\mathbf{P}^{(i)}\mathbf{A}^{(j)} + \mathbf{A}^{(j)T}\mathbf{P}^{(i)} - \mathbf{P}^{(i)}\left(\mathbf{B}^{(j)}\mathbf{R}^{-1}\mathbf{B}^{(v)T}\mathbf{P}^{(k)}\right) - \left(\mathbf{B}^{(j)}\mathbf{R}^{-1}\mathbf{B}^{(v)T}\mathbf{P}^{(k)}\right)^T\mathbf{P}^{(i)}\right) < 0$ which leads to the constraints (8.69) – (8.71). Thus (8.68) is a stable control law if the constraints are satisfied. □

Remark 1. *If $\mathbf{u}_1(\mathbf{x},t) = u^*$ is a policy that gives minimum cost J^* in (8.44) and $V(\mathbf{x}(t))$ in (8.58) can be represented as the value function of the system (8.45), then $V^*(\mathbf{x})$ is also minimum and $V^*(\mathbf{x}) \leq V(\mathbf{x})$. When the critic is modeled as TSK fuzzy network, parametrized in terms of matrix $\mathbf{P}^{(l)}$, the value of $\mathbf{P}^{(l)}$ plays an important role in the stability of the closed loop system. In this work, we have found a set of initial $\mathbf{P}^{(l)}$ matrices by solving (8.69) – (8.71) for which the closed loop system will remain stable. Simultaneously, we have found the range of $\mathbf{P}^{(l)}$ matrices during critic update, for which the closed loop system will remain stable. The given structure and constraints on $\mathbf{P}^l$ matrices ensure stability of the closed loop system. Since $\mathbf{u}_1$ is an admissible policy, it can be used as the first policy in the policy iteration scheme. In essence, this work for the first time provides a comprehensive method to design a critic*

Policy Iteration and SNAC 301

as well as keeps its parameter updates in the stable zone. Please note, $\mathbf{u}_1$ is selected as the first policy in the PI scheme, since it makes the closed loop system Lyapunov stable. Thus the algorithm provides a way to solve for the initial stable control policy.

8.3.4.4 Design of Fitness Function

Fitness function assigns a fitness value to each genome and the reproductive ability of a genome is decided by the fitness value. Selection and preservation of good genomes depend on the fitness function design. Algorithm 4 shows the design of fitness function which is used in the Genetic Algorithm to select initial weights.

Algorithm 4 Fitness function for initial weight selection

$\mathbf{P}^{(l)} \leftarrow \hat{\mathbf{P}} + \mathbf{D}^{(l)}$, $\mathcal{P}_0 \leftarrow 0$
for each $l \in L$ **do**
　　$\Lambda_{min} \leftarrow min\ eigenvalue\ of\ \mathbf{P}^{(l)}$
　　if $\Lambda_{min} < 0$ **then**
　　　　$\mathcal{P}_0 \leftarrow \mathcal{P}_0 - \Lambda_{min} \times \mathcal{N}_0$ [$\mathcal{N}_0$ is a large positive number]
　　end if
end for
if $\mathcal{P}_0 > 0$ **then**
　　return $\mathcal{P}_0$
else
　　for each $l \in L$, $i \in L_1$ and $j \in L_2$ **do**
　　　　$\Lambda_{max} \leftarrow max\ eigen\ value\ of\ \mathcal{M}$ [$\mathcal{M}$ represents the left hand side of equation (8.69) - (8.71)]
　　end for
　　$\mathcal{P}_1 \leftarrow max\ of\ \Lambda_{max}$
　　return $\mathcal{P}_1$
end if

8.3.5 Learning Near-Optimal Controller

This section presents the steps that eventually learn a near-optimal controller when the critic is iteratively updated. In general, a critic network is used to approximate the value function or the co-state vector $\lambda(\mathbf{x})$ on the optimal state trajectory. In our work we use SNAC to approximate $J(\mathbf{x})$ as $V(\mathbf{x})$ in (8.58) that represents cost-to-go in (8.74) for current state of the system.

$$V_c(\mathbf{x}(t), t_c) = \int_{t_c}^{t_c+T} \left(\mathbf{x}^T(t)\mathbf{Q}\mathbf{x}(t) + \mathbf{u}^T(\mathbf{x})\mathbf{R}\mathbf{u}(\mathbf{x}) \right) dt$$
$$+ V_c(\mathbf{x}(t), t_c + T). \qquad (8.74)$$

Lemma 8.3. V_c can be considered as a value function of nonlinear system (8.40) for a given cost function (8.41) where $\mathbf{u}$ is any stable control law and V_c is positive.

Proof. its complete derivative along the system state trajectory can be described as

$$\dot{V}_c(\mathbf{x},t) = \frac{\partial V_c(\mathbf{x},t)}{\partial t}\frac{dt}{dt} + \left(\frac{\partial V_c(\mathbf{x},t)}{\partial \mathbf{x}}\right)^T \frac{d\mathbf{x}}{dt}$$

$$= \frac{\partial V_c(\mathbf{x},t)}{\partial t} + \left(\frac{\partial V_c(\mathbf{x},t)}{\partial \mathbf{x}}\right)^T [\mathcal{A}\mathbf{x} + \mathcal{B}\mathbf{u}] \quad (8.75)$$

Again, since V_c represents cost (8.41), derivative of cost-to-go is given by

$$\dot{V}_c(\mathbf{x},t) = -\varphi(\mathbf{x},\mathbf{u}) \quad (8.76)$$

This leads (8.75) to

$$\frac{\partial V_c(\mathbf{x},t)}{\partial t} = -\varphi(\mathbf{x},\mathbf{u}) - \left(\frac{\partial V_c(\mathbf{x},t)}{\partial \mathbf{x}}\right)[\mathcal{A}\mathbf{x} + \mathcal{B}\mathbf{u}] \quad (8.77)$$

It follows from the above fact that (8.77) satisfies the Hamilton-Jacobi-Bellman equation (8.43) with a stable control law. Thus $V_c(\mathbf{x}(t))$ in (8.74) is a value function for system (8.45). □

Lemma 8.4. *Given a nonlinear system (8.40) with associated cost (8.41) while $V_0(\mathbf{x})$ is the value function related to any stabilizing control policy $\mathbf{u}_0$ and $V_1(\mathbf{x})$ is the value function related to control policy $\mathbf{u}_1$ where $\mathbf{u}_1$ is computed as in (8.68), it can be stated that $V_1(\mathbf{x}) < V_0(\mathbf{x})$.*

Proof. Using equation (8.76), the following can be written

$$\dot{V}_0(\mathbf{x}) = -(\mathbf{x}^T \mathbf{Q}\mathbf{x} + \mathbf{u}_0^T \mathbf{R}\mathbf{u}_0) \quad (8.78)$$
$$\dot{V}_1(\mathbf{x}) = -(\mathbf{x}^T \mathbf{Q}\mathbf{x} + \mathbf{u}_1^T \mathbf{R}\mathbf{u}_1) \quad (8.79)$$

Combining above equations, it can be written that,

$$\dot{V}_1(\mathbf{x}) = \dot{V}_0(\mathbf{x}) + \mathbf{u}_0^T \mathbf{R}\mathbf{u}_0 - \mathbf{u}_1^T \mathbf{R}\mathbf{u}_1$$
$$= \dot{V}_0(\mathbf{x}) + (\mathbf{u}_0 - \mathbf{u}_1)^T \mathbf{R}(\mathbf{u}_0 - \mathbf{u}_1) \quad (8.80)$$

Since $(\mathbf{u}_0 - \mathbf{u}_1)^T \mathbf{R}(\mathbf{u}_0 - \mathbf{u}_1) > 0$ *for* $\mathbf{u}_1 \neq \mathbf{u}_0$, (8.80) leads to the inequality

$$\dot{V}_1(\mathbf{x}(t)) \geq \dot{V}_0(\mathbf{x}(t))$$

Integrating both sides of the above equation, we get

$$\int_{t_0}^{\infty} \dot{V}_1(\mathbf{x}(t))dt \geq \int_{t_0}^{\infty} \dot{V}_0(\mathbf{x}(t))dt$$

Policy Iteration and SNAC

$$V_1(\mathbf{x}(\infty)) - V_1(\mathbf{x}(t_0)) \geq V_0(\mathbf{x}(\infty)) - V_0(\mathbf{x}(t_0)) \tag{8.81}$$

Since $V_1(\mathbf{x}(\infty)) = V_0(\mathbf{x}(\infty)) = 0$, equation (8.81) can be rewritten as

$$V_1(\mathbf{x}(t_0)) \leq V_0(\mathbf{x}(t_0)) \tag{8.82}$$

□

A solution to the optimal control law $\mathbf{u}^*$ can be obtained analytically by solving the optimal cost J^* from HJB equation (8.43) for the system which is explicitly modeled and simple. However, no analytic solution exists for systems with unknown dynamics. We exploit a learning technique by which we train the critic network to approximate optimal cost. The initial controller in *Lemma:* 8.2 eventually learns the optimal control law as we update the weights of the critic network through policy iteration. The critic learns to approximate the optimal cost as the training progresses. The policy iteration scheme is presented in the following.

$$V_i(\mathbf{x}(t_0)) = \int_{t_0}^{t_0+T} (\mathbf{x}^T \mathbf{Q} \mathbf{x} + \mathbf{u}_i^T \mathbf{R} \mathbf{u}_i) \, dt + V_i(\mathbf{x}(t_0+T)) \tag{8.83a}$$

$$\mathbf{u}_{i+1}(\mathbf{x}) = -\frac{1}{2}\mathbf{R}^{-1}\boldsymbol{\mathcal{B}}^T \frac{\partial V_i}{\partial \mathbf{x}} \tag{8.83b}$$

Here, subscript i denotes iteration number. At ith iteration, critic network estimates $V_i(x(t))$ in (8.83a) and policy is renewed to $\mathbf{u}_{i+1}(\mathbf{x})$ (8.83b) in $(i+1)^{th}$ iteration.

Theorem 8.2. *There exists an optimal control law $\mathbf{u}^*$ given in (8.84) for a nonlinear system (8.40) with completely unknown dynamics and associated cost (8.41)*

$$\mathbf{u}^* = -\frac{1}{2}\mathbf{R}^{-1}\boldsymbol{\mathcal{B}}^T \frac{\partial V^*}{\partial \mathbf{x}} \tag{8.84}$$

where the optimal cost V^* is approximated by a critic network. The optimal weights of the critic network can be obtained through policy iteration scheme given in (8.83). Here $V^* = V^i$ where iteration $i \to \infty$

Proof. It is evident from *Lemma* 8.4 that if the control policy follows (8.83b), the cost value is smaller than the previous control policy i.e. $V_i \geq V_{i+1}$ where $i = 0, 1, 2, ..., \infty$. Therefore, the following relation can be drawn

$$V_0 \geq V_1 \geq V_2 \geq ... \geq V_n \geq V_{n+1} \geq ... V_\infty \tag{8.85}$$

Since, $V(\mathbf{x})$ is a decreasing function with an updated control policy, eventually V_i falls on the optimal trajectory and converges to V^* as $i \to \infty$. □

Remark 2. *In this scheme, $\partial V/\partial \mathbf{x}$ is calculated analytically from (8.58) and $\mathcal{B}$ is fuzzy aggregated form of system input matrix. According to Lemma 8.4, the cost associated with $\mathbf{u}_i$ is always lower than the cost associated with $\mathbf{u}_{i-1}$ and any $\mathbf{u}_i$ is a stable policy for system (8.45) as long as it follows (8.69) to (8.71) as in Lemma 8.2. Hence, the critic network can be updated each time whenever a new policy is available. Essentially, Lemma 8.2 gives the flexibility to update the weights in real-time as it always ensures stable policy.*

8.3.5.1 Update of Critic Network

Using *Bellman's Principle of Optimality* the cost-to-go can be represented as summation of cost accumulated in time t_0 to T and cost-to-go at time T while staying on optimal trajectory. Thus the incremental cost can be given by the following

$$\Delta V(\mathbf{x}(t_0)) = \int_{t_0}^{t_0+T} (\mathbf{x}^T \mathbf{Q}\mathbf{x} + \mathbf{u}^{*T}\mathbf{R}\mathbf{u}^*)\, dt \qquad (8.86)$$

Since, the critic network is used to approximate the cost-to-go of the system at a given time, the difference of cost estimated by critic network at $t_0 th$ and $(t_0 + T)th$ instant should be same as $\Delta V(\mathbf{x}(t_0))$ in (8.86) when the network is trained properly. Hence, the weight matrices of the network is learned in such a way, so that $\|\Delta V(\mathbf{x}(k)) - (V(\mathbb{P}(k)) - V(\mathbb{P}(k+T)))\|$ is minimized. This must be noted that the constraints should be maintained while minimizing the error norm. Therefore, the error minimization problem can be given as the following

$$\begin{aligned}\underset{\mathbf{P}^{(l)}}{\text{minimize}} \quad & \|\Delta V(\mathbf{x}(k)) - (V(\mathbb{P}(k)) - V(\mathbb{P}(k+T)))\| \\ \text{subject to} \quad & \Upsilon_c \text{ holds}, \quad c = 1,2,3\end{aligned} \qquad (8.87)$$

Here, Υ_c represents constraints given in equations (8.69) to (8.71) and $l = 1, 2, ..., L$. The methodology is summarized in Algorithm 6. The block schematic of the overall control scheme is shown in Fig. 8.3.

Remark 3. *Training of the critic network is performed on a finite number of samples, collected when the system is influenced by policy $\mathbf{u}_i$. The incremental cost ΔV carries the information related to system dynamics. The algorithm tries to find a new set of $\mathbf{P}^l$ matrices in ith iteration of PI scheme that minimizes the difference between actual and network predicted incremental cost while enforcing Lyapunov stability to the system. With each update, the critic network weights are directed toward optimality. Genetic algorithm (GA) proves to be useful to search for more optimal $\mathbf{P}^l$; the fitness function of the GA implements (9.23). Since, GA returns weight matrices which are more optimal and stable by Lemma 8.2, one can update the network while keeping the system running thereby enabling online update of the critic network.*

The approach presented here is also computationally efficient since it uses a small rule base. The number of rules in the rule base is equal to the number of fuzzy zones of a state in the state space (if a state x_j is divided in F_z fuzzy zones, the number of rules in the rule base is also F_z). Thus the computational complexity of a n states system is $\mathcal{O}(n \times F_z)$ in this method whereas the burden is $\mathcal{O}(n \times (F_z)^n)$ for existing methods [241] [242] where all the possible combinations of fuzzy zones are considered in the rule base, which makes this algorithm at least $(F_z)^{n-1}$ times faster than the existing methods. Moreover, a common set of fuzzy zones are used in calculation of $\mathcal{B}$ and in approximation of the optimal value function in (8.44), which results less computations and fast update of the network. However, the computational burden in GA is not addressed in this paper, since GA is used here as a toolbox and its computational burden will not be reflected on real-time implementation as it works offline.

Algorithm 5 Fitness function for PI based training

$\mathbf{P}^{(l)} \leftarrow \hat{\mathbf{P}} + \mathbf{D}^{(l)}$, $\mathcal{P}_0 \leftarrow 0$, $\mathcal{P}_1 \leftarrow 0$
for each $l \in L$ **do**
   $\Lambda_{min} \leftarrow min\ eigenvalue\ of\ \mathbf{P}^{(l)}$
   if $\Lambda_{min} < 0$ **then**
     $\mathcal{P}_0 \leftarrow \mathcal{P}_0 - \Lambda_{min} \times \mathcal{N}_0$ [$\mathcal{N}_0$ is a large positive number]
   end if
end for
if $\mathcal{P}_0 > 0$ **then**
   return $\mathcal{P}_0$
else
   for each $l \in L$, $i \in L_1$ and $j \in L_2$ **do**
     $\Lambda_{max} \leftarrow max\ eigen\ value\ of\ \mathcal{M}$ [$\mathcal{M}$ represents the left hand side of equation (8.69) - (8.71)]
     if $\Lambda_{max} > 0$ **then**
       $\mathcal{P}_1 \leftarrow \mathcal{P}_1 + \Lambda_{max} \times \mathcal{N}_1$ [$\mathcal{N}_1 < \mathcal{N}_0$, is a large positive number]
     end if
   end for
   if $\mathcal{P}_1 > 0$ **then**
     return $\mathcal{P}_1$
   end if
   $\mathcal{P}_2 \leftarrow \|\Delta V(\mathbf{x}(k)) - (V(\mathbb{P}(k)) - V(\mathbb{P}(k+T)))\|$
   return $\mathcal{P}_2$
end if

8.3.5.2 Fitness Function for PI Based Training

A fitness value is assigned to each genome by the fitness function, based on the output of the function. The fitness function that is used in this phase of learning, by the GA is in fact some augmentation of the previous fitness function. Algorithm 5 shows the design of the fitness function.

Algorithm 6 Steps for discovering near-optimal weights.

1: Collect an input-output data set $\mathcal{L} \in \mathbb{U}$ with permissible random control torque or using any existing controller.
2: Learn TSK fuzzy model (8.45) of L rules that represent the original nonlinear model (8.40) from the data set. The algorithm can be found in [237]
3: Find out $\hat{\mathbf{P}}$ and $\boldsymbol{\delta}^{(l)}$ satisfying the inequality conditions given in (8.69) to (8.71), *where, l=1, 2, ..., L* while $\hat{\mathbf{P}}$ and $\boldsymbol{\delta}^{(l)}$ follow the definitions given in (8.62) and (8.63).
4: Assign $\mathbf{P}^{(l)}$ as initial or starting weight matrix of the critic network *where, l=1, 2, ..., L*.
5: Take any initial position and velocity state $\mathbf{x_0} = \mathbf{x}(t_0)$ within the discourse of learning.
6: Get the fuzzy input matrix $\mathcal{B}$ from TSK fuzzy model.
7: Calculate $\boldsymbol{\lambda}^*$ using the output of the critic network.
8: Evolve the original system (8.40) using (8.83b).
9: Store $\mathbf{x}(t_0)$, $\mathbf{x}(t_0 + T)$ and $\Delta V(\mathbf{x}(t_0))$ in (8.86) with time stamp t_0.
10: Collect critic network output for $\mathbf{x}(t_0)$ and $\mathbf{x}(t_0 + T)$ as approximated cost-to-go at $t_0 th$ and $(t_0 + T)th$ instant.
11: Store the difference $v = V(\mathbb{P}(k)) - V(\mathbb{P}(k+T))$ with time stamp t_0.
12: Collect $\mathcal{N}_0$ number of such points.
13: Search new set of $\mathbf{P}^{(l)}$, $l = 1, 2, .., L$ that minimizes $\|\Delta V(\mathbf{x}(k) - v\|$ over $\mathcal{N}_0$ points as given in (9.23).
14: Select those $\mathbf{P}^{(l)}$ where constraints (8.69) to (8.71) are satisfied.
15: Renew the policy with new set of $\mathbf{P}^{(l)}$ and evolve the system.
16: Repeat from Step 5.

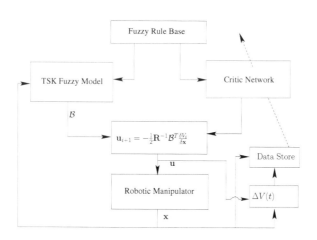

FIGURE 8.3: Overall control scheme.

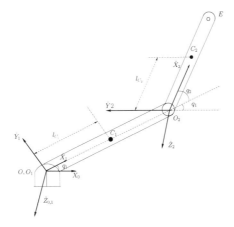

FIGURE 8.4: A manipulator with 2-DOF.

8.3.6 Examples

8.3.6.1 Simulated Model

In this section, the simulation results are presented. Let us consider a manipulator of two DOFs given in Fig. 8.4 [243]. Where the mass of link 1 & 2 are given by m_1 and m_2 respectively and the inertia matrices at the center of mass of both the links are given by

$$I_{C_1} = \begin{bmatrix} I_{xx1} & 0 & 0 \\ 0 & I_{yy1} & 0 \\ 0 & 0 & I_{zz1} \end{bmatrix}, \text{ and } I_{C_2} = \begin{bmatrix} I_{xx2} & 0 & 0 \\ 0 & I_{yy2} & 0 \\ 0 & 0 & I_{zz2} \end{bmatrix}$$

For the given manipulator configuration the torque model of the system can be represented as

$$\begin{bmatrix} m_{11} & m_{12} \\ m_{21} & m_{22} \end{bmatrix} \begin{bmatrix} \ddot{q}_1 \\ \ddot{q}_2 \end{bmatrix} + \begin{bmatrix} c_{11} & c_{12} \\ c_{21} & c_{22} \end{bmatrix} \begin{bmatrix} \dot{q}_1 \\ \dot{q}_2 \end{bmatrix} = \begin{bmatrix} u_1 \\ u_2 \end{bmatrix} \quad (8.88)$$

where m_{ij} and $c_{ij}, i = 1, 2; j = 1, 2$ are the elements of **M** and **C** respectively. $\mathbf{q} = [q_1 \quad q_2]^T$ and $\mathbf{u} = [u_1 \quad u_2]^T$ is the control torque at the joints. The dynamic model in Euler-Lagrange form is derived for WAM arm in a two DOF configuration. Joints two and four have been considered active and the rest of the joints are set to zero to reduce computational burden while deriving the model. It should be noted here that the gravity related torque is not considered here. It is assumed that the gravity term is compensated externally.

The dynamic model (8.88) of the manipulator is presented here to generate an input-output data set for system identification. The system model information is not used during the controller design. Training data can be generated using the existing controller. Though random input-output data points can also be used for system identification. It should be noted that the data set should cover all operating regions of the manipulator workspace. The design

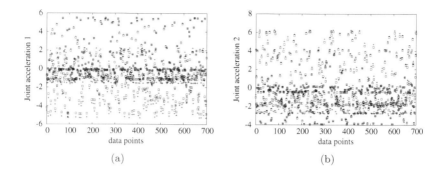

FIGURE 8.5: Validation with test data set: The empty squares represent desired acceleration and the filled squares are predicted by the model.

methodology is described in three steps. **Step 1.** The nonlinear manipulator model is identified as a TSK fuzzy model (8.45) in state space form where each joint position and velocity are the system states. Thirty thousand data points with sampling rate of 5ms are used to learn the system dynamics. We represent the fuzzy model by seven rules. The Gaussian functions are chosen as fuzzy set in the rule base. GA finds proper mean and variance of the fuzzy sets. GA runs for 500 generations with 150 population size. The procedure is given in [237]. Fig 8.5 shows the estimated output from TSK fuzzy model for testing data.

Step 2. The identified TSK fuzzy model is used to select the initial stable weights of the critic network. The procedure involves finding stable Lyapunov $\mathbf{P}^{(l)}$, $l = 1, 2, .., L$ matrices. Algorithm 4 shows the fitness function design which is used by GA to find initial weights. GA searches for suitable $\mathbf{P}^{(l)}$ and $\mathbf{K}$ for which equation (8.64) to (8.66) satisfy. Algorithm 4 is coded as the fitness function of the search problem. For this example, we take $\mathcal{N}_0 = 10^{25}$. The controller in (8.49) is a stabilizing controller where $\boldsymbol{\lambda}^*$ is associated with stable $\mathbf{P}^{(l)}$ matrices and $R = \mathbb{I} \in \Re^2$. Fig. 8.6 shows the state evolution when stabilizing controller is used. Results are presented for both the cases of known and unknown system dynamics.

Since we have a mathematical model of the system, we can use it to evaluate the performance of the algorithm. In this experiment we also learn the optimal controller using the information provided by (8.88). In the Fig. 8.6, trajectory 'a' is associated with known model information and trajectory 'b' is associated with unknown model parameters. It is evident that almost similar controller performance can be obtained using this technique without knowing the system dynamics.

Step 3. Finally, the critic weights are updated by improving the policy to approximate optimal cost-to-go from a given state. In this example we select $T = 0.5 \ sec$, $\mathbf{R} = \mathbb{I} \in \Re^2$ and $\mathbf{Q} = 10 * \mathbb{I} \in \Re^4$ where $\mathbb{I}$ represents identity matrix. The manipulator is operated to collect data points for optimal training

Policy Iteration and SNAC

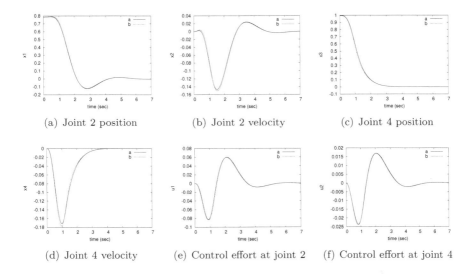

FIGURE 8.6: Performance of the initial stabilizing controller with both the cases of known and unknown system dynamics: Trajectory "a" is the case when the exact model dynamics is known and trajectory "b" is related to the to the case when the dynamics is unknown.

of the critic network. The system is left to evolve from some initial joint positions to desired positions under the influence of controller (8.83b). All the states, control inputs and associated cost are stored with corresponding time stamp at each T time. The critic network is updated when 1000 such data points are collected. GA finds another set of $\mathbf{P}^{(l)}$ matrices that minimize the error norm. Fitness function for this part of training is given in Algorithm 5 where $\mathcal{N}_1 = 10^{15}$. The policy is renewed by the solution provided by GA and the process continues. The results are shown after five such updates.

Fig. 8.7 shows time evolution of system states and controller output after the critic is iteratively trained with updated policy. Since states are penalized more during the training with optimal control law, the driving input is relaxed to increase in magnitude. All the simulations are done for seven seconds and with same initial points, so that a good comparison can be made and the conclusion can be obtained easily. Fig. 8.8 depicts convergence of cost on optimal trajectory. In the Fig. 8.8, cost trajectory 'c' denotes the cost that is generated by the initial control law before the critic is trained to estimate optimal cost. cost trajectory 'a' indicates the cost when the original input matrix $\mathbf{B}$ in (8.40) is used to generate control input in (8.83b) and cost trajectory 'b' shows the cost when the control input is generated using fuzzy estimated input matrix $\mathcal{B}$ from (8.45). It's evident from the results that optimal performances in both the cases are comparable.

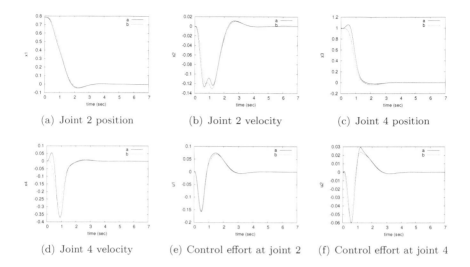

(a) Joint 2 position (b) Joint 2 velocity (c) Joint 4 position

(d) Joint 4 velocity (e) Control effort at joint 2 (f) Control effort at joint 4

FIGURE 8.7: Performance of controller after policy update: Trajectory "a" is the case when the exact model dynamics is known and trajectory "b" is related to the to the case when the dynamics is unknown.

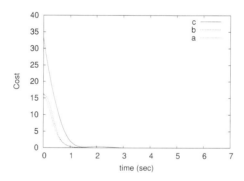

FIGURE 8.8: Comparison of cost accumulated in three cases: "a" represents cost after the critic is trained with known system dynamics; "b" represents cost after the critic is trained with unknown system dynamics, and "c" is the cost with the initial critic parameters.

8.3.6.2 Example using Real Robot

In earlier section we test our algorithm on a simulated model of the commercial robotic manipulator Barrett WAM. Now we test the algorithm on the real robotic manipulator given in Fig. 8.9. Robotic manipulators are inherently nonlinear and involve multi-body dynamics. Dynamic model of a general n-$Degree\ of\ Freedom\ (DOF)$ manipulator can be represented by the following

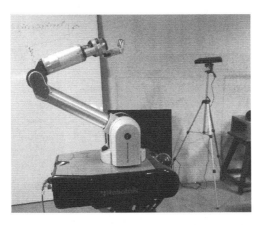

FIGURE 8.9: Experimental setup: Barrett Arm.

nonlinear equation:

$$\mathbf{M}(\theta)\ddot{\theta} + \mathbf{C}(\theta,\dot{\theta})\dot{\theta} + \mathbf{g}(\theta) = \boldsymbol{\tau}. \tag{8.89}$$

Here $\mathbf{M}(\theta) \in \Re^{n \times n}$ represents the mass matrix which is real and positive definite for a n-joints manipulator, $\mathbf{C}(\theta,\dot{\theta}) \in \Re^{n \times n}$ contains centrifugal and Coriolis force related terms, $\mathbf{g}(\theta) \in \Re^n$ represents the gravitational pull on the arms, $\boldsymbol{\tau} \in \Re^n$ represents the controlling torque applied to the joints and $\theta \in \Re^n$ represents the joint angles of the arm.

Considering the fact that the gravitational force on the arm can be compensated for quite efficiently, and hence the manipulator could be made unaffected by the gravitational pull. Assuming the joint angles and joint velocities as states of the system, equation (8.89) can be written in the following state space form,

$$\begin{aligned}\dot{\mathbf{x}} &= \begin{bmatrix} 0 & 1 \\ 0 & -\mathbf{M}^{-1}(\theta)\mathbf{C}(\theta) \end{bmatrix}\mathbf{x} + \begin{bmatrix} 0 \\ \mathbf{M}^{-1}(\theta) \end{bmatrix}[\boldsymbol{\tau} - \mathbf{g}(\theta)] \\ &= \mathbf{A}\mathbf{x} + \mathbf{B}\mathbf{u} \end{aligned} \tag{8.90}$$

Here $\mathbf{A} \in \Re^{2n \times 2n}$, $\mathbf{B} \in \Re^{2n \times n}$ and $\mathbf{x} = [\boldsymbol{\theta} \ \dot{\boldsymbol{\theta}}]^T$ defines the system states and $\mathbf{u} = (\boldsymbol{\tau}-\mathbf{g})$ represents effective torque at the joints after gravity compensation.

The algorithm is implemented on two joints of the robot, and hence the manipulator is represented as a 2-link configuration. Let us redefine the system states as $x_1 = \theta_1$, $x_2 = \dot{\theta}_1$, $x_3 = \theta_2$ and $x_4 = \dot{\theta}_2$ where θ_1 and θ_2 represent joint positions of $2nd$ and $4th$ joint respectively.

We formulate the optimal control problem of a two degree of freedom manipulator as the following. Given a manipulator with the performance index

$$J = \int_{t_0}^{\infty} (\mathbf{x}^T \mathbf{Q} \mathbf{x} + \mathbf{u}^T \mathbf{R} \mathbf{u}) \, dt \tag{8.91}$$

where $\mathbf{R} \in \Re^{2\times 2}$, $\mathbf{Q} \in \Re^{4\times 4}$, design a optimal controller that minimizes the infinite-horizon cost (8.91).

Overall control scheme is given in Fig. 8.3. The methodology is discussed in the following steps.

Step 1. The manipulator dynamics are represented as a TSK fuzzy model, based on the input-output data set that is collected for model identification. Joint positions, joint velocities and applied joint torques are considered as the input to the TSK fuzzy model and joint acceleration is taken as the output of the fuzzy model (8.45) where joint positions and joint velocities are the states of the system. A data set of 35,000 random data points is generated with in the range of $\begin{bmatrix} 0.8 & 0.5 & 0.8 & 0.5 \end{bmatrix}^T$ and $\begin{bmatrix} -0.8 & -0.5 & -0.523 & -0.5 \end{bmatrix}^T$ in radian and rad/sec. The sampling rate of the system is 5 ms. The state space is divided in nine fuzzy zones which gives a rule base of nine rules. The membership functions are chosen to be Gaussian functions. We search for proper mean and variance of the Gaussian functions using genetic algorithm toolbox in Matlab. The detailed procedure can be found in [237]. The affine term is not considered due to the requirement of our problem. Table 8.1 contains all the means and variances for nine rules, searched by GA. The population size is taken as 250 and the GA runs for 1000 generations.

$$\mathbf{A}^{(1)} = \begin{bmatrix} 0 & 1 & 0 & 0 \\ 0 & -67.59 & 0 & 3.36 \\ 0 & 0 & 0 & 1 \\ 0 & -8.20 & 0 & -11.28 \end{bmatrix}, \quad \mathbf{B}^{(1)} = \begin{bmatrix} 0 & 0 \\ 3.36 & 5.59 \\ 0 & 0 \\ -11.28 & 1.06 \end{bmatrix}$$

$$\mathbf{A}^{(2)} = \begin{bmatrix} 0 & 1 & 0 & 0 \\ 0 & 39.79 & 0 & -45.81 \\ 0 & 0 & 0 & 1 \\ 0 & 11.87 & 0 & -27.18 \end{bmatrix}, \quad \mathbf{B}^{(2)} = \begin{bmatrix} 0 & 0 \\ -45.81 & 5.87 \\ 0 & 0 \\ -27.18 & 1.26 \end{bmatrix}$$

TABLE 8.1: Mean and standard deviation of fuzzy sets associated to nine rules that represent the manipulator dynamics.

	Mean				SD			
	x1	x2	x3	x4	x1	x2	x3	x4
Rule 1	0.272	-0.091	-0.072	0.062	0.290	0.532	0.349	0.220
Rule 2	0.179	0.527	0.478	0.203	0.633	0.478	0.339	0.175
Rule 3	2.149	-0.356	0.468	0.077	1.683	0.263	0.434	0.608
Rule 4	-0.411	0.316	0.506	0.202	0.632	0.176	0.475	0.422
Rule 5	-0.041	0.328	0.684	-0.081	0.340	0.537	0.413	0.228
Rule 6	0.280	0.030	-0.261	-0.202	0.149	0.362	0.476	0.148
Rule 7	0.309	-0.015	0.805	-0.462	0.723	0.415	0.448	0.132
Rule 8	-0.117	-0.089	0.437	0.092	0.036	0.475	0.571	1.139
Rule 9	-0.774	0.394	0.723	-0.314	0.255	0.076	0.659	0.822

$$\mathbf{A}^{(3)} = \begin{bmatrix} 0 & 1 & 0 & 0 \\ 0 & -11.62 & 0 & -69.33 \\ 0 & 0 & 0 & 1 \\ 0 & -18.13 & 0 & 6.53 \end{bmatrix}, \quad \mathbf{B}^{(3)} = \begin{bmatrix} 0 & 0 \\ -69.33 & 5.11 \\ 0 & 0 \\ 6.53 & 0.47 \end{bmatrix}$$

$$\mathbf{A}^{(4)} = \begin{bmatrix} 0 & 1 & 0 & 0 \\ 0 & -96.62 & 0 & 11.51 \\ 0 & 0 & 0 & 1 \\ 0 & -17.74 & 0 & 5.6 \end{bmatrix}, \quad \mathbf{B}^{(4)} = \begin{bmatrix} 0 & 0 \\ 11.51 & 5.48 \\ 0 & 0 \\ 5.6 & 0.85 \end{bmatrix}$$

$$\mathbf{A}^{(5)} = \begin{bmatrix} 0 & 1 & 0 & 0 \\ 0 & -65.79 & 0 & 19.67 \\ 0 & 0 & 0 & 1 \\ 0 & -5.53 & 0 & -16.16 \end{bmatrix}, \quad \mathbf{B}^{(5)} = \begin{bmatrix} 0 & 0 \\ 19.67 & 5.38 \\ 0 & 0 \\ -16.16 & 0.73 \end{bmatrix}$$

$$\mathbf{A}^{(6)} = \begin{bmatrix} 0 & 1 & 0 & 0 \\ 0 & -12.96 & 0 & -0.02 \\ 0 & 0 & 0 & 1 \\ 0 & 7.77 & 0 & -28.19 \end{bmatrix}, \quad \mathbf{B}^{(6)} = \begin{bmatrix} 0 & 0 \\ -0.02 & 5.88 \\ 0 & 0 \\ -28.19 & 1.25 \end{bmatrix}$$

$$\mathbf{A}^{(7)} = \begin{bmatrix} 0 & 1 & 0 & 0 \\ 0 & -0.11 & 0 & 4.97 \\ 0 & 0 & 0 & 1 \\ 0 & -1.43 & 0 & -1.91 \end{bmatrix}, \quad \mathbf{B}^{(7)} = \begin{bmatrix} 0 & 0 \\ 4.97 & 5.50 \\ 0 & 0 \\ -1.91 & 0.91 \end{bmatrix}$$

$$\mathbf{A}^{(8)} = \begin{bmatrix} 0 & 1 & 0 & 0 \\ 0 & -23.93 & 0 & 28.79 \\ 0 & 0 & 0 & 1 \\ 0 & 0.48 & 0 & 1.31 \end{bmatrix}, \quad \mathbf{B}^{(8)} = \begin{bmatrix} 0 & 0 \\ 28.79 & 5.62 \\ 0 & 0 \\ 1.31 & 1.00 \end{bmatrix}$$

$$\mathbf{A}^{(9)} = \begin{bmatrix} 0 & 1 & 0 & 0.00 \\ 0 & -4.38 & 0 & -5.80 \\ 0 & 0 & 0 & 1.00 \\ 0 & -1.08 & 0 & -8.10 \end{bmatrix}, \quad \mathbf{B}^{(9)} = \begin{bmatrix} 0 & 0 \\ -5.80 & 5.39 \\ 0 & 0 \\ -8.10 & 0.79 \end{bmatrix}$$

Step 2. Weights of the critic network are initialized in such a way that the controller (8.49) is an admissible policy to the system (8.45). In our approach, we find out Lyapunov $\mathbf{P}^{(l)}$, $l = 1, 2, .., L$ ($L = 9$ in the experiment) as in step 3 of Algorithm 6, for which closed loop system is asymptotically stable. This technique successfully avoids learning of initial value function and guarantees stability of closed loop system by satisfying equation (8.69) to (8.71). The fuzzy zones those are identified during the TSK modeling, are also used to represent critic network. GA finds out Lyapunov $\mathbf{P}^{(l)}$, $l = 1, 2, .., L$ related to those specified fuzzy zones. In our work, GA toolbox in Matlab is used where each genome carries a set of $\mathbf{P}^{(l)}$, $l = 1, 2, .., L$ and the reproductive ability of the genome is decided by its fitness value. It should be noted that in our experiment the fuzzy rule base has only nine rules as nine fuzzy zones, were identified in the states space during TSK modeling. In this experiment, **R** is considered as identity matrix and is kept constant during the experiment. The gravity compensation algorithm that comes with robot operation library, keeps the joints at floating condition. However, it does not eliminate the static friction in the joints, which affects the learning of the model dynamics. In our experiment, the static friction term is learned during training to boost up gravity compensation in order to tackle static frictional force. **Step 3.** This step involves learning the optimal network parameters to approximate cost-to-go on optimal state trajectory. In this experiment we select $T = 0.8\ sec$, and $\mathbf{Q} = 10 * \mathbb{I} \in \Re^4$ where $\mathbb{I}$ represents identity matrix. The manipulator is operated within the discourse of learning for collecting data. Starting from an initial joint position within the manipulator workspace, we let the states to reach its desired values. The control torque is computed using equation (8.83b). Required data for optimal learning of the critic network is collected at each T interval as described in step 9 and 11 of Algorithm 6. The training is performed on 300 such points using (9.23) and the policy in (8.83b) is renewed after the critic network is updated. GA is used for this part of training where each genome contains a new set of $\mathbf{P}^{(l)}$ and the reproductive ability is evaluated based on (9.23). The results are shown after seventy-five such updates. Fig. 8.10 and Fig. 8.11 shows the state evolutions and related torques of joint one and two respectively from initial state $\begin{bmatrix} -0.78 & 0 & -0.52 & 0 \end{bmatrix}^T$ to $\begin{bmatrix} 0 & 0 & 0 & 0 \end{bmatrix}^T$ under the influence of controller (8.68). It should be noted that the a reference trajectory is provided from initial position to desired position

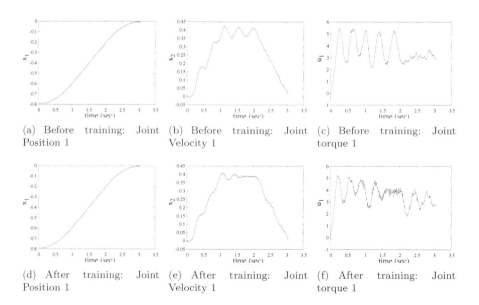

FIGURE 8.10: States and torque profiles of Joint Position.

Policy Iteration and SNAC

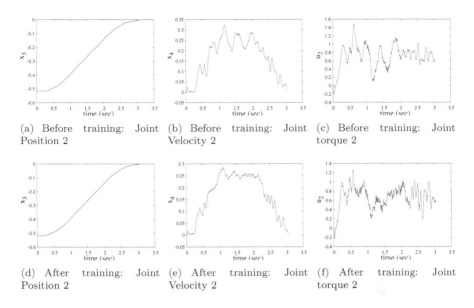

FIGURE 8.11: States and torque profiles of Joint Position 2.

by creating trapezoidal velocity profile to tackle huge initial acceleration which could damage the robotic arm severely. First row of the figures show states and torques before the critic was trained by policy iteration and second row contains the results after the iterative training. Since the states were penalized more than the control action by selecting $\mathbf{Q} = 10 * \mathbb{I}$ and $\mathbf{R}$ as *Identity*, the velocity profile of both the joints becomes smoother than the untrained case. Thus by selecting $\mathbf{Q}$ and $\mathbf{R}$, system state response can be customized based on the requirement.

Fig. 8.14 presents cost-to-go w.r.t time on the state trajectory. Here, *Initial cost* represents the cost-to-go, estimated by the critic network before the critic was iteratively trained. The time evolution of the near-optimal cost is given by *Near-optimal cost*. It is quite evident that the infinite horizon cost predicted by the critic is much lower that the untrained case, which essentially means that the critic network parameters converge toward optimal trajectory with such updates.

Fig. 8.12 and 8.13 show convergence of critic network parameters to the optimal trajectory. In the experiment, the parameters of the critic network have been updated seventy-five times. It is interesting to know that the critic network parameters at instant during the update, constitute a stable controller which is more optimal than the previous update.

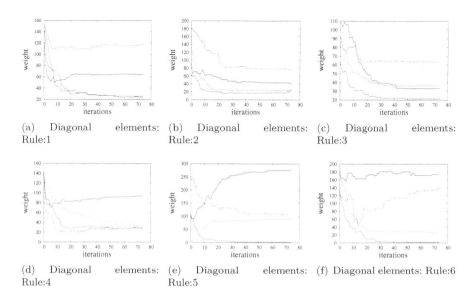

FIGURE 8.12: Convergence of weights to the optimal trajectory.

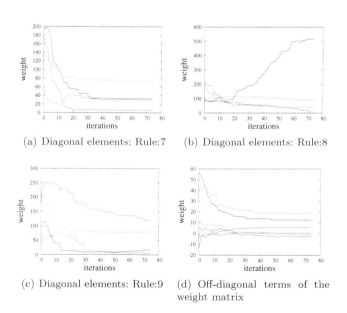

FIGURE 8.13: Convergence of weights to the optimal trajectory.

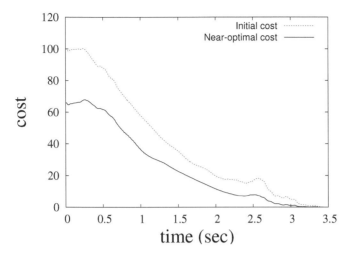

FIGURE 8.14: Comparison of costs.

8.4 Summary

We have seen the challenges of designing an optimal controller for a nonlinear system. AC methodologies help to overcome these challenges. In that connection, we have applied the adaptive critic both in discrete and continuous time systems. A TSK fuzzy model has been used as the critic network. The optimal controller for a nonlinear system has been obtained as a fuzzy cluster of the optimal controllers of the local linear models, establishing the relationship between the local linear dynamics and the global nonlinear dynamics. In discrete time SNAC, the costate vector has been approximated to design the optimal controller. Optimal control problem of continuous time systems with unknown dynamics is solved using policy iteration with SNAC architecture. The work analyzes the effect of critic network parameter variation on stability of the closed loop system. The algorithm learns stable weight matrices of a fuzzy critic network. The computational complexity is greatly reduced by incorporating a novel critic architecture. The continuous time dynamic model of the system is represented in state feedback form using TSK fuzzy model. Adaption of such a system identification technique leads to a small rule base which proves to be helpful for real-time implementation of the proposed algorithm. The algorithm also provides a way to solve for the initial stable control policy in the PI scheme. The critic weights are learned in such a way, that (8.49) is always a stable controller during the parameter update. The unconventional fuzzy Lyapunov function eases the stability analysis of the closed loop system and leads to simple stability criteria. It is shown analytically that the system is asymptotically stable through out the learning phase. In the process of learning optimal control policy, the weights eventually move toward optimal trajectory after each iteration, which is in harmony with Theorem 8.2. The algorithm is validated through real-time experiments on a commercial robotic manipulator whose dynamic model is unknown and involves highly nonlinear dynamics. Real-time experimental results, supported by the analytical proofs, suggest that the proposed algorithm can discover near-optimal critic network parameters in the presence of multiple solutions in the PI scheme for a nonlinear continuous time system where dynamic information is missing. However, the model uncertainties that may arise due to TSK fuzzy model representation will be dealt as a future scope of this work.

9
Imitation Learning

9.1 Introduction

Industrial evolution toward automation has opened the door to intelligent robots in order to reduce human labor at its minimum level. In the era of automation, robots have not only discarded the human intervention in repetitive tasks, also they are being considered as suitable replacement of human workers for complex jobs where decision making abilities related to motion planning are essential according to task requirements. Such a futuristic setup may also require sharing of workspace between humans and robots in a cooperative manner. Hence, robots which are able to learn skills from the human co-worker by only watching them accomplishing a task, would be quite useful for this purpose. As the modern intelligent robots come with complex architecture (both in hardware and software), challenges of operating these robots are also emerging. A demand of specially trained manpower is becoming high since the robots are needed to be programmed manually for different applications. To deal with the situation, current research concentrates on simplified skill learning for the robots. The skill learning is realized in the form of a robotic motion planner that plans/controls robot's motion in the workspace. The research problem here is to design the motion planner such that the user's expertise can be easily induced/encoded into the motion planner. In this chapter we present such a motion encoding system. The motion encoding system is a kinematic controller which works in the *outer loop* of the control hierarchy of the robotic manipulator.

In order to develop an efficient motion encoding system, the following criteria are desirable from an user's perspective: 1) *Reproduction of tasks*: The motion encoding system is capable of reproducing the demonstrated tasks (by task here we mean a particular motion profile). 2) *Stability and robustness to perturbations*: It is very important to design a system that can command a stable direction while meeting the first criteria and at the same time it is robust to perturbations. The perturbations can be spatial/temporal or external/internal. *Robustness to perturbation* feature will ensure that even if the end-effector trajectory is deviated from the desired path due to collision or something else, the end-effector still reaches the target point. 3) *Adaptivity*: The encoding system is adaptive to the changes in a dynamic environment, i.e., the motion model is capable of providing immediate directional command to the robot based on the current feedback from the environment. This is very important as the motion planner is expected to handle uncertainties in the environment. 4) *Generalization and re-usability*: the advantage of a generic motion encoding system is that it can be reused any where in the workspace. It avoids manual hard coding when ever a new task arrives. 5) *Effortless learning*: the model learning is effortless, i.e., the skills of a non-expert (in the field of robotics) user can be easily transferred to the encoding model.

Earlier approaches of learning robotic motions are mostly based on the principle of dividing a task in two parts [90]: a) *path planning* and b) *trajectory tracking*. *Path planning* involves generation of end-effector trajectory in the Cartesian space or joint space. As the robotic manipulator interacts with the environment in the Cartesian space, the Cartesian space motion planner demands significant attention for desired and safe robot operation. A Cartesian space motion planner or kinematic controller should be capable of generating suitable and feasible end-effector trajectory for the manipulator meeting all the specified criteria. The *trajectory tracking* algorithms use only local information from the environment as feedback to the robot to generate required control signal to execute the trajectory. The trajectory tracking controller generally provides joint level torques involving robot's dynamics. This enables the robot to track the desired trajectory [244]. However, the earlier approaches available for motion learning have failed to transfer skills from human to robots. In contrast, imitation learning provides flexibility to acquire skills from the human expert. Earlier we have discussed various elements of imitation learning. In this chapter, we'll discuss about various imitation learning / learning by demonstrations techniques. We'll mainly concentrate on IL through kinesthetic teaching.

9.2 Dynamic Movement Primitives

It is believed that complex motor actions can be represented as a sequence of simpler actions or action primitives. These action primitives are the building blocks that constitute higher level complex motions. The mathematical representation of these building blocks are known as dynamic movement primitives (DMPs) [92,114]. The main idea behind DMPs is they fuse a stable dynamical system and a nonlinear function follows some interesting trajectories. There are two categories of DMP: a) discrete and b) rhythmic. Discrete DMPs are generally used for learning point to point motions, whereas, rhythmic DMPs are good for learning tasks which involve repetitive motions such as walking. In discrete DMP, the base system is asymptotically stable, however, in the second category, a limit cycle is used as the base system. Here, we'll mainly consider the discrete type.

Dynamic Movement Primitives

9.2.1 Mathematical Formulations

The DMPs employ a second order dynamics which learns the acceleration profile from the given demonstrations. The base system is a PD controller that provides all the trajectories a stable behavior and is given as follows:

$$\ddot{\mathbf{x}} = k_p(\mathbf{g} - \mathbf{x}) - k_d \dot{\mathbf{x}} \tag{9.1}$$

where, $\mathbf{g}$ and $\mathbf{x}$ are the goal and current states. k_p and k_d are two positive scalers which are the gains of the PD controller. Being a stable system, (9.1) will always yield a stable trajectory and the properties of the trajectory can be manipulated to certain extent by playing with the gains. However, this system will always end up with a trivial trajectory. Real-life demonstrations comprise complex trajectories which cannot be encoded into the base system. To obtain a non-trivial trajectory from such models, we need an extra component that can hold the non-trivial characteristics of the trajectory. In this pursuit, the base system (9.1) is modulated by using a nonlinear function $\mathbf{f}$ and is given as follows:

$$\ddot{\mathbf{x}} = k_p(\mathbf{g} - \mathbf{x}) - k_d \dot{\mathbf{x}} + \mathbf{f} \tag{9.2}$$

Here, $\mathbf{f}$ is a kind of forcing function that forces the PD signal to follow a desired trajectory. The nonlinear function $\mathbf{f}$ is defined over a canonical system which is represented by

$$\dot{\tau} = -\alpha\tau \tag{9.3}$$

where, α is a positive scaler. The canonical system is a normalized state which starts from 1 and exponentially reaches to zero. The function f is defined as bellow

$$\mathbf{f}(\tau) = \sum_{i=1}^{N} w_i \psi_i \tau (\mathbf{g} - \mathbf{x_0}) \tag{9.4}$$

$\mathbf{x_0}$ in the above equation is the initial state. ψ_i is basis function weighted by w_i and is given by

$$\psi_i = exp(-\frac{1}{\sigma_i^2}(c_i - \tau)^2) \tag{9.5}$$

Here, we can see that the Gaussian function has been chosen as the basis function where, c_i is the mean and σ_i is the variance. The Gaussian function is defined over the trajectory of τ. As the value of τ decreases, the basis functions are activated based on their locations. The forcing function uses the normalized basis function multiplied with the spatial scaling. One can observe that the forcing term is diminishing as the time progresses. The canonical system trajectory starts at 1.0 and ends at 0.

9.2.1.1 Choice of Mean and Variance

τ in the canonical system (9.3) decays exponentially from its initial value. Therefore, for a set of equally spaced means, all the basis functions are activated during the early stage of evolution of τ. This would not have happened, if τ reached linearly to its target. To alleviate this problem, the means and the associated variances are chosen in time and projected back to τ. The steps are given in Algorithm 7.

Algorithm 7 Selection of mean and variance

1: Decide the number of basis functions (BFs) and tolerance ϵ.
2: Get the desired means (c_d) in time by equally segmenting the total time of evolution (t_f) by BFs.
3: i^{th} mean in τ can be given by

$$c_i = exp(\frac{-log(\epsilon) * c_d}{t_f}) \qquad (9.6)$$

4: The associated variance σ_i can be given by

$$\sigma_i = \frac{(BFs)^{1.5}}{c_i} \qquad (9.7)$$

9.2.1.2 Spatial and Temporal Scaling

One important advantage of DMP based motion planner is the learned trajectory can be scaled spatially and temporally. The spatial scaling property enables the trajectory to be expanded or contracted based on the distance between the initial and target/goal positions. If the goal position is far away from the initial position (irrespective of the initial and goal positions in the demonstrations), then the executed trajectory by the motion planner must reach the target/goal while maintaining the same demonstrated profile; or if the goal is very close to the initial position, then also the planner must maintain the demonstrated profile in its planned trajectory. In DMP model the spatial scaling is taken care by the term $(\mathbf{g} - \mathbf{x_0})$ in the forcing function (see equation 9.4). As the distance between the initial and goal position changes, the forcing function is spatially scaled accordingly.

Similarly, the temporal scaling property enables the robot to reach the a target point with different speeds while tracking the same trajectory. This property is achieved by introducing a scaling term s_t in the DMP model

Dynamic Movement Primitives

(9.2). Now the DMP model and the canonical system is given by the following equations.

$$\ddot{\mathbf{x}} = s_t^2 \left(k_p(\mathbf{g} - \mathbf{x}) - k_d \dot{\mathbf{x}} + \mathbf{f}\right) \tag{9.8}$$

$$\dot{\tau} = s_t(-\alpha\tau) \tag{9.9}$$

Here in (9.8) we use s_t^2 since the DMP dynamics is a second order differential equation. By selecting s_t between 0 to 1, we can achieve different speeds with same trajectory.

9.2.2 Example

Here, we shall discuss the implementation of the DMP based approach of motion learning through an example. Let us consider a 4−DOF manipulator which learns ball-hitting motions using a bat mounted as its end-effector. A human expert provides the kinesthetic demonstrations by guiding the robot to hit the ball. The joint positions are recorded during the hitting motion. Assume that the recorded joint positions are represented as $\boldsymbol{\theta}$ and the dataset to train the DMP model is represented by $\mathcal{D} = \{\boldsymbol{\theta}^t, \dot{\boldsymbol{\theta}}^t, \ddot{\boldsymbol{\theta}}^t\}_{t=1}^K$. $\dot{\boldsymbol{\theta}}^t$ and $\ddot{\boldsymbol{\theta}}^t$ in the data set are the joint velocities and acceleration respectively at t^{th} instant. The joint velocity and acceleration can be achieved by differentiating the joint positions and velocities respectively.

The joint acceleration in the dataset can be regarded as the desired acceleration which the the DMP model must generate. Hence, we can write

$$\ddot{\boldsymbol{\theta}} = k_p(\mathbf{g} - \boldsymbol{\theta}_0) - k_d \dot{\boldsymbol{\theta}} + \mathbf{f}$$

As we have discussed earlier that the forcing function $\mathbf{f}$ in fact encodes the trajectory. Thus the above equation can be rewritten as

$$\mathbf{f} = \ddot{\boldsymbol{\theta}} - \left(k_p(\mathbf{g} - \boldsymbol{\theta}_0) - k_d \dot{\boldsymbol{\theta}}\right) = \mathbf{f}_{ref}$$

$\mathbf{f}_{ref}$ is the target function for $\mathbf{f}$. Therefore the we need to learn w_i such that $\sum_t \|f_{ref}^t - \sum_i \psi_i w_i \tau(t)(g - \theta_0)\|$ is minimized.

We have chosen fifty Gaussian kernels as basis functions. 3,589 time steps were considered with 2 ms sampling time. The tolerance is chosen as 0.01. First we decided the means (c_i) and sigmas (σ_i) for the total time 7.18 seconds. The weights are then learned using least square optimization. After the training of the DMP model, we generate a hitting motion trajectory by selecting random initial and goal positions, which are not included in the demonstration. The results are plotted in Figure: 9.1. We can see the trajectories have been scaled according to the goal and initial positions.

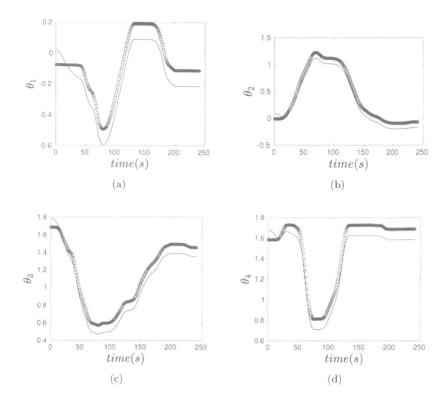

FIGURE 9.1: Four joint angles have been plotted here. The trajectory with circle (thick line) is the demonstrated joint motion profile during the hitting motion. The other trajectory (thin line) is generated by the DMP model.

9.3 Motion Encoding using Gaussian Mixture Regression

The point-to-point motion of the robotic system is represented as a continuous time nonlinear system and is given as follows:

$$\dot{\mathbf{x}} = \mathbf{f}(\mathbf{x}) \quad (9.10)$$

where $\mathbf{x} \in \Re^n$ is the state of the robotic system and $\mathbf{f}$ is the map. The function $\mathbf{f}(\mathbf{x})$ bears all characteristics of the trajectory that is executed by the motion model. The parameters of the nonlinear function are learned through an optimization process. In the IL paradigm, the map $\mathbf{f}$ is learned from the demonstrations, where the robot is shown reaching motions of particular pattern by the a human demonstrator. The map $\mathbf{f}$ in (9.10) can be represented as a nonlinear regressive function. Gaussian mixture regression has been very successful in modeling demonstrations, where the parameters of the mixture model are optimized using expectation maximization algorithm. Given a data set $\mathscr{D} = \{[\mathbf{x}_{i,j}, \dot{\mathbf{x}}_{i,j}]_{j=1}^{N_i}\}_{i=1}^{D}$, where $\mathbf{x}_{i,j}$ is the position state of the robot in ith instant of jth demonstration, the joint probability distribution of the demonstrations can be estimated using L Gaussian probability distribution functions and is given by

$$P(\mathbf{x}, \dot{\mathbf{x}}|\mu, \Sigma) = \sum_{k=1}^{L} P(k) P(\mathbf{x}, \dot{\mathbf{x}}|k), \qquad (9.11)$$

where $P(k)$ is termed as the prior and $P(\mathbf{x}, \dot{\mathbf{x}}|k)$ is the conditional probability density function which is given by

$$P(\mathbf{x}, \dot{\mathbf{x}}|k) = G_k\left(\mathbf{x}, \dot{\mathbf{x}}; \mu^k, \Sigma^k\right)$$
$$= \frac{\exp{-\frac{1}{2}\left(([\mathbf{x};\dot{\mathbf{x}}]-\mu^k)^T (\Sigma^k)^{-1} ([\mathbf{x};\dot{\mathbf{x}}]-\mu^k)\right)}}{\sqrt{2\pi^{2n}|\Sigma^k|}}, \qquad (9.12)$$

where μ^k and Σ^k are given by

$$\mu^k = \begin{bmatrix} \mu^k_{\mathbf{x}} \\ \mu^k_{\dot{\mathbf{x}}} \end{bmatrix} \text{ and } \Sigma^k = \begin{bmatrix} \Sigma^k_{\mathbf{xx}} & \Sigma^k_{\mathbf{x}\dot{\mathbf{x}}} \\ \Sigma^k_{\mathbf{x}\dot{\mathbf{x}}} & \Sigma^k_{\dot{\mathbf{x}}\dot{\mathbf{x}}} \end{bmatrix}. \qquad (9.13)$$

The posterior probability $P(\dot{\mathbf{x}}|\mathbf{x})$ gives the desired mean estimate using (9.12)

$$\mathbf{f}(\mathbf{x}) = \sum_{k=1}^{K} \frac{P(k)P(\mathbf{x}|k)}{P(\mathbf{x})} \left(\mu^k_{\dot{\mathbf{x}}} + \Sigma^k_{\dot{\mathbf{x}}\mathbf{x}} {\Sigma^k_{\mathbf{xx}}}^{-1} (\mathbf{x} - \mu^k_{\mathbf{x}}) \right) \qquad (9.14)$$

Comparing (9.13) with (9.24) with zero control input, the following can be written.

$$\mathbf{A}^k = \Sigma^k_{\dot{\mathbf{x}}\mathbf{x}} {\Sigma^k_{\mathbf{xx}}}^{-1}$$
$$\mathbf{b}^k = \mu^k_{\dot{\mathbf{x}}} - \Sigma^k_{\dot{\mathbf{x}}\mathbf{x}} {\Sigma^k_{\mathbf{xx}}}^{-1} \mu^k_{\mathbf{x}}$$
$$\sigma_k = \frac{P(k)P(\mathbf{x}|k)}{P(\mathbf{x})} \qquad (9.15)$$

Hence, (9.10) can be rewritten as

$$\dot{\mathbf{x}} = \sum_{k=1}^{K} \sigma_k (\mathbf{b}_k + \mathbf{A}_k \mathbf{x}) \qquad (9.16)$$

Here, each $(\mathbf{b}^k + \mathbf{A}^k \mathbf{x})$ is the local representation of the demonstrated motion profiles and the overall encoding is given by the contribution of each local subsystem when weighted by σ_k. However, model (9.14) is not stable by default. To ensure the motion trajectory reaches to the target, additional measures have to be taken. In other words, the dynamical system (9.14) must have an stable equilibrium point. Addressing this challenge has led to various algorithms using the above mentioned technique in recent times. Here, we shall discuss some of such important methods.

9.3.1 SED: Stable Estimator of Dynamical Systems

SED is a Gaussian mixture regression based motion encoding technique to learn point-to-point motion from user demonstrations. This technique was proposed by Khansari-Zadeh et al. [245]. The model ensures asymptotic stability of the predicted end-effector trajectory.

SED uses the dynamical system structure as given in (9.16). In general, even for simple motions, the learned motion model is not stable. It means that the model does not know where to stop while unfolding in time. As the characteristics of a dynamical system depends on its parameters, one needs to be careful while selecting those parameters in order to ensure stability. SED is motivated to that idea where the learning algorithm ensures that the learned model obeys the Lyapunov stability criteria.

9.3.1.1 Learning Model Parameters

As we have discussed earlier that the characteristics of the dynamical system depends on its parameters. In case of (9.16), $\mathbf{A}_k$ and $\mathbf{b}_k$ decide how the system unfold in time. The following theorem [245] presents the constraints on $\mathbf{A}_k$ and $\mathbf{b}_k$ during update.

Theorem 9.1. *The system* (9.16) *is globally asymptotically stable at target* $\mathbf{x}^* \in \Re^d$ *if*

$$\begin{cases} (i) \; \mathbf{b}_k + \mathbf{A}_k \mathbf{x}^* = \mathbf{0} \\ (ii) \; \mathbf{A}_k + (\mathbf{A}_k)^T < \mathbf{0}, \quad \forall k = 1 \cdots K \end{cases} \quad (9.17)$$

where, $(\mathbf{A}_k)^T$ *is the transpose of* $\mathbf{A}_k$ *and "*$< \mathbf{0}$*" implies the negative definiteness of the left hand side.*

Proof. The above theorem directly comes from the stability analysis of system (9.16). Consider a Lyapunov function $V(\mathbf{x})$

$$V(\mathbf{x}) = \frac{1}{2}(\mathbf{x} - \mathbf{x})^T (\mathbf{x} - \mathbf{x}) \quad (9.18)$$

V in the above equation in fact represents the error energy during the state evolution. For an asymptotically stable system, the error energy eventually diminishes as the state reaches to the target. Hence we have,

$$\dot{V}(\mathbf{x}) = (\mathbf{x} - \mathbf{x})^T \sum_{k=1}^{K} \sigma_k (\mathbf{b}_k + \mathbf{A}_k \mathbf{x}) \quad (9.19)$$

$$= (\mathbf{x} - \mathbf{x})^T \sum_{k=1}^{K} \sigma_k (\mathbf{A}_k (\mathbf{x} - \mathbf{x}^*) + \mathbf{A}_k \mathbf{x}^* + \mathbf{b}_k) \quad (9.20)$$

$$= \sum_{k=1}^{K} \sigma_k (\mathbf{x} - \mathbf{x})^T \mathbf{A}_k (\mathbf{x} - \mathbf{x}^*) \quad \text{(from (9.17))} \quad (9.21)$$

$$< 0 \quad \text{(from (9.17))} \quad (9.22)$$

□

i.e., the motion model (9.16) is an asymptotically stable system in the sense of Lyapunov. Theorem 9.1 says, during the search of the motion model parameters, one has to ensure the constraints given by (9.17) are maintained. The model parameters are selected through an optimization process where a cost is minimized. Optimization of model parameters is in fact determining $\boldsymbol{\theta} = \{\pi^1 \cdots \pi^K, \boldsymbol{\mu}^1 \cdots \boldsymbol{\mu}^K, \boldsymbol{\Sigma}^1 \cdots \boldsymbol{\Sigma}^K, \}$. Maximizing Log-Likelihood function in the constrained environment is an efficient way to determine the the model parameters.

9.3.1.2 Log-likelihood Cost

Maximization of log-likelihood is equivalent to minimization of negative log-likelihood. Hence the cost function of the optimization problem is given by

$$\begin{aligned}
\underset{\boldsymbol{\theta}}{\text{minimize}} \quad & -\frac{1}{N \sum_{n=1}^{N} T^n} \sum_{n=1}^{N} \sum_{m=1}^{T^n} \log P(\mathbf{x}^{m,n}, \dot{\mathbf{x}}^{m,n} | \boldsymbol{\theta}) \\
\text{subject to} \quad & \mathbf{b}_k + \mathbf{A}_k \mathbf{x}^* = \mathbf{0} \\
& \mathbf{A}_k + (\mathbf{A}_k)^T < \mathbf{0} \\
& \boldsymbol{\Sigma}^k > 0 \qquad \forall k = 1 \cdots K \\
& 0 < \pi^k \leq 1 \\
& \sum_{k=1}^{K} \pi^k = 1
\end{aligned} \qquad (9.23)$$

where $P(\mathbf{x}^{m,n}, \dot{\mathbf{x}}^{m,n} | \boldsymbol{\theta})$ is the joint probability of the demonstration given the parameters $\boldsymbol{\theta}$. The last three constraints are added to ensure that the model is a mixture of Gaussians. The constrained optimization problem can be solved a as a nonlinear programming (NLP).

9.4 FuzzStaMP: Fuzzy Controller Regulated Stable Movement Primitives

Fuzzy controller regulated stable movement primitives (FuzzStaMP) is another attempt to learn movement primitives in the imitation learning framework. It overcomes the limitation of the SED technique. As we have seen that the SED method imposes an equality constraint involving the target point $\mathbf{x}^*$. Satisfying an equality constraint is sometimes difficult from the optimization algorithm's perspective. The methods also learns a model which is valid for $\mathbf{x}^*$. In FuzzStaMP, the motion model is represented as a weighted summation of local linear models with a fuzzy controller. FuzzStaMP combines the statistical learning and the fuzzy reasoning for better encoding of demonstrations. The model is given as follows:

$$\dot{\mathbf{x}} = \sum_{k=1}^{L} \sigma_k (\underbrace{\mathbf{b}^k + \mathbf{A}^k \mathbf{x}}_{\mathbf{f}^k(\mathbf{x})} + \mathbf{B}^k \mathbf{u}_k) \qquad (9.24)$$

Here, (9.24) has the form of standard state space representation of a nonlinear dynamical system in closed loop, where $\mathbf{f}^k(\mathbf{x})$ bears the local properties of the demonstrated motion profiles and rest is the control input $\mathbf{u} \in \Re^n$ multiplied by system input matrix $\mathbf{B}$ of appropriate dimensions. $\mathbf{B}^1 = \mathbf{B}^2 = ... = \mathbf{B}^k = \mathbb{I} \in \Re^{n \times n}$ is chosen in this case. σ_k is the weightage of the kth local model. Parameters $\mathbf{A}^k$ and $\mathbf{b}^k$ are learned for the demonstration data as discussed in section 9.3. $\mathbf{u}$ is chosen as a TSK fuzzy controller of the following form: l^{th} rule:

IF x is G_l **THEN**
$$\mathbf{u}_l = \mathbf{k}_0^l + \mathbf{K}^l \mathbf{x} \qquad (9.25)$$

Here, G_l is a Gaussian probability distribution function that creates the lth fuzzy rule for the controller. The selection of the parameters $\mathbf{k}_0 \in \Re^n$ and $\mathbf{K} \in \Re^{n \times n}$ in consequent part is made in a constrained optimization process to achieve the desired performance of the model. FuzzStaMP comes with two variants based on strictness of the constraints during parameter update. First a conservative motion model is presented which is known as C-FuzzStaMP. Then a relaxed version of the algorithm is presented which is known as R-FuzzStaMP.

9.4.1 Motion Modeling with C-FuzzStaMP

C-FuzzStaMP model is conservative in a sense that the learning algorithm imposes strict asymptotic stability constraints during the update of the model parameters. Let us represent the system (9.10) as a fuzzy system and is given as follows:

$$\dot{\mathbf{x}} = \sum_{i}^{L} \sigma_i \mathcal{A}_i \mathbf{x} \qquad (9.26)$$

where $\mathcal{A}_i \in \Re^{n \times n}$ and $\mathbf{x} \in \Re^n$ represent system matrix and the state of the system (9.26). σ_i gives the firing strength of the ith rule. The following theorem [238] can be used to deal with the asymptotic stability of the system (9.26).

Theorem 9.2. *The equilibrium of the continuous fuzzy system (9.26) is globally asymptotically stable if there exists a common positive definite matrix* $\mathbf{P}$ *such that*

$$\mathcal{A}_i^T \mathbf{P} + \mathbf{P} \mathcal{A}_i < 0, \quad i = 1, 2, ..., L \qquad (9.27)$$

That means, there should be a common $\mathbf{P}$ matrix of appropriate dimension for the system (9.26) to be asymptotically stable. But it is sometimes difficult to find a common $\mathbf{P}$ for all the subsystems specially when the motion dynamics is complex. A fuzzy Lyapunov function in fact helps to deal with the situation.

9.4.1.1 Fuzzy Lyapunov Function

A fuzzy Lyapunov function is introduced to learn stable motion. Lyapunov function is described as energy like function. Thus, a Lyapunov function candidate is defined as (9.28) [246] which shares the same rule antecedent. The fuzzy Lyapunov function is given as follows:

$$V(\mathbf{x}(\mathbf{t})) = 2 \int_0^\mathbf{x} \mathbf{\Omega}(\mathbf{s}) \cdot \mathbf{ds} \qquad (9.28)$$

which is essentially a line integral function over the trajectory of any state $\mathbf{x}$ of the system. Here, $\mathbf{s} \in \Re^n$ and $\mathbf{\Omega}(\mathbf{s}) = [\Omega_1(\mathbf{s}) \;\; \Omega_2(\mathbf{s}) \;\; ... \;\; \Omega_n(\mathbf{s})]^T$.

Let us choose a fuzzy Lyapunov $\mathbf{P}$ matrices such that
l^{th} rule:

IF $\mathbf{x}$ is G_l **THEN**
$$\mathbb{P}^{(l)}(\mathbf{x}) = \mathbf{P}^{(l)} \qquad (9.29)$$

For $l=1,2,...,L$.

Here, $\mathbf{P}^{(l)} \in \Re^{n \times n}$ is symmetric positive definite matrix and From (9.29), we can write

$$\mathbf{\Omega}(\mathbf{x}) = \mathbb{P}(\mathbf{x})\mathbf{x} = \sum_{l=1}^{L} \sigma^{(l)} \mathbf{P}^{(l)} \mathbf{x} \qquad (9.30)$$

where, $\sigma^{(l)}$ is the normalized rule membership of lth rule. The following theorem [239] defines the structure of $\mathbf{P}^{(l)}$.

Theorem 9.3. $V(\mathbf{x})$ is a Lyapunov function candidate if there exists a $\mathbf{P}^{(l)} = \hat{\mathbf{P}} + \mathbf{D}^{(l)} > 0$ such that

$$\hat{\mathbf{P}} = \begin{bmatrix} 0 & \cdots & p_{1n} \\ \vdots & \ddots & \vdots \\ p_{1n} & \cdots & 0 \end{bmatrix}, \quad \mathbf{D}^{(l)} = \begin{bmatrix} d_{11}^{(l)} & \cdots & 0 \\ \vdots & \ddots & \vdots \\ 0 & \cdots & d_{nn}^{(l)} \end{bmatrix} \qquad (9.31)$$

The off diagonal terms of $\mathbf{P}^{(l)}$ matrices will remain same for all the rules. The diagonal elements construct different $\mathbf{P}$ for each rule. It can be shown that (9.28) is a generic form of conventional Lyapunov candidate function $V(\mathbf{x}) = \mathbf{x}^T \mathbf{P} \mathbf{x}$, where $\mathbf{P} = \hat{\mathbf{P}} + \mathbf{D}$ and $\mathbf{D} = \mathbf{D}^{(1)} = \mathbf{D}^{(2)} = ... = \mathbf{D}^{(L)}$.

Using equation (9.28), $\dot{V}$ of the system (9.24) with $\mathbf{u} = 0$, can be written as

$$\dot{V} = \mathbf{x}^T \left(\sum_{i=1}^{L} \sum_{j>i}^{L} \sigma_i \sigma_j \mathcal{V}_2^{i,j} + \sum_{i=1}^{L} \sigma_i^2 \mathcal{V}_1^i \right) \mathbf{x}$$

$$+ 2\mathbf{x}^T \sum_{i=1}^{L} \sigma_i^2 \left(\mathbf{P}^i \mathbf{b}^i \right) + 2\mathbf{x}^T \sum_{i=1}^{L} \sum_{j>i}^{L} \sigma_i \sigma_j \left(\mathbf{P}^j \mathbf{b}^i + \mathbf{P}^i \mathbf{b}^j \right) \quad (9.32)$$

where

$$\mathcal{V}_1^i = \left(\mathbf{A}^i \right)^T \mathbf{P}^i + \mathbf{P}^i \mathbf{A}^i \quad (9.33)$$

$$\mathcal{V}_2^{i,j} = \left(\mathbf{A}^i \right)^T \mathbf{P}^j + \left(\mathbf{A}^j \right)^T \mathbf{P}^i + \mathbf{P}^i \mathbf{A}^j + \mathbf{P}^j \mathbf{A}^i \quad (9.34)$$

Lemma 9.1. $\dot{V}$ *in (9.32) is strictly negative outside the ball* α *around the origin of system (9.24) with* $\mathbf{u} = 0$ *if there exist* $\mathbf{P}^l > 0$, $l = 1, 2, ..., L$ *that agree with the following constrains*

$$\lambda_{max}(\mathcal{V}_1^i) + \frac{2}{\alpha} \|\mathbf{P}^i \mathbf{b}^i\| < 0, \quad i = 1, ..., L \quad (9.35)$$

$$\lambda_{max}(\mathcal{V}_2^{i,j}) + \frac{2}{\alpha} (\|\mathbf{P}^i \mathbf{b}^j\| + \|\mathbf{P}^j \mathbf{b}^i\|) < 0, \quad i < j = 1, ..., L. \quad (9.36)$$

where $\lambda_{max}(.)$ *represents the maximum eigenvalue and* α *is a positive scalar.*

Proof. From the properties of matrices, it can be written that

$$\|\mathbf{x}\|^2 \lambda_{min}(\mathcal{V}_1^i) \leq \mathbf{x}^T \mathcal{V}_1^i \mathbf{x} \leq \|\mathbf{x}\|^2 \lambda_{max}(\mathcal{V}_1^i) \quad (9.37)$$

$$\mathbf{x}^T \left(\mathbf{P}^i \mathbf{b}^i \right) \leq \|\mathbf{P}^i \mathbf{b}^i\| \|\mathbf{x}\| \quad (9.38)$$

and similarly,

$$\|\mathbf{x}\|^2 \lambda_{min}(\mathcal{V}_2^{i,j}) \leq \mathbf{x}^T \mathcal{V}_2^{i,j} \mathbf{x} \leq \|\mathbf{x}\|^2 \lambda_{max}(\mathcal{V}_2^{i,j}) \quad (9.39)$$

$$\mathbf{x}^T \left(\mathbf{P}^j \mathbf{b}^i + \mathbf{P}^i \mathbf{b}^j \right) \leq \left(\|\mathbf{P}^i \mathbf{b}^j\| + \|\mathbf{P}^j \mathbf{b}^i\| \right) \|\mathbf{x}\| \quad (9.40)$$

$\dot{V}$ in (9.32) is negative for any $\mathbf{x}$ outsize the ball α if (9.35) and (9.36) are true. □

The following minimization problem searches $\mathbf{P}^i$, $i = 1, ..., L$ which make the fuzzy Lyapunov function as in (9.28) for the system (9.24).

$$\begin{aligned}
\underset{\mathbf{P}^i}{\text{minimize}} \quad & \rho_1, \rho_2 \\
\text{subject to} \quad & \mathcal{V}_1^i - \rho_1 \mathbb{I} < 0 \\
& \mathcal{V}_2^{i,j} - \rho_2 \mathbb{I} < 0 \\
& \mathbf{P}^i > 0, \quad i = 1, ..., L
\end{aligned} \quad (9.41)$$

Here, **I** is the identity matrix of appropriate dimensions and ρ_1 and ρ_2 are small scalars. The parameters $\mathbf{A}^k$ and $\mathbf{b}^k$ in (9.24) are learned from the demonstration data in an unconstrained optimization process. it is highly probable that the model learned through GMR is unstable and hence, it is possible that there does not exist such $\mathbf{P}^i$, $i = 1, ..., L$ that satisfy (9.35) and (9.36). In that scenario the intention is to find $\mathbf{P}^i$s such that the L.H.S of (9.35) and (9.36) have a value close to zero. These $\mathbf{P}^i$s subsequently help to design an associated stabilizing controller which has the least impact on motion regeneration.

9.4.1.2 Learning Fuzzy Controller Gains

After solving the LMIs in (9.41), the asymptotic stability of the system (9.24) can be ensured by choosing $\mathbf{K}^i$ and $\mathbf{k0}^i$ properly. The following theorem defines the global asymptotic stability of the motion model

Theorem 9.4. *The motion model given in (9.24) is globally asymptotically stable if the fuzzy controller gains $\mathbf{K}^i$ and $\mathbf{k0}^i$ in (9.25) are chosen as the following.*

$$\lambda_{max}(\mathcal{V}_{K1}^i) + \frac{2}{\alpha} \|\mathbf{P}^i (\mathbf{b}^i - \mathbf{k0}^i)\| < 0, \quad i = 1, ..., L \quad (9.42)$$

$$\lambda_{max}(\mathcal{V}_{K2}^{i,j}) + \frac{2}{\alpha}(\|\mathbf{P}^i (\mathbf{b}^j - \mathbf{k0}^j)\| + \|\mathbf{P}^j (\mathbf{b}^i - \mathbf{k0}^i)\|) < 0,$$
$$i < j = 1, ..., L. \quad (9.43)$$

and

$$\mathbf{k0}^i = \mathbf{b}^i, \quad \forall i \in L_\alpha \quad (9.44)$$

where

$$\mathcal{V}_{K1}^i = (\mathbf{A}^i - \mathbf{K}^i)^T \mathbf{P}^i + \mathbf{P}^i (\mathbf{A}^i - \mathbf{K}^i) \quad (9.45)$$

$$\mathcal{V}_{K2}^{i,j} = (\mathbf{A}^i - \mathbf{K}^i)^T \mathbf{P}^j + (\mathbf{A}^j - \mathbf{K}^j)^T \mathbf{P}^i + \mathbf{P}^i (\mathbf{A}^j - \mathbf{K}^j)$$
$$+ \mathbf{P}^j (\mathbf{A}^i - \mathbf{K}^i) \quad (9.46)$$

and L_α is the set of rules which are fired inside the ball α around origin.

Proof. According to Lemma 9.1 $\dot{V}$ is always negative outside the ball α around the origin if the constrains (9.35) and (9.36) are satisfied. It is possible to find $\mathbf{K}^i$ and $\mathbf{k0}^i$ such that (9.42)-(9.44) are true. Therefore, $\dot{V}$ of the system (9.24) is always negative on or outside the ball α around the origin. Since, $\frac{2}{\alpha} \|\mathbf{P}^i (\mathbf{b}^i - \mathbf{k0}^i)\| \geq 0$ and $\frac{2}{\alpha}(\|\mathbf{P}^i (\mathbf{b}^j - \mathbf{k0}^j)\| + \|\mathbf{P}^j (\mathbf{b}^i - \mathbf{k0}^i)\|) \geq 0$, the optimization algorithm finds $\mathbf{K}^i$ and $\mathbf{k0}^i$ such that, $\mathcal{V}_{K1}^i < 0$ and $\mathcal{V}_{K2}^{i,j} < 0$ while satisfying (9.42)-(9.44). Therefore, $\dot{V}$ takes the following form inside the ball α around origin.

$$\dot{V} = \mathbf{x}^T \left(\sum_{i=1}^{L} \sum_{j>i}^{L} \sigma_i \sigma_j \mathcal{V}_{K2}^{i,j} + \sum_{i=1}^{L} \sigma_i^2 \mathcal{V}_{K1}^{i} \right) \mathbf{x}$$
$$< 0, \quad \forall \mathbf{x} \neq 0$$
$$= 0, \quad \mathbf{x} = 0 \tag{9.47}$$

Hence, system (9.24) is globally asymptotically stable. □

The fuzzy controller gains $\mathbf{K}^i$ and $\mathbf{k0}^i$ are searched in the following optimization problem:

$$\begin{aligned}
& \underset{\mathbf{K}^i, \mathbf{k0}^j}{\text{minimize}} && \mathcal{E}_{model}, \mathcal{E}_{reg} \\
& \text{subject to} && \lambda_{max}(\mathcal{V}_{K1}^i) + \frac{2}{\alpha} \|\mathbf{P}^i (\mathbf{b}^i - \mathbf{k0}^i)\| < 0, \quad i \in L_r \\
& && \lambda_{max}(\mathcal{V}_{K2}^{i,j}) + \frac{2}{\alpha} (\|\mathbf{P}^i (\mathbf{b}^j - \mathbf{k0}^j)\| + \\
& && \|\mathbf{P}^j (\mathbf{b}^i - \mathbf{k0}^i)\|) < 0, \quad i < j \in L_r. \\
& && \mathbf{k0}^i = \mathbf{b}^i, \quad i \in L_\alpha
\end{aligned} \tag{9.48}$$

where the terms $\mathcal{E}_{model}$ and $\mathcal{E}_{reg}$ are defined as

$$\mathcal{E}_{model} = \frac{1}{D} \sum_{i}^{D} \frac{1}{N_i} \sum_{j}^{N_i} \|\dot{\mathbf{x}} - \hat{\dot{\mathbf{x}}}\| \tag{9.49}$$

$$\mathcal{E}_{reg} = \frac{1}{L} \sum_{i \in L_r} \|\mathbf{K}^i\| + \frac{1}{L - l_\alpha} \sum_{j \notin L_\alpha} \|\mathbf{k0}^j\| \tag{9.50}$$

Here, l_α is the number of element in L_α and L_r is the set of rules in the rule base. D represents the number of demonstrations and N_i represents the number of samples in ith demonstration. The optimization problem is solved using GA, where the parameters $\mathbf{K}^i$ and $\mathbf{k0}^i$ are searched in a constrained search space. $\mathcal{E}_{reg}$ is called the weight regularization cost which we shall discuss later.

Theorem 9.4 helps to design a globally asymptotically stable motion model that learns to generate stable human-like motion for a robotic manipulator. Fuzzy controller gains $\mathbf{K}^i$ and $\mathbf{k0}^i$ are searched based on the fuzzy Lyapunov function, defined in (9.28), such that the motion model has always a negative $\dot{V}$ for nonzero states. However, mere finding of the controller gains satisfying the constrains (9.42) to (9.44) may lead to poor performance of the motion model in regard to reconstruction of demonstrated motion profile, though the executed trajectory is globally asymptotically stable. Therefore, the selected gains should be from the feasible region of the search space, where it also minimizes the model reconstruction error. It is interesting to note here that if the existence of $\mathbf{P}^i$ matrices is discovered in the minimization problem (9.41), then the solution to $\mathbf{K}^i$ and $\mathbf{k0}^i$ can be simply set to zero. This means the GMR model that is learned from the demonstrations, is itself asymptotically stable. This will reduce the computation time of the learning problem.

9.4.1.3 Design of Fitness Function

The design of the fitness function is such that the parameters $\mathbf{K}^i$ and $\mathbf{k0}^i$ evolve in a direction, where the function approximation error is minimized while satisfying the constraints (9.42) to (9.44). The detailed design of the fitness function is given in algorithm 8.

Algorithm 8 Design of fitness function

1: **for** Each generation **do**
2:   **for** Each chromosome **do**
3:     Set $\mathbf{k0}^l|_{l \in L_\alpha} = \mathbf{b}^l$.
4:     Formulate $\mathcal{V}_{K1}^i$ and $\mathcal{V}_{K2}^{i,j}$, $i, j = 1, ..., L$, $i > j$ as in (9.45) and (9.46).
5:     Check all the constraints (9.42) to (9.44) and if constraints are satisfied, set $flag = 1$.
6:     **if** ($flag$) **then**
7:       Calculate model error as given in (9.49).
8:       Calculate weight regularization cost as in (9.50).
9:
10:       **return** $\mathcal{E}_{model} + \rho \mathcal{E}_{reg}$ (Here ρ is a positive scalar that sets the priority between two costs.)
11:     **end if**
12:   **end for**
13: **end for**

9.4.1.4 Example

Learn a C-FuzzStaMP motion encoding system from the real demonstrations $\mathscr{D} = \{[\mathbf{x}_{i,j}, \dot{\mathbf{x}}_{i,j}]_{j=1}^{N_i}\}_{i=1}^D$, where $\mathbf{x}_{i,j}$ is the position state of the robot end-effector in ith instant of jth demonstration.

An initial model is learned as per (9.14) that can encode point to point motion of the robotusing GMR of 5 Gaussians. Parameters of the model as in (9.15) are learned using expectation maximization (EM) algorithm. A fuzzy Lyapunov function is searched for the motion model learned in (9.14). The associated Lyapunov function is searched by searching $\mathbf{P}^i$, $i = 1, ..., L$ in a minimization problem as given in (9.41). The rule base of the fuzzy Lyapunov function is created by assigning the weighting function of the motion model as in (9.14) as the antecedent part of the rule base. The problem of finding $\mathbf{P}^i$ is solved using semi definite programming in YALMIP [247]. A fuzzy controller that shares the same rule antecedent part, is learned in another optimization process as in (9.48). The controller parameters $\mathbf{K}^i$ and $\mathbf{k0}^i$ are searched using GA where the fitness function is implemented as in Algorithm 8. The penalty ρ

Algorithm 9 Steps for execution
1: Training: Show user demonstration to the robot and create database $\mathscr{D} = \{[\mathbf{x}_{i,j}, \dot{\mathbf{x}}_{i,j}]_{j=1}^{N_i}\}_{i=1}^{D}$, where $\mathbf{x}_{i,j}$ is the position state of the robot end-effector in a shifted coordinate system whose origin coincides with the target point of the demonstration.
2: Learn the underlying nonlinear function $\mathbf{f}(\mathbf{x})$ as in (9.14) from the database $\mathscr{D}$ using EM.
3: Learn the Lyapunov function ($\mathbf{P}^i$) and associated fuzzy controller gains $\mathbf{K}^i$ and $\mathbf{k0}^i$.
4: Execution:Select an initial end-effector state and compute the velocity command from the motion model (9.24).
5: Use inverse kinematics to get the joint space solutions.
6: Evolve the system until the trajectory reaches the target state.

---

is chosen as 0.01 since our main priority is to minimize the function approximation error. The ball α is chosen as $\alpha = 0.1$ for this experiment. Please note that the parameter α is trade off in model performance between the regions inside and outside the ball. A bigger α gives better model performance outside the ball, where as a smaller α may distort the model outside the ball but gives a smaller area inside the ball. The chromosome size is taken as 500 and the parameters are evolved for 1000 generations. GA searches for $nL(n+1) - n(L-l_\alpha)$ variables, where $n = 3$ and l_α is the number of rules in L_0.

In all the figures, the human demonstration is shown in grey trajectories and executed trajectory by the motion model is given in black color. Figure 9.2 and Figure 9.3 are the plots of executed trajectories for an initial state starting inside the **RoD**. (Associated trajectory in each coordinate is given in Figure 9.4). The initial end-effector position is taken as $[0.4\ 0.47\ 0.074]^T$

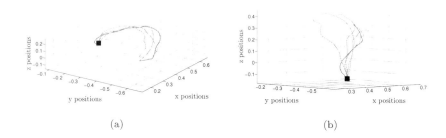

FIGURE 9.2: The grey trajectories represent the human demonstrations and the black one is the output of the learned motion model which is to be globally asymptotically stable. The initial state and the equilibrium state (black square) is chosen in side RoD for this experiment. The C-FuzzStaMP model is learned with (a) 3 demonstrations, and (b) 5 demonstrations.

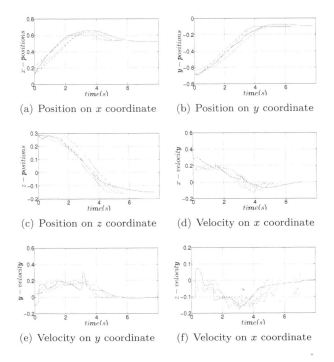

FIGURE 9.3: Projection plots of position and velocity of Figure 9.2(a). In this experiment, the initial state is chosen inside the *RoD*. The output trajectory is spatially similar with the demonstrated trajectories as the learned motion primitives are properly fired.

and the target position is taken as $[0.7 \ -0.18 \ -0.17]^T$. Figure 9.5(a) and Figure 9.6 show how the C-FuzzStaMP model behaves when the initial position is chosen as $[-0.32 \ 0.53 \ -0.4]$ which is a state outside the **RoD** with the target point $[0.7 \ -0.18 \ -0.17]^T$. (The associated error trajectories in each coordinate are given in Figure 9.7). The motion model generates trajectory as it was taught by the demonstrator when the end-effector started within the region of learning however, a different pattern is executed when the trajectory starts outside **RoD**. It happens as the motion primitives are not properly fired, since the region where the trajectory starts, is unexplored to the motion model.

9.4.2 Motion Modeling with R-FuzzStaMP

The C-FuzzStaMP scheme presented in the previous section sometimes may fail to encode subtle features of the demonstrations specially when the demonstrated trajectory is not simple. In such scenario, the executed trajectory is not precise reproduction of the demonstration, even though the target position (equilibrium point) of the trajectory is globally asymptotically stable

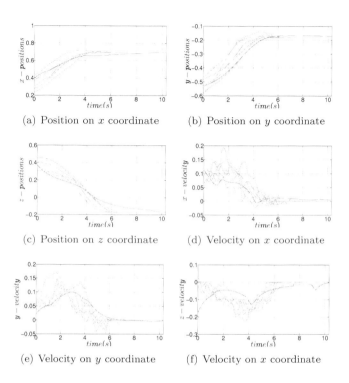

(a) Position on x coordinate
(b) Position on y coordinate
(c) Position on z coordinate
(d) Velocity on x coordinate
(e) Velocity on y coordinate
(f) Velocity on x coordinate

FIGURE 9.4: Projection plots of position and velocity of Figure 9.2(b). In this experiment, the initial state is chosen inside the *RoD*. The output trajectory is spatially similar with the demonstrated trajectories as the learned motion primitives are properly fired.

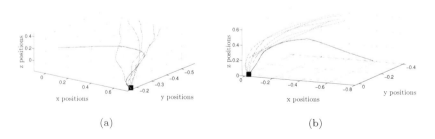

FIGURE 9.5: The C-FuzzStaMP model is used to generate trajectory starting outside **RoD**. (a) is the end-effector positions in the workspace, and (b) is evolution of the error state in $3D$. The reason for the output of the motion model being different from the demonstrations is, the learned local models are not fired properly as the states start outside the domain of learning.

FuzzStaMP: Fuzzy Controller Regulated Stable Movement Primitives

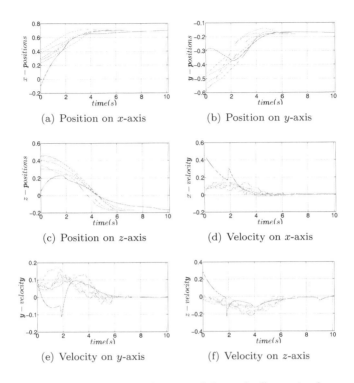

FIGURE 9.6: Positions and the velocities of the end-effector in the experiment using C-FuzzStaMP model with initial state starting outside RoD.

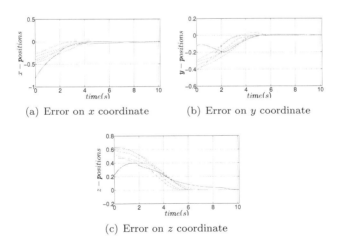

FIGURE 9.7: Position error state of the robot end-effector in each coordinate is presented in the figure.

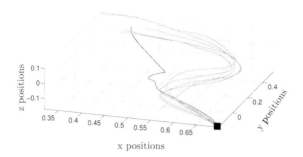

FIGURE 9.8: A C-FuzzStaMP model is learned from complex demonstrations. The initial end-effector position is chosen in RoD, where the local models have maximum firing. It is evident that the conservative model performs poorly as the optimization process compromises between the function approximation capability and the global asymptotic stability of the motion model.

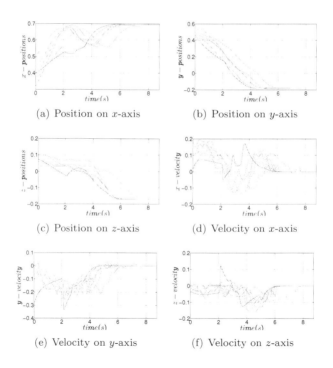

(a) Position on x-axis

(b) Position on y-axis

(c) Position on z-axis

(d) Velocity on x-axis

(e) Velocity on y-axis

(f) Velocity on z-axis

FIGURE 9.9: Positions and velocities of complex task execution by the C-FuzzStaMP model. The grey trajectories represent the original demonstrations and the black one is the output of the learned motion model.

FuzzStaMP: Fuzzy Controller Regulated Stable Movement Primitives 339

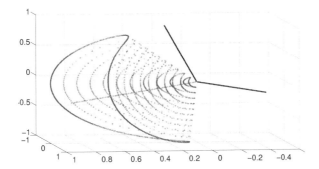

FIGURE 9.10: The grey and black lines represent the eigen vectors of **M** corresponding to positive and negative eigen values respectively. The region in black is spanned extremely on the subspace of two eigen vectors of different sign. $m(\mathbf{x})$ is positive inside the region. The replica of the region (is not shown here) also represent the same in the opposite direction of positive eigen vector.

(see Figures 9.8 and 9.9). The reason for such behavior is that, in order to learn a globally asymptotically stable motion model by imposing strict constraints, the poles of the closed loop system is pushed far negative on the real axis, such that, $\dot{V}$ of the system (9.24) is negative at any circumstances over the entire state space. Theoretically, the region of stability of closed loop system is expanded over the whole state space. In this process, local information encoding by the motion model is deprioritized over the global asymptotic stability and hence the performance is poor in the local perspective. In many of the applications of learning by demonstrations, such globally asymptotically stable model is redundant and moreover, a robot's workspace may not include the entire workspace. Considering the fact, a locally stable model is sufficient for producing stable trajectories if we can restrict the execution inside the stable region. R-FuzzStaMP is the relaxed version of the previous model, where the stability constraints are redefined to encode subtle feature of the demonstrations. Relaxing the stability constraints in fact broadens the solution space which helps minimizing the trajectory reconstruction error. To make the motion model more generalized, i.e., making it free from the restriction that model's operation is limited within the certain region of the workspace, a state translation technique is use. The methodology presented in this section can encode more complex demonstrations and makes the model globally valid.

9.4.2.1 Stability Analysis of the Motion System

For the system (9.24), the state evolution through time remains bounded in the region, where $\dot{V}$ of the system is negative. The states become unbounded, when $\dot{V}$ of the system is positive. Henceforth, we call the region stable, where $\dot{V}$ is negative; we call the region unstable otherwise.

Stable Region in the State Space

Stable region is the region of the state space where the time derivative of the Lyapunov function is negative, i.e., the states are bounded. For a negative definite matrix $\mathbf{M} \in \Re^{n \times n}$, the following can be written

$$m(\mathbf{x}) = \begin{cases} \mathbf{x}^T \mathbf{M} \mathbf{x} < 0, & \forall \mathbf{x} \neq 0 \\ 0, & \text{at } \mathbf{x} = 0 \end{cases} \quad (9.51)$$

The entire state space can be regarded as the stable region in this case. But such conclusions cannot be drawn when $\mathbf{M}$ is sign indefinite. However, the region in the state space can be identified for which the sign of the function $m(\mathbf{x})$ is known. Any state $\mathbf{x}$ falling on the eigen vector of matrix $\mathbf{M}$, is maximally scaled after the dot product with $\mathbf{M}$ to the direction of the eigen vector. Therefore, for a symmetric sign indefinite matrix $\mathbf{M}$ the function $m(\mathbf{x})$ changes its sign (goes +ve to −ve or vice versa) in the region between eigen vectors associated with the eigen values of different signs. The region expands up to certain angle with the eigen vectors in $\Re^n$, where the sign of the $m(\mathbf{x})$ is the sign of the corresponding eigen value. This can be visualized in Figure 9.10. In fact, the region can be represented by a cone in $\Re^2$ for a 2D-space. However for $n > 2$, the boundary of the stable/unstable region is not of a cone due to the asymmetricity in the dominance of eigen values. The minimum and maximum span of the stable/unstable region happens on the subspace created by two eigen vectors corresponding to two eigen values of different sign. Thus, $m(\mathbf{x})$ is positive for $\mathbf{x}$ around the positive eigen vector/s. Relating this phenomenon to $\dot{V}$ in (9.32), we can say $\mathbf{x}$ will go away from the equilibrium point around the positive eigen vector/s, if we assume the bias term to be zero for this moment.

Maximum Span of Stable/Unstable Region

The span of the stable/unstable region depends on the dominance of the eigen values of $\mathbf{M}$ if we consider (9.51). The maximum span occurs for the unstable region on the plane created by the eigen vectors of maximum positive and negative eigen values. The span can be computed as follows.

Let $\mathbf{x}_0$ is on the subspace created by eigen vectors associated with maximum positive and negative eigen values of $\mathbf{M}$ in (9.51), such that $m(\mathbf{x}_0) = 0$. Also assume that the maximum positive and negative eigen values are λ_{max}^+ and λ_{max}^- respectively and corresponding eigen vectors are $v_{\lambda+}$ and $v_{\lambda-}$. Then $\mathbf{x}_0$ can be written as the linear combination of $v_{\lambda+}$ and $v_{\lambda-}$.

$$\mathbf{x}_0 = \cos(\theta) v_{\lambda+} + \sin(\theta) v_{\lambda-}, \quad \forall \theta \in [0, \frac{\pi}{2}] \quad (9.52)$$

Therefore,

$$\mathbf{x}_0^T \mathbf{M} \mathbf{x}_0 = \lambda_{max}^+ \cos^2(\theta) + \lambda_{max}^- \sin^2(\theta), \quad \forall \, \|\mathbf{x}_0\| = 1 \quad (9.53)$$

We are looking for the $\mathbf{x}_0$ for which $\mathbf{x}_0^T \mathbf{M} \mathbf{x}_0 = 0$. Therefore,

$$\lambda_{max}^+ \cos^2(\theta) + \lambda_{max}^- \sin^2(\theta) = 0$$

$$\frac{sin(\theta)}{cos(\theta)} = \sqrt{-\frac{\lambda^+_{max}}{\lambda^-_{max}}}$$

$$\theta = arctan\left(\sqrt{-\frac{\lambda^+_{max}}{\lambda^-_{max}}}\right) \tag{9.54}$$

The maximum span of the unstable region is defined by the angle $\mathcal{S}$ between $\mathbf{x}_0$ and the corresponding positive eigen vector, i.e.,

$$\mathcal{S} = cos^{-1}\left(\frac{\mathbf{x}_0^T \lambda^+_{max}}{\|\mathbf{x}_0\|\|\lambda^+_{max}\|}\right) \tag{9.55}$$

Therefore,

$$\mathbf{x}^T \mathbf{M} \mathbf{x} > 0, \quad \forall \mathbf{x} \quad \mathcal{S}_{\mathbf{x}_0} < \mathcal{S} \tag{9.56}$$
$$\mathbf{x}^T \mathbf{M} \mathbf{x} < 0, \quad \forall \mathbf{x} \quad \mathcal{S}_{\mathbf{x}_0} > \mathcal{S} \tag{9.57}$$

where $\mathcal{S}_{\mathbf{x}_0}$ is the absolute value of the angle between $\mathbf{x}_0$ and the corresponding positive eigen vector. It is interesting to note that $\frac{\lambda^+_{max}}{\lambda^-_{max}} < 0$ since there does not exist such $\mathbf{x}$ for which $\mathbf{x}^T \mathbf{M} \mathbf{x} = 0$ when $\mathbf{M}$ is positive definite or negative definite. The region may be termed as unstable region[1].

Effect of Bias Term on Stability

The bias term $\mathbf{b}^k$ acts as a disturbance to the model (9.24). The effect of the disturbance will differ in terms of stability depending upon the region of the state space, where the model is currently operated. Here the effect of bias component is analyzed in terms of the sign of $\dot{V}$ in (9.32). From (9.32), we can see $\mathbf{b}^k$ contributes to $\dot{V}$ through the term $\mathbf{P}^i \mathbf{b}^i$ (where $i, j = 1, 2, .., L$). Thus, in this analysis the component $\mathbf{P}^i \mathbf{b}^i$ is treated as the bias term. Its effect is the strongest for a state near to the origin of the system since the contours of stable region shrinks significantly in this region. Therefore, any small disturbance that makes positive contribution to $\dot{V}$, can lead to instability of the system.

Definition 9.1. *A wining eigen vector is defined as the closest eigen vector to the mean demonstration for a local model.*

Definition 9.2. *Region of demonstrations (RoD) is a connected subset containing all the positions in the demonstrations and its neighborhood considering the robot's workspace as set.*

Definition 9.3. *Region of operation (RoO) is a connected subset containing all the positions in the region, where the robot is operated.*

---

[1] The bias term contributes positively to $\dot{V}$ only when it has an angle $< \frac{\pi}{2}$ with the state $\mathbf{x}$.

Lemma 9.2. *The bias term $\mathbf{P}^i\mathbf{b}^j$ associated with the active local model has a negative contribution to $\dot{V}$ in (9.32) if the bias or disturbance falls on the negative of the wining eigen vector of the corresponding models.*

Proof. For any state $\mathbf{x}$,

$$\mathbf{x}^T\mathbf{P}^i\mathbf{b}^j = \|\mathbf{x}\|\|\mathbf{P}^i\mathbf{b}^j\|cos(\theta) \leq 0, \quad \frac{\pi}{2} \geq \theta \leq \frac{3\pi}{2} \quad (9.58)$$

where θ is the angle between $\mathbf{x}$ and $\mathbf{P}^i\mathbf{b}^j$. Regions of the state space, where the states are at least $\frac{\pi}{2}$ radians away from the bias term, have zero or negative contribution to $\dot{V}$. Since the wining eigen vector is the closest eigen vector to the neighborhood states (locally), any vector having an angle of π radians with the wining eigen vector, will have an angle within the range $\frac{\pi}{2}$ and $\frac{3\pi}{2}$ with those states. Hence, $\mathbf{P}^i\mathbf{b}^j$ falling on negative of the wining eigen vector, will always have negative contribution to $\dot{V}$. □

Lemma 9.2 is an ideal scenario. In practice, we try to maximize the angle between the bias term and the wining eigen vector of the firing local model without effecting the trajectory regeneration accuracy of the motion model significantly. The idea is to put the bias term at least $\frac{\pi}{2}$ radians away from the **RoD**. It is observed that the bias terms of the learned local models are generally stay $\frac{\pi}{2}$ radians away from the **RoD**, since the demonstrations are itself stable motion.

9.4.2.2 Design of the Fuzzy Controller

The fuzzy controller plays a significant role in motion stability. The gains $\mathbf{K}^i$ and $\mathbf{k0}^i$ in (9.24) are chosen such that it can shrink the unstable region. The parameter $\mathbf{k0}^i$ is chosen to handle the bias term as per the requirement of stability. The fuzzy controller makes the zeroth local models (models that include the equilibrium state of the system) asymptotically stable by reducing the unstable region to zero, since the region near the origin is less spacious in regard to the stable region. Therefore, the fuzzy controller is given by

r^{th} *rule:*

IF $\mathbf{x}$ is G_r **THEN**
$$\mathbf{u}_r = \mathbf{k}_0^r + \mathbf{K}^r\mathbf{x} \quad (9.59)$$
where
$$\mathbf{k0}^r|_{r \in L_0} = \mathbf{b}^r \quad (9.60)$$

where L_0 is the set of rules that include the origin. The following theorem ensures the local stability of the motion model

Theorem 9.5. *The motion model (9.24) is locally stable with the controller (9.59) if* $\mathbf{RoO} \subseteq \mathbf{RoD}$ *while the following constraints are true.*

$$\mathcal{S}^i < \mathcal{S}^i_{RoD} \quad (9.61)$$

$$\mathcal{S}^{i,j} < \mathcal{S}^{i,j}_{RoD} \tag{9.62}$$

$$\frac{\pi}{2} + \xi_{\mu^i} \leq \psi^i_{\mathbf{Pb}} \leq \frac{3\pi}{2} + \xi_{\mu^i} \tag{9.63}$$

$$\frac{\pi}{2} + \xi_{\mu^{i,j}} \leq \psi^{i,j}_{\mathbf{Pb}} \leq \frac{3\pi}{2} + \xi_{\mu^{i,j}} \tag{9.64}$$

$$\left(\mathbf{A}^k - \mathbf{K}^k\right)^T \mathbf{P}^k + \mathbf{P}^k \left(\mathbf{A}^k - \mathbf{K}^k\right) < 0 \tag{9.65}$$

$$\left(\mathbf{A}^k - \mathbf{K}^k\right)^T \mathbf{P}^l + \left(\mathbf{A}^l - \mathbf{K}^l\right)^T \mathbf{P}^k + \mathbf{P}^k \left(\mathbf{A}^l - \mathbf{K}^l\right)$$
$$+ \mathbf{P}^l \left(\mathbf{A}^k - \mathbf{K}^k\right) < 0 \tag{9.66}$$

where $i < j \notin L_0$ and $k < l \in L_0$

Here, $\mathcal{S}^{(\cdot)}$ is the maximum span of the unstable region of the learned model for the corresponding rules. $\mathcal{S}^{(\cdot)}_{RoD}{}^2$ is the minimum angle between the demonstrations for that rule and the corresponding positive eigen vector. $\psi^{(\cdot)}_{\mathbf{Pb}}$ is the angle between the bias term $\mathbf{P}^{(\cdot)}\mathbf{b}^{(\cdot)}$ and the mean demonstration $\mu^{(\cdot)}_{\mathbf{x}}{}^3$ for the corresponding rules. $\xi_{\mu^{(\cdot)}}$ is the maximum angle between the $\mu^{(\cdot)}_{\mathbf{x}}$ and the demonstration for the corresponding local model or rule.

Proof. $\mathcal{S}^{(\cdot)}_{RoD}$ can be interpreted as the boundary of the region of demonstration for the robot, i.e., there is no state that has an absolute angle around the unstable eigen vector, less than $\mathcal{S}^{(\cdot)}_{RoD}$ in **RoD**. Since we are considering local stability, let us assume for now that the region of operation $\mathbf{RoO} \subseteq \mathbf{RoD}$. The controller gains are selected such that $\mathcal{S}^{(\cdot)} < \mathcal{S}^{(\cdot)}_{RoD}$. As $\mathcal{S}^{(\cdot)}$ is the maximum angle (absolute) or span of the unstable region for the local models / rules, using (9.57), the following can be written.

$$\mathbf{x}^T \left[\left(\mathbf{A}^i - \mathbf{K}^i\right)^T \mathbf{P}^i + \mathbf{P}^i \left(\mathbf{A}^i - \mathbf{K}^i\right)\right] \mathbf{x} < 0 \quad i \notin L_0, \; \forall \mathbf{x} \in \mathbf{RoO} \tag{9.67}$$

$$\mathbf{x}^T \left[\left(\mathbf{A}^i - \mathbf{K}^i\right)^T \mathbf{P}^j + \left(\mathbf{A}^j - \mathbf{K}^j\right)^T \mathbf{P}^i + \mathbf{P}^i \left(\mathbf{A}^j - \mathbf{K}^j\right)\right.$$
$$\left. + \mathbf{P}^j \left(\mathbf{A}^i - \mathbf{K}^i\right)\right] \mathbf{x} < 0, \quad i < j \notin L_0, \quad \forall \mathbf{x} \in \mathbf{RoO} \tag{9.68}$$

Using Lemma 9.2, (9.63), (9.64) and the above equations with (9.32), it can be said that

$$\dot{V} < 0, \quad i,j \notin L_0, \quad \forall \mathbf{x} \in \mathbf{RoO} \tag{9.69}$$

---

[2]$\mathcal{S}^{(\cdot)}_{RoD}$ is computed from the neighboring data associated with the local model. It is defined as

$$\mathcal{S}^{(\cdot)}_{RoD} \triangleq \min_{x \in \mathbf{RoD}} \angle \left(\mathbf{x}, \mathbf{v}^{(\cdot)}_{eig}\right)$$

where $\mathbf{v}^{(\cdot)}_{eig}$ is the corresponding eigen vector.

[3]$\mu^i_{\mathbf{x}}$ is the mean demonstration for the ith local model and can be regarded as the mean of the ith Gaussian projected on the position space. Similarly, $\mu^{i,j}_{\mathbf{x}}$ is the mean demonstration for ith and jth local models, which can be found from the means of the corresponding Gaussians weighted by their priors.

If the controller parameters $\mathbf{K}^i$ and $\mathbf{k0}^i$ are selected as in (9.65), (9.66) and (9.60) for rules in L_0,

$$\dot{V} < 0, \quad k, l \in L_0 \tag{9.70}$$

Hence, the motion model (9.24) is stable in $\mathbf{RoO} \subseteq \mathbf{RoD}$. □

Theorem 9.5 relaxes the constrains on learning of the motion model (9.24) as imposed by Theorem 9.4. The previous theorem makes the entire state space asymptotically stable which may not be required in most of the applications. The goal of this theorem is to learn a locally asymptotically stable model, i.e., to discover a region in the state space where the the $\dot{V}$ is always negative. The local region is considered here is the region of demonstrations. Theorem 9.5 ensures that the predicted trajectory by the motion model (9.24) is stable when the system is operated inside the region of demonstration, i.e., $\mathbf{RoO} \subseteq \mathbf{RoD}$.

Controller Parameters Selection:

The controller parameters $\mathbf{K}^i$ and $\mathbf{k0}^i$, $i = 1, ...L$ are learned in an constrained optimization process, where an objective cost is minimized. The objective function is chosen here such that minimization of such function would help reconstruct the demonstrated motion profile. $\mathscr{E}_{model}$ represents the objective function in the minimization problem. The constraints are taken from theorem 9.5 and is given by equation (9.60) to (9.66). The minimization problem is given in the following.

$$\underset{\mathbf{K}^i, \mathbf{k0}^j}{\text{minimize}} \quad \mathscr{E}_{model} \tag{9.71}$$

$$\text{subject to} \quad \mathcal{S}^i < \mathcal{S}^i_{RoD} \tag{9.72}$$

$$\mathcal{S}^{i,j} < \mathcal{S}^{i,j}_{RoD} \tag{9.73}$$

$$\frac{\pi}{2} + \xi_{\mu^i} \le \psi^i_{\mathbf{Pb}} \le \frac{3\pi}{2} + \xi_{\mu^i} \tag{9.74}$$

$$\frac{\pi}{2} + \xi_{\mu^{i,j}} \le \psi^{i,j}_{\mathbf{Pb}} \le \frac{3\pi}{2} + \xi_{\mu^{i,j}} \tag{9.75}$$

$$\left(\mathbf{A}^k - \mathbf{K}^k\right)^T \mathbf{P}^k + \mathbf{P}^k \left(\mathbf{A}^k - \mathbf{K}^k\right) < 0 \tag{9.76}$$

$$\left(\mathbf{A}^k - \mathbf{K}^k\right)^T \mathbf{P}^l + \left(\mathbf{A}^l - \mathbf{K}^l\right)^T \mathbf{P}^k +$$
$$\mathbf{P}^k \left(\mathbf{A}^l - \mathbf{K}^l\right) + \mathbf{P}^l \left(\mathbf{A}^k - \mathbf{K}^k\right) < 0 \tag{9.77}$$

$$\mathbf{k0}^l|_{l \in L_0} = \mathbf{b}^l \tag{9.78}$$

where $i < j \notin L_0$ and $k < l \in L_0$

Here, $\mathscr{E}_{model}$ is the model reconstruction error as given in (9.49). The above minimization problem is a constrained optimization problem and can be solved using evolutionary approaches. We use genetic algorithm (GA) to solve the problem. In GA, a chromosome represents a set of controller parameters $\mathbf{K}^i$ and $\mathbf{k0}^i$, $i = 1, ...L$. In each generation, many such chromosomes are checked for their fitness which is evaluated by their cost minimization capability. The evaluation is done using a fitness function.

Weight Regularization:

The above minimization problem does not include an weight regularization term, which may result the model to end up with high controller gain. Regularization is used to prevent over fitting problem of data-driven models. In other words, regularization of weight improves model generalization. The objective of the learning problem is accurate prediction of the end-effector velocity of the robot for a given position. Generally the learning is performed on a small subset of the input and output space. Additionally, the subset includes random noise due to the measurement errors. Unbounded search in the function space increases complexity in the model and leads to a over fitted model that works well on the training subspace, but fails to perform efficiently when the input does not belong to the training set. The regularization cost is given in the following using (9.50).

$$\mathcal{R}(\mathbf{K}, \mathbf{k0}) = p_{reg}\mathcal{E}_{reg} \tag{9.79}$$

Here p_{reg} is a positive scalar that acts as a trade off between the regularization cost and the main objective. The regularization cost is added to the main objective and the optimization problem is solved.

Design of the Fitness Function:

The efficient design of fitness function creates selection pressure on chromosome that can evolve toward the favorable region of the search space. The main objective is to find $\mathbf{K}^i$ and $\mathbf{k0}^i$, $i = 1, ...L$ that give minimum motion regeneration error (9.49) while the solution is constrained by (9.72) to (9.78). The hard constraints are solved by modifying (9.72) to (9.78) in the following manner.

$$\mathcal{S}^i - \mathcal{S}^i_{RoD} + \epsilon_1 \leq 0 \tag{9.80}$$

$$\mathcal{S}^{i,j} - \mathcal{S}^{i,j}_{RoD} + \epsilon_2 \leq 0 \tag{9.81}$$

$$\psi^i_{\mathbf{Pb}} - \frac{3\pi}{2} - \xi_{\mu^i} + \epsilon_3 \leq 0 \tag{9.82}$$

$$\frac{\pi}{2} + \xi_{\mu^i} - \psi^i_{\mathbf{Pb}} + \epsilon_4 \leq 0 \tag{9.83}$$

$$\psi^{i,j}_{\mathbf{Pb}} - \frac{3\pi}{2} - \xi_{\mu^{i,j}} + \epsilon_5 \leq 0 \tag{9.84}$$

$$\frac{\pi}{2} + \xi_{\mu^{i,j}} - \psi^{i,j}_{\mathbf{Pb}} + \epsilon_6 \leq 0 \tag{9.85}$$

$$\left(\mathbf{A}^k - \mathbf{K}^k\right)^T \mathbf{P}^k + \mathbf{P}^k \left(\mathbf{A}^k - \mathbf{K}^k\right) + \epsilon_7 \mathbb{I} \leq 0 \tag{9.86}$$

$$\left(\mathbf{A}^k - \mathbf{K}^k\right)^T \mathbf{P}^l + \left(\mathbf{A}^l - \mathbf{K}^l\right)^T \mathbf{P}^k + \mathbf{P}^k \left(\mathbf{A}^l - \mathbf{K}^l\right) + \mathbf{P}^l \left(\mathbf{A}^k - \mathbf{K}^k\right) + \epsilon_8 \mathbb{I} \leq 0 \tag{9.87}$$

$$\mathbf{k0}^l \big|_{l \in L_0} = \mathbf{b}^l \tag{9.88}$$

where $i < j \notin L_0$ and $k < l \in L_0$

Here, ϵ_c, $c = 1, ..., 8$ are positive scalars and $\mathbb{I}$ represents the identity matrix of appropriate dimensions. Each chromosome is first checked for its feasibility. A set of parameters from the feasible region is then checked for the motion regeneration accuracy. Chromosomes with minimum cost participate in reproduction and generate another set of parameters for the next generation.

The Algorithm 10 presents the fitness function for learning the controller gains.

Algorithm 10 Fitness function

1: **for** Each generation **do**
2:   **for** Each chromosome **do**
3:     Set $\mathbf{k0}^l|_{l \in L_0} = \mathbf{b}^l$.
4:     Formulate $\mathcal{V}_{K1}^i$ and $\mathcal{V}_{K2}^{i,j}$, $i, j = 1, ..., L$, $i > j$ as in (9.45) and (9.46).
5:     Calculate the maximum span of the region of instability $\mathcal{S}^i$, $\mathcal{S}^{i,j}$, $i, j = 1, ..., L$, $i > j$ as in (9.55).
6:     Calculate $\psi_{\mathbf{Pb}}^i$, $\psi_{\mathbf{Pb}}^{i,j}$, ξ_{μ^i}, and $\xi_{\mu^{i,j}}$.
7:     Check all the constraints (9.80) to (9.87) and if constraints are satisfied, set $flag = 1$.
8:     **if** $(flag)$ **then**
9:       Calculate model error as given in (9.49).
10:       Calculate weight regularization cost as in (9.79).
11:
12:       **return** $\mathcal{E}_{model} + \mathcal{R}(\mathbf{K}, \mathbf{k0})$
13:     **end if**
14:   **end for**
15: **end for**

9.4.3 Global Validity and Spatial Scaling

Earlier in this section, we have presented the motion model to be stable and valid when $\mathbf{RoO} \subseteq \mathbf{RoD}$. But in practical applications, the motion model would be more useful if it is stable and valid over the entire robot workspace. We adopt an algorithmic approach to include the entire workspace in $\mathbf{RoO}$. Let us assume that $\mathcal{I}_G$ is the spatial indexing of MPs in terms of demonstrations. $\mathcal{I}_G^k$, $k = 1, ..., L$ is an element of $\mathcal{I}_G$, represents the index of MP that is activated in the kth region of $\mathbf{RoD}$. Region 1 ($k = 1$) is the subspace where the demonstrations start. Region 2 ($k = 2$) comes second and so on. Similarly, region L ($k = L$) is the end of demonstrations, which includes origin. These local models are associated with Gaussian distributions which are learned from the demonstrations. To maintain the similar characteristics of the demonstration in the executed trajectory, the execution also needs to maintain the order of the motion primitives. The spatial scaling properties in the executed trajectories can only be retained by maintaining the spatial order of the motion primitives.

Let us define two parameters β and κ to represent the **RoD**, where each element β_k is associated with kth motion primitive representing the maximum angle of the demonstration with the mean demonstration projected on the position space and κ_k is the maximum scaling factor of the farthest point in the **RoD** with respect to mean demonstration associated with kth motion primitive. β_k and κ_k are given as follows.

$$\beta_k(\mathbf{x}_{ang}) = cos^{-1}\left(\frac{\mathbf{x}_{ang}^T \mu_x^k}{\|\mathbf{x}_{ang}\|\|\mu_x^k\|}\right) \tag{9.89}$$

$$\kappa_k(\mathbf{x}_{dist}) = \frac{\|\mathbf{x}_{dist}\|}{\|\mu_x^k\|} \tag{9.90}$$

where $\mathbf{x}_{ang}$ is the farthest state in the demonstration in terms of the angle with the mean demonstration μ_x^k and $\mathbf{x}_{dist}$ is the farthest state with maximum length in the demonstration.

Rotational Transformation Matrix:

Let the mean demonstration is $\boldsymbol{m}$ and a given state $\boldsymbol{x}$ and also $\|\boldsymbol{m}\| = 1$ and $\|\boldsymbol{x}\| = 1$. Also assume the $\boldsymbol{x}$ is rotated by angle θ from $\boldsymbol{m}$. Since the rotation occurs only on a $2D$ plane with normal $\boldsymbol{m} \times \boldsymbol{x}$, a rotation matrix can be written in the following form

$$\mathbf{R}_{ang} = \begin{bmatrix} cos(\theta) & -sin(\theta) & 0 \\ sin(\theta) & cos(\theta) & 0 \\ 0 & 0 & 1 \end{bmatrix} \tag{9.91}$$

As the angle between the vectors are unknown, (9.91) can be rewritten as

$$\mathbf{R}_{ang} = \begin{bmatrix} \boldsymbol{m} \cdot \boldsymbol{x} & -\|\boldsymbol{m} \times \boldsymbol{x}\| & 0 \\ \|\boldsymbol{m} \times \boldsymbol{x}\| & \boldsymbol{m} \cdot \boldsymbol{x} & 0 \\ 0 & 0 & 1 \end{bmatrix} \tag{9.92}$$

$\mathbf{R}_{ang}$ in (9.92) represents the rotation matrix from $\boldsymbol{m}$ to $\boldsymbol{x}$ in the coordinate frame $[\boldsymbol{i}\ \boldsymbol{j}\ \boldsymbol{k}]$ and is given by

$$\boldsymbol{i} = \boldsymbol{m}; \quad \boldsymbol{j} = \frac{\boldsymbol{m} - (\boldsymbol{m} \cdot \boldsymbol{x})\boldsymbol{m}}{\|\boldsymbol{m} - (\boldsymbol{m} \cdot \boldsymbol{x})\boldsymbol{m}\|}; \quad \boldsymbol{k} = \boldsymbol{x} \times \boldsymbol{m} \tag{9.93}$$

The basis transformation matrix for the orthogonal basis $[\boldsymbol{i}\ \boldsymbol{j}\ \boldsymbol{k}]$ is given by

$$\mathbf{T}_b = \begin{bmatrix} \boldsymbol{m} & \frac{\boldsymbol{m} - (\boldsymbol{m}\cdot\boldsymbol{x})\boldsymbol{m}}{\|\boldsymbol{m} - (\boldsymbol{m}\cdot\boldsymbol{x})\boldsymbol{m}\|} & \boldsymbol{x} \times \boldsymbol{m} \end{bmatrix}^{-1} \tag{9.94}$$

The rotational transformation matrix in the base coordinate system can be written by

$$\mathbf{T}_r = \mathbf{T}_b^{-1} \mathbf{R}_{ang} \mathbf{T}_b \qquad (9.95)$$

Here $\mathbf{T}_r$ represents the rotational transformation matrix between vector $\boldsymbol{m}$ to $\boldsymbol{x}$ in the base frame such that $\mathbf{T}_r \boldsymbol{m} = \boldsymbol{x}$. It is interesting to note that $\mathbf{T}_b$ being a orthonormal matrix, $\mathbf{T}_b^{-1} = \mathbf{T}_b^T$.

Generation of Trajectory anywhere in the State Space:

By relaxing the stability constraints, we have learned a locally asymptotically stable system. The method works as the following: each properly learned motion primitive pushes the trajectory toward the next local model / motion primitive in the **RoD** with the order given in $\mathcal{I}_G$. During the trajectory evolution one motion primitive is chosen as the *motion primitive in charge* (MPiC) at any given state in the state space. MPiC is the local model which has the maximum weight for a given state. Any state that starts outside **RoD**, is translated in **RoD**. The state evolves in side **RoD** and translated back to its original domain. At each instant, it is checked whether $\beta(\mathbf{x})$ and $\kappa(\mathbf{x})$ of the state is within β_{MPiC} and κ_{MPiC}[4]. Corrective actions are taken when it fails. The procedure is explained in detail in Algorithm 11.

9.4.3.1 Examples

For this example, we shall consider complex demonstrations where the trajectories approach multiple directions before reaching to the target state. Let us take those demonstrations where the C-FuzzStaMP model failed.

An initial dynamical model (9.15) is first learned from the demonstrations using EM algorithm. As this initial model is not stable, the R-FuzzStaMP takes feedback based corrective measures to render the closed loop motion model as a stable system. The corrective actions are generated by the fuzzy controller which is learned using evolutionary optimization as given in Algorithm 10.

We have seen earlier that the simple demonstrations are easy to encode but subtle features of complex demonstrations are not efficiently captured in the model since the model parameters are chosen from a fully constrained space. EM algorithm is iterated over 5 such trajectories and the parameters in (9.15) are identified for 5 Gaussians. $\mathbf{P}^i$, $i = 1, ..., L$ are searched for a fuzzy Lyapunov function as given by the minimization problem (9.41). This is done as the previous experiment. The fuzzy controller gains $\mathbf{K}^i$ and $\mathbf{k0}^i$, $i = 1, ..., L$ are selected using GA, where the fitness function is designed as the Algorithm 10. GA toolbox of Matlab is used for this purpose. The number of variables need to be searched is $nL(n+1) - n(L - l_0)$, where $n = 3$ and l_0 is number of rules in L_0, which is 1 in this case. $\epsilon_c = 0.0001$, $c = 1, .., 8$ are chosen for

---

[4] β_{MPiC} and κ_{MPiC} are in fact $\beta_k(\mathbf{x}_{ang})$ and $\kappa_k(\mathbf{x}_{dist})$, where k=MPiC.

Algorithm 11 Globally stable trajectory and spatial scaling

1: Let the initial position is $\mathbf{x}$ after translation of the target position to the origin of the state space.
2: Set MPiC as the first model from the list $\mathcal{I}_G$.
3: Calculate $\beta_{\mathbf{x}}(\mathbf{x})$ and $\kappa_{\mathbf{x}}(\mathbf{x})$ for $\mathbf{x}$ based on the MPiC.
4: **if** $(\beta_{\mathbf{x}}(\mathbf{x}) > \beta_{MPiC}$ or $\kappa_{\mathbf{x}}(\mathbf{x}) > \kappa_{MPiC})$ **then**
5:     Define the translated state $\hat{\mathbf{x}} = \mu_{\mathbf{x}}^{MPiC}$, where $\mu_{\mathbf{x}}^{MPiC}$ is the mean of the local model chosen as the MPiC or in other words it is the mean demonstration related to MPiC local model.
6: **else**
7:     Define the translated state $\hat{\mathbf{x}} = \mathbf{x}$.
8: **end if**
9: Get the rotational transformation $\mathbf{T}_r$ and the scaling factor $\mathbf{T}_\kappa$ between the translated state $\hat{\mathbf{x}}$ and the initial state $\mathbf{x}$.
10: **while** $(\|\hat{\mathbf{x}}\| > 0)$ **do**
11:     Calculate $\dot{\hat{\mathbf{x}}}$ using the motion model (9.24).
12:     Evolve the state as the following

$$\hat{\mathbf{x}} = \hat{\mathbf{x}} + dt * \dot{\hat{\mathbf{x}}}, \qquad (9.96)$$

    where dt is the sampling time.
13:     The current state $\mathbf{x}$ in the original coordinate frame can be obtained as the following

$$\mathbf{x} = \mathbf{T}_\kappa \mathbf{T}_r \hat{\mathbf{x}} \qquad (9.97)$$

14:     Update MPiC.
15:     Calculate $\beta_{\mathbf{x}}(\mathbf{x})$ and $\kappa_{\mathbf{x}}(\mathbf{x})$ for $\mathbf{x}$ based on the MPiC.
16:     **if** $(\beta_{\mathbf{x}}(\mathbf{x}) > \beta_{MPiC}$ or $\kappa_{\mathbf{x}}(\mathbf{x}) > \kappa_{MPiC})$ **then**
17:         Set MPiC as the next model in the list $\mathcal{I}_G$.
18:         Update the translated state as $\hat{\mathbf{x}} = \mu_{\mathbf{x}}^{(MPiC)}$.
19:         Update the rotational transformation $\mathbf{T}_r$ and the scaling factor $\mathbf{T}_\kappa$ between the translated state $\hat{\mathbf{x}}$ and the current state $\mathbf{x}$.
20:     **end if**
21: **end while**

constraint handling. The chromosome size is taken as 500 and the parameters are evolved for 1,000 generations. Figure 9.11 and 9.12 show the trajectory plots of this experiment.

Performance inside **RoD**

To test the model's performance, we first assign a task to the robot with an end-effector initial state inside **RoD**. As in Figure 9.11, the model learns the demonstrations very well. The pattern in the demonstrated motion profiles

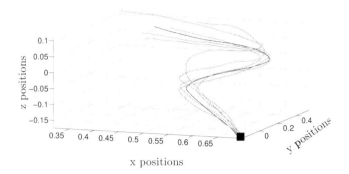

FIGURE 9.11: R-FuzzStaMP model is learned from complex demonstrations as the technique presented in Section 9.4.2. The initial end-effector position is chosen in *RoD*, where the local models have maximum firing. It is evident that performance of the R-FuzzStaMP model has a large improvement over the C-FuzzStaMP model.

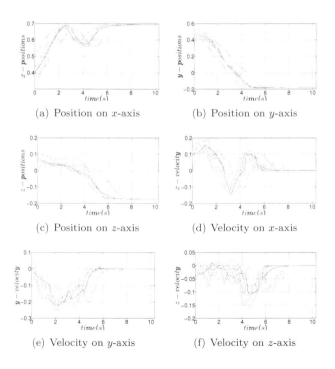

FIGURE 9.12: Positions and velocities of complex task execution by the R-FuzzStaMP model are here. The grey trajectories represent the original demonstrations and the black one is the output of the learned motion model.

is perfectly replicated. Figure 9.12 shows how the position and velocity are evolving on each coordinate.

Performance outside **RoD**

It is interesting to see the models performance outside the demonstrated region since it shows the generalization capability of model in the global perspective. In this case, the model generates trajectories in a region of the workspace where the demonstrator never explored. As Figure 9.13 shows, the end-effector trajectory exactly follows the demonstrated profile. In Figure 9.13(a) only the initial position is taken outside **RoD** and in Figure 9.13(b), both the initial and target positions are chosen outside **RoD**. The associated trajectories in each coordonate are given in Figures 9.14 and 9.15 respectively.

Spatial Scaling

The R-FuzzStaMP framework is able to spatially scale up and scale down the end-effector trajectory as per the requirement of the task. Let us first examine the performance of the motion model in scaling up the learned profiles. The initial and target states are chosen such that the length of the initial error state is significantly bigger than the demonstrations. The motion model executes an expanded trajectory generated by the Algorithm 11. Figure 9.16(a) shows the trajectory plots of the experiment. It is evident that the motion model has generated a bigger pattern than the demonstrated profile. Figure 9.17a-9.17c show how the trajectory in each axis has scaled up.

Another set of initial and target state is chosen, where the length of the initial error state is substantially smaller than the demonstrations. The end-effector motion is executed as earlier and the trajectory plots are given in the Figure 9.16(b) and Figure 9.17d-9.17f. It can be observed that the executed trajectory maintains the similar pattern as the demonstrations but in a smaller version. It is interesting to note that the special scaling on trajectories are performed outside **RoD**.

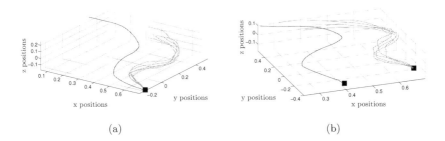

FIGURE 9.13: The end-effector trajectories in the workspace in complex task execution by the R-FuzzStaMP is shown. In this experiment, the initial end-effector state starts in the region outside **RoD**. (a) is the plot of trajectory when the target point is chosen from the demonstrations and (b) presents the trajectory when the target point is arbitrarily chosen outside **RoD**.

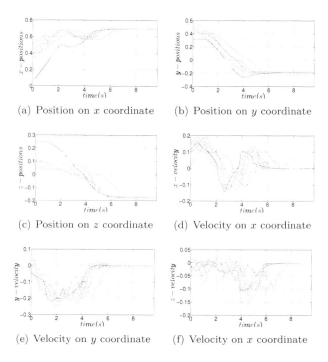

FIGURE 9.14: Position and velocity of the end-effector in each axis during complex task execution by the R-FuzzStaMP motion model is shown here. The equilibrium position is kept same as seen during the demonstrations.

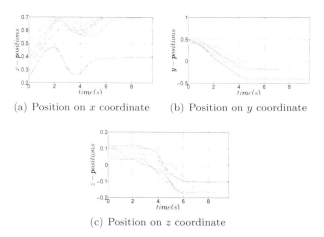

FIGURE 9.15: Position of the end-effector in each axis during complex task execution by the R-FuzzStaMP motion model is shown here. In this experiment, a different equilibrium position is chosen, which is not seen during the demonstrations.

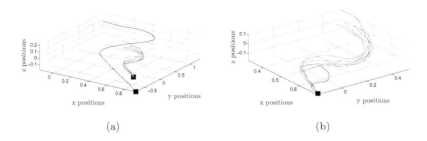

(a)   (b)

FIGURE 9.16: End-effector trajectories are spatially scaled depending on the length of the initial state. Here, grey trajectories are the demonstrated to the robot and black trajectories are executed by the motion model. (a) The trajectory is spatially scaled to create a larger pattern, and (b) shows a smaller pattern than the demonstrations.

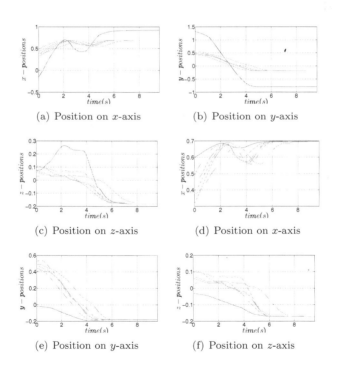

(a) Position on x-axis   (b) Position on y-axis

(c) Position on z-axis   (d) Position on x-axis

(e) Position on y-axis   (f) Position on z-axis

FIGURE 9.17: Position error evolution of the end-effector by the motion model is shown here. (a)-(c) refer to Figure 9.16(a) and (d)-(f) refer to the Figure 9.16(b).

9.5 Learning Skills from Heterogeneous Demonstrations

In the previous section we have seen how the demonstrated trajectories can be encoded in a dynamical regressive model known as FuzzStaMP. The asymptotic stability of the motion model has been ensured by imposing stability constraints during model parameter estimation. FuzzStaMP models are designed to encode demonstrations of a single task. However, many practical applications of robots require executing multiple task-trajectories. To be able to do that, a robot needs multiple motion encoders which are trained with different demonstrated skills. The kinematic controller must offer the features such as *multitasking capability, additional information processing*. *Multitasking capability* is another term for generic re-usability of the encoding system. A motion estimator is capable of multitasking, i.e., the motion model should is able to execute multiple task profiles based on the environmental input. The source of environmental input is either sensory information or the user's instruction to the robot. *Additional information processing* feature is in fact enables the controller to process additional sensory information (from the environment) and act accordingly to guide the robot.

The motion encoding algorithm discussed in this section is useful for learning multiple skills in a single model. Let us introduce the following definitions which are frequently used in the rest of this section.

Definition 9.4. *A* dynamical system *is represented as a tuple* (S, T, R) *where S is the state space, T is a set of times and R is the rule that governs the state evolution through time, such that $R : S \times T \to S$ which means $S \times T$ and S are domain and co-domain of R.*

Definition 9.5. *The* energy dissipation rate *(EDR) is defined as the rate at which the energy of the states of a dynamical system changes while evolving through time.*

Definition 9.6. $\Gamma(x, \tau) \in \{0, 1\}$ *is a switching function and is defined as the following:*

$$\Gamma(x, \tau) = sign\left(\frac{\tau - x}{|x - \tau|} + 1\right) \quad for \ x \neq \tau \tag{9.98}$$
$$= 1 \quad at \ x = \tau$$

assuming $sign(0) = 0$ and τ is the limiting value.

Definition 9.7. *The* task-trajectory *is defined by the end-effector trajectory which is generated during a task execution.*

Definition 9.8. *Given the robot end-effector trajectory in a human demonstration, the* task-equilibrium *state is defined as the end position B of the task trajectory that starts from arbitrary position A.*

Learning Skills from Heterogeneous Demonstrations 355

Let $\mathbf{x} \in \Re^n$ and $\dot{\mathbf{x}} \in \Re^n$ are respectively position and velocity vectors associated with the end-effector of a robotic arm in the Cartesian coordinate frame. Consider $\boldsymbol{\nu}$ as an arbitrary target position of the end-effector, where the user wants the end-effector to reach while executing an intended task/motion profile. Given repeated human demonstrations, corresponding motion profiles are captured as the temporal evolutions both in terms of $\mathbf{x}$ and $\dot{\mathbf{x}}$ of the end-effector. Let $\mathscr{D}_f = \left\{ [\mathbf{x}_{ij}, \dot{\mathbf{x}}_{ij}]_{j=1}^{N_i} \right\}_{i=1}^{D}$ is the database that contains all such temporal evolutions during D demonstrations given by a human demonstrator. These temporal data can be modeled as

$$\dot{\mathbf{x}}(t) = \mathbf{f}(\mathbf{x}(t)) \tag{9.99}$$

such that $\boldsymbol{\nu}$, the desired target position, is the equilibrium state. The mapping $\mathbf{f}(\mathbf{x}(t))$ can be learned using Gaussian mixture regression or any other function approximation technique. In Figure 9.18, we categorize the tasks performed by the robot, based on the task trajectories. A task, performed by the robot, when makes the end-effector move from the position A to B, has a single task-equilibrium state $\boldsymbol{\nu}_B$ with single-task trajectory as shown in Figure 9.18(a). Whereas, the task, performed by the robot, that makes the end-effector move from position A to B and C to B with different motion profiles, is an example of multi-task profile having a single task-equilibrium state $\boldsymbol{\nu}_B$ as shown in Figure 9.18(b). It is also possible that the task requires movement of the end-effector from position A to B and C to D, which is the case of multiple task-equilibrium states ($\boldsymbol{\nu}_B$ and $\boldsymbol{\nu}_D$) with multi-task profiles as shown in Figure 9.18(c). In the literature, system (9.99) has been used to model single-task trajectories having a single task-equilibrium state as shown in Figure 9.18(a). In this chapter, we propose to use system (9.99) to model multi-task trajectories/profiles with multiple task-equilibrium states. Essentially, a single dynamic system model (9.99) is expected to capture both multi-task profiles as well as multiple equilibrium states - this is the novel contribution of this work that is not available in the literature. However, the system model (9.99)

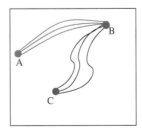
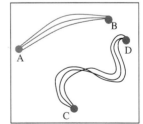

FIGURE 9.18: Task categorization: (a) single-task demonstrations with single task-equilibrium state $\boldsymbol{\nu}_B$, (b) multi-task demonstration with single task-equilibrium state $\boldsymbol{\nu}_B$, and (c) multi-task demonstration with multiple task-equilibrium state $\boldsymbol{\nu}_B$ and $\boldsymbol{\nu}_D$.

requires some structural refinements to accommodate additional information from the environment and stable dynamical behavior. In the first step, the model (9.99) is recasted as

$$\dot{\mathbf{x}}(t) = \mathbf{f}(\mathbf{x}(t), \boldsymbol{\xi}), \qquad (9.100)$$

where, $\mathbf{x}$ is redefined as the position vector in a translated coordinate system whose origin coincide with the target position $\boldsymbol{\nu}$ of a particular demonstration. $\boldsymbol{\xi}$ is a real valued parameter vector that contains information from the environment in the robot's workspace.

When the system (9.100) is learned using any function approximation technique, it is not guaranteed that the system will be dynamically stable. In addition, the learned system (9.100) can include spurious attractors that will generate trajectories without regarding human demonstrations. Thus to make desired equilibrium points globally stable, in the second step, an external guiding signal is introduced:

$$\dot{\mathbf{x}} = \mathbf{f}(\mathbf{x}(t), \boldsymbol{\xi}) + \mathbf{u}_g(\mathbf{x}(t)), \qquad (9.101)$$

where, $\mathbf{u}_g(\mathbf{x}(t))$ is called the guiding input that stabilizes the system (9.100). However, $\mathbf{u}_g$ is so designed that the EDR of the evolving dynamical system matches with the EDR of the demonstrations. This ensures faithful generations of human-like trajectories. The energy function for the system is defined as

$$\hat{E}_{\mathbf{x}} = \frac{1}{2}\mathbf{x}^T\mathbf{x} \qquad (9.102)$$

The change in energy with respect to time can be given as:

$$\dot{\hat{E}}_{\mathbf{x}} = \mathbf{x}^T \dot{\mathbf{x}}$$
$$= \mathbf{x}^T \mathbf{f}(\mathbf{x}(t), \boldsymbol{\xi}) + \mathbf{x}^T \mathbf{u}_g(\mathbf{x}(t)) \qquad (9.103)$$

EDR of the system (9.101) is thus represented by (9.103). By defining $E_{\mathbf{x}}$ as the energy associated with the demonstration states, EDR of the demonstration states $\dot{E}_{\mathbf{x}}$ is computed from the recorded position and velocity vectors of the end-effector. In this chapter, we propose a learning scheme that ensures that EDR of the system follows the EDR of the demonstration. The intention behind this proposed *learning from demonstrations* is to produce motions such that the end-effector moves from an initial position to the target position as per the demonstrations; which also includes the fact that, the state energy dissipation of the end-effector is similar to the demonstrations.

Assumption 9.1. *There exists an underlying function β that represents the EDR of the demonstrations and can be identified as weighted combination of local models as*

$$\beta = \sum_{l=1}^{\mathbb{L}} \sigma^l(\mathcal{M}^l), \qquad (9.104)$$

where σ^l is the weight or confidence of the lth local model $\mathcal{M}^D$.

Let us consider a database $\mathscr{D}_\beta = \left\{[\mathbf{x}_{ij}, \boldsymbol{\xi}_i, \dot{E}_{\mathbf{x}_{ij}}]_{j=1}^{N_i}\right\}_{i=1}^{D}$ that contains the EDR of D demonstrations given to the robot, where $E_{\mathbf{x}_{ij}}$ represents the energy of the ith sample in jth demonstration. $\mathscr{D}_\beta$ is modeled in $\beta(\mathbf{x}, \boldsymbol{\xi})$[5] which is a nonlinear function that encodes the EDR of all the demonstrations with arbitrary precision. By designing $\mathbf{u}_g$, the evolution of the dynamical system can be guided so that the dissipation rate of energy of the evolving system matches $\beta(\mathbf{x}, \boldsymbol{\xi})$. However, the learning of $\beta(\mathbf{x}, \boldsymbol{\xi})$ is crucial for determining appropriate $\mathbf{u}_g$ that keeps the states on the desired path. Thus, the following relationship is learned along the trajectory of the system.

$$\beta(\mathbf{x}, \boldsymbol{\xi}) = \begin{cases} \dot{E}_\mathbf{x}, & \forall \mathbf{x} \neq \mathbf{0} \\ 0, & \text{at } \mathbf{x} = \mathbf{0} \end{cases} \quad (9.105)$$

Here, $\dot{E}_\mathbf{x}$ refers to the energy dissipation rate in the demonstrations.

9.5.1 Stability Analysis

The dynamical system $\mathbf{f}(\mathbf{x}, \boldsymbol{\xi})$ that encodes motions, and the function $\beta(\mathbf{x}, \boldsymbol{\xi})$ that represents energy dissipation rate, are learned from the demonstrations. Regression models like Gaussian mixture regression (GMR), support vector regression (SVR), locally weighted projection regression (LWPR), etc., can be used to learn the function mapping. However, since the learned dynamical system (9.100) is not inherently stable, the guiding input $\mathbf{u}_g$ is introduced to maintain the closed loop stability. The guiding input is designed in such a way, that it helps the system (9.101) to stay on the intended trajectory while avoiding spurious attractors.

Lemma 9.3. *System in (9.101) exhibits motion that has energy dissipation rate $\beta(\mathbf{x}, \boldsymbol{\xi})$ as in (9.105) when the guiding input $\mathbf{u}_g$ is selected as*

$$\mathbf{u}_g = \frac{\mathbf{x}\left[\beta(\mathbf{x}, \boldsymbol{\xi}) - \mathbf{x}^T \mathbf{f}(\mathbf{x}, \boldsymbol{\xi})\right]}{\vartheta(\mathbf{x})} - e^{-\mathscr{K}\mathbf{x}^T\mathbf{x}} \mathbf{f}(\mathbf{x}, \boldsymbol{\xi}) \quad (9.106)$$

where, $\vartheta(\mathbf{x})$ is a positive scalar and is given as $\vartheta(\mathbf{x}) = \mathbf{x}^T\mathbf{x} + \epsilon$ with ϵ being a very small positive value at $\mathbf{x} = \mathbf{0}$ and is zero for nonzero $\mathbf{x}$. Here $\mathscr{K}$ is a large positive scalar.

Proof. Replacing $\mathbf{u}_g$ in (9.103) the EDR of the evolving system (9.101) can be written as

$$\dot{E}_\mathbf{x} = \mathbf{x}^T \left[\mathbf{f}(\mathbf{x}, \boldsymbol{\xi}) + \frac{\mathbf{x}\left[\beta(\mathbf{x}, \boldsymbol{\xi}) - \mathbf{x}^T \mathbf{f}(\mathbf{x}, \boldsymbol{\xi})\right]}{\vartheta(\mathbf{x})}\right]$$
$$- e^{-\mathscr{K}\mathbf{x}^T\mathbf{x}} \mathbf{x}^T \mathbf{f}(\mathbf{x}, \boldsymbol{\xi})$$

---
[5] Please note, the functions are in fact $\beta(\mathbf{x}, \boldsymbol{\xi}, \boldsymbol{\theta}_\beta)$ and $\mathbf{f}(\mathbf{x}, \boldsymbol{\xi}, \boldsymbol{\theta}_f)$ where $\boldsymbol{\theta}_\beta$ and $\boldsymbol{\theta}_f$ are the parameters of the models. We omit $\boldsymbol{\theta}_\beta$ and $\boldsymbol{\theta}_f$ at this stage as these are not explicitly required. $\boldsymbol{\theta}_\beta$ and $\boldsymbol{\theta}_f$ will reappear in the later part of this chapter.

$$= \left(1 - e^{-\mathscr{K}\mathbf{x}^T\mathbf{x}}\right)\mathbf{x}^T\mathbf{f}(\mathbf{x},\boldsymbol{\xi}) + \beta(\mathbf{x},\boldsymbol{\xi}) - \frac{\mathbf{x}^T\mathbf{x}\mathbf{x}^T\mathbf{f}(\mathbf{x},\boldsymbol{\xi})}{\vartheta(\mathbf{x})} \quad (9.107)$$

In (9.107) the term $exp(-\mathscr{K}\mathbf{x}^T\mathbf{x}) \approx 0$ for large $\mathscr{K}$ and nonzero $\mathbf{x}$; at $\mathbf{x} = \mathbf{0}$, $exp(-\mathscr{K}\mathbf{x}^T\mathbf{x}) = 1$. Thus (9.107) can be rewritten as

$$\dot{\hat{E}}_\mathbf{x} = \beta(\mathbf{x},\boldsymbol{\xi}) \quad (9.108)$$

which suggests that when the trajectory reaches $\mathbf{x} = \mathbf{0}$, the error energy $\hat{E}$ of the system (9.101) becomes zero and since $\dot{\hat{E}}$ is also zero at $\mathbf{x} = \mathbf{0}$ (as $\beta(\mathbf{x},\boldsymbol{\xi}) = 0$ according to (9.105)) and also $\dot{\mathbf{x}} = \mathbf{0}$ as $\mathbf{u}_g$ nullifies the affect of $\mathbf{f}(\mathbf{x},\boldsymbol{\xi})$ at the equilibrium, the system stays at $\mathbf{x} = \mathbf{0}$. □

Remark 9.1. *Lemma 9.3 defines an external signal which helps the closed loop system to attain the desired energy dissipation rate $\beta(\mathbf{x},\boldsymbol{\xi})$. Please note that with application of the control law (9.106), $\dot{\mathbf{x}}$ becomes zero at $\mathbf{x} = \mathbf{0}$, which is an important requirement for the equilibrium state to be stable.*

We are required to show that the closed loop system (9.101) is asymptotically stable so that, the trajectory reaches $\mathbf{x} = \mathbf{0}$ and stays there for ever. According to Lemma 9.3 application of $\mathbf{u}_g$ causes the energy of the system (9.101) to dissipate at a rate $\beta(\mathbf{x},\boldsymbol{\xi})$. In practice, there could be two situations related to the energy function of the system during the demonstrations. To analyze the demonstration, the workspace is divided in three regions (S_1, S_2 and S_3) based on the energy dissipation rate. The region S_1 represents the neighborhood of the initial state in the workspace. S_2 is the intermediate region in a demonstration and S_3 is the neighborhood of the target state of the demonstration. These are explained case wise in the following.

Case 1: Let us assume $||\mathbf{x}||^2$ is monotonically decreasing for all the demonstrations as in Figure 9.19(a). In this case, the demonstration starts at $\mathbf{x}_0$ in S_1 and the energy of the system decreases monotonically until it reaches the target $\boldsymbol{\nu}$ in S_3. In all the demonstrations, the energy function is a decreasing monotone. Thus, the energy function of the evolving system (9.101) can be considered as the Lyapunov function $V(\mathbf{x})$. The derivative of the Lyapunov function can be written using (9.108) as

$$\dot{V} = \dot{\hat{E}}_\mathbf{x} = \begin{cases} \beta(\mathbf{x},\boldsymbol{\xi}) < 0, & \forall \mathbf{x} \neq \mathbf{0} \\ 0, & at \quad \mathbf{x} = \mathbf{0} \end{cases} \quad (9.109)$$

Thus the equilibrium state of the dynamical system (9.101) is asymptotically stable and the system has a Lyapunov function with very simple structure. *Case 2:* The other scenario that needs to be considered is when demonstrations include motions where $||\mathbf{x}||^2$ is diverging in certain regions of the workspace as in Figure 9.19(b). In this case, a demonstration starts in region S_1 and the state energy decreases monotonically before reaching region S_2. In S_2, the states move away from the equilibrium state and enters in region S_3. Energy

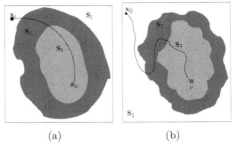

FIGURE 9.19: (a) Demonstration with monotonically decreasing error energy, (b) Demonstration with non-monotonous error energy dissipation.

is descrescent in S_3 as states move toward the equilibrium state in this region. Thus, in the evolving system (9.101), we have $\beta \leq 0$ in S_1, $\beta \geq 0$ in S_2 and $\beta < 0$ in S_3. Since the algorithm learns $\beta(\mathbf{x}, \boldsymbol{\xi})$ from stable demonstrations which ultimately converge to the equilibrium state $\mathbf{x} = \mathbf{0}$; application of $\mathbf{u}_g$, defined in (9.106), will exhibit similar energy dissipation rate that of demonstrations. If the system is operated within the domain of learning, the state $\mathbf{x}$ will eventually reach $\mathbf{x} = \mathbf{0}$ since the EDR follows β. The application of $\mathbf{u}_g$ also causes $\dot{\mathbf{x}} = \mathbf{0}$ at $\mathbf{x} = \mathbf{0}$. Which suggests, the state $\mathbf{x}$ stays at $\mathbf{x} = \mathbf{0}$ once it reaches here. Thus, $\mathbf{x} = \mathbf{0}$ is a stable equilibrium state. However, finding a Lyapunov function for such a system may not be trivial, since it has a complicated structure [248, 249].

Example of non-monotonic Lyapunov function

Let us take an autonomous system

$$\dot{\mathbf{x}} = \begin{bmatrix} -3 & -7 \\ 2 & 0 \end{bmatrix} \mathbf{x} \tag{9.110}$$

where, $\mathbf{x} \in \Re^2$ and the Lyapunov function of the system is $V = 0.5 * \mathbf{x}^T \mathbf{P} \mathbf{x}$ and the state trajectory is given in Fig: 9.20(a) & 9.20(a). Instead of searching for a Lyapunov function, if we simply consider the energy function along the state trajectory as the Lyapunov function (i.e. $\mathbf{P}$ is chosen as identity), the value of such positive definite function along the trajectory of the system is non-negative as given in Fig: 9.20(c). It can be seen that the function first decreases over a period of time, then it increases for some time and again decreases.

Now consider

$$\mathbf{P} = \begin{bmatrix} 3.7736 & 1.1787 \\ 1.1787 & 14.4693 \end{bmatrix} \tag{9.111}$$

The trajectory of V along the state trajectory, given in Fig: 9.20(d), shows that V is monotonically decreasing. This Lyapunov function has a more complex

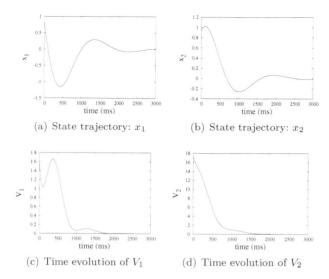

FIGURE 9.20: An example of linear autonomous system.

structure than the earlier as it provides monotonic response. Finding such Lyapunov functions is described by the following theorem [248].

Theorem 9.6. *Consider the continuous time dynamical system (9.101). If there exists scalars $\tau_1 \geq 0$ and $\tau_2 \geq 0$, and a three times differentiable Lyapunov function $V(\mathbf{x})$, such that*

$$\tau_2 \dddot{V}(\mathbf{x}) + \tau_1 \ddot{V}(\mathbf{x}) + \dot{V}(\mathbf{x}) < 0 \qquad (9.112)$$

$\forall \mathbf{x} \neq 0$, *then for any $\mathbf{x}(0)$, $V(\mathbf{x}) \to 0$ as $t \to \infty$ and the origin of (9.101) is globally asymptotically stable.*

It can be noted that the original Lyapunov theorem can be achieved by setting $\tau_1 = 0$ and $\tau_2 = 0$. It is also intuitive that if the higher order derivatives of the Lyapunov function is bounded and it is negative for $\mathbf{x} \neq 0$, eventually it will result $\dot{V}(\mathbf{x})$ to be negative. Thus, Theorem 9.6 tells that even though the quadratic Lyapunov function does not satisfy the condition $\dot{V}(\mathbf{x}) < 0$, one can find parameters τ_1 and τ_2 along with the parameters of a standard Lyapunov function to ensure stability. Theorem 9.6 essentially relaxes the search process for the Lyapunov function. Stability criterion is further relaxed in [249] and the search process for the parameters is made easier by convexifying sufficient condition for stability using higher order derivatives of Lyapunov function and is given as

$$W(\mathbf{x}) = \dddot{V}_3(\mathbf{x}) + \dot{V}_2(\mathbf{x}) + V_1(\mathbf{x}) \qquad (9.113)$$

Learning Skills from Heterogeneous Demonstrations 361

where, the Lyapunov stability criteria is stated as

$$W(0) = 0 \tag{9.114}$$
$$W(\mathbf{x}) = \ddot{V}_3(\mathbf{x}) + \dot{V}_2(\mathbf{x}) + V_1(\mathbf{x}) > 0 \quad \forall \mathbf{x} \neq 0 \tag{9.115}$$
$$\dot{W}(\mathbf{x}) = \dddot{V}_3(\mathbf{x}) + \ddot{V}_2(\mathbf{x}) + \dot{V}_1(\mathbf{x}) < 0 \quad \forall \mathbf{x} \neq 0 \tag{9.116}$$

It is further shown that when different functions $V_1(\mathbf{x})$ and $V_2(\mathbf{x})$ are used for stability analysis, exploration of first and second derivative of the functions alone is not vacuous. This implies that the stability criteria (9.114)-(9.116) can be relaxed as:

$$W(0) = 0 \tag{9.117}$$
$$W(\mathbf{x}) = \dot{V}_2(\mathbf{x}) + V_1(\mathbf{x}) > 0 \quad \forall \mathbf{x} \neq 0 \tag{9.118}$$
$$\dot{W}(\mathbf{x}) = \ddot{V}_2(\mathbf{x}) + \dot{V}_1(\mathbf{x}) < 0 \quad \forall \mathbf{x} \neq 0 \tag{9.119}$$

It is interesting to note here that there is no sign condition on $V_1(\mathbf{x})$ and $V_2(\mathbf{x})$ individually. In order to analyze stability of the system (9.101), we select $V_1(\mathbf{x})$ and $V_2(\mathbf{x})$ as quadratic functions parameterized by $\mathbf{P}_1$ and $\mathbf{P}_2$ respectively. $\mathbf{P}_1$ here is chosen to be an identity matrix of appropriate dimension, such that $V_1(\mathbf{x})$ represents the energy of the system as given in (9.102).

9.5.1.1 Asymptotic Stability in the Demonstrated Region

Let us design a Lyapunov candidate for the system (9.101) according to (9.118) and is given as

$$W(\mathbf{x}) = \mathbf{x}^T \mathbf{P}_2 \dot{\mathbf{x}} + E_\mathbf{x} \tag{9.120}$$

where, $E_\mathbf{x} = V_1(\mathbf{x}) = \mathbf{x}^T \mathbf{P}_1 \mathbf{x}$ with $\mathbf{P}_1 = \mathcal{I} \in \Re^{n \times n}$ ($\mathcal{I}$ is identity matrix) and $\mathbf{P}_2 \in \Re^{n \times n}$.

The asymptotic stability of the motion learning system in the region $\mathcal{W}^\mathcal{D}$[6] is defined by the following theorem.

Theorem 9.7. *$W(\mathbf{x})$ in equation (9.120) is a Lyapunov function of the system (9.101) in the region $\mathcal{W}^\mathcal{D}$ subjected to the following constraints:*

$$\mathbf{x}^T \mathbf{P}_2 \dot{\mathbf{x}} + E_\mathbf{x} > 0 \tag{9.121}$$
$$\dot{\mathbf{x}}^T \mathbf{P}_2 \dot{\mathbf{x}} + \mathbf{x}^T \mathbf{P}_2 \ddot{\mathbf{x}} + \beta(\mathbf{x}, \boldsymbol{\xi}, \boldsymbol{\theta}_\beta) < -\eta_0, \tag{9.122}$$
$$\forall \mathbf{x} : \{\mathbf{x} \in \mathcal{W}^\mathcal{D}, \mathbf{x} \neq \mathbf{0}\}$$

There exist parameters $\boldsymbol{\theta}_\beta$ and a matrix $\mathbf{P}_2 \in \Re^{n \times n}$ for which the guiding input $\mathbf{u}_g$ in Lemma 9.3 stabilizes the system (9.101) within the domain of

---
[6]Here, $\mathcal{W}^\mathcal{D}$ refers to the region in the robot's workspace where the demonstrations are given. It is also called the domain of learning as the motion model, learned from the demonstrated data, is only valid in this region of the workspace.

learning and also maintains the desired EDR $\beta(\mathbf{x}, \boldsymbol{\xi}, \boldsymbol{\theta}_\beta)$ along the system's trajectory.

Here η_0 is is a positive number that decays as the states move closer to the equilibrium point $\mathbf{x} = 0$ and becomes zero as the states reach to the equilibrium point. It is given as follows:

$$\eta_0 = \mathsf{c}_0(1 - e^{-\mathsf{d}_0\|\mathbf{x}\|}) \tag{9.123}$$

where, c_0 and d_0 are small positive constants and can be selected experimentally.

Proof. First part of the proof deals with $W(\mathbf{x})$ in (9.120) being a valid Lyapunov function. We search for a $\mathbf{P}_2$ for which $W(\mathbf{x}) > 0$ in the region $\mathcal{W}^\mathcal{D}$, $\mathbf{x} \neq 0$ and at the equilibrium state $\mathbf{x} = 0$,

$$\mathbf{x} = 0 \;\; and \;\; \dot{\mathbf{x}} = 0 \;\; as \;\; \mathbf{u}_g = -\mathbf{f}(\mathbf{x}, \boldsymbol{\xi})$$

which gives, $W(0) = 0$. Therefore, $W(\mathbf{x})$ is a valid Lyapunov candidate function.

The derivative of the Lyapunov function w.r.t time is given by

$$\dot{W} = \dot{\mathbf{x}}^T \mathbf{P}_2 \dot{\mathbf{x}} + \mathbf{x}^T \mathbf{P}_2 \ddot{\mathbf{x}} + \beta(\mathbf{x}, \boldsymbol{\xi}, \boldsymbol{\theta}_\beta) \tag{9.124}$$

Again at $\mathbf{x} = \mathbf{0}$, $\dot{\mathbf{f}}(\mathbf{x}, \boldsymbol{\xi}) = \frac{\partial \mathbf{f}}{\partial \mathbf{x}} \dot{\mathbf{x}} = 0$ and $\dot{\mathbf{u}}_g = 0$, which gives $\ddot{\mathbf{x}} = 0$. Here, $\beta(\mathbf{x}, \boldsymbol{\xi}, \boldsymbol{\theta}_\beta)$ is a nonlinear map that is learned from the demonstrations. The parameter $\boldsymbol{\theta}_\beta$ is tuned over all the data points in a minimization problem such that $\beta(\mathbf{x}, \boldsymbol{\xi}, \boldsymbol{\theta}_\beta)$ represents the EDR of the demonstrations and simultaneously, a $\mathbf{P}_2$ is also searched for which constraints (9.121)-(9.122) are true. Hence, using (9.124)

$$\dot{W} < 0, \quad \forall \mathbf{x} : \{\mathbf{x} \in \mathcal{W}^\mathcal{D}, \; \mathbf{x} \neq \mathbf{0}\}$$
$$\dot{W} = 0 \quad at \quad \mathbf{x} = \mathbf{0}$$

Therefore, the system (9.101) is asymptotically stable in $\mathcal{W}^\mathcal{D}$. □

Remark 9.2. *During the training of the motion encoding system, the function $\beta(\mathbf{x}, \boldsymbol{\xi}, \boldsymbol{\theta}_\beta)$ is learned such that it obeys (9.122). The optimization algorithm finds $\boldsymbol{\theta}_\beta$, such that, there exists $\mathbf{P}_2$ for an exponentially decaying positive scalar η_0. The choice of the parameter η_0 is a trade off between the robustness on system stability and the ability of the motion model to closely mimic the demonstrations. The parameters of the motion model are learned from the locally generated demonstrations. Thus, a globally asymptotically stable system can only be learned when the demonstrations cover the whole state space. It is interesting to note that there is no sign constraint on $\mathbf{P}_2$ which essentially relaxes the search process.*

9.5.1.2 Ensuring Asymptotic Stability outside Demonstrated Region

The learning approach presented in the earlier section ensures stability in the region $\mathcal{W}^\mathcal{D}$. However, ensuring global stability requires the region $\mathcal{W}^\mathcal{D}$ to be spanned over the entire state space. Providing demonstrations covering the entire state, which is not practically feasible. Hence, during run time when a state arrives from the non-demonstrated region, the models generate erroneous output which may result to an unstable state. To avoid this problem, the control input $\mathbf{u}_g$ (9.106) is modified. This modification ensures a stable motion of the robot and helps to reach the equilibrium state.

The EDR of the demonstrations is modeled as the weighted summation of the local models, where the weights come from the firing strength of the local models. We use this firing strength as the confidence or reliability of the models. In this work, all these models are created under Assumption 9.1 where the firing strength of the local models is calculated using Gaussian distribution as follows

$$\psi^l = exp\left(-\frac{1}{2}(\mathbf{x} - \boldsymbol{\mu}^l)^T {\boldsymbol{\Sigma}^l}^{-1}(\mathbf{x} - \boldsymbol{\mu}^l)\right), \quad (9.125)$$

where, $\mathbf{x}$ is the state, $\boldsymbol{\mu}^l$ and $\boldsymbol{\Sigma}^l$ are the mean and covariance matrix of the lth Gaussian distribution function related to lth local model. The desired EDR is selected based on the maximum firing strength ψ_{max} which is the confidence of the local model closest to the current state $\mathbf{x}$. Hence, ψ_{max} actually tells whether a state has occurred outside the demonstrated region $\mathcal{W}^\mathcal{D}$. It is defined as:

$$\psi_{max}(x) \triangleq \max_{l \in [1...\mathbb{L}]}(\psi^l) \quad (9.126)$$

where $\mathbb{L}$ is the total number of local models.

Thus the guiding input in (9.106) is modified as the following:

$$\mathbf{u}_g = \frac{\mathbf{x}\left[(1 - \Gamma(\psi_{max}, \tau))\beta(\mathbf{x}, \boldsymbol{\xi}) - \Gamma(\psi_{max}, \tau)\bar{\eta}\right]}{\vartheta(\mathbf{x})}$$

$$- \frac{\mathbf{x}\mathbf{x}^T \mathbf{f}(\mathbf{x}, \boldsymbol{\xi})}{\vartheta(\mathbf{x})} - exp(-\mathscr{K}\mathbf{x}^T\mathbf{x})\mathbf{f}(\mathbf{x}, \boldsymbol{\xi}) \quad (9.127)$$

where, $\Gamma(\psi_{max}, \tau)$ is a switching function as defined in (9.98) with the switching limit τ is a constant positive scalar. If the maximum firing strength over all the local models drops bellow τ (which implies that the robot trajectory is away from the demonstrated region), $\Gamma(\cdot)$ modifies the input signal such that the state trajectory is pushed toward the equilibrium state. $\bar{\eta}$ has the same functional form as given in (9.123). Thus, outside demonstrated region the effective control signal becomes:

$$\mathbf{u}_g = \frac{\mathbf{x}\left[-\bar{\eta} - \mathbf{x}^T \mathbf{f}(\mathbf{x}, \boldsymbol{\xi})\right]}{\vartheta(\mathbf{x})} - e^{-\mathscr{K}\mathbf{x}^T\mathbf{x}}\, \mathbf{f}(\mathbf{x}, \boldsymbol{\xi}) \quad (9.128)$$

and the EDR of the system (9.101) takes the form as follows:

$$\dot{E}_{\mathbf{x}} = -\frac{\mathbf{x}^T \mathbf{x} \bar{\eta}}{\vartheta(\mathbf{x})} < 0, \quad \forall \mathbf{x}: \{\mathbf{x} \notin \mathcal{W}^{\mathcal{D}}\}$$
$$= 0 \quad at \; \mathbf{x} = 0 \tag{9.129}$$

Remark 9.3. $\dot{E}_{\mathbf{x}}$ *in (9.129) of the closed loop system (9.101) is negative outside the $\mathcal{W}^{\mathcal{D}}$ region. Thus, during the execution, if the state appears outside the demonstrated region, the control law makes it sure that the energy of the state decays monotonically such that the equilibrium state is reached.*

9.5.2 Learning Model Parameters from Demonstrations

Proper learning of the functions $\mathbf{f}(\mathbf{x}, \boldsymbol{\xi}, \boldsymbol{\theta}_f)$ and $\beta(\mathbf{x}, \boldsymbol{\xi}, \boldsymbol{\theta}_\beta)$ is important since they hold the characteristics of the demonstrated motion profiles and also they are responsible for stabilization of the system (9.101). Here, we present learning architectures using three regression techniques as listed in the following:

1. Gaussian mixture regression (GMR)
2. Locally weighted projection regression (LWPR)
3. Support vector regression (SVR)

The detailed learning methodology using above mentioned regression techniques is given as follows.

9.5.2.1 Motion Modeling using GMR

Gaussian mixture regression is a powerful tool for function approximation. The dynamical system in (9.100) and the energy dissipation rate in (9.105) are approximated using GMR. However, to deal with the instability issues, careful measures need to be taken while learning the function $\beta(\mathbf{x}, \boldsymbol{\xi}, \boldsymbol{\theta}_\beta^{gmr})$.

Regression model for $\mathbf{f}(\mathbf{x}, \boldsymbol{\xi}, \boldsymbol{\theta}_f^{gmr})$

The joint probability distribution of the demonstration dataset $\mathscr{D}_f = \{[\mathbf{x}_{i,j}, \boldsymbol{\xi}_i, \dot{\mathbf{x}}_{i,j}]_{j=1}^{N_i}\}_{i=1}^{D}$ is given by

$$P(\mathbf{x}, \boldsymbol{\xi}, \dot{\mathbf{x}}|\mu, \Sigma) = \sum_{k=1}^{K} P(k) P(\mathbf{x}, \boldsymbol{\xi}, \dot{\mathbf{x}}|k), \tag{9.130}$$

where $P(k)$ is termed as the prior and $P(\mathbf{x}, \boldsymbol{\xi}, \dot{\mathbf{x}}|k)$ is the conditional probability density function which is given by

$$P(\mathbf{x}, \boldsymbol{\xi}, \dot{\mathbf{x}}|k) = G^k \left(\mathbf{x}, \boldsymbol{\xi}, \dot{\mathbf{x}}; \mu^k, \Sigma^k\right)$$
$$= \frac{\exp^{-\frac{1}{2}\left(([\mathbf{x};\boldsymbol{\xi};\dot{\mathbf{x}}] - \mu^k)^T (\Sigma^k)^{-1} ([\mathbf{x};\boldsymbol{\xi};\dot{\mathbf{x}}] - \mu^k)\right)}}{\sqrt{2\pi^{(2n+n_\xi)} |\Sigma^k|}}, \tag{9.131}$$

Learning Skills from Heterogeneous Demonstrations 365

where n_ξ is the dimension of ξ. μ^k and Σ^k are given by

$$\mu^k = \begin{bmatrix} \mu^k_{\mathbf{x}} \\ \mu^k_{\xi} \\ \mu^k_{\dot{\mathbf{x}}} \end{bmatrix} \text{ and } \Sigma^k = \begin{bmatrix} \Sigma^k_{\mathbf{xx}} & \Sigma^k_{\mathbf{xx}^*} & \Sigma^k_{\mathbf{x}\dot{\mathbf{x}}} \\ \Sigma^k_{\mathbf{xx}^*} & \Sigma^k_{\xi\xi} & \Sigma^k_{\xi\dot{\mathbf{x}}} \\ \Sigma^k_{\mathbf{x}\dot{\mathbf{x}}} & \Sigma^k_{\xi\dot{\mathbf{x}}} & \Sigma^k_{\dot{\mathbf{x}}\dot{\mathbf{x}}} \end{bmatrix}. \tag{9.132}$$

The posterior probability $P(\dot{\mathbf{x}}|\mathbf{x}, \xi)$ gives the desired mean estimate using (9.131)

$$\mathbf{f}(\mathbf{x}, \xi, \boldsymbol{\theta}^{gmr}_f) = \sum_{k=1}^{K} \frac{P(k)P(\mathbf{x}, \xi|k)}{P(\mathbf{x}, \xi)} \left(\mu^k_{\dot{\mathbf{x}}} + \begin{bmatrix} \Sigma^k_{\dot{\mathbf{x}}\mathbf{x}} & \Sigma^k_{\dot{\mathbf{x}}\xi} \end{bmatrix} \right.$$
$$\left. \begin{bmatrix} \Sigma^k_{\mathbf{xx}} & \Sigma^k_{\mathbf{x}\xi} \\ \Sigma^k_{\mathbf{x}\xi} & \Sigma^k_{\xi\xi} \end{bmatrix} \begin{bmatrix} \mathbf{x} - \mu^k_{\mathbf{x}} \\ \xi - \mu^k_{\xi} \end{bmatrix} \right) \tag{9.133}$$

Approximation of EDR

The joint probability distribution of the demonstration dataset $\mathscr{D}_\beta = \left\{ [\mathbf{x}_{i,j}, \xi_i, \dot{E}_{\mathbf{x}_{ij}}]_{j=1}^{N_i} \right\}_{i=1}^{D}$ is given by

$$P(\mathbf{x}, \xi, \dot{E}_\mathbf{x} | \mu, \Sigma) = \sum_{k=1}^{K} P(k) P(\mathbf{x}, \xi, \dot{E}_\mathbf{x} | k), \tag{9.134}$$

where $P(k)$ is termed as the prior and $P(\mathbf{x}, \xi, \dot{E}_\mathbf{x}|k)$ is the conditional probability density function. Similarly, the posterior mean of the EDR is given by

$$\beta(\mathbf{x}, \xi, \boldsymbol{\theta}^{gmr}_\beta) = \sum_{k=1}^{K} \frac{P(k)P(\mathbf{x}, \xi|k)}{P(\mathbf{x}, \xi)} \left(\mu^k_{\dot{E}_\mathbf{x}} + \begin{bmatrix} \Sigma^k_{\dot{E}_\mathbf{x}\mathbf{x}} & \Sigma^k_{\dot{E}_\mathbf{x}\xi} \end{bmatrix} \right.$$
$$\left. \begin{bmatrix} \Sigma^k_{\mathbf{xx}} & \Sigma^k_{\mathbf{x}\xi} \\ \Sigma^k_{\mathbf{x}\xi} & \Sigma^k_{\xi\xi} \end{bmatrix} \begin{bmatrix} \mathbf{x} - \mu^k_{\mathbf{x}} \\ \xi - \mu^k_{\xi} \end{bmatrix} \right) \tag{9.135}$$

The parameter $\boldsymbol{\theta}^{gmr}_\beta$ in (9.105) is learned in two steps. The first step is called pre-training where expectation maximization (EM) is used to learn the initial values of $\boldsymbol{\theta}^{gmr}_\beta$. In the next step, the parameters are fine-tuned to ensure stability of the closed loop system (9.101). A cost function is minimized in the optimization process. We select Negative Log-Likelihood as the cost function which is minimized to ensure propoer learning of $\beta(\mathbf{x}, \xi, \boldsymbol{\theta}^{gmr}_\beta)$. The steps are given in the Algorithm 12.

Cost function

The parameters of $\beta(\mathbf{x}, \xi, \boldsymbol{\theta}^{gmr}_\beta)$ are searched in a direction where the likelihood of the random variable maximizes over all the data points. However, since the maximization of the likelihood function $p(\mathbf{x}, \boldsymbol{\xi}, \dot{E}_{\mathbf{x}} | \boldsymbol{\theta}_\beta^{gmr})$ suffers from numerical underflow problem due to small likelihood values, log-likelihood is maximized instead. Moreover, from the calculus's perspective, natural logarithm is a monotone transformation. Thus, the log-likelihood function is given by

$$\mathcal{L}(\mathbf{x}, \boldsymbol{\xi}, \dot{E}_{\mathbf{x}}, \boldsymbol{\theta}_\beta^{gmr}) = \log \ p(\mathbf{x}, \boldsymbol{\xi}, \dot{E}_{\mathbf{x}} | \boldsymbol{\theta}_\beta^{gmr}) \qquad (9.136)$$

The function $\mathcal{L}(\mathbf{x}, \boldsymbol{\xi}, \dot{E}_{\mathbf{x}}, \boldsymbol{\theta}_\beta^{gmr})$ in (9.136) is maximized in the following optimization problem where the parameter $\boldsymbol{\theta}_\beta^{gmr}$ is learned during the process.

$$\underset{\boldsymbol{\theta}_2}{\text{minimize}} \quad -\frac{\sum_{i=1}^{D} \sum_{j=1}^{N_i} \mathcal{L}(\mathbf{x}_{i,j}, \boldsymbol{\xi}_i, \dot{E}_{\mathbf{x}_{i,j}}, \boldsymbol{\theta}_\beta^{gmr})}{\sum_{i=1}^{D} N_i} \qquad (9.137)$$

subject to   Constraints $(9.121) - (9.122)$

The procedure is described in Algorithm 12.

Algorithm 12 Steps to learn parameter $\boldsymbol{\theta}_\beta^{gmr}$.

---

1. Prepare $\mathscr{D}_\beta = \left\{ [\mathbf{x}_{ij}, \boldsymbol{\xi}_i, \dot{\mathbf{x}}_{ij}, \ddot{\mathbf{x}}_{ij}, E_{\mathbf{x}_{ij}}, \dot{E}_{\mathbf{x}_{ij}}]_{j=1}^{N_i} \right\}_{i=1}^{D}$ during the demonstrations. Here, D represents the number of demonstrations and N_i is the number of samples in each demonstration.
2. Run EM over the dataset $\bar{\mathscr{D}}_\beta = \left\{ [\mathbf{x}_{ij}, \boldsymbol{\xi}_i, \dot{E}_{\mathbf{x}_{ij}}]_{j=1}^{N_i} \right\}_{i=1}^{D}$ to get an initial estimate for all $\boldsymbol{\pi}_\beta^k$, $\boldsymbol{\mu}_\beta^k$ and $\boldsymbol{\Sigma}_\beta^k$.
3. **initialize:** $\boldsymbol{\pi}_{\beta 0} \leftarrow \emptyset$ $\boldsymbol{\mu}_{\beta 0} \leftarrow \emptyset$, $\boldsymbol{\Sigma}_{\beta 0} \leftarrow \emptyset$
    for k=1:K do
        $\boldsymbol{\pi}_{\beta 0} \leftarrow [\boldsymbol{\pi}_{\beta 0}; \boldsymbol{\pi}_\beta^k]$
        $\boldsymbol{\mu}_{\beta 0} \leftarrow [\boldsymbol{\mu}_{\beta 0}; \boldsymbol{\mu}_\beta^k]$
        for i=1:d+1 do
            for j=i:d+1 do
                $\boldsymbol{\Sigma}_{\beta 0} \leftarrow [\boldsymbol{\Sigma}_{\beta 0}; \boldsymbol{\Sigma}_\beta^k{}_{i,j}]$
            end for
        end for
    end for
4. Set initial values $\hat{\boldsymbol{\theta}} \leftarrow [\boldsymbol{\pi}_{\beta 0}^T \ \boldsymbol{\mu}_{\beta 0}^T \ \boldsymbol{\Sigma}_{\beta 0}^T]$.
5. Run optimization algorithm over $\mathscr{D}_\beta$.
6. Find new $\hat{\boldsymbol{\theta}}$ that minimizes the cost (9.137) while maintaining the constraints (9.121)-(9.122).
7. **Return:** $\boldsymbol{\theta}_\beta^{gmr} \leftarrow \hat{\boldsymbol{\theta}}$

9.5.2.2 Motion Modeling using LWPR

LWPR is an incremental learning framework for nonlinear function approximation in high dimensional space. The algorithm is widely used to learn the functional relationship between the input and output data where the input has irrelevant and redundant dimensions. The algorithm assumes that there exists a lower dimensional distribution of the input space. The idea is to learn piece wise local linear models from a given nonlinear function. The overall function is represented as the weighted sum over the local models. The weightage of the kth local region is given by [250],

$$w_k = \exp\left(-0.5(\mathbf{x} - \mathbf{c}_k)^T \mathbf{D}_k (\mathbf{x} - \mathbf{c}_k)\right)$$

where, $\mathbf{x} \in \Re^n$ is the input of the system and $\mathbf{c}_k \in \Re^n$ is the center of kth local region, called receptive field (RF) which is parameterized by a distance metric $\mathbf{D}_k$.

Regression model for $\mathbf{f}(\mathbf{x}, \boldsymbol{\xi}, \boldsymbol{\theta}_f^{lwpr})$

The underlying function of the demonstrations is learned using LWPR technique and is given as follows.

$$\mathbf{f}(\mathbf{x}, \boldsymbol{\xi}, \boldsymbol{\theta}_f^{lwpr}) = \sum_{k=1}^{K_f} \left(w^k \mathbf{b_0}^k + \mathbf{A}^k \mathbf{s}^k\right) \tag{9.138}$$

Here, $\mathbf{b_0}^k$ is the bias term in the kth local model and is given by the weighted mean of the velocities $\dot{\mathbf{x}}$, seen during the demonstrations. For a n dimensional state vector, $\mathbf{A}^k \in \Re^{n \times R}$ is the regression parameter for the projections $\mathbf{s}^k \in \Re^R$ in k^{th} local model, where R is the number of projections in the input space.

Approximation of EDR

The energy dissipation rate is approximated using LWPR and can be written as

$$\boldsymbol{\beta}(\mathbf{x}, \boldsymbol{\xi}, \boldsymbol{\theta}_\beta^{lwpr}) = \sum_{k=1}^{K_\beta} \left(w^k \boldsymbol{\beta_0}^k + \mathbf{B}^k \mathbf{v}^k\right) \tag{9.139}$$

Here, $\boldsymbol{\beta_0}^k$ is the bias term in the kth local model and is given by the weighted mean of the energy dissipation rates, seen during the demonstrations. $\mathbf{B}^k \in \Re^R$ is the regression parameter for the projections $\mathbf{v}^k \in \Re^R$ in k^{th} local model, where R is the number of projections in the input space.

The parameters $\boldsymbol{\theta}_f^{lwpr}$ and $\boldsymbol{\theta}_\beta^{lwpr}$ are learned in two separate optimization processes. In pre-training, the initial models are learned where the distance metric $\mathbf{D}_k$ is updated using stochastic gradient descent. In the next phase, the parameters are fine-tuned in a constrained optimization process, where a cost function is minimized.

Cost function

The optimization problem is formulated to minimize a cost function using leave-one-out cross validation error of each local model and is given by

$$J_k(\mathbf{D}_k) = \frac{1}{W_k} \sum_{i=1}^{D} \sum_{j=1}^{N_i} w_{k_{i,j}} \left(\dot{E}_{\mathbf{x}_{i,j}} - \beta_{k_{i,j,-j}} \right)^2 \quad (9.140)$$

$$+ \frac{\gamma}{d} \sum_{i,j=1}^{d} (D_k)_{ij}^2$$

where $W_k = \sum_{i=1}^{D} \sum_{j=1}^{N_i} w_{k_{i,j}}$ is a normalization term. Here, D denotes the number of demonstrations and N_i is number of samples in i^{th} demonstration. The second term in the cost function is the regularization term weighted by γ. Thus, the optimization problem is given as

$$\begin{aligned}\underset{\mathbf{D}_k}{\text{minimize}} \quad & J_k(\mathbf{D}_k) \text{ in } (9.140) \\ \text{subject to} \quad & \text{Constraints } (9.121) - (9.122)\end{aligned} \quad (9.141)$$

The detailed description of the learning methodology using LWPR is given in Algorithm 13.

Algorithm 13 Steps to learn parameter $\boldsymbol{\theta}_\beta^{lwpr}$.
---
1. Perform step 1 of Algorithm 12
2. Run LWPR algorithm [250] over the dataset $\bar{\mathscr{D}}_\beta = \left\{ [\mathbf{x}_{ij}, \boldsymbol{\xi}_i, \dot{E}_{\mathbf{x}_{ij}}]_{j=1}^{N_i} \right\}_{i=1}^{D}$
   to get an initial estimate for all $\mathbf{c}_k$, $\mathbf{D}_k$, $\boldsymbol{\beta_0}^k$, $\mathbf{B}^k$ and $\mathbf{v}^k$.
3. **initialize:** $\hat{\boldsymbol{\theta}} \leftarrow \emptyset$
   for k=1:K_β do
     $\hat{\boldsymbol{\theta}} \leftarrow [\hat{\boldsymbol{\theta}} \quad \mathbf{c}_k^T \quad diag(\mathbf{D}_k)]$
   end for
4. Run optimization algorithm over $\mathscr{D}_\beta$.
5. Recalculate $\boldsymbol{\beta_0}^k$, $\mathbf{B}^k$ and $\mathbf{v}^k$ for new $\hat{\boldsymbol{\theta}}$.
6. Find new values $\hat{\boldsymbol{\theta}}$ that minimizes the cost (9.141) while maintaining the constraints (9.121)-(9.122)
7. **Return:** $\boldsymbol{\theta}_\beta^{lwpr} \leftarrow \hat{\boldsymbol{\theta}}$

9.5.2.3 Motion Modeling using ϵ-SVR

Support vector regression is a powerful tool for nonlinear function approximation. The algorithm learns the support vectors from the input data space. The dynamical system and the energy dissipation rate are learned as a regression problem from the demonstration data. For a given data set $\mathcal{D} = \{(\mathbf{x}^1, y^1), ...(\mathbf{x}^Z, y^Z)\}$ where $\mathbf{x} \in \Re^n$ and $y \in \Re$, the underlying function $g(\mathbf{x}, \mathbf{w})$ is learned by optimizing the following cost function over the training dataset [251].

$$\mathcal{L}_\epsilon(y^i, g(\mathbf{x}^i, \mathbf{w})) = \begin{cases} 0, & \forall |y^i - g(\mathbf{x}^i, \mathbf{w})| \leq \epsilon \\ |y^i - g(\mathbf{x}^i, \mathbf{w})| - \epsilon, & \text{otherwise} \end{cases} \quad (9.142)$$

Here, ϵ is a small positive scalar that gives the boundary of deviation of $g(\mathbf{x}^i, \mathbf{w})$ from the target y^i. The problem is solved as a constrained optimization problem and the following cost is maximized

$$J(\alpha^i, \alpha_*^i) = \sum_{i=1}^{l} \alpha_*^i(y^i - \epsilon) - \alpha^i(y^i + \epsilon)$$

$$- \frac{1}{2} \sum_{i=1}^{l} \sum_{j=1}^{l} (\alpha_*^i - \alpha^i)(\alpha_*^j - \alpha^j) \mathcal{K}(\mathbf{x}^i, \mathbf{x}^j) \quad (9.143)$$

where, α^i and α_*^i are the Lagrange multipliers and $\mathcal{K}(\mathbf{x}^i, \mathbf{x}^j)$ is a kernel for the input space. Therefore, the optimization problem is posed as

$$\begin{aligned}
\underset{\alpha_i^i, \alpha_*^i}{\text{maximize}} \quad & J(\alpha^i, \alpha_*^i) \\
\text{subject to} \quad & 0 \leq \alpha^i, \alpha_*^i \leq C \quad i = 1, 2, ..., l \\
& \sum_{i=1}^{l} (\alpha^i - \alpha_*^i) = 0
\end{aligned} \quad (9.144)$$

where $C > 0$ is the penalty parameter that determines trade off between the flatness of the learned function and the deviation from the target state.

Regression model for $\mathbf{f}(\mathbf{x}, \boldsymbol{\xi}, \boldsymbol{\theta}_f^{svr})$

The function that preserves the characteristics of the demonstrations $\mathscr{D}_f = \{[\mathbf{x}_{i,j}, \boldsymbol{\xi}_i, \dot{\mathbf{x}}_{i,j}]_{j=1}^{N_i}\}_{i=1}^{D}$ is learned using ϵ-SVR and is given by

$$f_j(\mathbf{x}, \boldsymbol{\xi}, \boldsymbol{\theta}_{f_j}^{svr}) = \sum_{i=1}^{K_{f_j}} (\bar{\alpha}_j^i - \bar{\alpha}_{*j}^i) \mathcal{K}_{f_j}(\bar{\mathbf{x}}^i, \bar{\mathbf{x}}) + \bar{b}_j^f; \quad j = 1, ..., n \quad (9.145)$$

where, K_{f_j} is the number of support vectors, $\bar{\alpha}_j^i$ and $\bar{\alpha}_{*j}^i$ are the optimal Lagrange multipliers and $\boldsymbol{\theta}_{f_j}^{svr}$ represents the parameters of the model and $\bar{\mathbf{x}}$ is the concatenation of $\mathbf{x}$ and $\boldsymbol{\xi}$. Here, $\bar{b}_j^f$ is given by

$$\bar{b}_j^f = \dot{x}_j^v - \sum_{i=1}^{T_N} (\bar{\alpha}_j^i - \bar{\alpha}_{*j}^i) \mathcal{K}_{f_j}(\bar{\mathbf{x}}^i, \bar{\mathbf{x}}^v) + \epsilon; \quad \text{s.t.} \quad 0 < \alpha_*^v < C_f$$

where, T_N is the total number of samples and C_f is a penalty parameter.

Approximation of EDR

The relation between the present states of the system and energy dissipation rate of the demonstrations is learned as regression function. ϵ-SVR is used to learn the map from the dataset $\mathscr{D}_\beta = \left\{[\mathbf{x}_{ij}, \boldsymbol{\xi}_i, \dot{E}_{\mathbf{x}_{ij}}]_{j=1}^{N_i}\right\}_{i=1}^{D}$ and is given by

$$\beta(\mathbf{x}, \boldsymbol{\xi}, \boldsymbol{\theta}_\beta^{svr}) = \sum_{i=1}^{K_\beta}(\bar{\mu}^i - \bar{\mu}_*^i)\mathcal{K}_\beta(\bar{\mathbf{x}}^i, \bar{\mathbf{x}}) + \bar{b}^\beta \tag{9.146}$$

where, K_β is the number of support vectors, $\bar{\mu}^i, \bar{\mu}_*^i$ are the optimal Lagrange multipliers and $\boldsymbol{\theta}_\beta^{svr}$ represents the parameters of the model. Here, $\bar{b}^\beta$ is given by

$$\bar{b}^\beta = \dot{E}_\mathbf{x}^v - \sum_{i=1}^{T_N}(\bar{\mu}^i - \bar{\mu}_*^i)\mathcal{K}_\beta(\bar{\mathbf{x}}^i, \bar{\mathbf{x}}^v) + \epsilon; \quad s.t. \ 0 < \mu_*^v < C_\beta$$

where, T_N is the total number of samples and C_β is a penalty parameter.

The kernel is taken as the Gaussian function such that $\mathcal{K}(\bar{\mathbf{x}}, \bar{\mathbf{x}}') = e^{-\frac{(\bar{\mathbf{x}}-\bar{\mathbf{x}}')}{\gamma}}$, where $\bar{\mathbf{x}}$ and $\bar{\mathbf{x}}'$ are the samples and $\gamma > 0$ is the variance parameter of the Gaussian kernel. Algorithm 14 provides step by step procedure of learning the EDR.

Algorithm 14 Steps to learn parameter $\boldsymbol{\theta}_\beta^{svr}$.

---

1. Perform step 1 of Algorithm 12
2. Run SVR algorithm [251] over the dataset $\bar{\mathscr{D}}_\beta = \left\{[\mathbf{x}_{ij}, \boldsymbol{\xi}_i, \dot{E}_{\mathbf{x}_{ij}}]_{j=1}^{N_i}\right\}_{i=1}^{D}$ to get an initial estimate for $C_\beta, \gamma_\beta, \bar{\mu}$ and $\bar{\mu}_*$
3. **initialize:** $\hat{\boldsymbol{\theta}} \leftarrow C_\beta, \gamma_\beta, \bar{\mu}^i, \bar{\mu}_*^i$
4. Run optimization algorithm over $\mathscr{D}_\beta$
5. Find new $\hat{\boldsymbol{\theta}}$ that minimizes the cost (9.144) while maintaining the constraints (9.121)-(9.122)
6. **Return:** $\boldsymbol{\theta}_\beta^{svr} \leftarrow \hat{\boldsymbol{\theta}}$

The model $f(\mathbf{x}, \boldsymbol{\xi}, \boldsymbol{\theta}_f^{svr})$ and the pre-model of $\beta(\mathbf{x}, \boldsymbol{\xi}, \boldsymbol{\theta}_f^{svr})$ in SV based regression are learned using LibSVM [252] library which is publicly available on their web page.

9.5.2.4 Complete Pipeline

We have seen three regression techniques for encoding the demonstrations and also formulated the learning conditions for the models. In the following Algorithm 15 we shall describe the entire procedure including the execution on a robotic manipulator.

Learning Skills from Heterogeneous Demonstrations 371

Algorithm 15 Steps for encoding heterogeneous profiles

0: Collect data samples during demonstrations as shown in Figure 9.21 and prepare the datasets $\mathscr{D}_f$ and $\mathscr{D}_\beta$
0: Learn $\mathbf{f}(\mathbf{x},\boldsymbol{\xi},\boldsymbol{\theta}_f)$ and $\beta(\mathbf{x},\boldsymbol{\xi},\boldsymbol{\theta}_\beta)$ from the dataset using any of the methods described above and keep memory of the environmental information $\boldsymbol{\xi}_i$ in $\bar{\boldsymbol{\xi}}$. In our experiment, $\boldsymbol{\xi}_i$ is the concatenation of target position $\boldsymbol{\nu}_i$ (can be referred as object's picking position) and initial positions for ith demonstration.
0: Get the initial position and desired target $\boldsymbol{\nu}$ within the domain of learning in task space.
0: Select the nearest $\boldsymbol{\xi}$ from the store $\bar{\boldsymbol{\xi}}$ and consider it as the environmental input $\boldsymbol{\xi}$ to the models.
0: Evaluate velocity command as in (9.101) using the learned models.
0: Get the desired position of the end-effector and apply inverse kinematics to obtain corresponding joint positions.
0: Send the joint command to the robot.
0: Repeat until the target is reached.

9.5.3 Spatial Error Calculation

We introduce an error metric that gives a measure of the performance of the motion encoding system. Since the generated trajectory differs in reaching time with the demonstration, the mean square error cannot be applied directly to check the error at each sample point. Moreover, only observing the difference between states will not provide the entire picture regarding the performance of the encoding system. Hence, we develop an algorithm that essentially calculates two types of errors: one is related to the positions and other is related to the velocity of the overall trajectory. Let us define few variables: $\mathbf{T}_D^{\mathbf{x}}$ and $\mathbf{T}_D^{\dot{\mathbf{x}}}$ contain all points in the demonstrated trajectory; $\mathbf{T}_h^{\mathbf{x}}$ and $\mathbf{T}_h^{\dot{\mathbf{x}}}$ contain all data points of the trajectory generated by the motion encoding system; $e^{\mathbf{x}}$ and $e^{\dot{\mathbf{x}}}$ are the errors related to position and velocity respectively. (The unit of the errors are same as position and velocity respectively). Algorithm 16 describes the computation of spatial error.

9.5.4 Examples

Let us see some examples of learning multiple task profiles in a single model using the approaches we have learned in the previous section. In this example we shall use 7 DOF robotic manipulator. First, the manipulator is given kinesthetic demonstrations of heterogeneous task. The data in terms of end-effector position, velocity, target point and EDR in the Cartesian space is filled in the datasets $\mathscr{D}_f = \{[\mathbf{x}_{i,j}, \boldsymbol{\xi}_i, \dot{\mathbf{x}}_{i,j}]_{j=1}^{N_i}\}_{i=1}^{D}$ and $\bar{\mathscr{D}}_\beta = \left\{[\mathbf{x}_{ij}, \boldsymbol{\xi}_i, \dot{E}_{\mathbf{x}_{ij}}]_{j=1}^{N_i}\right\}_{i=1}^{D}$.

Algorithm 16 Steps to calculate spatial error.
1: **for each data in** $\mathbf{T}_h^x$ **do**
2:     Get the nearest data $\mathbf{x}_i$ in $\mathbf{T}_D^x$ forward in time.
3:     Store it in $\mathbf{O_x}$ such that, $\mathbf{O} \leftarrow [\mathbf{O} \ \mathbf{x}_i]$
4:     Extract associated $\dot{\mathbf{x}}_i$ from $\mathbf{T}_D^{\dot{x}}$ and store it in $\mathbf{O}_{\dot{\mathbf{x}}}$ such that, $\mathbf{O}_{\dot{\mathbf{x}}} \leftarrow [\mathbf{O}_{\dot{\mathbf{x}}} \ \dot{\mathbf{x}}_i]$
5: **end for**
6: Calculate $\mathbf{E_x} = (\mathbf{O_x} - \mathbf{T}_h^x)/mean(abs(\mathbf{O_x}))$.
7: Calculate $\mathbf{E}_{\dot{\mathbf{x}}} = (\mathbf{O}_{\dot{\mathbf{x}}} - \mathbf{T}_h^{\dot{x}})/mean(abs(\mathbf{O}_{\dot{\mathbf{x}}}))$.
8: Get the mean square value along the rows of $\mathbf{E_x}$ and $\mathbf{E}_{\dot{\mathbf{x}}}$ such that,
9: $e_j^x = meansqr(\mathbf{E_x}[j,:])$
10: $e_j^{\dot{x}} = meansqr(\mathbf{E}_{\dot{\mathbf{x}}}[j,:])$
10: **Return** $mean(e^x)$ and $mean(e^{\dot{x}})$.

Then we use the previously described algorithms to learn the parameters of the motion model.

9.5.4.1 Example of Monotonic and Non-monotonic State Energy

The robot is introduced with two kinds of demonstration. Half of the demonstrations are such that the energy of the evolving system is always decreasing, i.e., the EDR is always negative. This kind of situation occurs when a robot simply picks an object and places it away from the picking point. For rest of the demonstrations, the energy increases in certain region of the workspace - in this case the EDR is sometimes positive as well. These type of motion profiles are common when a robot picks an object and place it into a deep bucket where the target point is not far away from the picking point. The system models as given in (9.101) with the control input (9.127) are learned from the datasets $\mathcal{D}_f$ and $\mathcal{D}_\beta$ using three regression techniques described earlier. The learned models are tested by letting the dynamical system evolve for the specified tasks. The performance of these models are compared with the original demonstrations for those tasks. The training dataset contains ten demonstrations of which five are such that the EDR is always negative and in the rest, the EDR attains positive values in certain region of the workspace. An initial set of Gaussian parameters for the EDR model is first learned from the data. The optimization algorithm then finds new parameters in the neighborhood of the initial values satisfying the stability constraints. Matlab's Optimization toolbox (fmincon) with interior-point algorithm is used to find stable parameters. The learned model is used for end-effector trajectory generation. An initial end-effector position (marked by the black circle in Figure 9.21) is chosen within the domain of learning from which the robot moves toward the target position (marked by black square in Figure 9.21). The motion encoding model provides new end-effector position at each control loop and related joint positions are computed using IKfast [253]. The PID controller of the robot produces required torque for the joints. Figure 9.21 presents the trajectories

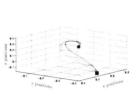 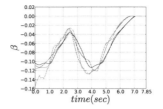

(a) End-effector positions   (b) Rate of change of energy

FIGURE 9.21: In this experiment the EDR is always negative. Output of all the three models (GMR, LWPR and SVR) with the demonstrations are shown here. The trajectory starts from the black circle and reaches to the target marked by black square.

when the EDR is always negative during the task execution and Figure 9.23 presents the trajectories with positive EDR. Figure 9.22 and 9.24 present corresponding position and velocities on each coordinate in both the cases.

GMR models:

In this experiment, we first use Gaussian mixture regression to model $\mathbf{f}(\mathbf{x}, \boldsymbol{\xi}, \boldsymbol{\theta}_f)$ and $\beta(\mathbf{x}, \boldsymbol{\xi}, \boldsymbol{\theta}_\beta)$ (as in Algorithm 12). 11 Gaussians are used to capture the data distribution for both the models. Standard EM algorithm is used to estimate the Gaussian parameters. The output trajectory of the GMR model is given in Figure 9.21 with the associated demonstration for that target position. Figure 9.23 represents the output of the model when the actual demonstration has positive EDR.

LWPR models:

LWPR algorithm is useful for incremental learning. Here, the models $\mathbf{f}(\mathbf{x}, \boldsymbol{\xi}, \boldsymbol{\theta}_f)$ and $\beta(\mathbf{x}, \boldsymbol{\xi}, \boldsymbol{\theta}_\beta)$ (as in Algorithm 13) are learned offline, using the entire dataset. The algorithm learns 102 local models to represent $\mathbf{f}(\mathbf{x}, \boldsymbol{\xi}, \boldsymbol{\theta}_f)$ and 86 local models to represent $\beta(\mathbf{x}, \boldsymbol{\xi}, \boldsymbol{\theta}_\beta)$. The output trajectory of the LWPR model is given in Figure 9.21. Figure 9.23 represents the output of the model when the actual demonstration had positive EDR.

SVR models:

Support vector regression technique is also used to learn the underlying dynamical systems that represent the demonstrations. First, the initial model parameters of $\mathbf{f}(\mathbf{x}, \boldsymbol{\xi}, \boldsymbol{\theta}_\beta)$ and $\beta(\mathbf{x}, \boldsymbol{\xi}, \boldsymbol{\theta}_\beta)$ are learned using standard SVR algorithm. These initial parameters are then used to learn the final model that satisfy stability criterion. The procedure is elaborated in the Algorithm 14. The end-effector related trajectories by the SVR model are shown in Figure 9.21 where the EDR remains negative during the task. Figure 9.23 shows a different task where the EDR reaches positive values during the task execution.

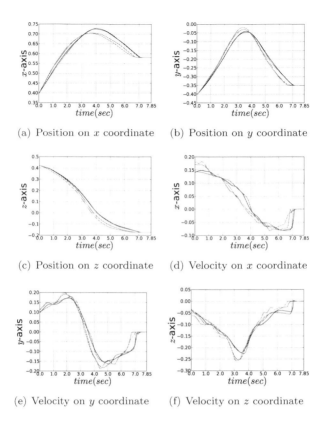

FIGURE 9.22: The trajectory of the original demonstration in Cartesian coordinate is such that the energy of the states monotonically decreases to zero, thereby making β always negative. Here, three trajectories, produced by GMR, LWPR and SVR models along with associated demonstrations are presented. The trajectory starts from the black circle and reaches to the target marked by black square. (a) (b), (c), (d), (e), (f) show the positions and velocities of the end-effector on x, y, and z coordinate respectively. Positions and velocities are in m and m/sec.

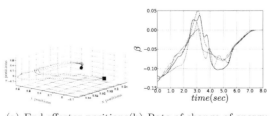

(a) End-effector positions (b) Rate of change of energy

FIGURE 9.23: In this experiment the EDR attains positive values. Trajectories are produced by GMR, LWPR, and SVR models.

9.5.4.2 Example of Multitasking with Single and Multiple Task-equilibrium

Let us divide multitasking in two categories in a dynamical system's perspective. Category one contains a set of tasks where the end-effector of a manipulator starts from different locations in the workspace and reaches at a single task-equilibrium. And the other category includes tasks where the end-effector starts from a fixed / different initial location(s) in the workspace and ends at different task-equilibrium positions while executing different type of motions.

For this example, the robot is given twenty demonstrations which include four type of tasks. First ten demonstrations represent two tasks with one task-equilibrium and different initial positions and the rest of the demonstrations are about two more tasks where the end-effector starts from different initial locations and ends at different task-equilibrium positions. Essentially, in this experiment the robot is demonstrated complex task profiles with multiple initial and target locations. The demonstrations are logged in datasets $\mathscr{D}_f$ and $\mathscr{D}_\beta$ and the algorithm learns the motion encoding model using three regression techniques. To test the model an initial position is chosen within the domain of learning, from which the end-effector starts moving to the desired location.

Multi-task trajectories with single task-equilibrium:

In this part let us evaluate the performance of the algorithms in executing multiple tasks (multi-task trajectories) for a single task-equilibrium position. *Task-1* and the *task-2* are shown in the figures. The end-effector starts from black circles and ends at black square (task-equilibrium) which is the target for that task. Figures 9.25 and 9.26 shows performance of the Gaussian mixture regression models. The results related to LWPR and support vector regression models are given in Figures 9.27 and 9.29 respectively. The associated trajectories in each coordinate are given in Figures 9.28 and 9.30 respectively. These results show that the model is able to generate two different task-trajectories while reaching for the single target. In addition, the model generated trajectories are very closely following the human demonstrations.

Multi-task trajectories with Multiple task-equilibrium:

The demonstrated tasks are learned using three regression techniques mentioned earlier. The task execution starts at black circles and ends at squares in black (task-equilibrium). Figures 9.31 and 9.32 shows performance of the Gaussian mixture regression models. The results related to LWPR and support vector regression models are given in Figures 9.33 and 9.35 respectively. The associated trajectories in each coordinate are given in Figures 9.34 and 9.36 respectively. The figures show that multi-task trajectories with multiple task-equilibria are efficiently learned using a single dynamical system based motion model.

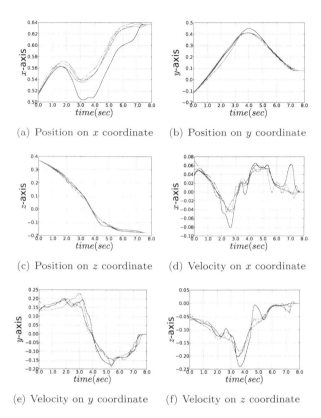

(a) Position on x coordinate (b) Position on y coordinate

(c) Position on z coordinate (d) Velocity on x coordinate

(e) Velocity on y coordinate (f) Velocity on z coordinate

FIGURE 9.24: The trajectory of the original demonstration in Cartesian coordinate is such that the energy of the states increases in certain regions of the workspace, thereby making β positive during task execution. Here, three trajectories, produced by GMR, LWPR and SVR models along with associated demonstrations are presented. The trajectory starts from the black circle and reaches to the target marked by black square. (a), (b), (c), (d), (e), (f) show the positions and velocities of the end-effector on x, y, and z coordinate respectively. Positions and velocities are in m and m/sec.

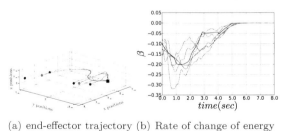

(a) end-effector trajectory (b) Rate of change of energy

FIGURE 9.25: Multi-task trajectories by GMR model with single task-equilibrium and different initial positions for the *task-1* and *task-2* with the associated demonstrations have been shown. Two demonstrations out of five demonstrations for each task have been presented here.

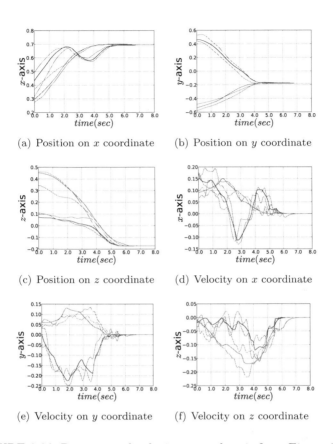

FIGURE 9.26: Position and velocity on each axis from Figure 9.25.

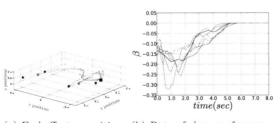

(a) End-effector positions (b) Rate of change of energy

FIGURE 9.27: Multi-task trajectories by LWPR model with single task-equilibrium position. *Task-1* and *task-2* are shown here with the associated demonstration.

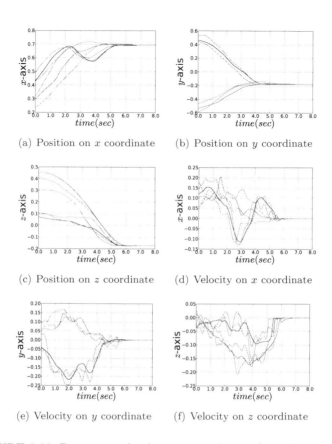

FIGURE 9.28: Position and velocity on each axis from Figure 9.27.

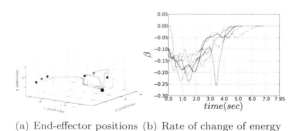

(a) End-effector positions (b) Rate of change of energy

FIGURE 9.29: Multi-task trajectories by SVR model with single task-equilibrium position. *Task-1* and *task-2* trajectories, generated by the SVR model are shown here.

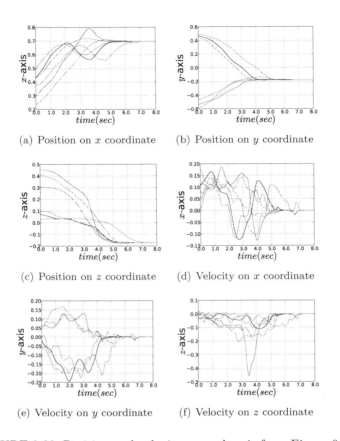

FIGURE 9.30: Position and velocity on each axis from Figure 9.29.

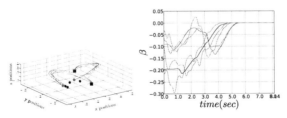

(a) Position on x coordinate (b) Rate of change of energy

FIGURE 9.31: Multi-task trajectories by GMR model with different task-equilibria.

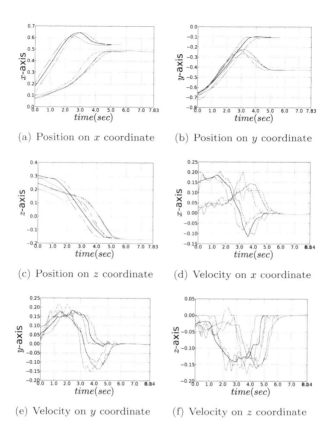

FIGURE 9.32: Projections of positions and velocities on individual axis from Figure 9.31 are given in (a), (b), (c), (d), (e), (f) where the model outputs are plotted alongside human demonsrtions. The motion encoding system generates trajectories from different initial points and ends at different targets. Tasks are selected based on the targets.

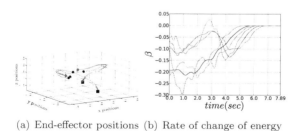

(a) End-effector positions (b) Rate of change of energy

FIGURE 9.33: This figure depicts the performance of the motion encoding model in an multitasking environment using LWPR technique.

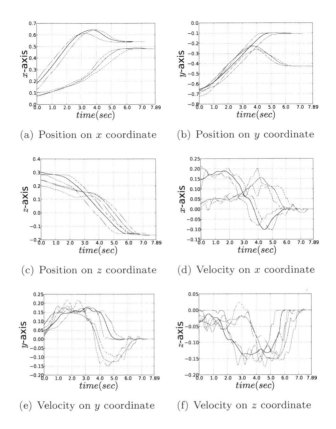

FIGURE 9.34: Projections of positions and velocities on individual axis from Figure 9.33 are given in (a), (b), (c), (d), (e), (f). The motion encoding system generates trajectories from different initial points and ends at different targets. Tasks are selected based on the targets.

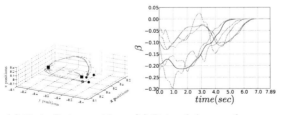

(a) End-effector positions (b) Rate of change of energy

FIGURE 9.35: Multi-task trajectories by SVR model with different task-equilibrium positions.

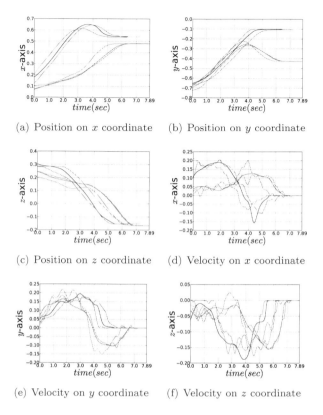

FIGURE 9.36: Projections of positions and velocities on individual axis from Figure 9.35 are given in (a), (b), (c), (d), (e), (f). The motion encoding system generates trajectories from different initial points and ends at different targets. Tasks are selected based on the targets.

9.5.5 Summary

In this chapter we have discussed various imitation learning techniques based on kinesthetic teaching where the expert (teacher) provides demonstrations of certain task while holding the robotic manipulator and guiding it through the task trajectory. The dynamic movement primitives (DMPs) presented here, have been shown to learn task in joint space and the other methods which use Gaussian mixture model, learn the task in the Cartesian space. The task trajectory is asymptotically stable in DMP models as it uses an inbuilt PD controller. Parameters of DMP based model are learned from single demonstration in an unconstrained optimization process which make them computationally efficient. But the main disadvantage of this technique is that is has a strong time dependency and cannot exploit multiple demonstrations. Multiple demonstrations are useful in many cases as the quality of demonstrations cannot be ensured by a single demonstration. Another problem with DMP based motion encoders is that they are generally learned for every singe dimension of the task space / joint space. It is assumed that each dimension / DoF is decoupled from each other. However, it is not true for complex tasks. Hence the efficiency of DMP based system is little lesser that the the Gaussian mixture model-based imitation learning methods.

Asymptotic stability of the models imitating human motion has been an important concern of this chapter. SED motion encoders which are great at generating asymptotically stable trajectories have been presented in this chapter. The parameters of the SED models are learned in a constrained optimization process. The constraints are formulated using Lyapunov stability criteria. The reader should note that when we learn the parameters of the model in a constrained search space, the model accuracy in regards to input-output data mapping is hampered. Handling equality constraints while learning a highly accurate model is very difficult. FuzzStaMP models alleviates this problem by combining statistical learning and Fuzzy logic. Since the stability is a great concern for DS based motion estimators, we have presented rigorous stability analysis of the motion system. A fuzzy Lyapunov function is searched for stability analysis as it is difficult to find a common Lyapunov function for all demonstrations. An associated fuzzy controller is designed using GA. The motion learning architecture C-FuzzStaMP presented in this chapter is conservative in nature since it imposes strict stability constraints during the controller gain selection and thereby giving more weightage on learning a globally asymptotically stable motion model. This approach sometimes may ignore complex features of the demonstrated motion profiles. However its performance is good in case of simple motion learning. By relaxing the asymptotic stability constraints on learning, R-FuzzStaMP learns the complex attributes of the demonstrations. Rather than making the $\dot{V}$ in (9.32) negative over the entire workspace, the algorithm ensures that the $\dot{V}$ is negative at least inside **RoD**. During the execution of the motion model, whenever a new state appears outside the **RoD**, which is quite typical, the state is translated inside the **RoD**. The translated state is made to evolve through time and translated back to its original domain. This also endows the motion model to spatially scale the trajectory as required. The spatial scaling is realized by using a space transformation technique, which prevents exaggeration of the scaled trajectory.

The multitask learning framework exploits additional information to enhance multitasking capability of the proposed technique. Three algorithms are presented to describe learning using different regression techniques namely GMR, LWPR and SVR. The model $\mathbf{f}(\mathbf{x}, \boldsymbol{\xi})$ holds the characteristics of the demonstrations, where $\boldsymbol{\xi}$ bears the information from the environment in the robot's workspace. The information can be either in the form of sensory feedback to the robot or the instructions to the robot, provided by the user. In this work, the target (task-equilibrium) positions of the end-effector is represented as $\boldsymbol{\xi}$ which is in fact the camera feedback of an object to be grasped. The guiding input $\mathbf{u}_g$ proves to be quite helpful in avoiding spurious attractors and at the same time it also pushes the system on the desired path where the EDR matches with the demonstrations. Instead of constraining the learning of $\mathbf{f}(\mathbf{x},\boldsymbol{\xi})$, a matrix $\mathbf{P}_2$ in (9.120) is searched during the parameter estimation of $\beta(\mathbf{x},\boldsymbol{\xi})$, which essentially relaxes the stability constraints on parameter estimation, since there is no sign constraint on $\mathbf{P}_2$.

Task requirements may vary based on various environmental conditions. Since we have considered target positions of the objects (to be grasped) as environmental input in our experiments, we have defined tasks based on target positions, by demonstrating different motion profiles to the robot. In that perspective, multitasking has been generalized by dividing it in two different categories. In the first category, the robot is demonstrating two tasks with a single task-equilibrium position and two more different tasks based on target position are demonstrated in the second category. Essentially, a single motion model learns and execute multi-task trajectories for different task-equilibrium positions.

10
Visual Perception

10.1 Introduction

Recent success in the area of object recognition, segmentation, and graspable region identification using deep learning techniques has intrigued robotics researchers to bring new improvements in visual perception techniques. These enormous improvements in both the deep learning techniques and high performance computing devices, like GPU machines have motivated many organizations to conduct various events, such as Amazon Picking Challenge (APC) in the year 2015-2016, Amazon Robotics Challenge (ARC) in 2017, and Scene Parsing Challenge during 2015-2017 in order to encourage the research in visual perception for real-time robotics applications. One such application, includes automation of warehouses, where thousands of wheeled mobile robots are deployed to move objects within the warehouses. In existing scenarios, it still requires several hundred people in each warehouse to do things like pulling items from shelves and placing them into packaging boxes to be shipped to the user. Robots that can automatically pick and place items would boost the efficiency of operation by reducing the reliance on human workers which is very expensive in highly competitive e-commerce markets with very narrow profit margins. A perception module can be developed to accurately recognize a specific instance of the objects to be picked and keep a record of it for further use. Until recently, such vision-based techniques have been considered as very challenging task and it is due to the current progresses in the field of deep learning, that makes the implementation of such a complicated task possible. This chapter mainly focuses on the vision-based deep learning techniques for automatic recognition of an object class and accurate detection of object boundary for localizing graspable region. The existing deep learning-based state-of-the-art object detection and segmentation techniques, such as Faster RCNN [254,255], Single-shot-detection [256], PSPNet [257], and Mask-RCNN [258] are designed to perform well only when they are rigorously trained using a large set of images. However, all these deep learning techniques are mostly supervised in nature, i.e, they need a large amount of manually labeled ground-truth or image-masks for training the detection model. Manual generation of annotated image sets of such large size need a lot of human intervention, which is very tedious and are often prone to error due to fatigue. Taking this above challenge into consideration, this chapter first demonstrates some techniques of automatically generating annotated images and then is followed by deep learning-based object detection and instances segmentation for pick and place task. It is desirable to have some prerequisites, such Artificial Neural Networks (ANNs), linear algebra, and basics of machine learning and optimization techniques, for the readers to understand all the aforesaid deep learning techniques. It may not be in the scope of this chapter to cover all these prerequisites. However, inclusion of some of the basics, like a brief overview of the Artificial Neural Networks (ANNs), Convolutional Neural Networks (CNNs), and its sub-blocks and a brief evolutionary history of CNNs, we have tried to make this chapter more readable and easier to understand even for a newbie. We thus start with an introduction of deep learning techniques and the prerequisites needed to implement a deep network. The Section 10.3.1 and Section 10.3.2 then demonstrates two different deep learning-based automatic annotation techniques for warehouse applications, and finally it presents a deep learning-based object detection and semantic segmentation technique for objects in a densely populated, completely unconstrained environment.

10.2 Deep Neural Networks and Artificial Neural Networks

Deep Neural Networks (DNNs) have created new bench mark in almost all the vision-based applications. Some of these include object instances and category recognition, semantic segmentation, and graspable region identification. On the other hand, the conventional classification techniques, like Support Vector Machines [259] which had sustained almost a decade, is in no way closer to any of the current deep learning techniques for the performance measures in terms of recognition accuracy. The improvements made in recognition results using DNNs compared to the conventional techniques are mostly in the order of 2, which is astonishing. Certainly, a basic question sticks in our mind, what is unique in DNNs which makes them so different from Artificial Neural Networks (ANNs)? Deep learning is one kind of supervised machine learning technique where the model is provided with a big training set of examples to learn a complex non-learning function. Each example in the training set is a pair of input and output from the function. The more complex the task is, the bigger the training set has to be. Deep learning is based on the concept of ANNS that is designed to mimic the way the human brain learns. This section provides a clear explanation about the distinct features present in DNNs. However, before that we familiarize the readers with some prerequisites for better understanding of the underlying concepts. We will hence start with a brief history and explanation of those neural networks.

10.2.1 Neural Networks

Neural networks (NN) or Artificial Neural Network (ANN) is an information processing technique inspired from the way our biological nervous system works. It is the structural combination of the neurons which makes the ANN a unique processing system. It is composed of a large number of interconnected neurons. As the brain adjusts synaptic connections between the neurons, a similar fundamental strategy is also applied in ANN to learn the parameters during the training process. Warren McCulloch, a neurophysiologist, and logician Walter Pits first introduced the concept of artificial neuron in the year 1943 [260]. However, due to unavailability of high processing devices at that time, the researchers were not able to explore more in this direction, and the work was almost stagnant for a couple of years. The advancements in computational devices have motived researchers to explore more in the area of ANNs and lots of improvements have been made during the last few decades [261]. A few remarkable advantages of using ANNs are given in the following.

1. The ANN, has a remarkable ability of deriving vital information or meaning from very complicated or imprecise data. The ANN can be used to extract patterns and detect trends which are usually very complex to be noticed by either bare eyes or through other computerized techniques.

2. *Adaptive learning*: An ANN can learn a model-based on the training data.

3. *Self-Organization*: An ANN has the ability to create its own organization or representation of the information it gets while learning or updating the parameters.

4. *Real-time operation*: Computationally efficient devices like, GPU can be used for real time operation of complex ANN models.

Let us give a brief architectural overview of a simplest ANN with only one neuron, where the input feature vector is $\mathbf{x} \in \mathbb{R}^4$. The labeled training example is given as (x_i, y_i). The NN defines a complex nonlinear form of hypothesis $\mathbf{h}_{\mathbf{w},b}$ to approximate the output function. Figure 10.1 shows an example of such ANN structure. The neuron is a computational unit which takes $\mathbf{x} = [x_1, x_2, x_3, x_4]$ as input feature vector and a bias term $b = +1$. The neuron outputs $\mathbf{h}_{\mathbf{w},b}$ is given as

$$\mathbf{h}_{\mathbf{w},b}(\mathbf{x}) = \mathbf{f}(\mathbf{W}^T \mathbf{x}) \tag{10.1}$$

$$= \mathbf{f}\left(\sum_{i=1}^{3} \mathbf{W}_i x_i + b\right) \tag{10.2}$$

where $f: \mathbb{R} \to \mathbb{R}$ is known as the *activation function* which maps input and output via (generally) a nonlinear function. Mostly, it is chosen as activation function. The is given as

$$f(z) = \frac{1}{1 + e^{-z}} \tag{10.3}$$

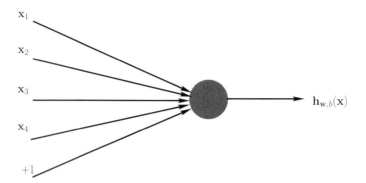

FIGURE 10.1: An example of a ANN with a single neuron.

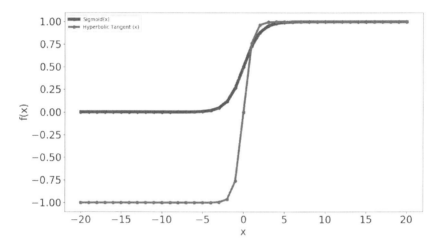

FIGURE 10.2: Example of activation functions: Sigmoid function and tanh function.

Hyperbolic tangent or tanh function is also used as activation function. The tanh function is defined as

$$f(z) = \tanh(z) = \frac{e^z - e^{-z}}{e^z + e^{-z}} \tag{10.4}$$

Figure 10.2 gives examples of sigmoid function and tanh function respectively. The tanh function is re-scaled version of the sigmoid function. The range of outputs of the sigmoid and tanh functions are $[0, 1]$ and $[-1, 1]$. A sample code written in python for plotting activation functions, such as sigmoid and hyperbolic tangent is also given below in this section.

```python
# Required Python Packages
import numpy as np
import matplotlib.pyplot as plt
FONT_SIZE = 20
plt.rc('xtick', labelsize=FONT_SIZE)
plt.rc('ytick', labelsize=FONT_SIZE)
def sigmoid(inputs):
    y1 = [1 / float(1 + np.exp(- x)) for x in inputs]
    return y1

def tanh(inputs):
    y2 = [(1-np.exp(-2*x)) / float(1 + np.exp(- 2*x)) for x
        in inputs]
    return y2

x = range(-20, 21)
sig_a = sigmoid(x)
tanh_a = tanh(x)

fig = plt.figure()
plt.xlabel('x', fontsize=FONT_SIZE)
plt.ylabel('f(x)', fontsize=FONT_SIZE)

ax  = fig.add_subplot(111)
ax.plot(x, sig_a, c='b', label='Sigmoid(x)', marker ='o',
    linewidth=6.0)
ax.plot(x, tanh_a, c='r', label='Hyperbolic Tangent (x)' ,
    marker ='o', linewidth=4.0)
leg = plt.legend()
plt.show()
```

Many simple neurons are combined together to give a structure of NN. Output of one neuron can be the input for another neuron. One such architecture is Multi-layer perceptron (MLP). The following section gives a detailed architectural overview of MLP.

10.2.1.1 Multi-layer Perceptron

The multi-layer perceptron (MLP) [261] is a feed forward NN which can input feature vector to the output class using a nonlinear function. It has multiple layers of neurons. Each neuron is associated with an activation function at its output. The training method used in MLP is supervised and the learning approach is called backpropagation method. MLP is a improvement over the standard linear perceptron which can classify the data that are not linearly separable [261]. Figure 10.3 gives a basic network architecture of an MLP

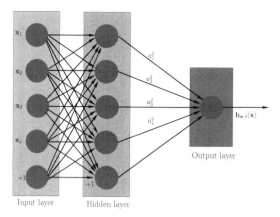

FIGURE 10.3: A basic architecture of a multilayer perceptron.

with only three layers. The leftmost layer of the network is called *input layer* which connects the input features $x_i \in (x_1, x_2, x_3, x_4)$ to each neuron. The middle layer of the neurons is called *hidden layer* as its values are not directly observed in the training set. The outermost layer is known as *output layer*. The output layer connects all the activation outputs of the hidden layer.

$$a_1^{(2)} = f(W_{11}^{(1)}x_1 + W_{12}^{(1)}x_2 + W_{13}^{(1)}x_3 + W_{14}^{(1)}x_4 + b_1^{(1)}) \qquad (10.5)$$
$$a_2^{(2)} = f(W_{21}^{(1)}x_1 + W_{22}^{(1)}x_2 + W_{23}^{(1)}x_3 + W_{24}^{(1)}x_4 + b_2^{(1)}) \qquad (10.6)$$
$$a_3^{(2)} = f(W_{31}^{(1)}x_1 + W_{32}^{(1)}x_2 + W_{33}^{(1)}x_3 + W_{34}^{(1)}x_4 + b_3^{(1)}) \qquad (10.7)$$
$$a_4^{(2)} = f(W_{41}^{(1)}x_1 + W_{42}^{(1)}x_2 + W_{43}^{(1)}x_3 + W_{44}^{(1)}x_4 + b_4^{(1)}) \qquad (10.8)$$

where a_i^l denotes the activation of neuron i in layer l, $W_{ij}^{(l)}$ gives the parameters or the weights associated with the connection between neuron j in layer l and neuron i in layer $l+1$ and the term $b_i^{(l)}$ is the bias associated with the neuron i in the layer $l+1$. The output of the entire network is given by the hypothesis $h_{W,b}$ which can be expressed as

$$h_{W,b}(x) = a_1^{(3)} = f(W_{11}^{(2)}a_1^{(2)} + W_{12}^{(2)}a_2^{(2)} + W_{13}^{(2)}a_3^{(2)} + W_{14}^{(2)}a_4^{(2)} + b_1^{(2)}) \qquad (10.9)$$

Backpropagation algorithm:

Backpropagation (BP) is an abbreviation for *backward propagation of errors*. The BP method calculates the gradient of a cost function with respect to all the weights associated with the network. The gradients with respect to the parameters (weights) are given to the optimization method (usually

Algorithm 17 Backpropagation algorithm

1: % Given a training data (x,y) there are broadly two step: First run *forward pass* to calculate all the activation output in entire network including the final output $h_{W,b}(x)$ and then calculate the *error term* $\delta_i^{(l)}$ for each neuron i in layer l. It gives how much the neuron is responsible for any error in the output.
2: Carry out a feed-forward pass to compute all the activations in layer L_2, L_3 till L_{nl}.
3: For each output neuron i in layer nl calculate the cost

$$\delta_i^{(nl)} = \frac{\partial}{\partial z_i^{(nl)}} \frac{1}{2} \|y - h_{W,b}(x)\|^2 = -(y_i - a_i^{(nl)})f'(z_i^{(nl)}). \qquad (10.10)$$

4: **for** $l = nl-1, to\ 2$ **do**
5:
6:     **for** neuron $i = 1$ to sl **do**
7:

$$\delta_i^{(l)} = \left(\sum_{j=1}^{sl+1} W_{ji}^{(l)} \delta_j^{(l+1)} \right) f'(z_i^{(l)}). \qquad (10.11)$$

8:     **end for**
9: **end for**
10: Compute the partial derivatives as

$$\frac{\partial}{\partial W_{ij}^{(l)}} J(W,b;x,y) = a_j^{(l)} \delta_i^{(l+1)}, \qquad (10.12)$$

$$\frac{\partial}{\partial b_i^{(l)}} J(W,b;x,y) = \delta_i^{(l+1)}. \qquad (10.13)$$

---

gradient descent method) to update the weights, which in an attempt minimizes the cost function. To train a feature vector $x \in \mathbb{R}^n$ with output vector $y \in \mathbb{R}^m$, we need training examples (x^i, y^i). Given m training examples, i.e., a set of $\{(x^{(1)}, y^{(1)}), \ldots, (x^{(m)}, y^{(m)})\}$ the ANN is trained using batch gradient descent method. The batch gradient descent computes the gradient using the whole dataset.

For instance (x,y), the cost function for a particular sample can be given as

$$J(W,b;x,y) = \frac{1}{2} \|h_{W,b}(x) - y\|^2 \qquad (10.14)$$

The overall cost function for the entire training set (for batch gradient descent) with weight regularization term is given as

$$J(W,b) = \left[\frac{1}{m}\sum_{i=1}^{m} J(W,b;x^{(i)},y^{(i)})\right] + \frac{\lambda}{2}\sum_{l=1}^{n_l-1}\sum_{i=1}^{s_l}\sum_{j=1}^{s_l+1}(W_{ji}^{(l)})^2, \quad (10.15)$$

$$= \left[\frac{1}{m}\sum_{i=1}^{m}\left(\frac{1}{2}\|h_{W,b}(x^{(i)}) - y^{(i)}\|^2\right)\right] + \frac{\lambda}{2}\sum_{l=1}^{n_l-1}\sum_{i=1}^{s_l}\sum_{j=1}^{s_l+1}(W_{ji}^{(l)})^2. \quad (10.16)$$

The term in the above equation (10.16) is an average sum of squared error and the second term is a regularization term also known as weight decay term. The term tends to decrease the magnitude of the weights to prevent overfitting. λ is the weight decay parameter, which controls the relative importance of the two terms. The goal of the backpropagation algorithm is to minimize $J(W,b)$ which is a function of the parameters W and b. As $J(W,b)$ is non-convex in nature, use of simple gradient descent may lead to trapping into local minima and thus, batch gradient descent is suggested for use. The parameters update rules are as follows:

$$W_{ij}^{(l)} = W_{ij}^{(l)} - \alpha \frac{\partial}{\partial W_{ij}^{(l)}} J(W,b), \quad (10.17)$$

$$b_i^{(l)} = b_i^{(l)} - \alpha \frac{\partial}{\partial b_i^{(l)}} J(W,b). \quad (10.18)$$

where α is the learning rate which decays as the number of iteration increases. The backpropagation algorithm is applied in this stage to efficiently calculate the partial derivatives. Using the equation 10.16, the partial derivatives for overall cost function are given as

$$\frac{\partial}{\partial W_{ij}^{(l)}} J(W,b) = \left[\frac{1}{m}\sum_{i=1}^{m} \frac{\partial}{\partial W_{ij}^{(l)}} J(W,b;x^{(i)},y^{(i)})\right] + \lambda W_{ij}^{(l)} \frac{\partial}{\partial b_i^{(l)}} J(W,b), \quad (10.19)$$

$$\frac{\partial}{\partial b_i^{(l)}} J(W,b) = \frac{1}{m}\sum_{i=1}^{m} \frac{\partial}{\partial b_i^{(l)}} J(W,b;x^{(i)},y^{(i)}). \quad (10.20)$$

10.2.1.2 MLP Implementation using Tensorflow

Since all neural networks need a sufficiently larger amount of data to train the network, it becomes a concern in the neural network community to optimize the network for faster implementation. In the year 2015, *Google* group open-sourced a new library called *Tensorflow* for numerical computations, which is computationally the most efficient library presently available. In this chapter, most of the example codes are given in python based tensorflow.

Readers can follow basic tensorflow APIs to get accustomed with the tensorflow implementation. In this section, we are providing an implementation of MLP using tensorflow for better understanding of the architecture and back-propagation. In machine learning, every newbie is recommended to start working with *MNIST* dataset that has a total of 70,000 images of size 28×28 (flattened to 784 dimensional vector) containing handwritten digits. Let's say, we have a MLP with two hidden layers, one input layer and one output layer. As we have seen from the data that the input is of dimension 784 while the output has 10 nodes ($0-9$, total numbers). The network parameters can be defined as follows.

```
#Initialize the network parameters and learning parameters
n_hidden1 = 256
n_hidden2 = 256
n_input = 784
n_output = 10
learning_constant = 0.001
number_epochs = 10000
batch_size = 200
```

where, the two hidden layers have 256 neurons each, input layer contains 784 nodes and the output layer has 10 nodes. The learning rate is defined as $\alpha = 0.01$. Number of epochs is set to 1000 and the batch size (number of images the network train in each epoch) to 100. Tensorflow implementation is very simple and compact. Initially, it defines the variables called tensor to store the data during runtime. The tensor X has size of the input vector and Y has the size of the output vector.

```
# create variable for input X and output Y
X = tf.placeholder("float", [None, n_input])
Y = tf.placeholder("float", [None, n_output])
```

Each neuron in the network is associated with a bias and each pair of neurons in the neighboring layer has a weight, which is defined as follows.

```
# Create variables for bias and weights at each layer.
# Bias at first hidden layer
b1 = tf.Variable(tf.random_normal([n_hidden1]))
# Bias at second hidden layer
b2 = tf.Variable(tf.random_normal([n_hidden2]))
# Bias at output layer
b3 = tf.Variable(tf.random_normal([n_output]))
#weights between first input data   layer to first hidden
    ↪ layer
w1 = tf.Variable(tf.random_normal([n_input, n_hidden1]))
```

```
#weights between first hidden layer   layer to second hidden
    ↪ layer
w2 = tf.Variable(tf.random_normal([n_hidden1, n_hidden2]))
#weights between second hidden layer to output layer
w3 = tf.Variable(tf.random_normal([n_hidden2, n_output]))
```

Once the weights w and biases b are assigned, the network becomes ready to perform multiplication followed by addition operation $x \cdot W + u$. It is further applied to a nonlinear activation function, like sigmoid or relu.

```
def multilayer_perceptron(input_data):
    # operation at hidden layer 1 for input data
    layer_1 = tf.nn.relu(tf.add(tf.matmul(input_data, w1),
        ↪ b1))
    # operation at hidden layer 2 for input data of layer 1
    layer_2 = tf.nn.relu(tf.add(tf.matmul(layer_1, w2), b2)
        ↪ )
    # operation at output layer  for input data of layer 2
    out_layer = tf.add(tf.matmul(layer_2, w3),b3)
    return out_layer
```

```
#loss function for training the mlp
loss_func = tf.reduce_mean(tf.nn.
    ↪ softmax_cross_entropy_with_logits(
                        logits=multilayer_perceptron(X),
                            ↪ labels=Y))
#optimize the losses using gradient descent
optimizer = tf.train.GradientDescentOptimizer(
    ↪ learning_constant)
                    .minimize(loss_func)
```

Once the network is defined and optimizer is set, the next step is to train the network in an iterative manner. All the aforesaid operations will be performed only when it is called within the tensorflow session. For that we need to create and run a session after initializing the parameters.

```
#Initialization
init = tf.global_variables_initializer()
#Create a session
with tf.Session() as sess:
    sess.run(init)
    #Perform training  operation for every epoch
    for epoch in range(number_epochs):
        #load training data for a batch
        batch_x, batch_y = mnist.train.next_batch(
            ↪ batch_size)
```

```
#Run the optimizer feeding the network with the
⤷ batch
sess.run(optimizer, feed_dict={X: batch_x, Y:
⤷ batch_y})
#Display the epoch (just every 100)
if epoch % 100 == 0:
    print("Epoch:", '%d' % (epoch))
```

Model evaluations can be done as follows.

```
# Evaluate the trained model using test data
pred = tf.nn.softmax(neural_network)   # Apply softmax to
⤷ logits
correct_prediction = tf.equal(tf.argmax(pred, 1), tf.argmax
⤷ (Y, 1))
# Calculate the prediction accuracy
accuracy = tf.reduce_mean(tf.cast(correct_prediction, "
⤷ float"))
print("Accuracy:", accuracy.eval({X: mnist.test.images, Y:
⤷ mnist.test.labels}))
```

10.2.2 Deep Learning Techniques: An Overview

Over the years, deep learning techniques have taken various shapes. They can be broadly categorized into two different kinds: unsupervised (generative models) and supervised approaches [254, 262, 263]. As there have been enormous research in the area of deep learning in recent years, it may not be practically possible to cover the entire literature in this chapter. We have limited it to basic building blocks of deep neural networks, such as Convolutional Neural Network (CNNs), which have been proven to be performing well in vision-based applications. We have also demonstrated a few examples of deep learning-based techniques for automatic annotation of objects, localization of a given instance, and accurate segmentation of objects. The following section gives a clear description for Convolutional Neural Networks.

10.2.2.1 Convolutional Neural Network (Flow and Training with Back-propogation)

Convolutional Neural Networks (CNN/ConvNets) are cascaded layers of neural networks mainly designed to classify images (name the object), localize the region of the object, and perform each object recognition within scenes. It has given eyes to the machine which enables it to perceive and predict like a human, at times even more powerful than what human eyes can actually perceive. The proficiency of the convolutional neural networks has woken up the world to dig a lot more into this area. It is used in all the vision-based applications.To name a few, detection and recognition of objects in a scene,

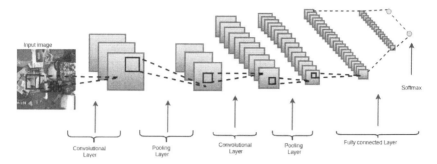

FIGURE 10.4: A basic CNN architecture is shown with different building blocks: Convolutional layer, max-pooling and fully connected layer.

identify faces, individuals, street signs, eggplants, diagnosis of medical images, remote image analysis, self-driving cars, robotics, drones, security, etc. Here we are going to explain how CNNs are different from ANNs and why they is so powerful.

A basic architectural overview of the CNN is shown in Figure 10.4. The CNNs are very similar to ordinary Artificial Neural Networks (ANNs). Like ANNs, the CNNs are also consisting of neurons which have learnable weights and biases. Each neuron in the CNN network receives input data that performs a dot product, often followed by nonlinear transformation. The entire network finally expresses a single differentiable score function using the raw image pixels at the input end to class recognition/detection scores at the final layer. It also has a loss function in the last (fully-connected) layer and all the tips/tricks which are mostly applied to NNs are applied to this network too. In contrast to ANNs, the CNN architecture mostly comprises repetitive blocks of neurons that are applied across 2D space (in case of an image input) or time (for audio signals, etc.). These blocks of neurons can be interpreted as 2D kernels to perform convolution operations incase of images which are repeatedly applied over each patch of a given input image. The same can be treated as 1D convolutional kernels for audio or speech data. As mentioned previously, the CNN consists of repeated blocks connected in a cascaded manner. One of the important differences of the CNN with other ANNs is that the weights of each block are shared. The weight gradients learned over various patches are averaged. The choice of using image patches, instead of the entire image as input, helps to make the system invariant to temporal or spatial changes for recognition of an object. On the other hand, the ANNs, such as MLP must be trained for all the possible spatial locations of an object within a scene in order to recognize it. Alongwith these differences, there are some other specially designed techniques in CNN that makes it entirely different from conventional ANNs. Some of those techniques are described later in this chapter. As shown in the Figure 10.4, three main layers are being used to build

Deep Neural Networks and Artificial Neural Networks 397

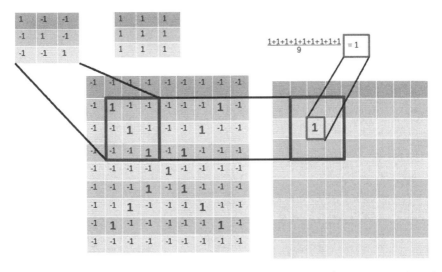

FIGURE 10.5: An example of a convolutional operation. A 3×3 kernel with ones are used to convolve an image patch.

a CNN architecture: Convolutional layer, pooling layer, and fully connected layer. Silent features of a basic CNN architecture are highlighted below:

1. Input layer: Raw image data, i.e, pixel values of an image is given as an input. The format of the input data is $[M \times N \times Z]$, where $M \times N$ is the width and height of the image and Z is the depth (3 for R,G and B channels).

2. Convolution layer: It computes output of neurons that are locally connected to an image patch. Given a K number of filters, an image patch of size $M \times N$ computes dot product with each of the K filters, resulting into a $M \times N \times K$ volume of convolved data if all the filters are used. An example for the convolution operation is shown in Figure 10.5.

3. Pooling layer: It basically performs downsampling of a convolved image along the spatial dimensions. The technique is explained with an example in Figure 10.6. The filter size is taken as 2×2 with the stride as 1. Max pooling performs the operation $max(x)$ within the 2×2 window.

   There are some other pooling approaches, such as min pooling, average pooling. However, max pooling has been widely accepted over other techniques, due to its better performance.

4. RELU layer: This is a nonlinear function which is operates on each of the output neurons. Element-wise activation function, such as $max\ (0,\ x)$ is mostly used leaving the size of the volume unchanged. Figure 10.7 gives an example of a RELU function using $max(0, x)$ operation. A sample code is also provided here to show a RELU function.

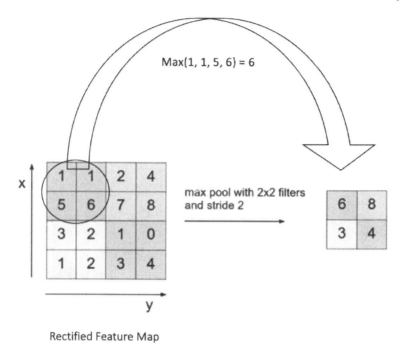

FIGURE 10.6: Max pooling, an illustration.

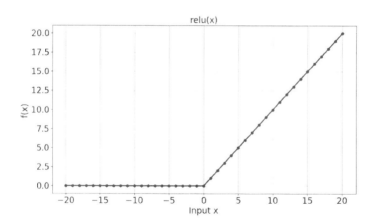

FIGURE 10.7: RELU, an illustration.

```
# A sample code to plot activation function relu
import numpy as np
import matplotlib.pyplot as plt
FONT_SIZE = 20
plt.rc('xtick', labelsize=FONT_SIZE)
plt.rc('ytick', labelsize=FONT_SIZE)

def relu_func(input, zero):
    y = np.max([zero, input], axis=0)
    return y
def line_graph(x, y, x_title, y_title, title):
    plt.plot(x, y, c='b', marker ='o', linewidth =2.5)
    plt.title(title, fontsize=FONT_SIZE)
    plt.xlabel(x_title, fontsize=FONT_SIZE)
    plt.ylabel(y_title, fontsize=FONT_SIZE)
    plt.grid()
    plt.show()

graph_x = range(-20, 21)
zero = np.zeros(len(graph_x))

graph_y = relu_func(graph_x, zero)

print "Graph X readings: {}".format(graph_x)
print "Graph Y readings: {}".format(graph_y)
line_graph(graph_x, graph_y, "Input x", "f(x) ", "relu(x
    )")
```

5. Fully connected layer (FC): This is the final layer of a CNN architecture. Each neuron of this layer gets connected to all the output neurons of the previous layer and computes class scores at the end. The parameters, such as weights and biases of the FC layer are trained with optimization techniques, like gradient descent in order to align the network performance with the labels given in the training image sets. An example of a fully connected layer is shown in Figure 10.8.

The fully connected layer shown in the Figure 10.8 has an input layer of features, a hidden layer, and an output layer. Each feature x_i in the input feature vector **x** is connected to all the neurons in the hidden layer and each output node is connected to all the neurons in the hidden layer.

10.2.3 Different Architectures of Convolutional Neural Networks (CNNs)

CNNs came into the limelight after they won the ImageNet challenge for the first time in 2012. Since than, in every year variants of CNNs have been

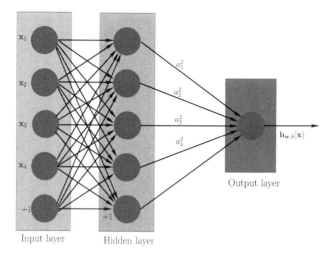

FIGURE 10.8: An illustration of a fully connected layer with only one hidden layer and one output node.

introduced with promising improvements in recognition accuracy over its previous counterparts. A few of those winning architectures which have created long lasting impacts are presented in this section.

LeNet-5 (1998)
LeNet-5 was a pioneering seven-level convolutional neural network developed by LeCun et al. in the year 1998 [264]. The network was able to successfully recognize handwritten numbers on cheques and was adopted by several banks for this task as digitized in 32×32 pixels images. Images with higher resolution demanded more convolutional layers with much higher computing resources, which was unavailable at that time. An architectural overview of the LeNet-5 is shown in the Figure 10.9.

AlexNet (2012)
Long after the LeNet, another CNN called AlexNet was invented in the year 2012. The improvements in GPU based computational power enabled such CNN networks with quite a large number of hidden layers to outperform all the prior competitors. And it won the ImageNet challenge reducing the top-5 error to 15.3%. The second place top-5 error rate, which was not a CNN variation, was around 26.2%. The architectural framework of the AlexNet was similar to LeNet [264]. The only difference was that AlexNet was much deeper with more filters in each layer and had more stacked CNN layers. The network has been split into two parts as it has used two parallel Nvidia Geforce GTX 580 GPUs. It was designed by the SuperVision group, consisting of Alex Krizhevsky, Geoffrey Hinton, and Ilya Sutskever. The architecture is shown in the Figure 10.10.

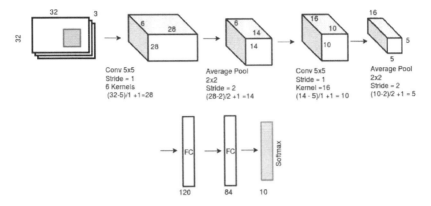

FIGURE 10.9: An architectural overview of LeNet [264].

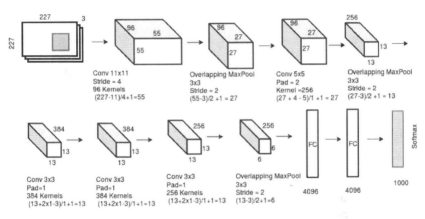

FIGURE 10.10: An architectural overview of AlexNet [265].

ZFNet (2013)
ZFNet [266], another CNN framework, become a winner of ILSVRC 2013. The network became famous for achieving a top-5 error rate of 14.8% which is half of the prior mentioned non-neural error rate. Not surprisingly, the ILSVRC 2013 winner was also a CNN which became known as ZFNet. It achieved a top-5 error rate of 14.8% which is now already half of the prior mentioned non-neural error rate. It was mostly an achieved by tweaking the hyper-parameters of AlexNet while maintaining the same structure with additional deep learning elements as discussed earlier in this essay. The architectural overview of ZFNet is shown in Figure 10.11.

GoogleNet/Inception(2014)
The winner of the ILSVRC 2014 competition was GoogleNet(Inception) [267] from Google. It achieved a top-5 error rate of 6.67%! This was very close

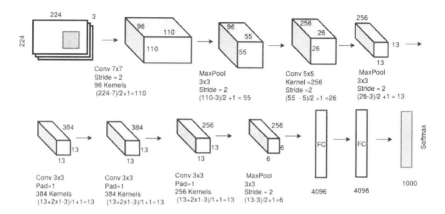

FIGURE 10.11: An architectural overview of ZFNet [266].

to human level performance which the organizers of the challenge were now forced to evaluate. As it turns out, this was actually rather hard to do and required some human training in order to beat GoogLeNets accuracy. After a few days of training, the human expert (Andrej Karpathy) was able to achieve a top-5 error rate of 5.1%. The network used a CNN inspired by LeNet but implemented a novel element which is dubbed an inception module. This module is based on several very small convolutions in order to drastically reduce the number of parameters. Their architecture consisted of a 22-layer deep CNN but reduced the number of parameters from 60 million (AlexNet) to 4 million. A *naive* inception module is shown in the Figure 10.12 (a). It performs convolution on an input, with three different sizes of filters (1x1, 3x3, 5x5). Apart from this, it also performs a max pooling operation. The outputs are then concatenated and given to the next inception module. In a successive work, dimensionality reduction is done by introducing a convolutional layer with 1x1 filter before the convolution operations using 3x3, 5x5 and after max-pooling operation using 3 x 3. The module is shown in Figure 10.12 (b).

VGG networks

The VGG Network [268] was introduced by the researchers at Visual Graphics Group at Oxford (hence the name VGG). This network is specially characterized by its pyramidal shape, where the bottom layers which are closer to the image, are wide and the top layers are deep. There are certain remarkable advantages of VGG network. One of the most significant advantage of the network is that, the architecture is very good for benchmarking on a particular task. Also, pre-trained networks for VGG are available freely on the internet, so it is commonly used out of the box for various applications. However, it also has an unavoidable disadvantage, that it is very slow to train if trained from scratch. Even using a latest GPU machine, the computation goes for weeks to properly train the network. A 16-layer VGG network is shown in Figure 10.13.

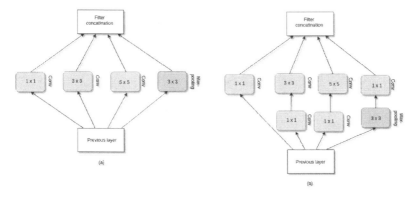

FIGURE 10.12: An architectural overview of naive inception modules ((a) without, and (b) with dimensionality reduction).

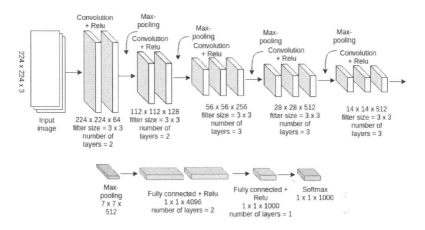

FIGURE 10.13: An architectural overview of VGGnet [268] with sixteen layers.

ResNet

Residual Networks (ResNet) [269] is one of the largest deep architectures which truly define how deep a deep learning architecture can be. One of the main advantages of the ResNet over the previous deep networks is that it has the ability to train extremely deep networks (150 layers or more) successfully while ensuring no vanishing gradient problems. It consists of multiple subsequent residual modules, which are the basic building blocks of the ResNet architecture. An architectural overview of the ResNet block is given in the Figure 10.14.

A residual module in the network has two options, i.e., either it can perform a set of functions on the input, or it can skip this step altogether. There is

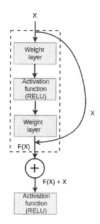

FIGURE 10.14: A representation of a ResNet module [269].

an architectural similarity of ResNet with GoogleNet [267], in a sense, the residual modules are stacked one over the other to form a complete end-to-end network. The ResNet has following features:

1. It uses standard Stochastic Gradient Descent (SGD) instead of newly introduced adaptive learning techniques, that claims to be performing better. This is done along with a reasonable initialization function which keeps the training intact.

2. The input is first divided into patches and then it is passed to the network. The main advantage of the ResNet in contrast of other aforesaid networks is that, we can use any number of residual layers, even hundreds or thousands and the performance gets improved with increasing number of layers, unlike other networks where it gets degraded with a higher number of layers. The residual blocks make sure that with the increase in layer, the vanishing gradient problem will not arise.

10.3 Examples of Vision-Based Object Detection Techniques

In this section we have presented some use cases of visual perception techniques applied to automate warehouses. Amazon is one such organization which has been constantly focusing on warehouse automation by organizing various events, such as Amazon Picking Challenge (APC) in the year 2015, 2016 and Amazon Robotic Challenge (ARC) in 2017. ARC 2017 was associated with three main tasks: the first task was to stow each object from a

Examples of Vision-Based Object Detection Techniques 405

overly populated tote into a partially crowded shelf; the second part involves picking a given set of object instances from a densely populated shelf and place it into a dispatching box, and the third part includes stow followed by pick. The tasks were associated with various vision-based challenges. A few of them include:

1. Generation of a training dataset such that it can handle new objects at the time of training.

2. An approach of updating the recognition module to train newly introduced objects within a short duration, so that the additional training set includes almost all the possible combination of the total objects.

Here we address the above challenges with some deep learning-based demonstration examples.

1. Automatic detection of ROI for any new object using Faster-RCNN based deep learning framework

2. Detection of object ROI for eighty-three different classes of object.

3. Automatic segmentation of any new object boundary and class-wise segmentation of objects.

All these examples are based on the objects given by Amazon for the Amazon Robotic Challenge (ARC) 2017. A few images of those objects are shown in the Figure 10.15.

10.3.1 Automatic Annotation of Object ROI

In the era of deep learning-based object recognition systems, where a large number of annotated images are required for training, manual annotation of each object is a challenging job. It is associated with a lot of human intervention, which is very tedious and often prone to error due to fatigue. Thus, development of an automatic annotation technique plays a very crucial role for effective recognition of object category or instances. This section demonstrates a technique for developing a completely automatic object annotation technique by using a few manually annotated images. A very small set of images of size 8000 is used to generate a big dataset of size 1,10,000 by using affine transformation, multiple color augmentation and clutters generation techniques, discussed later in the section. Color augmentations, affine transformations, and cluttering increases the size of the training dataset which is an essential requirement to train any deep convolutional neural network. Introduction of affine transformation and color augmentation also makes the model invariant change in color and object transformation due to rotation and scaling. Training the network with synthetically generated clutters has another significant advantage as it makes the model capable of annotating multiple objects at a time even in a densely cluttered environment. The generated big data is then applied to train Faster RCNN for binary classification

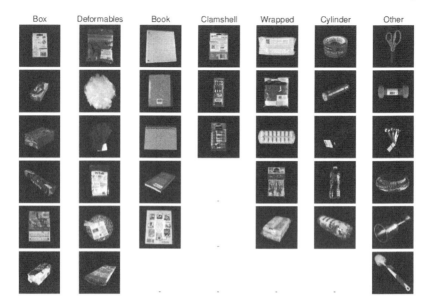

FIGURE 10.15: Example images of objects used in Amazon Robotics Challenge.

where the region proposal network suggests a set of bounding boxes followed by regression to best fit the bounding box for effective ground truth generation. The models are developed using Faster RCNN as well as R-FCN for the same pre-trained models. Semantic segmentation based techniques, such as PSPNet [257] and MaskRCNN [258] are not used for the aforesaid purpose as both the approaches need comparatively much more training time. An architectural overview of the annotation model is shown in the Figure 10.16.

1. Data augmentation technique to generate training set invariant to brightness variation, colors, scale, and orientations. Introduction of contrast enhancement and noises further increases the training data size.

2. Automatic generation of cluttered objects and the corresponding annotations using the manually annotated image set.

3. Faster RCNN and RFCN are fine-tuned using pre-trained models for two-class classification (either foreground or background) using a few manually annotated images and a large number of automatically generated augmented images.

4. The final model when tested on any new image generates bounding box significantly close to the ground truth.

5. The detection model is invariant to change in background.

Examples of Vision-Based Object Detection Techniques

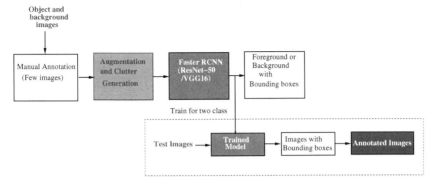

FIGURE 10.16: Proposed network architecture for automatic annotation.

10.3.1.1 Image Acquisition

The model uses a set of forty different objects provided by Amazon for the Amazon Robotic Challenge (ARC) 2017. Images are captured using different cameras like, Foscam, Realtek and webcam for forty different objects at various orientations. Images with multiple resolutions, such as (800×800), (600×600), (1320×1080), (540×480) are used in training set and testing sets. This makes the system capable to detecting new objects of any resolutions. Cameras are mounted on a rotating platform. Background images (where the objects are to be placed, tote in our case) are also captured at different directions.

10.3.1.2 Manual Annotation

The captured images are annotated manually to generate a training set for modeling a two-class classifier (foreground and background). A widely used software tool called LabelMe [270] is used for this task. Each training image is thus having a corresponding annotated image containing the segmentation region of the object in the image called mask image. We are able to manually annotate 8000 images, 200 of images from each of the forty objects.

10.3.1.3 Augmentation and Clutter Generation

Augmentation of images and synthetic generation of clutters are done to automatically generate sufficient data within a very short duration. Training data of significantly large size is a prime requirement for training any deep network. Another advantage of this approach is that it prevents the network from over fitting and makes the network more generic for detecting new objects even in an unknown environment. Data augmentation technique using affine transformation is also introduced to generate numerous cluttered data within a very short duration when images of individual objects with annotations are provided to the system.

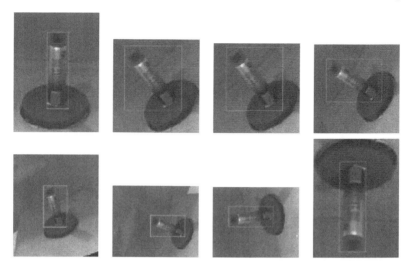

FIGURE 10.17: Figure shows an example of augmentation results when affine transformation is applied to an image.

The affine transformation is done by selecting ten combinations of rotation(anti clockwise) using θ, scaling with λ, horizontal translation by T_x and vertical translation by T_y. It hence generates ten new images for a given manually annotated image. The transformation matrix (H) is thus given as:

$$H = \begin{bmatrix} \lambda cos\theta & -\lambda sin\theta & T_x \\ \lambda sin\theta & \lambda cos\theta & T_y \\ 0 & 0 & 1 \end{bmatrix} \quad (10.21)$$

The annotation for the augmentated images are generated by using affine transformation of the corresponding original image's ground truth points $[xmin, ymin]$ and $[xmax, ymax]$.

Some example images of the augmentation results obtained after using affine transformation are shown in the Figure 10.17. It is to be noted that, the affine transformation is applied to the object after it is detected using the Faster RCNN. Color augmentation is also applied to every object around its ROI, obtained from the mask image. The augmentation is done by combining R,G,B channels. The three channels R, G and B are interchanged to generate six new images. Color agumentation results of few images are demonstrated in the Figure 10.18. The introduction of color augmentation technique increases training data-size to five times of its original size, which is very significant.

Using the given set of object images, various combination of cluttered images are generated within the boundary of the bin. Following method is used for clutter generation.

Examples of Vision-Based Object Detection Techniques 409

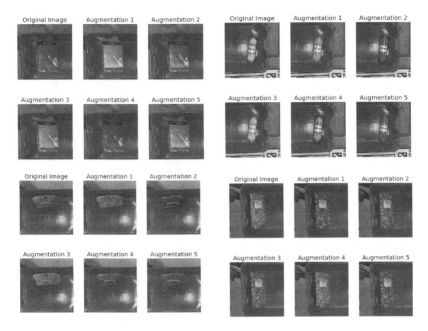

FIGURE 10.18: Example images showing color augmentation results. Color augmentation increase the data size five times to its original size.

1. The background image(tote image) is divided into grids. Cropped object images using manually generated mask are randomly pasted pasted on those grids.

2. The generated mask are associated with different binary value for different objects in order to distinctly obtain object ROI.

The color augmentation and affine transformation are followed by a clutter generation approach. Some of the resultant images generated after applying our clutter creation technique are shown in the Figure 10.19.

The generated clutter includes all the possible occlusion, brightness variation, orientation, scale and combination of all the forty objects. Finally, a total of 110,000 training images comprises of forty objects are generated after applying affine transformation and color augmentation on the 8,000 manually annotated images. For each of the forty objects, 200 images were captured to maintain a balanced data distribution.

10.3.1.4 Two-class Classification Model using Deep Networks

The training dataset of size 110,000 generated using 8,000 manually annotated images followed by color augmentation, affine transformation and cluttering generation is hence used to train the aforesaid detection framework

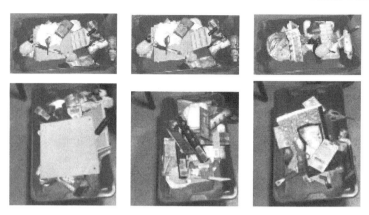

FIGURE 10.19: Example images shows synthetically generated cluttered objects using manually annotated images of single objects. Cluttered images are used to train the proposed deep learning framework for foreground and background detection.

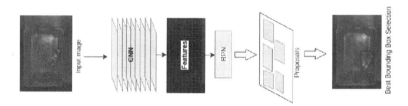

FIGURE 10.20: An architectural overview of the proposed network. The network first generates region proposals to get all possible foreground locations, and then it is trained to identify the correct ROI from the proposed ROIs.

for detecting the foreground ROI in an image. Object detection frameworks: Faster RCNN and R-FCN are used to fine-tune the VGG-16 and ResNet-101 respectively. The basic layout of the framework has already been explained in the block diagram shown in Figure 10.20. The recognition performance of the framework based on Faster RCNN is experimentally found to be much better than R-FCNN in terms of $mIOU$.

The network is trained in two stages, in the first stage the network is trained for generating region proposals to get all possible foreground locations, while in the final stage the network is trained to identify the correct ROI among proposed ROIs. The training can be done in either using alternate optimization technique for training each stage individually or by end-to-end training both stages at a time.

Examples of Vision-Based Object Detection Techniques 411

10.3.1.5 Experimental Results and Discussions

Some results of the annotation model when tested using an entirely new set of objects are shown in the Figures 10.21 and 10.22.

Table 10.1 gives an overall summary of the experimental results. Five different sets are used to validate the annotation performance of the proposed approaches. The performances are given in terms of mean Average Precision (mAP) which is standardized by Pascal VOC [271]. The observation shows that, the performance of the proposed ResNet-101 model is slightly higher than the Faster-RCNN based technique. However, the training time of the former is comparatively much higher than the later approach. The user can choose any of the network based on priorities. The validation is done by using the automatically generated dataset including new set of objects to train both Faster-RCNN and RFCN based based multi-class classification network. Pre-trained models Vgg16 and RestNet-101 are used for Faster RCNN (F-RCNN) and RFCN respectively. A mean Average Precision (mAP) of 99.19% is achieved by using F-RCNN based multi-class object detector and an mAP of 99.61% is achieved with RFCN based network. However, training time of the later approach is much higher than its previous counterpart. The model is trained by using a single GPU machine (Quadro M5000M). Training the entire dataset of size 1,10,000 takes around 8 hours for F-RCNN

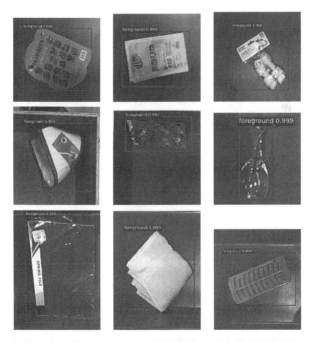

FIGURE 10.21: Automatic annotation results of few images when tested on same background using an entirely new set of objects. These objects were never shown to the model previously.

TABLE 10.1: Test results for new set of objects with multiple backgrounds. Brown(1) stands for set of object images taken using rotating platform and Brown(2) stands for the test set images taken from rack. The third column shows number of images in each of the test sets, fourth column gives corresponding count of new objects. Mean average Precision (mAP) for both Faster RCNN (F-RCNN) and RFCN based approaches are presented for the given test sets. Training has been done in two steps: first using object images with red background only. The second part uses augmented background. BG stands for background.

Test Set	Set BG	# Images	# objects	mAP % (Trained with Red BG)		mAP % (Trained with Augmented BG)	
				F-RCNN (VGG16)	RFCN (ResNet-101)	F-RCNN (VGG16)	RFCN (ResNet-101)
1.	Brown(1)	1760	13	38.62	40.14	92.34	94.36
2.	Brown (2)	3855	23	21.38	26.42	70.34	73.12
3.	Black (1)	6000	23	58.69	64.38	96.12	98.65
4.	White (1)	5880	23	41.21	46.43	96.05	96.78
5.	Red(1)	24900	83	98.96	99.05	-	-

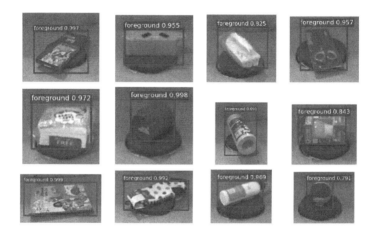

FIGURE 10.22: Automatic annotation results of few images when tested on new a background using an entirely new set of objects. These objects were never shown to the model previously.

and approximately 13 hours for RFCN based network. The precision values of individual objects when tested with new set of data equivalent to 20% of the training data size is presented in the Table 10.2. A few example images of the automatic ground truth detection results when objects are placed in varying degree of clutters are shown in Figure 10.23.

10.3.2 Automatic Segmentation of Objects for Warehouse Automation

This example presents a semantic segmentation network, inspired by PSP-Net [272] which is used for scene parsing applications where the macro objects are semantically segmented by exploiting high inter-class variability.

Examples of Vision-Based Object Detection Techniques 413

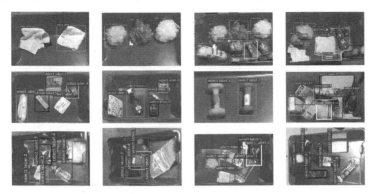

FIGURE 10.23: A few example images of the automatic ground truth detection results when objects are placed in varying degree of clutters. The annotation model detects ROI and end-user can write a label on each detected ROI for further categorization of objects. The clutters contain both a known set of objects as well as unknown objects.

The presented framework is a customized deep architecture, for a specific task of semantically segmenting objects mostly used in warehouses. In PSPNet, the ambiguities in class label is resolved using contextual information which is itself learned through training using a spatial pyramidal pooling [273] module. While it learns to segment a 'river' from a 'house,' it also learns that a house surrounded by a river could be a 'boat house.' Such contextual information may not be useful in warehouse automation where the objects belong to more or less one class (may be labeled as 'household retail objects' and are to be segmented based on their intra-class variability. For instance, a 'cup' being near to a 'book' does not make it a 'pencil holder.' All these three objects are to be segmented based on how different they are to each other. Since the extraction of such semantic contextual relationships requires more data and time, the standard PSPNet may not be a suitable for this use case.

10.3.2.1 Network Architecture

Our approach is based on a semi-supervised deep convolutional network for automatic generation of an annotated dataset using only few manually labeled training images, artificial clutter generation, and finally pixel-wise 40-class classification with the same framework.

An overview of the approach is shown Figure 10.24 as a flow diagram. A ResNet based deep learning framework is designed to make use of low-level features along with high level features.The same is used for both binary-classification and multi-class classification. A rectangular ROI detector based on Single Shot Detection (SSD) technique is used in parallel to the pixel-wise binary classification to generate final mask. This helps to eliminate semantic segmentation outside the object region and hence further enhances the

TABLE 10.2: Multi-class object detection results for eighty-three classes. The first column gives name of the object class followed by precision values in terms of percentage for both Faster RCNN (F-RCNN) and RFCN based multi-class object detectors. Vgg16 and RestNet-101 are used as pre-trained models for Faster RCNN (F-RCNN) and RFCN respectively. A mean Average Precision (mAP) of 99.19% is obtained in F-RCNN and the latter given an mAP 99.61%. However, training time of the latter approach is much higher than its previous counterpart.

Class	Precision % (F-RCNN)	Precision % (RFCN)	Class	Precision % (F-RCNN)	Precision % (RFCN)
allenkey-set	100	100	robot-dvd	98.17	100
augmentedReality-book	100	100	saffola-salt	100	100
barbie-book	100	100	selpak-tissue	100	100
bisleriSmall-bottle	95.1	95.35	semanticWeb-book	100	100
blackCap-bottle	97.53	97.67	teddyBear-toys	100	100
black-ball	100	100	tulsi-greenTea	100	100
black-tape	97.53	99.88	green-dumbell	100	100
blueCap-bottle	96.77	97.5	homefoil-aluminiumFoil	100	100
blueFeeding-bottle	87.75	90.5	introToRobotics-book	100	100
blueHandleToilet-brush	96.29	100	kiwiShoePolish-bottle	98.05	97.79
blue-dumbell	100	100	microfiber-clothWipes	100	100
blue-notebook	100	97.56	miltonBlue-bottle	100	100
brownBlack-TTbat	100	100	miltonSmall-bottle	100	100
brown-cup	100	100	multimediaOntology-book	100	100
camlin-colourPencil	100	100	nivea-deo	100	100
careTaker-swipes	100	100	origamiWhite-plates	100	100
circularBase-meshcup	100	100	paint-brush	94.28	98.25
cloth-clips	100	100	panteneShampoo-bottle	100	100
colgate4-toothbrushs	100	100	patanjali-toothpaste	100	100
colinBig-bottle	100	100	pink-scotchBite	100	100
deepBlue-bottle	99.16	98.34	plato-book	100	100
deepGreen-bottle	98.68	98.67	careTaker-napkin	100	100
dettol-bottle	100	100	green-battery	100	100
devi-coffeeBox	100	100	probablisticsRobotics-book	100	100
dove-soap	100	100	violet-bottle	100	100
fevicol-bottle	100	100	wet-wipes	100	100
fevikwik-tubePacket	100	100	whiteBoard-duster	100	100
fiama-loofa	100	100	whiteIron-brush	99.98	100
foundationRobotics-book	100	100	whiteVoiletCap-bottle	100	100
garnet-bulb	100	100	whiteWritten-cup	100	100
gillet-razor	90.6	95.92	whiteYellowCloth-brush	100	100
greenCapTransparent-bottle	100	100	whiteYellow-cup	100	100
redBig-scissor	97.6	100	white-cottonBalls	100	100
redBlack-socks	100	100	white-cup	100	100
redGreen-ball	100	100	white-gloves	100	100
redmi-mobilePhone	100	100	white-tape	100	100
redPlastic-spoons	98.23	100	wooden-brush	100	100
redWhiteSwiping-cloth	100	100	woolen-cap	100	100
red-bottle	95.05	100	yellowMagic-tape	100	100
red-feviquik	100	100	yellow-DSTape	100	100
restInPractice-book	97.7	100	yellow-ScrewDriver	100	100
roboticsManual-binder	94.34	100			

segmentation accuracy. Unlike PSPNet, the spatial pyramidal pooling module is replaced by a Feature Pyramid Network (FPN) that aggregates multi-level features obtained from successive layers of the standard ResNet network to provide better segmentation between the objects without needing additional data and time, making it faster and better compared to PSPNet for this application. First, the deep network is trained on a small set of manually annotated datasets to act a binary classifier that can segment foreground objects from its background. This binary classifier is used to automatically generate labeled

Examples of Vision-Based Object Detection Techniques   415

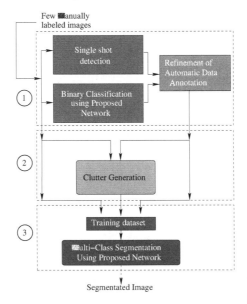

FIGURE 10.24: Flow diagram of the approach. (1) Semi-supervised technique to automatically annotate training data set with Single Shot Detection running in parallel to refine the generated masks; (2) Automatic data generation of cluttered environment; (3)Multi-class image segmentation trained on data generated in (1) and (2).

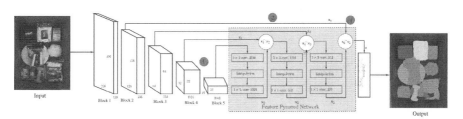

FIGURE 10.25: Overall flow diagram of the proposed network: Base network is Residual Network 50; major differences from Pyramid Scene Parsing Network: (1) No dilation is used convolutional layers of the last two blocks; (2) Feature Pyramid Network is used instead of Spatial Pyramid Pooling; (3) Each feature space is smoothened before concatenation.

templates. The second step involves generating clutters synthetically by superimposing individual templates. The third step involves training the proposed network using this machine generated dataset for multi-class segmentation.

The framework, specifically designed for objects in warehouses is an improvement over the state of the art technique, PSPNet, both in terms of faster computation and segmentation accuracy. Figure 10.25 gives an architectural overview of our proposed technique. Features from different layers are convolved followed by interpolation to generate equivalent dimensional feature vector and then concatenated in three different steps. The final concatenated feature vector **z** is then used for FCN based semantic segmentation.

10.3.2.2 Base Network

The base network used in the proposed framework is a popular deep architecture, residual network consisting of fifty convolutional layers (ResNet50) [274]. Unlike PSPNet, this network is not using dilation in any of the convolutional layers as shown in Figure 10.25. The main purpose of inclusion of dilation in PSPNet was to increase the receptive field which makes it easier for the network to detect large items [275]. As in the given use-case the distance of the camera from all the objects is almost same, this was not very useful. Since dilated convolution is much slower than normal convolution, this change helped the network to increase the computation speed.

As shown in Figure 10.25, the presented architecture has five blocks of Resnet 50. We use output of all the residual blocks {block 2, block 3, block 4, and block 5} to make the final prediction. This part of the network is called Feature Pyramid Network (FPN) [276] (Component 2 of Figure 10.25). These blocks have strides of {4, 8, 16, 32 } pixels with respect to the input image. We do not include the output from the first block, block 1 into the pyramid because of increased memory requirement for the same. We chose the output of the last layer of each stage as our reference set of feature maps because the deepest layer of each block should have the strongest features [276]. This ensured we used the information from primitive feature maps along with the higher level features. Also, in the proposed architecture, all feature maps were concatenated (Component 3 in Fig. 10.24) so as to retain more information since concatenation does not require matching of number of channels. Before concatenation of two feature maps, an interpolation layer is used to match sizes of the feature maps. Smoothing the feature map before interpolation by replacing each pixel with a linear combination of some of its neighbors is very important. Thus the proposed network uses convolution to slow spatial variations over abrupt changes before each interpolation layers. The motivation behind choosing FPN over Spatial Pyramid Pooling (SPP) [273] of PSPNet [272] were two fold: 1) The multi-scale feature map prediction was nearly twice as fast as SPP, and 2) the SPP module [273] of PSPNet increases accuracy by bringing in contextual information to differentiate between confusing categories. Since the objects in the bin or the stack are unrelated, contextual information does not add much information.

10.3.2.3 Single Shot Detection

Single Shot Multibox Detector (SSD) [256] is a state-of-the-art technique for object detection and localization. For 300 × 300 input, SSD achieves 74.3% mAP1 on VOC2007 test at 59 FPS on a Nvidia TitanX and for 512 × 512 input, SSD achieves 76.9% mAP, outperforming a comparable state-of-the-art

Experimental Results 417

Faster R-CNN model [256]. SSD is as accurate as slower techniques like Fast RCNN [254] and Faster RCNN [255] that perform explicit region proposals but is still faster because of the elimination of both bounding box proposals and the subsequent pixel or feature resampling stage [256]. Thus, we chose SSD for this network. We use SSD in combination with pixel-wise binary classification for final mask creation. Thus no foreground detection by PSPNet outside the bounding box provided by SSD is considered. This further enhanced the accuracy of ground truth generation.

10.3.3 Automatic Generation of Artificial Clutter

After obtaining images with their corresponding ground truths from the semi-supervised technique mentioned above, we generated synthetic clutters. This was essential for ARC 2017 as the network was expected to perform in a cluttered environment similar to that in a ware house. Using the dataset created above with single object per image, we created images of a cluttered environment consisting of four-sixteen items. Using the binary masks, more objects could be artificially placed in the bin creating a cluttered environment along with its ground truth. Objects were placed with some augmentation like rotation, scaling, etc. Images with a clutter of different permutation and combinations of items along with varying pose and orientation of that item were created. This data augmentation results in a more robust system.

10.3.4 Multi-Class Segmentation using Proposed Network

The final dataset consisting of images with cluttered environment as well as single object images was passed to the deep convolutional neural network for segmentation task, that is a pixel-wise 40 class classification. For the picking or stowing task of ARC 2017, one of the most important tasks to be performed is image segmentation, a 40-class pixel-wise labeling of images. The proposed architecture is trained on single object images as well on the artificially created clutters. Here we have not used SSD in parallel because 1) our model showed high accuracy without SSD so an additional computational overhead can be avoided; and 2) instance segmentation was not required in this task since the ARC 2017 dataset had only a single instance of every object at any given time.

10.4 Experimental Results

10.4.1 System Description

This vision system takes in RGB images from multiple views to train the algorithm for ground truth generation followed by clutter creation. Then these images are given to a deep convolutional neural network that outputs segmented images for the robot to complete the picking and stowing tasks. The following components helped in completing the task of labeling of data-set and training for the new items within thirty minutes in ARC 2017:

10.4.1.1 Server

As the architecture is residual networks with fifty convolutional layers, there was a requirement of good GPU power to ensure that the training is completed within the thirty-minute constraint posed by Amazon in the Challenge 2017. Thus, training was run on a server with multiple GPU configurations. The server used was NVIDIA Quadro P6000 with two 24GB GPUs. Half an hour training on this server with a batch size of ten produced promising results in the pick and place task of ARC 2017.

10.4.2 Ground Truth Generation

Generation of data along with its corresponding ground truth was the first task. Using the rig, images of each item from the training list were obtained at varying scales, positions, and orientations. All images were obtained on a uniform red background to facilitate semantic background subtraction. First we took about 200 images of these items. These images were then manually annotated to be fed into our network. The data collected so far had only one object per image, that is, there was no clutter environment making it easy to manually annotate the images. Then these images along with their binary masks were passed to the proposed architecture for training the pixel-wise binary classifier. Apart from the softmax loss at the branch, an auxiliary loss is applied at the fourth block of the architecture to train the final binary classifier. Like PSPNet, we let the auxiliary loss backpropagate all previous layers. As shown in [272], this deeply supervised learning strategy for ResNet-based network optimizes the learning process. This auxiliary loss is not used in the testing phase. For training phase, it is assigned 0.4 weight in this network. Before feeding into the predictor, we make use of feature pyramid network of low-level features along with the high-level features. A batch size of 10 and e^{-3} learning rate was used for 1,200 iterations to fine-tune this network.

Once this training is done, 300 images per item of the training list provided in ARC 2017 were passed to this network. As output we obtained the binary masks of each of these items. Figure 10.26 shows masks generated by this network for some of the items. Now we had 300 images per item along with their ground truths. But our system was expected to perform the task of picking and stowing in a cluttered environment. Thus, in our next step we artificially created cluttered environments by picking and placing items with other items. This was made possible using the binary masks generated in the previous stage. Their corresponding ground truth data was created simultaneously in a similar manner. Figure 10.27 shows some images of the

Experimental Results 419

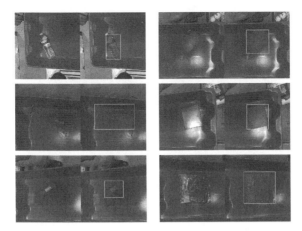

FIGURE 10.26: Figure illustrates a few examples of binary masks generated using the proposed network. It can be observed that even for transparent backgrounds the network can generate precise segmented regions.

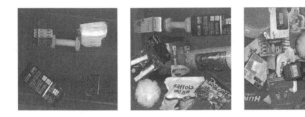

FIGURE 10.27: A few examples of synthetically generated cluttered environments are shown here.

cluttered environment generated. This resulted in the final data set consisting of 38,000 images of cluttered environments and 12,000 images with one object at a time. These 50,000 images were used to fine-tune our network for image segmentation in the next step.

10.4.3 Image Segmentation

For the picking and stowing task of ARC 2017, once the dataset along with its ground truth is obtained, a deep convolutional neural network was trained for image segmentation task. For this, images along with their ground truth were passed to the modified structure of ResNet [274]. The ADE20k model was used for initialization of the weights and then it was fine-tuned with the dataset

 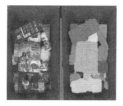

FIGURE 10.28: A few examples of the automatic segmentation results when tested in a real-world cluttered environment as shown here. It is important to note here that the training dataset contains only synthetically generated clutters.

TABLE 10.3: A comparison of the proposed network with the state-of-the-art technique: PSPNet is given here.

Criteria (per-image)	PSPNet	Proposed Architecture
Accuracy	94.031%	94.851%
Forward pass	108.066 ms	63.622 ms
Backward pass	172.048 ms	82.650 ms
Forward-Backward pass	280.460 ms	146.598 ms

generated. A batch size of ten and a learning rate of e^{-3} was used for 3,000 iterations. Softmax loss was also added on the fourth block as an auxiliary loss to optimize the learning process. The main branch loss along with the auxiliary loss were used to train the classifier. Also feature maps form all blocks are concatenated and sent to the classifier for the final prediction. Results produced by this architecture is shown in Fig. 10.28. Table 10.3 compares the performance of our architecture with PSPNet. Proposed architecture's average forward-backward pass is 1.91 times faster than PSPNet with a slightly better accuracy than PSPNet. An overall accuracy of 94.85% was produced by our network compared to 94.03 of PSPNet. Thus our system outperforms the architectures of PSPNet in terms of both speed and accuracy on the same dataset. Class-wise precision and recall produced by our network for forty items from the training-list provided by Amazon are plotted in Figure 10.29.

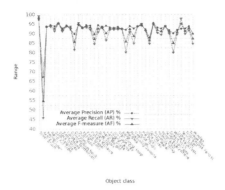

FIGURE 10.29: Multi-class object recognition results in terms of precision and recall. Test data contains entirely new set of objects.

10.5 Summary

This chapter is intended to provide some practical use cases of CNNs in robotics using only visual data. The chapter began with a brief introduction of CNNs and their advantages over the conventional ANNs followed by a detailed description of each of the sub-modules present in a standard CNN architecture for better understanding of the subsequent applications. While describing each and every module of the CNN architecture, we make sure that some small example codes written in python are provided in this chapter, which can help the reader understand the practical implementation of a CNN network. Later, in this chapter, we demonstrated two different applications of CNNs toward warehouse automation. Automation of warehouses demand accurate localization and segmentation of an object of interest which is further to be picked and placed by a robotic arm from/in a given location. These applications are involved with various visual challenges which make a conventional vision-based technique almost impossible to segment and localize an object of interest. We thus, demonstrate here two such CNN based techniques for accurate localization and segmentation of object present in a cluttered and partially occluded environment. The first application mainly dealt with detection of a rectangular bounding box for an object of interest, whereas, the second application was to precisely localize the object boundary for a more accurate pick and place of an object. Training a CNN network for multi-class object detection and semantic segmentation demands a large amount of annotated data which is very tedious to obtain manually. To this end, we developed two different automatic annotation techniques and demonstrated their results for different degrees of clutters with varying backgrounds and lighting conditions.

11

Vision-Based Grasping

11.1 Introduction

A robot that can manipulate its environment is much more useful than one that can only perceive. Such robots can act as active agents which will someday replace humans from all types of dull, dangerous, and dirty works completely, thereby, freeing them for more creative pursuits. Grasping is an important capability necessary for realizing this end. Solving the grasping problem ininvolves two steps. This first step uses a perception module to estimate the pose (position and orientation) of the object and hence, the gripper pose needed for picking it. Then, the second step uses a motion planner to generate necessary robot and gripper movement to make contact with the object. In this chapter, we are primarily interested in solving the first part of the problem, namely, finding graspable regions and suitable grasp poses (together known as graspable affordances) for a two-finger gripper. Visual grasping or vision-based grasping solves the grasping problem by using 2D RGB images or 3D RGBD images obtained from normal cameras and depth cameras respectively. Depending on whether the object model information is used or not, the approaches for grasping are divided into two categories: Model-based and Model-free methods.

In model-based grasping, we have the prior object information either in the form of 3D CAD models or training dataset comprising 3D (RGBD) image scans of the object and/or 2D perspective views of the objects. This information is used for estimating the pose of the target object in the scene. This object pose is, then used to determine the suitable gripper pose required for grasping the object.

Let us consider an example of grasping a cup placed on a table. Even though there are multiple ways to hold the cup, we would like to hold it by only one way - by its handle. The pose of the object in the scene is determined by fitting the target CAD model (or object scan) on the actual object visible in the scene. This is done by rotating and scaling the template CAD model (or its scan) on the actual object. Once the object pose is known, the corresponding gripper pose could be determined to hold the cup by its handle. This process of finding object pose is more commonly known as template matching method. Template matching methods may make use of visual features in addition to its geometric features to decide the object pose. While these methods are easy to

423

implement, these methods are not known to be robust. Recent advancement in machine learning, particularly deep learning, methods [278] have been shown to be more robust and accurate however, at an increase computational cost.

Once the object pose is known, a stable grasping region or pose can be calculated using the following three approaches:

1) In the first case, we try to find contact points on a 3D object surface that ensure a stable grasp. At each contact location, the object is subjected to normal/tangential forces and torsional moment about the normal. These forces or wrenches form wrench matrix which can be used to decide the stability of a grasp. The grasp is said to be stable if a small disturbance, on the object position or finger force, generates a restoring wrench that tends to bring the system back to its original configuration. Finding the contact points requires the precise knowledge of the physical properties of objects and gripper like mass distribution, friction coefficient, etc.

2) In the second case, we use simulators to predict the most suitable grasp regions. For example, if you have a 3D model of the object and the gripper, then in the simulator you can render the object and gripper, and calculate the stability score for each feasible grasp pose. This score can be based on stability against external disturbances, the area of contact between objects and gripper, or any other geometric constraint. But the idea is to assign some score to each grasp pose and then using that score, choose the best grasp pose for the object. This strategy is very simple, but it could be very time consuming for larger objects because we need to evaluate all the possible grasp configurations.

3) In the third case, we use machine learning techniques to predict the grasp regions. Generally, people train their network on synthetic data produced in step 2 and fine-tune the network to work in real situations. Similar work is done in [278] [279].

The above pipeline can be used for grasping isolated objects. Additional constraints need to be imposed in order to grasp objects in a clutter to avoid collision with neighboring objects around the target. There are several ways to achieve this. For instance, one can use the local geometric properties of the object and the gripper to constrain the motion of the robot to minimize collisions with neighboring objects. Another option would be to render the objects in a simulator using the computed object poses and then test all possible gripper pose to select one that leads to minimum collisions. It is also possible to generate such data by using actual robots as demonstrated in [280] [281] and use reinforcement learning to learn optimal motion planning required for grasping objects in a clutter. One of the model-based approach for computing grasping handles from visual data will be explained in Section 11.2 below.

In many practical cases, it is impossible to have prior object model information given the availability of large diversity of objects and in many cases, the decision to grasp an objects needs to be taken on the fly making model-based methods less suitable for such cases. Many of the recent approaches now try to solve the grasping problem independent of the object identity by directly using RGBD point cloud scans obtained using low cost depth and range sensors. The advantage of this approach lies in its simplicity which allows real-time implementation and does not require any time-consuming and data-intensive training phase common in most of the learning-based methods [282] [283] [284]. The details of one such model-free method for grasp pose detection will be described in Section 11.3 along with the implementation details and experimental results. The example which we will present in Section 11.3 is inspired by the approach presented in [285] [286] which uses surface curvature to localize graspable regions in the point cloud.

11.2 Model-Based Grasping

In this section we will give an example to estimate the grasp region given the clutter of objects. In this example, we assume that we have the object information in advance, and we will use a neural network to estimate the pose of the object. Please note that in this example we will use Mask RCNN [258] for semantic segmetation, i.e., assign a label to each pixel and then we will perform the axis assignment operation which gives the pose of the object.

In second step we will estimate the grasp regions. In this example we will use a two-finger gripper. The advantage of using a two-finger gripper is that its projection on image plane is a rectangle as shown in Figure 11.1. So to decide if a grasp region is valid or not we need to consider the 3D point cloud data corresponding to that rectanglar region in image plane. Each step will be explained in subsequent sections.

11.2.1 Problem Statement

Given an image I and its corresponding point cloud $P \subset R^3$ of a robot workspace where O number of objects are placed separately or in clutter, the goal is to find out the best pose (position and orientation of end-effector in image plane) for picking the object.

Assumptions:

1. Robot gripper in the image plane is represented as a rectangle with its shorter sides representing the fingers and longer sides representing the opening of gripper.

2. Grasp pose estimation means the final end-effector orientation to grasp the object. This assumption is valid as long as all the objects are placed within the task space of robot.

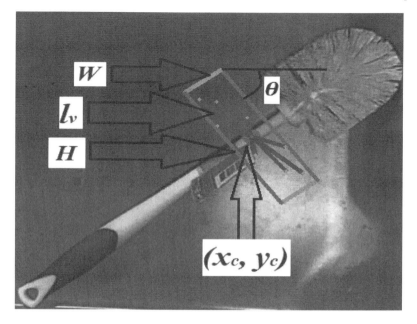

FIGURE 11.1: Projection of gripper in image plane.

3. Robot-camera extrinsic calibration is known to us.

11.2.2 Hardware Setup

The hardware consists of following main components:

1. Universal Robot 5 (UR5).

2. Kinect V1 and Foscam IP Cameras.

3. SAKE EZGripper- A two-finger single servo-controlled gripper. Grasp Force (0-35N).

ROS drivers are used to communicate with robot hardware and camera sensors. The real-world environment is created in Gazebo [287] for testing the algorithm. MoveIt [288] is used to plan the trajectories of robot using OMPL (Open Motion Planning Library) [289] plugin. Figure 11.2 show the environment created in Gazebo and RViz (ROS Visualization). OMPL contains so many path planning algorithms' implementation. In all our experiments, RRT (Rapidly-exploring Random Trees) is used.

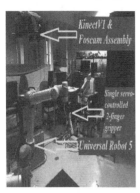  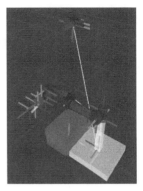

(a) Workspace Set Up   (b) Gazebo   (c) ROS Visualizer

FIGURE 11.2: Workspace simulated in Gazebo and Ros Visualizer.

FIGURE 11.3: Data augmentation on one training example.

11.2.3 Dataset

Amazon Robotics Challenge 2017 objects dataset is used to train a deep net. The dataset contains 600 × 600 RGB images of objects of forty classes with corresponding binary masks and bounding boxes. Every image contains exactly one object. The total dataset contains 12,000 images of 300 images per class. The dataset contains versatile type of objects.

11.2.4 Data Augmentation

As the number of images in dataset is 12,000 and Mask-RCNN has millions of parameters so to avoid the over-fitting, we did data augmentation and increased the dataset four times the previous dataset. Data augmentation includes increase in brightness, mean subtraction, scale the minimum dimension of image between 500 to 800 randomly and flip it horizontally. Figure 11.3 shows the effects of data augmentation to the original image.

11.2.5 Network Architecture and Training

For segmenting the objects, we have used ResNet50 architecture with its head modified as mentioned in the Mask R-CNN and Faster R-CNN [290] papers. Input to the network is a 600 × 600 RGB image, ground truth bounding box and corresponding binary mask. The loss function used for training is combination of log loss for classification, smooth L1 loss for bounding boxes (BBoxes) and average binary cross-entropy loss as described in Mask R-CNN paper. Classification loss and bounding boxes loss are described in a same way as described in the Fast R-CNN [254]. The detailed architecture is explained in Figure 11.4.

Weights are initialized with MS-COCO pre-trained model. The reason behind this initialization is to reduce the overfitting which is caused by presence of less data and the large number of learnable parameters. For fine tunning we set batch size of 4, learning rate 0.00025, weights are updated using SGD with momentum haveing momentum term 0.9081 and we run the network for 180,000 number of iterations.

11.2.6 Axis Assignment

After binary mask prediction, an *axis assignment* step is performed. In this step we try to find the major axis of the object so that we can grasp the object along its major axis. In [291], it is mentioned that an axis is the major axis if sum of perpendicular distances of contour point from that axis is minimum. So given the binary mask (output of network) which has N contour points and let k^{th} contour point has pixel value represented by (C_x^k, C_y^k). Average of these N contour points gives the value of $\bar{C}_x$ and $\bar{C}_y$ centroid co-ordinates in image plane. So our objective function is

$$min(\sum_{k=1}^{N}[(C_x^k - \bar{C}_x)*sin(\phi) - (C_y^k - \bar{C}_y)*cos(\phi)]^2)$$

$$\bar{C}_x = \frac{\sum_{k=1}^{N} C_x^k}{N}, \quad \bar{C}_y = \frac{\sum_{k=1}^{N} C_y^k}{N}$$

$$tan(2\phi) = 2*(\frac{\sum_{k=1}^{N}(C_x^k - \bar{C}_x)*(C_y^k - \bar{C}_y)}{\sum_{k=1}^{N}[(C_x^k - \bar{C}_x)^2 - (C_y^k - \bar{C}_y)^2]})$$

$\phi :=$ Angle of major axis with horizontal line.

After finding the major axis minor, axis is simply the perpendicular axis in the image plane as shown in Figure 11.5.

11.2.7 Grasp Decide Index (GDI)

This index takes into account, the collision of gripper fingers with the other objects [292]. It also ensures that the selected hand configuration corresponds

Model-Based Grasping

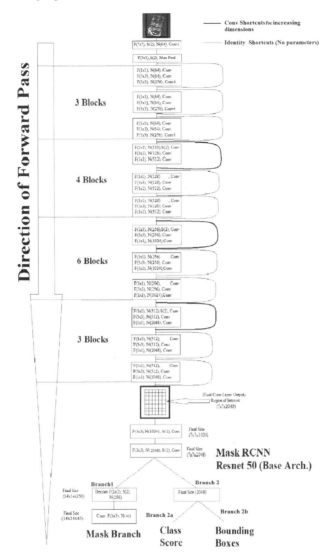

FIGURE 11.4: Network architecture for semantic segmentation.

to a least cluttered object. We define:

$$GDI = min(F \cap O)$$

Where, **F** be the fingers and **O** be the objects in the vicinity of target object.

Ideally GDI must be a null set [292]. However, just by looking at the image (Figure 11.6), it can't be inferred whether the gripper (rectangle) is colliding

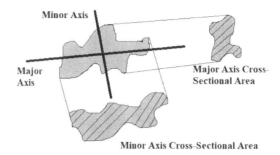

FIGURE 11.5: Major-minor axis assignment.

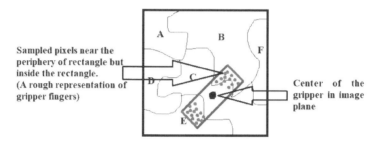

FIGURE 11.6: Pictorial representation of image with segmented objects and rectangle with sampled points near the periphery. A, B, C, D, E, F showing the segments of objects in the image plane.

with the surrounding objects. So, in order to cop up with this, we utilize depth information and denote the gripper fingers by densely sampling inside rectangle representation near its shorter edges in the image plane. Next, we use the z-values of every sampled pixel and redefine it as follows:

$$GDI = max(\Delta Z_r) \; \forall \; r_s, \Delta Z_r = mean(Z - Z_c), \Delta Z_r \; \epsilon \; r$$

$$(Z_i - Z_c) \; \epsilon \; R^+ \; 0 \leq i \leq S_p \; and \; Z \; \epsilon \; R^{S_p}$$

S_p Set of all sampled pixels,

Z_i Z-value (depth) of ith sampled pixel in S_p

Z Vector representation of all Z_is

ΔZ Mean of all positive deviation of S_p from center of the gripper after removing univariate point outliers

Z_c is rectangle (gripper in image plane) center pixel's z-value

r_s represents each rectangle and all rectangles in the image plane

Model-Based Grasping

In order to make the grasp with fingers, collision free, GDI (practically) must be as large as possible. This is also the ideal case in practice. It also corresponds to the case, when there is no clutter in the environment. In the case of dense clutter as the case in real-time grasping situations, the GDI will automatically adjust itself according to the surrounding z-values. In the case of clutter, the most dominant cases for GDI are tabulated in Table 11.1, the possibilities can even be more but these are the cases which affect the performance of grasping systems dominantly.

11.2.8 Final Pose Selection

When objects are placed in isolation, to approach the centroid of the object for grasping will lead to success in grasp. But in the case of clutter this approach is not good. So in case of clutter, we discretize major axis of the object segment and try to find out the best suited grasp. For every discrete point gripper configuration a corresponding to axis angle is drawn. Inside each configurations' periphery, pixels are sampled and on those pixels GDI is applied. The one with highest GDI is selected for grasp planning. The GDI is applied on those configurations which are ranked according to the large number of positive deviations in Z-values.

11.2.9 Overall Pipeline and Result

Following are the steps for estimating the grasp region in a cluttered environment:

1. Feed the image into Mask-RCNN and the network will produce the segmented output.
2. For each segment we will find major and minor axes.
3. For each major axis we will select multiple poses and for each pose we will calculate the GDI.
4. Finally we will select the pose which has maximum GDI score.

A complete pipeline has been shown in Figure 11.7

TABLE 11.1: Possible cases affecting Grasp Decide Index for different surroundings

Cases	GDI Sign
(diagram: Kinect View Direction, Z_d, Z_i, ΔZ_i)	GDI > 0 Favorable
(diagram: Kinect View Direction, Z_d, Z_i)	Partially Favorable or Unfavorable GDI ≤ 0 $\cup$ GDI ≥ 0 $\cup$ is used to show either of the sign is possible
(diagram: Z_d, Z_c, Kinect View Direction)	GDI < 0 Unfavorable

FIGURE 11.7: Overall pipeline.

11.3 Grasping without Object Models

In this section we will give a working example to estimate the grasp region given the clutter of objects without any prior information of objects.

11.3.1 Problem Definition

In this section, we look into the problem of finding graspable affordances for a two-finger parallel-jaw gripper in a 3D point cloud obtained from a single view of a range or RGBD sensor. The affordances for objects are to be computed in an extreme clutter scenario where many objects could be partially occluded. A graspable affordance is the six-dimensional grasp pose required for grasping (or holding) the object. The problem is solved by taking a geometric approach where the geometry of the robot gripper is utilized to simplify the problem.

Various geometrical parameters corresponding to the gripper and the object to be grasped are shown in Figure 11.8(a). The maximum hand aperture is the maximum diameter that can be grasped by the robot hand and is denoted by d. It should be greater than the diameter b of a cylinder encircling the object. It is further assumed that each finger of the gripper has a width w, thickness e and total length h. The minimum amount of length needed for grasping an object successfully is assumed to be l. There has to be sufficient clearance between objects so that a gripper can make contact with the target object without colliding with its neighbors. Let this minimum clearance needed between two objects be g and it should be more than the width of each finger, i.e., $g > w$ to avoid collision with non-target objects while making

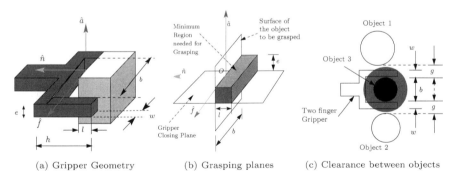

(a) Gripper Geometry   (b) Grasping planes   (c) Clearance between objects

FIGURE 11.8: Grasp configuration for a two-finger parallel jaw gripper. For a successful grasp, the clearance between objects must be greater than the width of each finger $(g > w)$. The gripper approaches the object along a direction opposite to the surface normal with its gripper closing plane coplanar with the minor axis $\hat{f}$ of surface segment as shown in (b). The objects 1 and 2 in (c) are obstacles which are to be avoided by the gripper while grasping the target object 3. The 6D pose detection problem becomes a 1-D linear search for a volume of $l \times b \times e$ along the major axis $\hat{a}$.

a grasping manoeuvre. This clearance is shown as blue ring in Figure 11.8(c). Each object surface is associated with three principal axes, namely, $\hat{n}$ normal to the surface and two principal axes - major axis $\hat{a}$ orthogonal to the plane of finger motion (gripper closing plane) and minor axis - $\hat{f}$ which is orthogonal to other two axes as shown in Figure 11.8(b). Readers can refer to [292] to understand some of the terms which have been used here without being defined to avoid repetitions.

The proposed grasp pose detection algorithm takes a 3D point cloud $\mathscr{C} \in \mathscr{R}^3$ and a geometric model of the robot hand as input and produces a six-dimensional grasp pose handle $H \subseteq SE(3)$. The six-dimensional grasp pose is represented by the vector $\mathbf{p} = [x, y, z, \theta_x, \theta_y, \theta_z]$, where (x, y, z) is the point where a closing plane of the gripper and object surface seen by the robot camera intersect; and $(\theta_x, \theta_y, \theta_z)$ is the orientation of the gripper handle with respect to a global coordinate frame. Searching for a suitable 6 DOF grasp pose is a computationally intensive task and hence, a practical approach is taken where the search space is reduced by applying several constraints. For instance, it is assumed that the gripper approaches the object along a plane which is orthogonal to the object surface seen by the robot camera. In other words, the closing plane of the gripper is normal to the object surface as shown in Figure 11.8(b). Since the mean depth of the object surface is known, the pose detection problem becomes a search for three-dimensional $(l \times b \times e)$ bands along the major axis $\hat{a}$ where l is the minimum depth necessary for holding the object. Hence, the grasp pose detection becomes a one-dimensional search problem once an object surface is identified.

Hence, the problem of computing graspable affordances or grasp pose detection boils down to two steps: (1) creating surfaces in 3D point clouds and, (2) applying geometric constrains of a two finger parallel jaw gripper to reduce the search space for finding suitable gripper hand pose. The details of the proposed method to solve these two problems is described in the next section.

11.3.2 Proposed Method

As explained in the previous section, the proposed method for finding graspable affordances involves two steps: (1) Creating continuous surfaces in the 3D point cloud and then, (2) applying geometrical constraints to search for suitable gripper poses on these surfaces. This is described next in the following subsections.

11.3.2.1 Creating Continuous Surfaces in 3D Point Cloud

The method involves creating several surface patches in the 3D point cloud using region growing algorithm [293] [294]. The angle between surface normals is taken as the smoothness condition and is denoted by symbol θ. The process starts from one seed point and the points in its neighborhood are added to the current region (or label) if the angle between the surface normals of new point and that of seed point is less than a user-defined threshold. Now the procedure is repeated with these neighboring points as the new seed points. This process continues until all points have been labeled to one region or the other. The quality of segmentation heavily depends on the choice of this threshold value. A very low value may lead to over segmentation and a very high value may lead to under segmentation. The presence of sensor noise further exacerbates this problem leading to spurious edges when only one threshold is used. This limitation of the standard region growing algorithm is overcome by introducing a concept called *edge points* and using a pair of thresholds instead of one. The use of two thresholds is inspired by a similar technique used in a Canny edge filter [295] [296] and is demonstrated to provide robustness against spurious edges. This modified version of the region growing algorithm is described next in the following section.

To begin, we first describe the concept of edge points and then, explain how a pair of two thresholds on smoothness condition can improve the performance of the standard region growing algorithm. Some of the notations which will be used for describing the proposed method are as follows. Also refer to Figure 11.9 for a better understanding of these notations. Let us consider a seed point $s \in \mathscr{C}$ with its own spherical neighborhood $\mathscr{N}(s)$ shown as a circle in Figure 11.9. It is further assumed that this neighborhood consists of m points $(p_i, i = 1, 2, \ldots, m)$ in the 3D point cloud. Mathematically, this neighborhood may be written as follows:

$$\mathscr{N}(s) = \{p_i \in \mathscr{C} \mid \|s - p_i\| \leq r\}; \ i = 1, 2, \ldots, m \qquad (11.1)$$

where r is an user-defined radius of the spherical neighborhood. Each neighboring point p_i has an associated surface normal N_i which makes an angle of θ_i with the normal associated with the seed N_s. As stated earlier, θ_i is the smoothness condition for the region growing algorithm. In this context, we define two thresholds θ_{low} and θ_{high} which are used for defining the region label for the neighboring point and creating new seed for further propagation. Let Q_s be the set of new seeds which will be used in the next iteration of

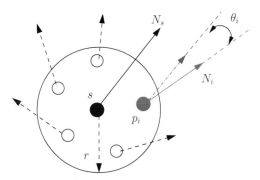

FIGURE 11.9: Defining an edge point. It is a point on the surface around which the surface normals are widely scattered in different directions.

the region growing algorithm. Before describing the modification to the standard region growing algorithm, it is necessary to introduce the concept of *edge points* which is defined as follows:

Definition 11.1 (Edge Point). *Let $R(s)$ be a set of those neighbors p_i of seed point s for which $\theta_i > \theta_{high}$. In other words,*

$$R(s) = \{p_i \in \mathcal{N}(s) \mid \theta_i > \theta_{high}, \ i = 1, 2, \ldots, m\} \quad (11.2)$$

Let C_R be the cardinality of the set $R(s)$, i.e., $C_R = |R(s)|$. Then, a seed point will be called as an edge *point if the following condition is satisfied:*

$$\frac{C_R}{m} > k; \quad 0 < k < 1.0 \quad (11.3)$$

The set of all edge points for a given point cloud $\mathscr{C}$ be denoted by the symbol $\mathscr{E}$ and $\mathscr{E} \subset \mathscr{C}$. ∎

The value of $k = 0.4$ is found to be empirically effective in providing better segmentation of surfaces as will be shown later in this section. Essentially, an edge point is a point on the edge of a surface where a majority of its neighbors will have surface normals scattered in all directions and for such a seed point, the neighboring points will have angles $\theta_i > \theta_{high}$ as mentioned above. One such edge point is shown in Figure 11.10 as point B. An edge point is different from a non-edge point in the sense that the latter lies away from an edge and its neighbors have surface normals more or less in the same direction. One such non-edge point is shown as point A in Figure 11.10. Even with sensor noise, the neighboring points around such a seed point will have surface normals with smaller values of angles with respect to the surface normal of the seed point, i.e., $\theta_i < \theta_{high}$.

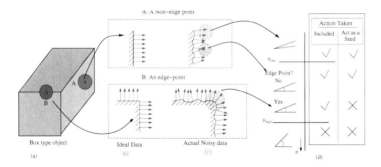

FIGURE 11.10: Update criteria of proposed region growing is illustrated here a cuboid. Two cases are shown here- one on a face (A) and other on the boundary (B). Noisy data leads to errors in normal directions as shown in (c) compared to ideal input data, (b) resulting in either spurious boundary or undetected boundary. Our method combines the boundary condition with two thresholds (d) in order to achieve better performance.

Now, in the region growing algorithm starting with the seed point s, the label $L\{p_i\}$ for a neighboring point $p_i \in \mathcal{N}(s)$ is defined as follows:

$$\begin{aligned}\text{if } \theta_i < \theta_{low}; &\quad \text{then,} \quad L\{p_i\} = L\{s\} \land p_i \to Q_s \\ \text{if } \theta_i > \theta_{high}; &\quad \text{then,} \quad L\{p_i\} \neq L\{s\} \land p_i \not\to Q_s\end{aligned} \quad (11.4)$$

where the notation $p_i \to Q_s$ indicates that the point p_i is added to the list of seed points which will be used by the region growing algorithm in the next iteration. However, if the angle between normals lies between the above two thresholds, i.e., $\theta_{low} < \theta_i < \theta_{high}$, the label to the neighboring point is assigned as follows:

$$\begin{aligned}\text{if } s \notin \mathcal{E} &\quad \text{then,} \quad L\{p_i\} = L\{s\} \land p_i \to Q_s \\ \text{if } s \in \mathcal{E} &\quad \text{then} \quad L\{p_i\} = L\{s\} \land p_i \not\to Q_s\end{aligned} \quad (11.5)$$

The above equation only states that while the neighboring point p_i is assigned the same label as that of the seed point s, it is not considered as a new seed point if the current seed is an edge point. This allows the region growing algorithm to terminate at the edges of each surface where there is a sudden and large change in the direction of surface normals thereby obtaining the natural boundaries of the objects. The above process for deciding labels for neighboring points is demonstrated pictorially in Figure 11.10. It is also shown how a pair of thresholds are effective in dealing with sensor noise, thereby eliminating spurious edges. The effect of this modified version of region growing algorithm on the object segmentation can be seen clearly in Figure 11.11. Figures 11.11(a) and (b) shows the case of segmentation obtained with only one threshold. In the first case, a lower threshold cut-off value θ_{low} is used while in the later, upper cut-off threshold θ_{high} is used. As discussed earlier, lower value of threshold leads to under-segmentation and may generate multiple patches even on the same surface. On the other hand, higher value of thresholds leads to over-segmentation where different surfaces of a rectangular box may be identified as a single surface patch. In contrast to these two cases, the use of two thresholds provide better segmentation leading to creation of two separate surfaces one for each face of the rectangular box.

This modified version of region growing algorithm allows us to find graspable affordances for rectangular box-type objects which were hitherto difficult. For instance, authors in [285] [286] find graspable affordances only for objects with cylindrical or spherical shapes as they relied on curve fitting methods. In [292] [297], the authors use a trained SVM to identify rectangular edges using HoG features and pre-defined hand poses were used for grasping objects at these detected regions. Compared to these approaches, the above proposed method is much simpler which does not require any training phase and can be implemented in real-time. More details about real-time implementation will be provided in the experiment section later in this chapter. The surfaces identified in this section are then used to find valid graspable regions on the object as described in the next section.

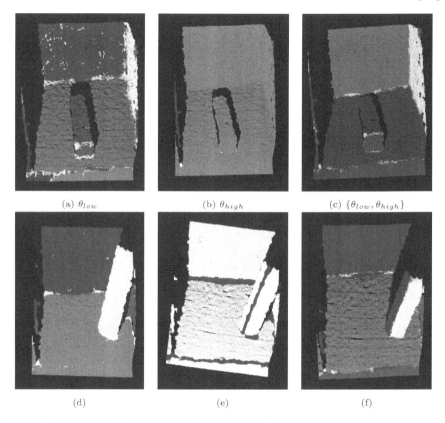

FIGURE 11.11: Effect of double thresholds on smooth conditions in region growing algorithm. (a) Using single threshold: θ_{low} leads to under-segmentation - multiple and discontinuous patches on the same surface (b) Using single threshold: θ_{high} leads to over-segmentation where different surfaces (having different normals) are merge together into a single surface. (c) using double boundary thresholds $\{\theta_{low}, \theta_{high}\}$ provides better surface segmentation compared to the case when only one threshold is used. (d) Shows the case when only one threshold is used. Two orthogonal surfaces of the object gets merged into one continuous surface. (e) Shows the edge points in blue color (f) Shows that the use of double thresholds lead to creation of two surfaces for the rectangular object.

11.3.3 Finding Graspable Affordances

Once the surface segments are created, the grasping algorithm needs to find suitable handles which could be used by the gripper for picking objects. This is otherwise known as the problem of grasp pose detection [297] which essentially aims at finding a 6 dimensional pose for the gripper necessary for making a stable grasping contact with the object. However, this is a computationally intensive task as one has to search in a 6-dimensional pose space. The searching procedure is broadly handled in two ways. In one approach, the object to be picked is matched with its CAD model. Once a match is found, then the geometric parameters of the object model is used to compute the 6 DOF gripper pose directly. As CAD models may not always be available, the objects are generally approximated with some basic shape primitives [298] [299] [300] or superquadric [301] models. While these methods take 3D point cloud as input, other methods can work with RGBD data. They generally take color and depth information as image and apply a sliding window based search with different scale to find valid grasping regions [283] [284]. We simplify this search problem at first by grouping similar type of points based on the boundaries obtained in the previous step and, by making some practical assumptions about the grasping task. As described earlier in section 11.3.1, the gripper is assumed to approach the object in a direction opposite to surface normal of the object. It is also assumed that the gripper closing plane coincides with the minor axis of the surface segment under consideration as shown in Figure 11.8(b). In this way, the 6D pose problem is solved in a single step and can be implemented in real-time. However, it is still necessary to identify suitable regions on the surface segments that can fit within the fingers of the gripper while ensuring that the gripper does not collide with neighboring objects. In other words, one still needs to search for a three-dimensional cube of dimension $l \times b \times e$ around the centroid of the object segment as shown in Figure 11.8(b). This requires carrying out a linear search along the three principal axes of the surface to find regions that meet this bounding box constraint. These regions are the graspable affordances for the object to be picked by the gripper. The details of the search process is described next in this section.

Let us assume that the region growing algorithm, described in the previous section, leads to the creation of S segments in the 3D point cloud $C \in \mathscr{R}^3$. As a first step, we extract the following parameters for each of these segments $s = 1, 2, \ldots, S$:

- The centroid of the segment: $\boldsymbol{\mu}_s = [\mu_x^s, \mu_y^s, \mu_z^s]$.

- The associated surface normal vector: $\hat{n}_s \in \mathscr{R}^3$.

- First two dominant directions obtained from Principle Component Analysis (PCA) and their corresponding lengths. These two axes correspond to vectors $\hat{a}$ and $\hat{f}$ respectively in Figure 11.8(b).

The search for suitable handles starts from the centroid μ_s of the surface and proceeds along the three principal axes, i.e., major axis $\hat{a}$, minor axis $\hat{f}$ and surface normal $\hat{n}$. In order to do this, the 3d point in the original point cloud corresponding to the surface segment under consideration s are projected onto these new axes $(\hat{f}, \hat{a}, \hat{n})$ as shown in Figure 11.12. So for every point $\vec{p}_O = (x_1, y_1, z_1)_O$ in the orginal coordinate system $(\hat{x}, \hat{y}, \hat{z}, O)$ that lies within a sphere of radius $d/2$ results in a vector $\vec{q}_{O'} = (f_1, a_1, n_1)_{O'}$ in the

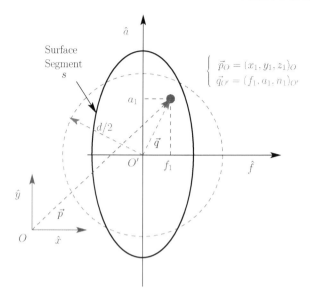

FIGURE 11.12: Scalar Projection of 3D cloud points to a new coordinate frame. The 3D point cloud points on the surface segment s represented by $\vec{p}_O$ within the sphere of radius $d/2$ is projected onto the axes of the new coordinate frame $(\hat{f}, \hat{a}, \hat{n})$ represented by the vector $\vec{q}_{O'}$. These projected points are used for finding suitable graspable affordances for the object. The axes $\hat{z}$ and $\hat{n}$ are perpendicular to the plane of the paper and point outward.

new coordinate system $(\hat{f}, \hat{a}, \hat{n}, O')$. The radius of the sphere is selected to be half of the maximum hand aperture of the gripper to be used for picking the object. The third axes $\hat{n}$ and $\hat{z}$ are normal to the surface of the chapter and hence is not displayed in the figure.

Through this scalar projection, the three dimensional search problem is converted into three one-dimensional search problems, which is computationally much simpler compared to the former. The search is first performed along the direction $\hat{a}$ and $\hat{n}$ respectively. All the points that lie within the radius $e/2$ around the centroid μ_s is considered to be a part of the gripper handle. Similarly, all points of the surface that lie within the radius of l along a direction of $-\hat{n}$ is considered to be part of the gripper handle. Please note that e and l are the width and the length of gripper fingers needed for holding the object. Once these two boundaries are defined, we get a horizontal patch of points extending along the minor axis $\hat{f}$ as shown in Figure 11.13 (c), (e) and (f). So now, we need to find the boundary along the minor axis to see if it would fit within the gripper finger gap. This is done by searching for a gap along the minor axis $\hat{f}$ which is at least bigger than a given user defined threshold which itself depends on the thickness of the gripper finger. The idea is that there should be sufficient gap between two objects to avoid collision with the neighboring objects. This is illustrated in Figure 11.13. The working of the

Grasping without Object Models 441

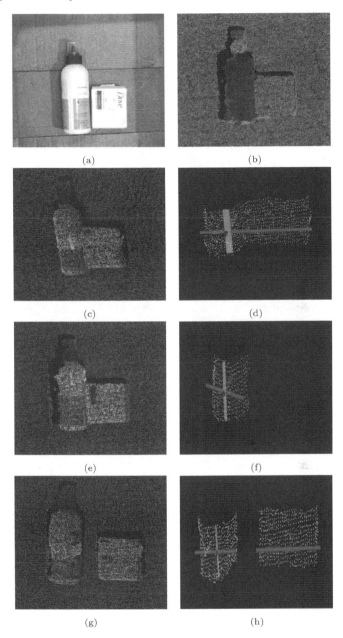

FIGURE 11.13: Searching for suitable grasping handle. (a) Actual picture for a case where objects are stacked very close to each other. (b) Segmented point cloud obtained after applying region growing on surface normals; (c) - (d) Suitable handle is not found as discontinuity is detected in the horizontal axis. (e) - (f) Suitable handle is found for another patch on the same object; (g)-(h) suitable handle found when objects are separate as a discontinuity is detected along the red axis

search process could be understood by analyzing this figure as explained in the following paragraph.

The figure 11.13(a) shows two objects which have been kept adjacent to each other such that their boundaries touch each other. The figure (b) shows the surface segments obtained using the proposed region growing algorithm. The objective is to find a suitable graspable handle for the cylinder object. The figure (c) shows the horizontal patch obtained using the linear search as explained above. Since there is no gap along the minor axis (shown in red), the region belonging to both the objects within the yellow band gets included into the graspable region. Total horizontal length of this band may exceed the maximum hand aperture d of the gripper making it an invalid grasping handle for the object. Now the next band of width e on the top of the last band is taken into consideration. In this case, a gap is found immediately around the boundary of the cylindrical surface along the minor axis as shown in Figure 11.13(e). Since this length along the red axis fits within the gripper handle, it will be considered as a valid handle for the object. The figures 11.13(g)-(h) shows the case when these two objects have been kept apart. In this case, the gap is found along the minor axis and hence the handle for the bottle is detected successfully without any further search. Hence the search process involves four steps:

1. Project all the points on the surface segment within a spherical radius of $d/2$ onto the axes $\hat{a}$, $\hat{f}$ and $\hat{n}$.

2. Fix boundary along the major axis $\hat{a}$ at a distance of $e/2$ on either side of the centroid the patch under consideration.

3. Fix boundary along the normal axis $-\hat{n}$ at a distance of l from the top surface.

4. Search for gap along the minor axis $\hat{f}$ on either side of the centroid. If this gap is greater than or equal to g, then search is stopped. The resulting patch is considered a valid grasping handle for the object if the total length of the patch along the minor axis is less than maximum hand aperture d of the gripper.

A new patch along the major axis either side of the centre patch is analyzed for validity in case the current one fails to satisfy the gripper constraints. So it is possible to obtain multiple handles on the same object, which is very useful, as the robot motion planner may not be able to provide a valid end-effector trajectory for a given graspable affordance. The gripper approaches the object at the centroid of the yellow patch shown in Figure 11.13(c) or (e) along the direction of surface normal (shown in blue color in Figure 11.13 (d) or (f) respectively, toward the object with its gripper closing plane coinciding with the minor axis (show in red color). As one can appreciate, the pose detection problem is solved by a simple method that converts a 6-D search problem into a simple 1-D search problem. This is much faster computer other method such as [301] that use complex optimization methods to arrive at the same conclusion. The proposed provides remarkable improvement over the state-of-the-art method [285] [286] which provides much inferior performance in a cluttered environment as will be shown in the next section.

11.3.4 Experimental Results

In this section, we provide results of various experiments performed to establish the usefulness of the proposed algorithm in comparison to the existing state-of-the-art methods. As explained before, our focus is to find suitable graspable affordances for various household items. The input to our algorithm is a 3D point cloud obtained from an RGBD or a range sensor and, the output is a set of graspable affordances comprising graspable regions and gripper poses required to pick the objects. We have particularly tested our algorithms on datasets obtained using Kinect [302], realsense [303] and Ensenso [304] depth sensors. An additional smoothing pre-processing step is applied to the Ensenso point cloud which is otherwise quite noisy compared to that obtained using either Kinect or realsense sensors. As we will demonstrate shortly, we have considered the grasping of individual objects in an extremely cluttered environment. The performance of the proposed algorithm is compared with other methods on four different datasets, namely, (1) Big bird dataset [305], (2) Cornell Grasping dataset [283], (3) ECCV dataset [306], (4) Kinect Dataset [307], (5) Willow garage dataset [308], (5) the TCS Grasping Dataset-1, and (6) TCS Grasping Dataset-2. The last two are created by us as a part of this work and are made available online [309] along with the program source code for the convenience of users. A snapshot of images for these two datasets is shown in Figure 11.14. The first TCS dataset contains 382 frames each having only single object in its view inside the bin of a rack where the view could be slightly constrained due to poor illumination. Similarly, the second dataset consists of forty frames with multiple objects in an extreme clutter environment. Each dataset contains RGB images, point cloud data (as .pcd files) and annotations in the text format. These datasets exhibit more difficult real world scenarios compared to what is available in the existing datasets. The algorithm is implemented on a Linux laptop with a i7 processor and 16 GB RAM.

11.3.4.1 Performance Measure

Different authors use different parameters to evaluate the performance of their algorithm. For instance, authors in [297] use *recall at high precision* as a measure while few others as in [283] use *accuracy* as a measure. In some cases, accuracy may not be a good measure for grasping algorithms because the number of true negatives in a grasping dataset is usually much more than the number of true positives. So, the accuracy could be high even when the number of true positives (actual handles detected) are less (or the precision is

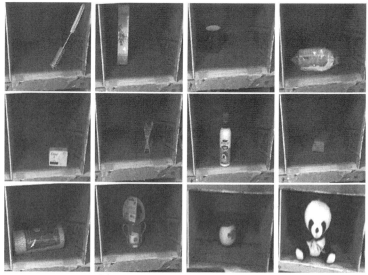
(a) Snapshot of TCS Grasp Dataset 1

(b) Snapshot of TCS Grasp Dataset 2

FIGURE 11.14: Snapshot of frames in TCS Grasping datasets 1 and 2. Each dataset consists of images and point cloud data files along with annotations in text files.

less). There are other researchers as in [292] [298] [301] who use *success rate* as a performance measure which is defined as the number of times a robot is able to successfully pick an object in a physical experiment. The success rate is usually directly linked to the precision of the algorithm as the false detections or mistakes could be detrimental to the robot operation. In other words, a grasping algorithm with high precision is expected to yield high success rate. The precision is usually defined as the fraction of total number of handles detected which are true. However, in a cluttered scenario, the precision may not always provide an effective measure to evaluate the performance of the grasping algorithm. For instance, it is possible to detect multiple handles for some objects and no handles at all for some others, without affecting the total precision score. In other words, the fact that no handles are detected for a set of objects may not have any effect on the final score as long as there are other objects for which more than one handle is detected.

In our case, the precision is considered to be 100% as any handle that does not satisfy the gripper and the environment constraints is rejected. In order to address the concerns mentioned above, we use *recall at high precision* as a measure of the performance of our algorithm which is defined as the fraction of total number of graspable objects for which *at least one valid handle* is detected. Mathematically, it can be written as

$$\text{recall \%} = \frac{\text{Number of objects for which at least one handle is detected}}{\text{Total number of graspable objects}} \times 100 \qquad (11.6)$$

The total number of graspable objects includes objects which could be actually picked up by the robot gripper in a real world experiment. It excludes the objects in the clutter which cannot picked up due to substantial occlusion. This forms the ground truth for the experiment. Note that the above definition is slightly different from the conventional definition of recall in the sense that the later may include multiple handles for a given object which are not considered in our definition. We analyze and compare the performance of our algorithm with an existing state-of-the-art algorithm using this new metric as described in the next section.

11.3.5 Grasping of Individual Objects

First, we demonstrate the performance of the proposed algorithm in picking individual objects. Table 11.2 shows the performance of the proposed algorithm on TCS dataset 1. This dataset has 382 frames along with the corresponding 3D point cloud data and annotations for ground truth. A snapshot of objects present in these dataset is shown in Figure 11.14. The performance of our proposed algorithm on this dataset is compared with Platt's algorithm reported in [285] [286]. As one can see in Table 11.2, the proposed algorithm is able to find graspable affordances for objects in more number of frames and hence it is more robust compared to the previous approach. On an average, our algorithm is able to detect handles in 94% of the frames compared to Platt's approach which can detect handles only for 51% of frames. This could be attributed to the fact that Platt's algorithm primarily relies on surface curvature to find handles and hence, cannot deal with rectangular objects with flat surfaces. They try to overcome this limitation in [292] by training a SVM classifier to detect valid grasp out of a number of hypotheses created using HoG features. Compared to this approach, our proposed method is much simpler to implement as it does not require any training and can be implemented in real-time. It also does not depend on image features which are

TABLE 11.2: Performance Comparison for TCS Grasping Dataset 1 - Individual Objects

Object	Total Number of frames	% of frames where a valid handle is detected	
		Platt's Method [286]	Proposed Method
Toothpaste	40	38	90
Cup	50	70	96
Dove Soap	40	25	100
Fevicol	40	75	92
Battery	50	36	98
Clips	21	45	90
CleaningBrush	40	30	90
SproutBrush	21	63	95
Devi Coffee	40	76	93
Tissue Paper	40	40	96
Total	382	51	**94**

more susceptible to various photometric effects. Some of the handles detected by our algorithm for individual objects are shown in Figure 11.15. Examples (a)-(c) shows few instances of simple objects where it is easier to find affordances while the (d)-(f) shows few difficult objects for which finding a suitable handle is challenging.

11.3.6 Grasping Objects in a Clutter

In this section, we demonstrate the performance of our proposed algorithm in a cluttered environment. A new dataset is created for this purpose. It is called 'TCS Grasp Dataset 2' and it contains forty frames, each one showing multiple objects in extreme clutter situation. The objects in the clutter have different shapes and sizes and, may exhibit partial or full occlusion. The performance of our algorithm on some of these frames are shown in Figure 11.16. The performance comparison with Platt's algorithm [286] [285] is shown in Figure 11.17. As one can see in Figure 11.16, the proposed algorithm is successful in finding graspable affordances for rectangular objects with flat surfaces such as books in addition to objects with curved surfaces. It also shows multiple handles detected for some of the objects. All those handles which do not satisfy the geometric constraints of the gripper are rejected and hence not shown in this figure. The maximum hand aperture considered for finding these affordances is 8 cm. In contrast, Platt's algorithm [286] [285] fails to detect any handles for flat rectangular objects as shown in Figure 11.17. Table 11.3

Grasping without Object Models

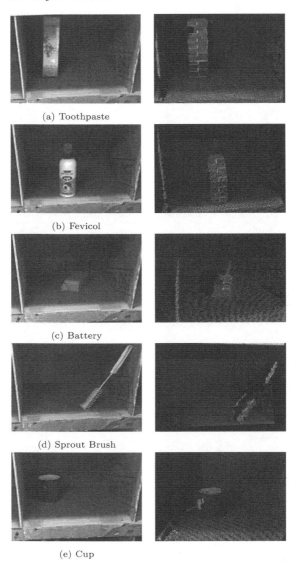

(a) Toothpaste

(b) Fevicol

(c) Battery

(d) Sprout Brush

(e) Cup

FIGURE 11.15: Finding graspable affordances for few objects inside a rack. The objects in (d), (e), and (f) show a few cases where it is difficult to find graspable affordances.

provides a more quantitative comparison between these two algorithms. It shows that the proposed algorithm is able to detect at least 86% of unique handles in the dataset compared to 36% recall achieved with Platt's algorithm. The performance of these two algorithms on various publicly available datasets is summarized in Table 11.4. Cornell Grasping Dataset [283] contains single

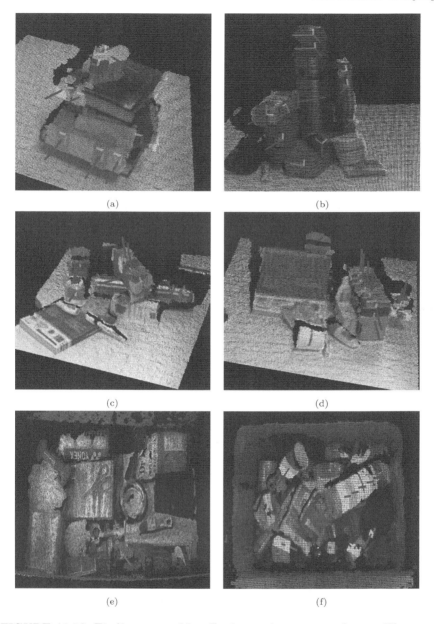

(a) (b) (c) (d) (e) (f)

FIGURE 11.16: Finding graspable affordances in extreme clutter. The proposed algorithm is capable of finding graspable affordances for rectangular objects as well as objects with curved surface. The maximum hand aperture (d) considered here is 8 cm.

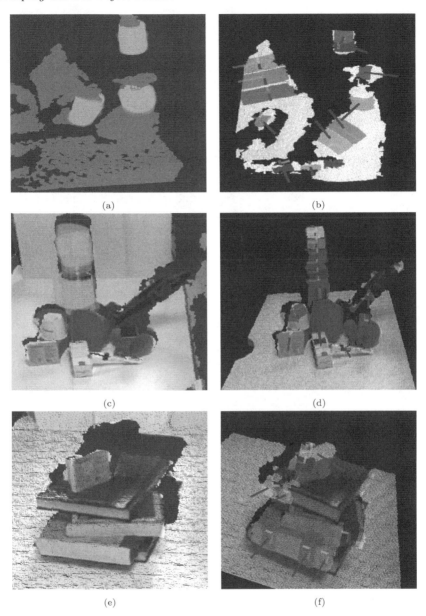

FIGURE 11.17: Visual comparison of the performance of the proposed algorithm with Platt's algorithm [285] on TCS Dataset 2. The cyan coloured patches on left hand side figures are the handles detected using Platt's algorithm. The patches on right side figures along with gripper pose show affordances obtained using the propose algorithm.

object per frame and grasping rectangle as ground truth. Their best result (93.7%) reported is in terms of accuracy whereas recall from our method is 96% at 100% precision. The Bird Bird dataset [305] consists of segmented individual objects and yields a maximum recall of 99%. This high level of performance is due to the fact that the object point cloud is segmented and processed for noise removal. This dataset, as such, does not include clutter and has been included in this section for the sake of completeness. The ECCV dataset [306], Kinect Dataset [307] and the Willowgarage dataset [308] have multiple objects in one frame and may exhibit low level of clutter. All of these datasets are created for either segmentation or pose estimation purposes, therefore ground truth for grasping is not provided. We have evaluated the performance (as reported in Table 11.4) using manual annotation. The extent of clutter in these datasets is not comparable to what one will encounter in a real world scenario. This is one of the reasons why we had to create our own dataset. As one can see in Figure 11.14(b), the TCS grasp dataset 2 exhibits extreme clutter scenario. As one can observe in Table 11.3, the

TABLE 11.3: Performance comparison for TCS dataset 2 - multiple objects in a cluttered environment

Frame No.	No. of graspable objects in the frame	Platt's Method [285] [286]		Proposed Method	
		max no. of handles detected	% Recall	max no. of handles detected	% Recall
#1	8	2	25	6	75
#3	8	3	38	6	75
#5	6	3	50	6	100
#7	7	2	28	7	100
#10	6	3	50	5	83
#12	7	2	28	7	100
#13	7	2	28	7	100
#16	8	1	13	6	75
#20	8	2	25	6	75
#23	9	2	22	8	89
#24	6	3	50	5	83
#26	5	3	60	3	60
#28	5	2	40	5	100
#30	6	2	33	6	100
#32	6	2	33	5	83
#37	5	1	20	5	100
#38	2	2	100	2	100
#39	4	2	50	3	75
#36	5	1	20	4	80
Total	118	40	33	102	86

Grasping without Object Models

TABLE 11.4: Performance comparison on various publicly available datasets

S. No.	Dataset	% Recall	
		Proposed Method	Platt's Algorithm [285] [286]
1	Big Bird [305]	99%	85% [297]
2	Cornell Dataset [283]	95.7%	93.7% [283]
3	ECCV [306]	93%	53%
4	Kinect Dataset [307]	91%	52%
5	Willow Garage [308]	98%	60%
6	TCS Dataset-1 [309]	94%	48%
7	TCS Dataset-2 [309]	85%	34%

proposed algorithm provides better grasping performance compared to the current state-of-the-art reported in literature.

11.3.7 Computation Time

The computational performance of the algorithm can be assessed by analyzing Table 11.5. This table shows the average computation time per frame for two TCS datasets. As one can observe, the bulk of the time is taken by the region growing algorithm which is the first step of our proposed method. This time is proportional to the size of the point cloud data. The second stage of our algorithm detects valid handles by applying geometric constraints on the surface segments found in the first step. This step is considerably faster compared to the first step. Many of the segments created in the first step are rejected in the second step to identify valid grasping handles as can be see in the fourth and fifth columns in this table. The computation time for each valid handle for the two datasets is 4 and 5 ms respectively.

TABLE 11.5: Average computation time per frame. All values are reported per frame basis and are averaged over all frames.

Dataset	# data in point cloud	Time for Region Growing algorithm (sec)	# segments detected	# valid handles detected	Handle detection time (sec)
TCS Dataset 1	37050	0.729	77	10	0.055
TCS Dataset 2	42461	0.82	182	43	0.171

The total processing time for a complete frame with around 40K data point is approximately 800 ms to 1 second. This is quite reasonable in the sense that the robot can process around sixty frames per second which is very good for most of the industrial applications. This time can be further reduced by detecting a particular ROI within the image thereby reducing the number of points to be processed in the frame. The computation time per frame can also be reduced significantly by downsampling the point cloud. There is a limit to the extent of downsampling allowed as it is directly linked to the quality and quantity of handles detected. For high speed applications, one may use FPGA or GPU based embedded computing platform.

11.4 Summary

This chapter focuses on finding graspable handles for objects in a clutter. We describe two methods - one that requires prior knowledge of object model and the other that does not require any such information. These two methods are named model-based and model-free methods respectively. In the first case, the grasp pose detection is carried out by performing semantic segmentation of the object followed by the object axis assignment. These two steps provide the object pose (position and orientation of the object) in the image plane. The object pose information is then used to predict multiple grasp poses along the major axis and select that the pose with maximum GDI score. In the second case, the graspable affordances (or suitable grasp poses) are computed by first creating surface segments by finding contiguous regions on the 3D point clouds based on some smoothness condition and then apply the geometric constraints of gripper to reduce the search space for possible gripper poses required for grasping the object. The full implementation details of each of the method is provided along with the experimental results for the benefit of readers. The interested readers are referred to appropriate literature for further reading.

12

Warehouse Automation: An Example

In this chapter, we provide details of a robotic system that can automate the task of picking and stowing objects from and to a rack in an e-commerce fulfillment warehouse. The system primarily comprises four main modules: (1) Perception module responsible for recognizing query objects and localizing them in the 3-dimensional robot workspace; (2) Planning module generates necessary paths that the robot end-effector has to take for reaching the objects in the rack or in the tote; (3) Calibration module that defines the physical workspace for the robot visible through the on-board vision system; and (4) Gripping and suction system for picking and stowing different kinds of objects. The perception module uses a faster region-based Convolutional Neural Network (R-CNN) to recognize objects. We designed a novel two-finger gripper that incorporates a pneumatic valve based suction effect to enhance its ability to pick different kinds of objects. The system was developed by IITK-TCS team for participation in the Amazon Picking Challenge 2016 event. The team secured a fifth place in the stowing task in the event. The purpose of this chapter is to share our experiences with students and practicing engineers and enable them to build similar systems. The overall efficacy of the system is demonstrated through several simulation as well as real-world experiments with actual robots.

12.1 Introduction

Warehouses are important links in the supply chain between the manufacturers and the end consumers. People have been increasingly adopting automation to increase the efficiency of managing and moving goods through warehouses [310]. This is becoming even more important for e-commerce industries like Amazon [311] that ships millions of items to its customers worldwide through its network of fulfillment centers. These fulfillment centers are sometimes as big as nine football fields [312] employing thousands of people for managing inventories. While these warehouses employ IoT and IT infrastructure [313, 314] to keep track of goods moving in and out of the facility, they still require the staffs to travel several miles each day in order to pick or stow products from or to different racks [312]. The problem related to the goods

FIGURE 12.1: Amazon plans to employ robots to pick and stow things from racks in retail warehouses.

movement was solved by the introduction of mobile platforms like KIVA systems [315] that could carry these racks autonomously to human 'pickers' who would then, pick things from these racks while standing at one place. These mobile platforms could then be programmed [316] [317] to follow desired paths demarcated using visual [318] or magnetic markers [319]. However, it still needs people to pick or stow items from or to these racks. Amazon hires several hundred people during holiday seasons, like Christmas or New Year, to meet this increased order demands. Given the slimmer operating margins, e-commerce industries can greatly benefit from deploying robotic 'pickers' that can replace these humans. This transition is illustrated in Figure 12.1. The left hand side of this figures shows the current state of affairs where a human picks or stows items from or to the racks, which are brought to the station by mobile platforms. The right hand side of this figure shows the future where robots will be able to do this task autonomously. In the later case, it won't be required to bring the racks to a picking station anymore if the robot arm is itself mounted on a mobile platform [320]. However, building such robots that can pick / stow items from / to these racks with the accuracy, dexterity and agility of a human picker is still far too challenging. In order to spur the advancement of research and development in this direction, Amazon organizes annual competition known as 'Amazon Picking Challenge' [321] every year since 2015. In this competition, the participants are presented with a simplified version of the problem where they are required to design robots that can pick and stow items autonomously from or to a given rack.

The picking task involves moving items from a rack and placing them into a tote while the stowing task involves moving items from the tote to the rack. The objects to be picked or stowed are general household items that vary greatly in size, shape, appearance, hardness, and weight. Since there is no constraint on how the products are organized on the rack or the tote, there are several possibilities of configuration one might encounter during the actual operation. This uncertainty that may arise due to factors like occlusion, variation in illumination, pose, viewing angle, etc., makes the problem of autonomous picking and stowing extremely challenging.

Introduction

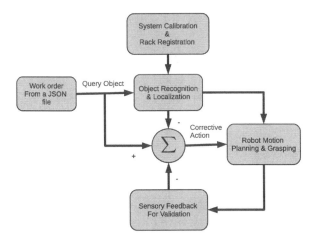

FIGURE 12.2: Schematic showing the important blocks of the system.

This chapter provides the details of the proposed system that can accomplish this task and share our experiences of participating in the APC 2016 event held in Leipzig, Germany. The proposed system primarily consists of three main modules: (1) Calibration, (2) Perception, (3) Motion Planning as shown in Figure 12.2. Some of the distinctive features of our implementation are as follows. In contrast to other participants, we took a minimalistic approach making use of minimum number of sensors necessary to accomplish the task. These sensors were mounted on the robot itself and the operation did not require putting any sensor in the environment. Our motivation has been to develop robotic systems that can work in any environment without requiring any modification to the existing infrastructure. The second distinctive feature of our approach was our lightweight object recognition system that could run on a moderate GPU laptop. The object recognition system uses a trained Faster RCNN based deep network [290] to recognize objects in an image. Deep network requires large number of training examples for higher recognition accuracy. The training examples are generated and annotated through a laborious manual process requiring considerable amount of time and effort. Moreover, larger training set requires larger time for training the network for a given GPU configuration. In a deviation to the usual trend, a hybrid method is proposed to reduce the number of training examples required for obtaining a given detection accuracy. Higher detection accuracy corresponds to tighter bounding box around the target object while lesser training examples will result in a bigger bounding box around the target object. The exact location for making contact with the object within this bounding box is computed using an algorithm that uses surface normals and depth curvatures to segment the target object from its background and finds suitable graspable affordance to facilitate its picking. In other words, the limitations of having smaller training set is overcome by an additional step which uses depth information to localize the targets within the bigger bounding box obtained from the RCNN network. This is another step which helps us in maintaining our minimalistic approach toward solving the problem. This approach allowed us to achieve accuracy of about $90 \pm 5\%$ in object recognition by training the RCNN network using only 5,000 images as opposed to other participants who used more than 20,000 images and high end GPU machines. The third distinctive feature of this chapter is the details that have been put in to explain the system integration process which, we believe, would be useful for students, researchers, and practicing engineers in reproducing and replicating similar systems for other applications.

In short, the major hallmarks of the proposed algorithm can be summarized as follows: (1) A novel hybrid perception method is proposed where depth information is used to compensate for the lesser size of dataset required for training a deep network based object recognition system. (2) The proposed system uses minimal resources to accomplish the complete task. It essentially uses only one Kinect sensor in an eye-in-hand configuration for all perception tasks in contrast to others [322] who used expensive camera like Ensenso [304]. (3) An innovative gripper design is provided that combines both suction as well as gripping action. (4) A detailed description of the system implementation is provided which will be useful for students, researchers, and practicing engineers. The performance of the proposed system is demonstrated through rigorous simulation and experiments with actual systems. The current system can achieve a pick rate of approximately two-three objects per minute.

12.2 Problem Definition

As described before the objective of this work is to replace humans for picking and stowing tasks in an e-commerce warehouse as shown in Figure 12.1. The schematic block diagram of our proposed system which can accomplish this objective is shown in Figure 12.2. The list of items to be picked or stowed is provided in the form of a JSON file. The system comprises a rack, a tote, and a 6 DOF robotic arm with appropriate vision system and end-effector for picking items from the rack or the tote.

The task is to develop a robotic system that can automatically pick items from a rack and put them in a tote and vice-versa. The reverse task is called the stowing task. The information about the rack as well as the objects to be picked or stowed are known a priori. The rack specified by APC 2016 guidelines had twelve bins arranged in a 4×3 grid. There were about forty objects in total which were provided to each of the participating teams.

In the pick task, the robot is expected to move items from the shelves of a rack to a tote. A subset of the forty objects (known a priori) were randomly distributed in these twelve bins. Each bin would contain minimum of one and maximum of ten items and the list of items at individual bins are known. Multiple copies of the same item could be placed in the same bin or in different bins. The bins may contain items which are partially occluded or in contact with other items or the wall of the bin. In other words, there is no constraint on how the objects would be placed in these bins. A *JSON* file is given prior to start the task which contain the details about which item is in which bin and what items are to be picked up from these bins. The task is to pick twelve specified items, only one from each of the bin in any sequence and put it into the tote.

In the stow task, the robot is supposed to move items from a tote and place them into bins on the shelf. The tote contained twelve different items, which are placed in such a way that some items are fully occluded or partially occluded by other items. The rest of the items are placed in the bins so that each bin can have minimum one item and maximum ten items. The task is to stow twelve items from the tote one by one in any sequence and put them into any bin.

The challenge was to get the robot to pick or stow autonomously as many items as it could within fifteen minutes. Different objects carried different reward points if they were picked or stowed successfully. A penalty was imposed on making mistakes such as picking or stowing wrong items, dropping them midway or damaging the items or the rack during robot operation, etc.

12.3 System Architecture

The schematic block diagram of the complete system is shown in Figure 12.2. The system reads the query items one by one from a JSON file. The JSON file also provides the bin location for each of these queried items. The robot has to pick these items from their respective bins. Since there could be several other objects in the bin, robot has to identify and localize the target object inside these bins. The system consists of the following three main components: (1) Calibration module, (2) Perception module, and (3) Motion planning module. The calibration module is used for defining the workspace of the robot with respect to the rack and the tote. It computes the necessary transformations needed for converting image features into physical real world coordinates. The perception module is responsible for recognizing queried items, localizing them in the bin, and finding the respective physical coordinates which can be used by robot for motion planning. The motion planning module generates necessary robot configuration trajectories and motion to reach the object, pick it using suction or gripping action, and move it to a tote. This module makes use of several sensors to detect the completion of the task. Once the task is completed, the system moves to the next item in the JSON query list.

The system is implemented using Robot Operating System (ROS) framework [323]. The operation of the complete system is divided into different modules each performing a specific task. Each of these modules are made available as a *node* which are the basic computation units in a ROS environment. These nodes communicate with each other using *topics*, *services* and *parameter servers*. Readers are advised to go through basic ROS tutorials available online in order to understand these concepts before proceeding further. Topics are unidirectional streaming communication channels where data is continuously published by the generating node and other nodes can access this data by subscribing to this topic. In this case, nodes are required to receive a response from other nodes; it can be done through *services*. The complete set of modules which are required for building the entire system is shown in Figure 12.3. These modules or *nodes* run on different computing machines which are connected to a common LAN. The dotted lines indicate service calls which execute a particular task on a demand basis. All these modules are controlled by a central node named "apc_controller." Simulation environment and RVIZ visualizer is also part of this system and is made available as an independent node.

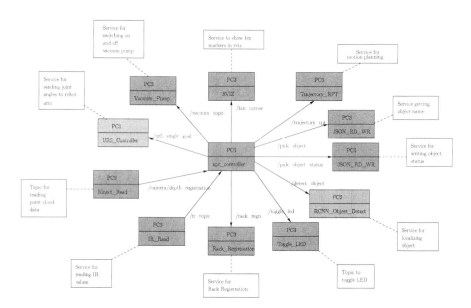

FIGURE 12.3: ROS architecture for pick and place application. Various nodes and topics run on three different computers (PC1, PC2 and PC3). The solid arrows indicate the topics which are either published or subscribed by a node. The dotted lines represent service calls.

12.4 The Methods

In this section, we provide the details of underlying methods for each of the modules described in the previous section.

12.4.1 System Calibration

The calibration step is needed to define the workspace of the robot as seen through a camera so that the robot can reach any visible location in the workspace. The calibration is an important step in all robotic systems that use camera as a sensor to perceive the environment. The purpose is to the transform the points visible in the camera plane to the physical Cartesian plane. A number of methods have been devised for calibration the normal RGB cameras [324] [325] which try to estimate the camera parameters so that the required transformation from pixel coordinates to 3D Cartesian coordinates could be carried out. The depth estimation has been simplified with the advent of RGBD camera such as Kinect [302] [326] which provides depth value for each RGB pixel of the image frame.

In this work, a Kinect RGBD camera is used in eye-in-hand configuration to detect as well as find the Cartesian coordinate of a query object with respect to its frame F_K. These coordinates are required to be transformed into robot base frame coordinate F_R so that it can be reached by the robot. In order to do this, it is necessary to know the transformation between the Kinect camera frame F_K and the robot end-effector frame F_E. The corresponding frames are shown in Figure 12.4. The transformation between the frames F_E

FIGURE 12.4: Cartesian coordinate frames for the robotic system. The transformation matrix between the robot base frame F_b and the end-effector frame F_e is known through robot forward kinematics. The transformation matrix between the Kinect frame F_k and the end-effector frame F_e is estimated in the calibration step.

and F_R is known through the forward kinematics of the robot manipulator. Hence the calibration step aims at finding this transformation between the robot end-effector frame F_e and the Kinect frame F_K as explained below.

Let us consider a set of points $\{P_K^i,\ i=1,2,\ldots,N\}$ which are recorded with respect to the Kinect frame F_K. The same set of points as recorded with respect to the robot base frame F_B is represented by $\{P_B^i,\ i=1,2,\ldots,N\}$. These latter points are obtained by moving the robot so that the robot end-effector touches these points which are visible in the Kinect camera frame. Since these two sets refer to the same set of physical locations, the relation between them may be written as

$$P_B^i = RP_K^i + \mathbf{t} \tag{12.1}$$

where $\{R, \mathbf{t}\}$ denotes the corresponding rotation and translation needed for the transformation between the coordinate frames. These equations are solved for $\{R, \mathbf{t}\}$ using least square method based on Singular Value Decomposition (SVD) [327] [328] as described below.

The centroid of these points is given by

$$\bar{P}_K = \frac{1}{N}\Sigma_{i=1}^N P_K^i$$
$$\bar{P}_B = \frac{1}{N}\Sigma_{i=1}^N P_B^i$$

and the corresponding covariance matrix is given by

$$C = \sum_{i=1}^N (P_K^i - \bar{P}_K)(P_B^i - \bar{P}_B)^T \tag{12.2}$$

Given SVD of covariance matrix $C = USV^T$, the rotation matrix R and translation vector $\mathbf{t}$ are given by

$$R = VU^T \tag{12.3}$$
$$\mathbf{t} = -R\bar{P}_K + \bar{P}_B \tag{12.4}$$

The RMS error between the actual points and the points computed using estimated $\{R, \mathbf{t}\}$ is shown in Figure 12.5 and the corresponding points are shown in Figure 12.6. The points are shown with respect to the robot base coordinate frame. The red points are the actual points and the yellow points are computed using estimated values of $\{R, \mathbf{t}\}$. It is possible to obtain an RMS error of 1 cm with as small as 8 points.

12.4.2 Rack Detection

Rack detection involves finding the corners of the rack and the bin centers automatically from an RGBD image recorded by the on-board Kinect camera.

The Methods

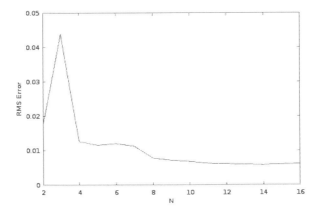

FIGURE 12.5: Plot of average RMS error (in meters) with the sample size N.

The bin corners information is useful for defining region of interest (ROI) for identifying objects within the bin. The bin corners and centers are also useful for planning motion to and inside the bins for picking objects. The bins in the rack are in form of a grid structure consisting of four vertical and five horizontal lines, and hence the bin corners can be identified by the intersection of vertical and horizontal lines. The vertical and the horizontal lines on the rack are detected using Hough line transform [329]. If $(x_1^v, y_1^v), (x_2^v, y_2^v)$ are end points of a vertical line and $(x_1^h, y_1^h), (x_2^h, y_2^h)$ are end points of a horizontal line then the equation to compute the intersection (x_i, y_i) of the two lines is

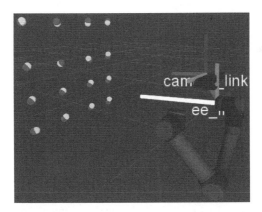

FIGURE 12.6: Checking the accuracy of calibration. Points in red color are the robot end-effector points collected prior to calibration. The yellow points are the points in the Kinect frame which are transformed into the robot base frame using estimated $\{R, \mathbf{t}\}$.

(a) Line detection (b) Bin Centres

FIGURE 12.7: Rack Detection from the RGBD point cloud. (a) Vertical and horizontal lines are detected using Hough line transform. Intersection of these vertical and horizontal lines provide corners for bins. (b) The bin centers are computed as the mean of bin corners.

given by

$$x_i = \frac{(x_1^v y_2^v - y_1^v x_2^v)(x_1^h - x_2^h) - (x_1^v - x_2^v)(x_1^h y_2^h - y_1^h x_2^h)}{(x_1^v - x_2^v)(y_1^h - y_2^h) - (y_1^v - y_2^v)(x_1^h - x_2^h)}$$

$$y_i = \frac{(x_1^v y_2^v - y_1^v x_2^v)(y_1^h - y_2^h) - (y_1^v - y_2^v)(x_1^h y_2^h - y_1^h x_2^h)}{(x_1^v - x_2^v)(y_1^h - y_2^h) - (y_1^v - y_2^v)(x_1^h - x_2^h)}$$

Once the corners are known, the bin center can be computed as the mean of its centers. Figure 12.7(a) shows the vertical and horizontal lines detected using an OpenCV [330] implementation for Hough transform. The intersection points computed using above equations are shown in Figure 12.7(b) where the bin corners are shown in red while the bin centers are shown in green. Note that only three middle horizontal lines and two outer vertical lines are required to be detected. The rest of the points can be estimated using the prior knowledge of rack geometry.

12.4.3 Object Recognition

Recognition and localization of an object in an image has been a fundamental and challenging problem in computer vision in past decades [331–334]. In the era of deep learning, CNN has been widely used for object recognition task, and it has shown outstanding performance [256, 290, 335, 336] as compared to the conventional hand-crafted feature based object recognition techniques [333, 334]. Techniques, like deformable parts models (DPM) [337] use a sliding window method where at every evenly spaced spatial locations the classifier is trained. The approach hence fails to progress further due to huge computational complexity. Eventually, in 2014 R-CNN was introduced by Girshick et al. [338], which uses region proposal methods to generate potential bounding boxes at the first stage. Then the classifier is trained on each of these proposed boxes. The bounding boxes are fine-tunned by post-processing followed by eliminating duplicate detection and re-evaluating the box based on objects in the scene. There are other variants of R-CNN with improved recognition accuracy and faster execution time. Some of theses are presented in [254, 290, 339].

In a recent work Redmon et al. proposes *you only look once* (YOLO) [339], where the object detection is transformed to a single regression problem. The approach improves the performance in terms of computational cost, however, the recognition accuracy is slightly inferior as compared to the Faster RCNN [290]. We use Faster RCNN as a base for our object recognition and localization task, as it localizes the objects in an image in real-time with very high recognition accuracy.

In APC 2016, object detection is considered to be a challenging problem due to varying illumination conditions, placement of the objects in different orientation, and depths inside the rack. In the case of stowing, the objects in the tote can be fully or partially occluded by other objects. These, resulted in a very complex object recognition task.

We have combined the deep learning approach and standard image processing techniques for robust object detection. We are using Faster RCNN based deep neural network to find the bounding box of the target object. A second step verification of target object in the bounding box provided by RCNN is performed using random forest classifier. We have done fine tuning of pretrained object detection model with our own dataset. RCNN layer architecture used for object detection is given in Figure 12.9. The details of the data preparation, training, and verification steps are given in the below sections.

Annotation: We have prepared two different training datasets for picking and stowing tasks. We have annotated 150 RGB images per object with different orientations and backgrounds for each task. A total of 6,000 images were annotated for each task.

Training models: To do object detection tasks, which include classification and localization, we are using VGG-16 layered classification network in combination with region proposal networks. RPN are basically fully covolutional network which takes an image as input and outputs a set of rectangular object proposals, each with an objectness score. It is a sixteen-layered classification network which consists of thirteen convolution layers and three fully connected layers. These RPNs share convolutional layers with object detection networks because it does not add significant computations at run time(10ms per image). We have fine tuned VGG-16 pretrained models of faster RCNN for our own dataset of 6,000 images for forty different objects. Snapshot of examples used for training the RCNN network is shown in Figure 12.8.

Object verification: We have added an additional step in object detection pipeline to verify objects in the window proposed by RCNN. It uses shape and color information to verify the presence of the object. Both shape and color informations are incorporated as a feature vector and a random forest is used to classify each pixel inside the object box. After finding the most probable region inside the window using random forest, we apply a meanshift algorithm

FIGURE 12.8: Snapshot of examples used for training the RCNN network.

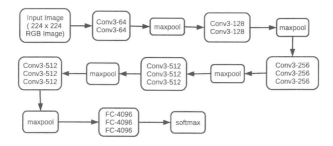

FIGURE 12.9: RCNN layer architecture used for object detection.

to obtain the suction point for that object. The details of the feature (shape and color) and classifier used are explained below:

Shape and color information as a feature: As we know, any 3D surface of the object is characterized by the surface normals at each point in the point cloud. The angle between neighboring normals at interest points on any suface can be used to the shape of any object. A shape histogram is created for each object model which is used as a shape feature vector. Similarly, we are incorporating color information in the feature using color and grayscale histogram of the objects.

Random forest: After computing histograms, all three histograms are concatenated as a 37 dimensional feature vector for each object. The training data is prepared by extracting features from pointcloud and RGB data for each object. A random forest classifier is trained for each object with one versus all strategy. In one versus all, the target object features are trained as positive class and rest all features are considered as negative class. The number of trees and depth of the trees used in the random forest are one hundred and thirty respectively.

The Methods

Algorithm 18 Algorithm for object detection technique

1: Calibrate and get the rack transformation matrix using kinect.
2: **for** each object i in the JSON file $i \leftarrow 1$ to N **do**
3:    Read JSON file. Get bin number and object identifier.
4:    Take RGB image of the bin and corresponding 3d Point Cloud according to transformation matrix.
5:    Using trained Faster R-CNN model get the ROI of the object in the RGB input image.
6:    Select the object ROI with the highest (score) probability
7:    Apply color and shape backprojection technique in the resultant object ROI using corresponding 3d Point Cloud.
8:    Classify each pixel inside the object ROI using Random Forest classifier based on combined shape and color information.
9:    Apply adaptive meanshift to find the most probable suction point.
10:   Find normal at the suction point and the centroid of the object to be picked.
11:   Instruct motion planner to move to the given position.
12:   Robot controller
13: **end for**

12.4.4 Grasping

Grasping involves two steps - finding a grasp pose for the target object then making the actual motion to make physical contact with the object. The first part is usually difficult and has attracted a lot of attention over the last couple of decades. There are primarily two approaches to solve the grasping problem - one of them makes use of known 3D CAD models [340] and the other one does not require these CAD models [341] [342] [343]. The latter method directly works on the partial depth point cloud obtained from a range sensor. Quite recently, researchers are exploring the use of deep learning networks to detect grasping directly from images [344] [345].

In this chapter, we follow the latter approach where we detect the graspable affordance for the recognized object directly from the RGBD point cloud obtained from the on-board Kinect camera. Figure 12.10 shows the schematic

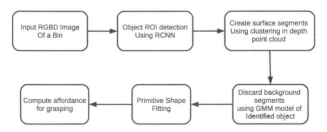

FIGURE 12.10: Schematic block diagram for computing grasping affordances for objects using RGBD images obtained from a Kinect Camera.

block diagram of the method employed for grasp pose detection. Input to this scheme is an RGBD point cloud of the bin viewed by the on-board robot camera. The bounding box of the query object is obtained by the RCNN based object recognition system. The bounding box returned by the RCNN module may have a bigger size than the object itself depending on the amount of training of the network used. This bounding box acts as the region of interest (ROI) for finding graspable regions. This bounding box may contain parts of the background as well other objects in the vicinity. Within this ROI, a clustering method combined with region growing algorithm [346] [347] is used to create several surface segments by identifying discontinuity in the space of surface normals [348] [349] [294]. Apart from having different surfaces for different objects and backgrounds, there can be multiple surface segments for the same object. Then the background segments are separated from the foreground target segments using a Gaussian Mixture Model (GMM) [350] [350] of the identified object using both color (RGB) and depth curvature information. Once the background segments are discarded, a primitive shape is identified for the object using empirical rules based on surface normals, radius of curvature, alignment of surfaces, etc. Once the shape is identified, the best graspable affordance for the object is computed using a modified version of the method presented in [350].

12.4.5 Motion Planning

In the case of industrial manipulators where one does not have access to internal motor controllers, motion planning refers to providing suitable joint angle position (or velocity) trajectories needed for taking the robot from one pose to another. In other words, motion planning becomes a path planning problem which is about finding a way to point poses between the current pose and the desired end-effector pose. The problem of generating collision free paths for manipulators with increasingly larger number of links is quite complex and has attracted considerable interest over last couple of decades. Readers can refer to [351] for an overview of these methods. These methods could be primarily divided into two categories - local and global. Local methods start from a given initial configuration and step toward final configuration by using local information of the workspace. Artificial potential field-based methods [352] [353] [354] are one such category of methods where the search is guided along the negative gradient of artificially created vector fields. On the other hand, global methods use search algorithms over the entire workspace to find suitable paths. Some of the examples of global methods are probabilistic roadmaps (PRM) [355] [356]and cell-decomposition based C-Space methods [357] [358]. Rapidly exploring random tree (RRT) [359] is one of the most popular PRM method used for path planning. Many of these state-of-the-art algorithms are available in the form of the open motion planning library (OMPL) [289] which has been integrated into several easy-to-use software packages like *Moveit!* [288], Kautham [360], and OpenRave [361].

The Methods

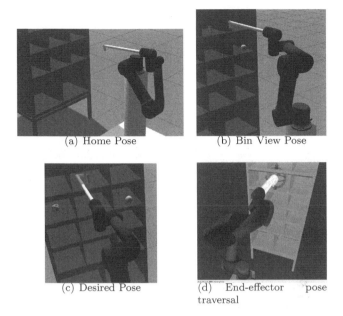

(a) Home Pose   (b) Bin View Pose

(c) Desired Pose   (d) End-effector pose traversal

FIGURE 12.11: Simulating motion planning using Moveit and Gazebo. (a) In idle state the robot stays at home pose; (b) on receiving queried bin number, the robot moves to the bin view pose where it takes an image of the bin; (c) the required desired for picking the can in picture is obtained after processing the image to identify target item; and (d) the end-effector trajectory from bin view pose to the desired pose is obtained using RRT motion planning algorithm available with Moveit.

In this chapter, we have used *Moveit!* package available with ROS [323] for building motion planning algorithms for UR5 robot manipulator. The simulation is carried out using the Gazebo [362] environment. Some of the snapshots of the robot are shown in Figure 12.11. The robot starts its operation from its home pose which is shown in Figure 12.11(a). The pose is so selected so that the entire rack is visible from the on-board Kinect camera (not shown in the picture). This image is used for system calibration process as described in Section 12.4.1 and 12.4.2 respectively. Once the bin number is obtained from the JSON query file, the robot moves to the bin view pose shown in Figure 12.11(b). At this pose, a close up picture of the bin is taken by the Kinect camera mounted on the wrist of the robot. Every bin has a pre-defined bin view pose which is selected so as to get a good view of the bin. The desired pose necessary for picking an item in the bin is obtained from the object recognition and grasping algorithm. One such desired pose is shown in Figure 12.11(c). The robot configuration trajectory generated by the motion planning

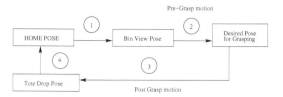

FIGURE 12.12: Sequence of steps for motion planning for a picking task.

algorithm is shown in Figure 12.11(d). It also performs collision avoidance by considering the rack (shown in green color) as an obstacle.

The sequence of steps involved in carrying out motion planning for the pick task is shown in Figure 12.12. It primarily involves four steps. The motion for segments 1 and 4 are executed using pre-defined joint angles as these poses do not change during the pick task. However, the motion planning for segment 2 (pre-grasp motion) and segment 3 (post-grasp motion) is carried out using RRT algorithm during the run-time. This is because the desired pose required for grasping the object will vary from one object to another and hence, the paths are required to be determined in the run-time. The online motion planning uses flexible collision library (FCL) [363] to generate paths for robot arm that avoid collision with the rack as well as the objects surrounding the target item. In order to avoid collision with the rack, the bin corners obtained from the rack detection module, described in Section 12.4.2, are used to define primitive boxes for each wall of the bin. These primitive boxes, shown in green color in Figure 12.13(a), are then treated as obstacles in the motion planning space. Similarly, the collision with other objects in the bin is achieved by creating 3D occupancy map called OctoMap [364] which converts point cloud into 3D voxels. This OctoMap feed is added to the Moveit

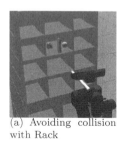

(a) Avoiding collision with Rack

(b) Avoiding collision with Objects using Octomap

FIGURE 12.13: Collision avoidance during motion planning. In (a), the green color shows the obstacle created using primitive shapes. In (b) Octomap is used to create 3D voxels for each object which are considered obstacles during motion planning.

The Methods

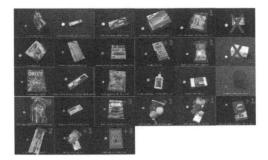

FIGURE 12.14: Typical items that were to be picked or stowed in the Amazon Picking Challenge.

motion planning scene and FCL is used for avoiding collision with the required objects. The OctoMap output of a 3D point cloud is shown in Figure 12.13(b).

12.4.6 End-Effector Design

The Amazon Picking Challenge focuses on solving the challenges involved in automating picking and stowing various kinds of retail goods using robots. These items include both rigid as well as deformable objects of varied shape and size. The maximum specified weight was about 1.5 Kgs. A snapshot of typical objects that were specified for the APC 2015 event [321] is shown in Figure 12.14. The authors in [365] provide a rich dataset for these objects which can be used for developing algorithms for grasping and pose estimation. It was necessary to design an end-effector which could grasp or pick all kinds of objects. We designed two kinds of end-effectors to solve this problem which are described below.

12.4.6.1 Suction-based End-effector

This end-effector essentially makes use of a vacuum suction system to pull the objects toward it and hold it attached to the end-effector. Such a system was successfully used by the TU-Berlin team [366] in the APC 2015 event where they came out as clear winners. A normal household cleaner could be used as the robot end-effector. It was sufficient only to make the nozzle end of the vacuum suction to reach any point on the object to be picked irrespective of its orientation. However, the suction can work only if it makes contact with the object with sufficient surface area necessary to block the cross-section of the suction pipe. One such system designed for our system is shown in Figure 12.15. The cross section of the suction pipe should be big enough to generate necessary force to lift the object. It cannot be used for picking small objects having smaller cross section area, for instance, a pen or a pencil or a metal dumbbell having narrow cylindrical surface. A more close-up view of the suction cup is shown in Figure 12.15(b). A set of IR sensors is used inside the

(a) Suction-based end-effector

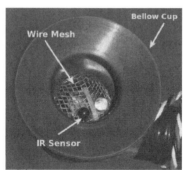
(b) Close-up view of suction cup

FIGURE 12.15: The end-effector using a suction cup for picking objects. The suction cup uses IR Sensor to detect if an object has been picked up successfully. The wire mesh prevents smaller or softer items getting sucked into the system.

bellow cup in order to detect the successful pick operation for a given object. A fine mesh is embedded inside the cup to prevent finer and soft materials like cotton or clothes from getting sucked into the tube and thereby, damaging the end-effector.

12.4.6.2 Combining Gripping with Suction

This particular design was employed by the MIT team [367] during the APC 2015 event. In this design, they combined a parallel jaw gripper with a suction system. They also used a compliant spatula to emulate scooping action. In this design, suction was used for picking only a very few items which could not be picked by the parallel jaw gripper and hence, it employed a single bellow cup capable of picking smaller items. Inspired by this design, we developed a similar hybrid gripper by combining suction cups with a two-finger gripper as shown in Figure 12.16. This gripper was designed to lift a weight of around 2 kgs. It uses a single actuator with rack pinion mechanism to achieve linear

The Methods 471

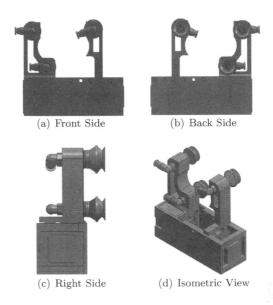

FIGURE 12.16: Novel gripper design that combines gripping with suction.

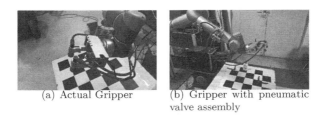

FIGURE 12.17: Actual gripper after fabrication and assembly.

motion between the fingers. The stationary finger houses two bellow cups while the moving finger houses one bellow cup. Hence, it is possible to pick bigger objects through suction by increasing the space between the fingers. The bellow cups are actuated by pneumatic valves that create suction by diverting pressurized air through them. The actual gripper assembly with pneumatic valves and pipes are shown in Figure 12.17(a) and (b) respectively. The working of the gripper is demonstrated in the experiment section.

12.4.7 Robot Manipulator Model

In order to carry out simulation for the actual system, one may need the forward kinematic model of the robot being used. This can be derived using the D-H parameters [368] of the robot. The D-H parameters for UR5 robot [369]

TABLE 12.1: D-H Parameters of UR5 robot

a (m)	d (m)	α (rad)	θ
0	0.0895	1.5708	θ_1
-0.425	0	0	θ_2
-0.3923	0	0	θ_3
0	0.1092	1.5708	θ_4
0	0.0947	-1.5708	θ_5
0	0.0823	0	θ_6

is shown in Table 12.1 and the corresponding axes for deriving these values are shown in Figure 12.18. The forward kinematic model thus obtained can be used for solving inverse kinematics of the robot manipulator, developing visual servoing and other motion planning algorithms. In the rest of this section, we describe three popular methods for solving inverse kinematics. The readers are referred to [368] [370] for more detailed treatment on the subject.

The forward-kinematic equation is given by the following equation:

$$\mathbf{x} = \boldsymbol{f}(\mathbf{q}) \qquad (12.5)$$

Let us assume that $\mathbf{q} \in \mathscr{R}^n$ and $\mathbf{x} \in \mathscr{R}^m$. For a redundant manipulator, $n > m$. By taking time-derivative on both sides of the above equation, we get

$$\dot{\mathbf{x}} = J(\mathbf{q})\dot{\mathbf{q}} \qquad (12.6)$$

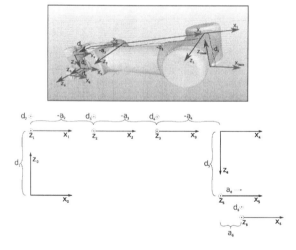

FIGURE 12.18: Axes for computing D-H parameters of UR5 robot manipulator.

The Methods

where J is the $m \times n$ dimensional Jacobian of the manipulator. The joint angles for a given end-effector pose $\mathbf{x}_d$ can be obtained using matrix pseudo-inverse as shown below:

$$\dot{\mathbf{q}} = J^\dagger(\mathbf{q})\dot{\mathbf{x}}_d \tag{12.7}$$

where $J^\dagger(\mathbf{q})$ represents the inverse of the Jacobian matrix J. If $(J^T J)$ is invertible, the pseudo-inverse is given by the Moore-Penrose inverse equation:

$$J^\dagger(\mathbf{q}) = (J^T J)^{-1} J^T \tag{12.8}$$

This is otherwise known as the least square solution which minimizes the cost function $\|\dot{\mathbf{x}} - J\dot{\mathbf{q}}\|^2$. The equation (12.8) is considered as a solution for an over-constrained problem where the number of equations (m) is less than the number of variables n and rank$(J) \le n$.

If (JJ^T) is invertible, then pseudo-inverse is the **minimum norm** solution of the least square problem given by the following equation:

$$J^\dagger = J^T (JJ^T)^{-1} \tag{12.9}$$

The equation (12.9) is considered to be a solution for an under-constrained problem where the number of equations m is less than the number of unknown variables n. Note that the equation (12.9) is also said to provide the *right pseudo-inverse* of J as $JJ^\dagger = I$. Note that, $J^\dagger J \in \mathscr{R}^{n \times n}$ and in general, $J^\dagger J \ne I$.

12.4.7.1 Null Space Optimization

Another property of the pseudo-inverse is that the matrix $I - J^\dagger J$ is a projection of J onto null space. Such that for any vector ψ that satisfies $J(I - J^\dagger J)\psi = 0$, the joint angle velocities could be written as

$$\dot{\mathbf{q}} = J^\dagger \dot{\mathbf{x}} + (I - J^\dagger J)\psi \tag{12.10}$$

In general, for $m < n$, $(I - J^\dagger J) \ne 0$, and all vectors of the form $(I - J^\dagger J)\psi$ lie in the null space of J, i.e., $J(I - J^\dagger J)\psi = 0$. By substituting $\psi = \dot{\mathbf{q}}_0$ in the above equation, the general inverse kinematic solution may be written as

$$\dot{\mathbf{q}} = J^\dagger \dot{\mathbf{x}} + (I - J^\dagger J)\dot{\mathbf{q}}_0 \tag{12.11}$$

where $(I - J^\dagger J)$ is a projector of the joint velocity vector $\dot{\mathbf{q}}_0$ onto $\mathscr{N}(J)$. The typical choice of the null space joint velocity vector is

$$\dot{\mathbf{q}}_0 = k_0 \left(\frac{\partial w(\mathbf{q})}{\partial \mathbf{q}} \right)^T \tag{12.12}$$

with $k_0 > 0$ and $w(\mathbf{q})$ is a scalar objective function of the joint variables and $\left(\frac{\partial w(\mathbf{q})}{\partial \mathbf{q}} \right)^T$ represents the gradient of w. A number of constraints could be imposed by using this objective function. For instance, the joint limit avoidance can be achieved by selecting the objective function as

$$w(q) = \frac{1}{n} \sum_{i}^{n} \left(\frac{q_i - \bar{q}_i}{q_{iM} - q_{im}} \right)^2 \quad (12.13)$$

where $\bar{q}_i$ is the middle value of joint angles while q_{iM} (q_{im}) represent maximum (minimum) value of joint angles. The effect of the null space optimizing on joint angle norm is shown in Figure 12.19(d). As one can see from this figure, the null space optimization for joint limit avoidance leads to a solution with smaller joint angle norm compared to the case when self motion is not used.

12.4.7.2 Inverse Kinematics as a Control Problem

The inverse kinematic problem may also be formulated as a closed-loop control problem as described in [371]. Consider the end-effector pose error and its time derivative be give as follows:

$$\mathbf{e} = \mathbf{x_d} - \mathbf{x}; \quad \dot{\mathbf{e}} = \dot{\mathbf{x}}_\mathbf{d} - \dot{\mathbf{x}} = \dot{\mathbf{x}}_\mathbf{d} - J\dot{\mathbf{q}} \quad (12.14)$$

By selecting the joint velocities as

$$\dot{\mathbf{q}} = J^{\dagger}(\dot{\mathbf{x}}_\mathbf{d} + K_p(\mathbf{x}_d - \mathbf{x})) \quad (12.15)$$

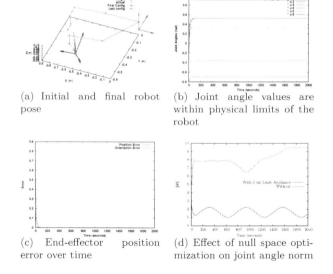

(a) Initial and final robot pose

(b) Joint angle values are within physical limits of the robot

(c) End-effector position error over time

(d) Effect of null space optimization on joint angle norm

FIGURE 12.19: Computing inverse kinematics for a given target pose using conventional methods. Figure (a) shows the initial and final robot pose along with the end-effector trajectory. The frame coordinates for desired and actual end-effector pose are also shown.

The Methods

the closed loop error dynamics becomes

$$\dot{\mathbf{e}} + K_p \mathbf{e} = 0$$

Hence the control law (12.15) stabilizes the closed loop error dynamics and the error will converge to zero if K_p is positive definite. The homogeneous part of the inverse kinematic solution in (12.10) could be combined with (12.15) in order to obtained a generalized closed loop inverse kinematic solution.

12.4.7.3 Damped Least Square Method

The pseudo-inverse method for inverse kinematics is given by

$$\Delta \mathbf{q} = J^\dagger \mathbf{e} \quad (12.16)$$

In damped least square method, the $\Delta \mathbf{q}$ is selected so as to minimize the following cost function

$$V = \|J \Delta \mathbf{q} - \mathbf{e}\|^2 + \lambda^2 \|\Delta \mathbf{q}\|^2 \quad (12.17)$$

This gives us the following expression for joint angle velocities:

$$\Delta \mathbf{q} = J^T (JJ^T + \lambda^2 I)^{-1} \mathbf{e} \quad (12.18)$$

The inverse kinematic solutions computed using these conventional methods are shown in Figure 12.19 and 12.20 respectively. Figure 12.19(a) shows

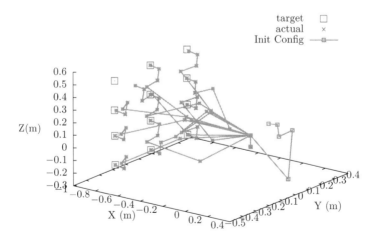

FIGURE 12.20: Robot pose for the bin centers of the rack obtained by solving inverse kinematics of the robot manipulator. The average error over 12 points is about 6 mm.

the inverse kinematic solution obtained for a given pose using null space optimization method that avoids joint limits as explained above. The corresponding joint angles are within their physical limits as shown in Figure 12.19(b). Figure 12.20 shows the joint configurations for reaching all bin centers of the rack.

12.5 Experimental Results

The actual system developed for accomplishing the task of automated picking and stowing is shown in Figure 12.21. The system comprises a 6 DOF UR5 robot manipulator with a suction based end-effector, a rack with twelve bins in a 3×4 grid. The end-effector is powered by a household vacuum cleaner. It uses a Kinect RGBD sensor in an eye-in-hand configuration for carrying out all perception tasks. As explained in Section 12.3, the entire system runs on three laptops connected to each other through ethernet cables. One of these laptops is a Dell Mobile Precision 7710 workstation with a NVIDIA Quadro M5000M GPU process with 8GB of GPU RAM. This laptop is used for running the RCNN network for object detection. The other two laptops have a normal Intel i7 processor with 16 GB of system RAM. The distribution of various nodes on the machines are shown in Figure 12.3. It is also possible to run the whole system on a single system having necessary CPU and GPU configuration required for the task. The videos showing the operation of the entire system

FIGURE 12.21: Experimental setup for automated pick and stow system.

Experimental Results

TABLE 12.2: Computation time for various modules of the robotic pick and place system.

S. No.	Component	Description	Time (seconds)
1	Reading JSON file	For ID extraction	0.01
2	Motion 1	Home position to Bin View Position	3.5
3	Object recognition	using trained RCNN model	2.32
4	Motion 2	Pre-grasp motion	9.6
5	Motion 3	Post-grasp motion	4.97
6	Motion 4	Motion from Tote drop to home position	3.41
	Total loop time for each object		23.81
7	Rack Detection		2.1
8	Calibration		13.1

using a suction end-effector [372] [373] and a two-finger gripper [374] is made available on internet for the convenience of the readers. The readers can also use the source codes [375] made publicly available under MIT license for their own use.

12.5.1 Response Time

The computation time for different modules of the robotic pick and place system is provided in Table 12.2. As one can see the majority of time is spent in image processing as well as in executing robot motions. Our loop time for picking each object is about 24 seconds which leads to a pick rate of approximately 2.5 objects per minute. The rack detection and system calibration is carried out only once during the whole operation and does not contribute toward the loop time.

(a) Gripping Action (b) Suction Action

FIGURE 12.22: The hybrid gripper in action. It uses two-finger gripper to pick objects that can fit into its finger span. Suction is used for picking bigger objects with flat surfaces.

12.5.2 Grasping and Suction

The working of our custom gripper is shown in Figure 12.22. The maximum clearance between the fingers is about 7 cm and it has been designed to pick up a payload of 2 kgs. The gripper can grasp things using an antipodal configuration [292] as shown in Figure 12.22(a). The suction is applied whenever it is not possible to locate grasping affordances on the object. The bellow cups are positioned normal to the surface of the object being picked as shown in Figure 12.22(b). For grasping, it is necessary to detect the grasp pose and compute the best graspable affordance for a given object. This is done by using the method as described in Section 12.4.4. Some of the results corresponding to the grasping algorithm is shown in Figure 12.23. As explained before, a GMM model comprising color (RGB) information and depth curvature information is effective in segmenting the target object from its background as shown in Figure 12.23(a) and 12.23(b) respectively. The outcome of the grasping algorithm is shown in Figure 12.23(c) and 12.23(d) respectively. Figure 12.23(c) shows the best graspable affordance for objects with different shapes while Figure 12.23(d) shows the graspable affordance of objects in a clutter.

12.5.3 Object Recognition

Experiments on object recognition are performed using our APC dataset with 6,000 images for forty different objects. The images are taken at different lighting conditions with various backgrounds. Pretrained VGG-16 model of the Faster R-CNN is fine tuned using 80% of the whole dataset and remaining 20% is used to validate the recognition performance. Figure 12.24 presents some object recognition results when tested with new images. Statistical analyses have been carried out on the validation set. We have achieved a mean Average Precision (mAP) of 89.9% for our validation set, which is a pretty

Experimental Results 479

(a) Segmenting 'Fevicol' tube from the clutter

(b) Use of GMM model using both color and depth curvature information

(c) Primitive shape fitting and identifying best graspable affordance for objects with different shapes

(d) Identifying shapes and computing graspable affordance in a clutter

FIGURE 12.23: Computing graspable affordance of target object in a cluttered environment. (a) Shows the use of GMM model comprising of RGB and depth curvature information in segmenting the target object from clutter. (b) Shows the GMM model used in (a). It shows the Gaussian corresponding to depth curvature provides better discrimination compared to colors in identifying the target. (c) Shows the detection of shape and best graspable affordance for isolated objects. (d) Shows the detection of shape and graspable affordance in a clutter.

good performance for such an unconstrained and challenging environment. The individual precision of randomly picked twenty-nine objects and their mAP are shown in Table 12.4. Observation shows that, when the objects are deformable, such as cherokee t-shirt and creativity stems, the precision is reasonably lower. In our case, the precisions are 74.7% and 73.65% respectively.

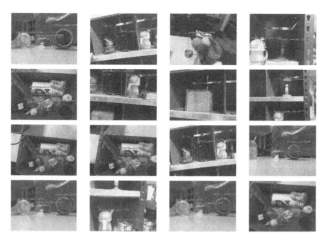

FIGURE 12.24: Output of RCNN after training. The objects are detected in different environments (different backgrounds). Each recognized object is provided a label and a bounding box.

TABLE 12.3: Experimental details for object recognition task using Faster-RCNN

System configuration	Training data size	Validation data size	Testing time	mAP
GPU NVIDIA Quadro M5000M	4800 samples	1200 samples	0.125 second	89.9%

The performance can be boosted if the size of the dataset is increased with a new set of images. Detailed information of the experimental setup are given in the Table 12.3. GPU system NVIDIA Quadro M5000M is used to train the Faster R-CNN VGG-16 model. Objects in an image are detected in just 0.125 second, which is in real-time. In order to compare the recognition performance of VGG-16, we trained and validated the given dataset using ZF model. Object recognition results using VGG-16 is observed to be slightly better than that of ZF model (mAP is 89.3% in case of ZF model). Average precision of individual objects for both the VGG-16 and the ZF model is shown in Figure 12.25.

12.5.4 Direction for Future Research

While the current system can carry out the picking and stowing tasks with reasonable accuracy and speed, a lot of work still needs to be done before it can be deployed in real world scenarios. Improving the system further forms

Experimental Results

TABLE 12.4: Mean average precision and per-class average precision

mAP	per-class average precision				
	barkely bones	bunny book	cherokee tshirt	clorox brush	cloud bear
	95.31	83.51	74.70	97.63	90.58
	command hooks	crayola 24 ct	creativity stems	dasani bottle	easter sippy cup
	93.52	90.57	73.65	91.21	91.13
89.9	elmers school glue	expo eraser	fitness dumbell	folgers coffee	glucose up bottle
	90.36	95.27	95.64	88.45	94.34
	jane dvd	jumbo pencil cup	kleenex towels	kygen puppies	laugh joke book
	95.43	96.53	81.24	84.35	93.41
	pencils	platinum bowl	rawlings baseball	safety plugs	scotch tape
	83.93	96.54	97.39	92.77	94.75
	staples cards	viva	white lightbulb	woods cord	
	90.84	81.46	87.62	85.01	

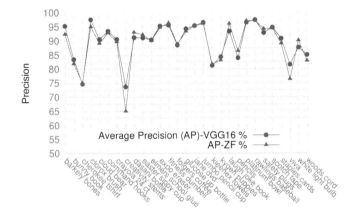

FIGURE 12.25: Plot showing average precision of individual objects obtained using Faster RCNN for both VGG-16 and ZF model.

the direction of our future work. Some of the ways of improving the system are as follows:

- The performance of the system relies on the performance of individual modules, particularly perception module for object recognition and grasping. One of the future directions would be to carry out research toward improving the performance of the perception module.

- One of the challenges of deep learning-based approaches for perception and grasping is the amount of samples required for training such models. Most of these training samples are created manually which is laborious and slow. One of our future directions would be to automate the process of data generation and explore deep learning models that can be trained incrementally [376] or through transfer learning [377] [378].

- The design of custom grippers that can pick all types of objects, including soft and deformable objects, still remains a challenge. Picking items from a closely packed stack of objects is another challenge which will be looked into as a part of our future research.

- The real-time performance of the system needs to be improved further without increasing the infrastructure cost. This can be done by parallelizing several modules, improving CPU utilization, and reducing network latency. The use of state machine based software architecture [379] such as ROS SMACH [380] [381] will be explored as a part of our future work.

- Even though the existing motion planning algorithms are quite mature, it is still not possible to deal with all possible cases of failure. One possible way to deal with these extreme cases would be to have a human in loop that intervenes only when a failure is reported by the system. The human operator can then teach the robot through demonstration [382] [383] to deal with such situations. The robot, in turn, can learn from such human inputs over a period of time to deal with such complex situations through methods such as deep reinforcement learning [384] [385] [386]. Some of these directions will be explored in the future.

- The real-world deployment of such systems will be explored through the use of cloud robotics platforms like Rapyuta [387].

12.6 Summary

This chapter presents the details of a robot-arm based automatic pick and place system for a retail warehouse. The two major tasks which are considered here include (1) *picking* items from a rack into a tote, and (2) *stowing* items from a tote to a rack. These two tasks are currently done manually by employing a large number of people in a big warehouse. This work was carried out as a part of our preparation for participating in the Amazon Picking Challenge 2016 event held in Leipzig, Germany. The problem is challenging from several perspectives. Identifying objects from visual images under conditions of varying illumination, occlusion, scaling, rotation, and change of shape (for deformable objects) is an open problem in the computer vision literature. This problem is solved in this work by the use of deep learning networks (RCNN) which gives reasonable accuracy suitable for real world operation. The average recognition accuracy is about $90 \pm 5\%$ for twenty-nine objects which is obtained by training the RCNN network using 4,800 training samples.

Part II

Mobile Robotics

13

Introduction to Mobile Robotics and Control

13.1 Introduction

As the name indicates, mobile robots have the capability to navigate from one place to another. They can be broadly classified as ground robots, surface water vehicles, underwater robots, and aerial robots. In this chapter we are focusing on ground robots. The ground robots can be classified as front wheel steering robots and differential-drive robots. In a front wheel steering robot, viz., an ordinary car type robot, the right and left wheels cannot be independently controlled. Hence, it can only be steered with a nonzero turning radius, depending on the physical dimensions of the robot. Whereas, in differential-drive robots, the left and right wheels can be independently controlled and they can do maneuvers with zero turning radius. Differential-drive robots can again be classified as holonomic and nonholonomic robots. In nonholonomic robots shown in Figure 13.1, lateral motion is not possible, and the nonholonomic constraints hold good for the same, whereas in holonomic vehicles, sideways motion is possible. In this chapter, we are dealing with two-wheel differential-drive nonholonomic mobile robots.

FIGURE 13.1: Differential-drive nonholonomic mobile robot.

13.2 System Model: Nonholonomic Mobile Robots

The nonlinear system model of a nonholonomic robot [388] is presented in this section. Using Figure 13.2, the mathematical model of the robot [389], [390] can be defined with respect to the center of gravity as follows:

$$\dot{x}_c = v \cos \theta \tag{13.1}$$
$$\dot{y}_c = v \sin \theta \tag{13.2}$$
$$\dot{\theta} = \omega \tag{13.3}$$
$$\dot{v} = F/m \tag{13.4}$$
$$\dot{\omega} = \tau/I \tag{13.5}$$

where $\mathbf{x} = [x, y]^T$ represents the position vector, θ denotes the turn angle; and v, ω, F, τ represent the velocity, turn rate, force, and torque respectively. Considering the off-axis point A as the handling point, as shown in Figure 13.2, the equivalent kinematic model [388, 391] of the differentially driven robot can be defined as

$$\begin{pmatrix} \dot{x} \\ \dot{y} \end{pmatrix} = \begin{pmatrix} \cos \theta & -L \sin \theta \\ \sin \theta & L \cos \theta \end{pmatrix} \begin{pmatrix} v \\ \omega \end{pmatrix} \tag{13.6}$$

Similarly, by differentiating (13.6), the dynamic model [388] of the robots can be obtained as

$$\begin{pmatrix} \ddot{x} \\ \ddot{y} \end{pmatrix} = \begin{pmatrix} -v\omega \sin \theta - L\omega^2 \cos \theta \\ v\omega \cos \theta - L\omega^2 \sin \theta \end{pmatrix} + \begin{pmatrix} \frac{1}{m} \cos \theta & -\frac{L}{I} \sin \theta \\ \frac{1}{m} \sin \theta & \frac{L}{I} \cos \theta \end{pmatrix} \begin{pmatrix} F \\ \tau \end{pmatrix} \tag{13.7}$$

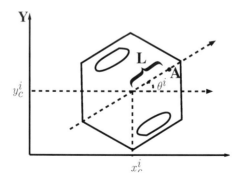

FIGURE 13.2: Differential-drive mobile robot.

13.3 Robot Attitude

For the navigation of the robots in outdoor environments, the position and orientation of them need to be monitored. The knowledge of coordinate frames and their transformations are extremely important in transforming the motion parameters from the robot fixed moving frame to a fixed frame or to another moving frame. Hence, this section deals with the coordinate frames and their transformations [392]. In the robot coordinate frame, $X-$ axis is aligned along the longitudinal axis of the robot, $Y-$axis along the lateral direction, and the $Z-$axis aligned along the upward direction to complete a right-handed co-ordinate system. The rotation about the longitudinal axis (X), lateral axis (Y), and the Z-axis are termed as roll (R), Pitch (P), and yaw (Y) respectively. In all the cases, the rotation in the anti-clockwise direction is taken as positive.

13.3.1 Rotation about Roll Axis

The frame rotation about the roll axis is shown in Figure 13.3, where the frame is rotated about the x-axis by an angle θ. In this case, (X_1, Y_1, Z_1) represents the coordinates in the original frame, and (X_2, Y_2, Z_2) represents coordinates in the rotated frame. The location of any arbitrary point as marked by "X" in the rotated frame can be expressed [392] in terms of the original frame as follows:

$$X_2 = X_1$$

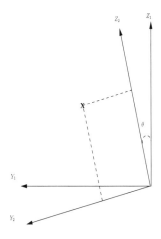

FIGURE 13.3: Rotation about X-axis.

$$Y_2 = Y_1 \cos\theta - Z_1 \sin\theta \qquad (13.8)$$
$$Z_2 = Y_1 \sin\theta + Z_1 \cos\theta$$

$$\begin{bmatrix} X_2 \\ Y_2 \\ Z_2 \end{bmatrix} = \begin{bmatrix} 1 & 0 & 0 \\ 0 & \cos\theta & -\sin\theta \\ 0 & \sin\theta & \cos\theta \end{bmatrix} \begin{bmatrix} X_1 \\ Y_1 \\ Z_1 \end{bmatrix} \qquad (13.9)$$

The rotation matrix for roll can be expressed as

$$R(\theta) = \begin{bmatrix} 1 & 0 & 0 \\ 0 & \cos\theta & -\sin\theta \\ 0 & \sin\theta & \cos\theta \end{bmatrix} \qquad (13.10)$$

Example 13.1. Consider a vector $x = \begin{bmatrix} 1 & 2 & 4 \end{bmatrix}^T$ defined in robot coordinate frame. Find the coordinates of the vector in a frame rotated by $\theta = \pi/3$ about the roll axis of the original frame.

Solution 13.1. The vector can be expressed in rotated frame as

$$\begin{bmatrix} X_2 \\ Y_2 \\ Z_2 \end{bmatrix} = \begin{bmatrix} 1 & 0 & 0 \\ 0 & \cos(\pi/3) & -\sin(\pi/3) \\ 0 & \sin(\pi/3) & \cos(\pi/3) \end{bmatrix} \begin{bmatrix} 1 \\ 2 \\ 4 \end{bmatrix}$$
$$= \begin{bmatrix} 1 \\ 2.464 \\ 3.732 \end{bmatrix}$$

13.3.2 Rotation about Pitch Axis

The rotation matrix for the frame rotation about Y-axis can be derived in the similar manner [392]. The respective rotation of the frame about the pitch axis is shown in Figure 13.5 The location of any arbitrary point in the rotated frame-2 can be expressed in terms of the original frame-1 as follows:

$$X_2 = X_1 \cos\psi - Z_1 \sin\psi$$
$$Y_2 = Y_1$$
$$Z_2 = -X_1 \sin\psi + Z_1 \cos\psi \qquad (13.11)$$

$$\begin{bmatrix} X_2 \\ Y_2 \\ Z_2 \end{bmatrix} = \begin{bmatrix} \cos\psi & 0 & -\sin\psi \\ 0 & 1 & 0 \\ -\sin\psi & 0 & \cos\psi \end{bmatrix} \begin{bmatrix} X_1 \\ Y_1 \\ Z_1 \end{bmatrix} \qquad (13.12)$$

The rotation matrix for Pitch can be expressed as

$$P(\psi) = \begin{bmatrix} \cos\psi & 0 & -\sin\psi \\ 0 & 1 & 0 \\ -\sin\psi & 0 & \cos\psi \end{bmatrix} \qquad (13.13)$$

Robot Attitude

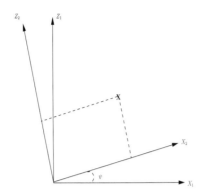

FIGURE 13.4: Rotation about Y-axis.

Example 13.2. Consider a vector $x = \begin{bmatrix} 1 & 2 & 4 \end{bmatrix}^T$ defined in a rotated frame, which has undergone a positive pitch rotation about the original robot frame, by an amount, $\psi = 2\pi/5$. Find the coordinates of the vector in the original robot coordinate frame.

Solution 13.2. The vector can be expressed in rotated frame as

$$\begin{bmatrix} X_1 \\ Y_1 \\ Z_1 \end{bmatrix} = \begin{bmatrix} \cos(2\pi/5) & 0 & -\sin(2\pi/5) \\ 0 & 1 & 0 \\ -\sin(2\pi/5) & 0 & \cos(2\pi/5) \end{bmatrix}^{-1} \begin{bmatrix} 1 \\ 2 \\ 4 \end{bmatrix}$$

$$= \begin{bmatrix} 1 & 0 & -1.73 \\ 0 & 1 & 0 \\ -1.73 & 0 & 1 \end{bmatrix} \begin{bmatrix} 1 \\ 2 \\ 4 \end{bmatrix}$$

$$= \begin{bmatrix} -5.92 \\ 2 \\ 2.27 \end{bmatrix}$$

13.3.3 Rotation About Yaw Axis

The respective rotation of the frame about the yaw axis is shown in Figure 13.5 The location of any arbitrary point in the rotated frame-2 can be expressed in terms of the original frame-1 as follows [392]:

$$X_2 = X_1 \cos \phi - Y_1 \sin \phi$$
$$Y_2 = X_1 \sin \phi + Y_1 \cos \phi$$
$$Z_2 = Z_1 \qquad (13.14)$$

$$\begin{bmatrix} X_2 \\ Y_2 \\ Z_2 \end{bmatrix} = \begin{bmatrix} \cos \phi & -\sin \phi & 0 \\ \sin \phi & \cos \phi & 0 \\ 0 & 0 & 1 \end{bmatrix} \begin{bmatrix} X_1 \\ Y_1 \\ Z_1 \end{bmatrix} \qquad (13.15)$$

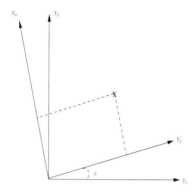

FIGURE 13.5: Rotation about Z-axis.

The rotation matrix for the rotation about the yaw axis can be expressed as

$$Y(\phi) = \begin{bmatrix} \cos\phi & -\sin\phi & 0 \\ \sin\phi & \cos\phi & 0 \\ 0 & 0 & 1 \end{bmatrix} \quad (13.16)$$

13.4 Composite Rotation

Consider a robot frame undergoing composite rotation, with roll followed by yaw and pitch. Then, the location of any point in the rotated frame can be obtained from the coordinates of the original frame as follows [392]:

$$\begin{bmatrix} X_2 \\ Y_2 \\ Z_2 \end{bmatrix} = P(\psi)Y(\phi)R(\theta) \begin{bmatrix} X_1 \\ Y_1 \\ Z_1 \end{bmatrix} \quad (13.17)$$

In this case, roll was the first rotation. Hence, it is first transformation matrix to operate on the coordinates, followed by yaw and pitch.

Sometimes, the situation arises where the frame is subjected to rotation as well as translation. Consider the case in which the the frame-B is rotated and displaced with respect to frame-A. Hence, $\mathbf{P}_T = [X_0, Y_0, Z_0]^T$ represents the location of origin of frame-A relative to frame-B. The coordinates of any point in frame-A can be transformed to frame-B using the following transformation.

Coordinate System

$$\begin{bmatrix} X_B \\ Y_B \\ Z_B \end{bmatrix} = R_{AB}(\theta, \psi, \phi) \begin{bmatrix} X_A \\ Y_A \\ Z_A \end{bmatrix} + \begin{bmatrix} X_0 \\ Y_0 \\ Z_0 \end{bmatrix} \qquad (13.18)$$

Example 13.3. Consider a vector $x = \begin{bmatrix} 1 & 2 & 4 \end{bmatrix}^T$ defined in robot coordinate frame.

Find the coordinates of the vector in a frame, which is rotated by $2\pi/3$ rad about the pitch axis, $\pi/5$ rad about the roll axis and $\pi/3$ rad about the yaw axis, and if the coordinates of the origin of the original frame with respect to the rotated frame be $\mathbf{P}^T = [10, 20, 30]^T$.

Solution 13.3. The vector can be expressed in rotated frame, B, as

$$\begin{bmatrix} X_B \\ Y_B \\ Z_B \end{bmatrix} = Y(\phi) R(\theta) P(\psi) \begin{bmatrix} X_A \\ Y_A \\ Z_A \end{bmatrix} + \begin{bmatrix} X_0 \\ Y_0 \\ Z_0 \end{bmatrix}$$

$$= \begin{bmatrix} c_{\frac{\pi}{3}} c_{\frac{2\pi}{3}} - s_{\frac{\pi}{3}} s_{\frac{\pi}{5}} s_{\frac{2\pi}{3}} & -c_{\frac{\pi}{5}} s_{\frac{\pi}{3}} & -c_{\frac{\pi}{3}} s_{\frac{2\pi}{3}} + s_{\frac{\pi}{3}} s_{\frac{\pi}{5}} c_{\frac{2\pi}{3}} \\ s_{\frac{\pi}{3}} c_{\frac{2\pi}{3}} + s_{\frac{\pi}{5}} c_{\frac{\pi}{3}} s_{\frac{2\pi}{3}} & c_{\frac{\pi}{5}} c_{\frac{\pi}{3}} & -s_{\frac{\pi}{3}} s_{\frac{2\pi}{3}} - s_{\frac{\pi}{5}} c_{\frac{\pi}{3}} c_{\frac{2\pi}{3}} \\ -c_{\frac{\pi}{5}} s_{\frac{2\pi}{3}} & s_{\frac{\pi}{5}} & c_{\frac{\pi}{5}} c_{\frac{2\pi}{3}} \end{bmatrix} \begin{bmatrix} 1 \\ 2 \\ 4 \end{bmatrix}$$

$$+ \begin{bmatrix} 10 \\ 20 \\ 30 \end{bmatrix}$$

$$= \begin{bmatrix} 5.162 \\ 18.218 \\ 28.858 \end{bmatrix}$$

13.5 Coordinate System

The knowledge of coordinate system is also equally significant in extracting the navigation data of the robot with respect to a known reference. This section gives a brief outline on Earth-Centered Earth-Fixed co-ordinate systems, one of the most commonly used coordinate systems in robot navigation.

13.5.1 Earth-Centered Earth-Fixed (ECEF) Co-ordinate System

In a ECEF coordinate system [392] as shown in Figure 13.6, the positive X-axis passes through the intersection of the equator with the prime meridian in Greenwich, Z-axis is aligned along the North pole, and Y-axis complete the right-handed coordinate system aligned along the equator, with the origin fixed at the center of Earth. For any arbitrary point on the surface of the Earth at a radial distance of r from the center, the positive longitude is measured

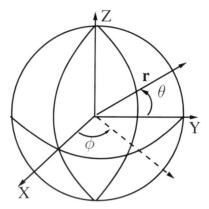

FIGURE 13.6: Earth-Centered Earth-Fixed (ECEF) co-ordinate system.

from the prime meridian in Greenwich in the Eastward direction, and the positive latitude is measured from the equator toward the positive Z-axis of the ECEF frame. The ECEF coordinates [392] can be expressed in terms of latitude (θ) and longitude(ϕ) as follows:

$$X = r \cos\theta \cos\phi$$
$$Y = r \cos\theta \sin\phi \qquad (13.19)$$
$$Z = r \sin\theta$$
$$(13.20)$$

The coordinates in any Earth-centered coordinate frame can be transformed to ECEF frame in terms of longitude and latitude using respective rotation matrices.

13.6 Control Approaches

A robust controller is needed for the smooth navigation of a robot tracking the desired trajectory in a cluttered environment, subjected to external disturbances. The next sections deal with various types of nonlinear control techniques, which can be applied to the wheeled mobile robots.

13.6.1 Feedback Linearization

A feedback linearization-based control law can make the closed loop dynamics of any system linear, by canceling out the nonlinearities in the system [191, 393].

Consider the system dynamics in input affine form

$$\dot{\mathbf{x}} = \mathbf{f}(\mathbf{x}) + \mathbf{g}(\mathbf{x})\mathbf{u} \qquad (13.21)$$

where $\mathbf{x} \in \Re^n$ is the state vector, $\mathbf{u} \in \Re^m$ is the control input, and $\mathbf{g}(\mathbf{x})$ is nonsingular. For the system states to track the reference trajectory given by $\mathbf{x}_d$, the feedback linearization (FL) based tracking control law can be designed as,

$$\mathbf{u} = \frac{1}{\mathbf{g}(\mathbf{x})}(-\mathbf{f}(\mathbf{x}) + \mathbf{K}\mathbf{e} + \dot{\mathbf{x}}_d) \qquad (13.22)$$

where $\mathbf{e} = \mathbf{x}_d - \mathbf{x}$ is the tracking error, $\mathbf{K} = \text{diag}\{k_1, k_2\}$, $k_1, k_2 > 0$ are the controller gains. The gains can be precomputed offline using trial and error method. As we plug in the FL-based control law in the system dynamics given by (13.21), the closed loop error dynamics becomes $\dot{\mathbf{e}} = -\mathbf{K}\mathbf{e}$, which is linear and stable.

Example 13.4. Consider a two-wheel differential-drive mobile robot with the kinematic model given by (13.6). Design and implement a feedback linearization-based stable controller for the robot, starting from $[0.5, 0.5]^T$ in the reference frame, to track a circular reference trajectory of radius 1 m centered at the origin, with a desired linear velocity (v) of 0.1 m/s.

Solution 13.4. The kinematic model of the robot can be defined as $\dot{\mathbf{x}} = \mathbf{g}(\mathbf{x})\mathbf{u}$, where $\mathbf{g}(\mathbf{x}) = \begin{pmatrix} \cos\theta & -L\sin\theta \\ \sin\theta & L\cos\theta \end{pmatrix}$. $\mathbf{g}(\mathbf{x})$ is nonsingular, since $|\mathbf{g}(\mathbf{x})| \neq 0$.

Based on the desired specifications, the required angular velocity can be obtained as $\omega = 0.1/1 = 0.1$ rad/s, and the reference trajectories are derived as $\mathbf{x}_d = [\cos 0.1t, \sin 0.1t]^T$ and $\dot{\mathbf{x}}_d = [-0.1\sin 0.1t, 0.1\cos 0.1t]^T$. The tracking error is defined as $\mathbf{e} = \mathbf{x}_d - \mathbf{x}$.

The control input, $\mathbf{u} = [v, \omega]^T$, can be designed as

$$\mathbf{u} = \frac{1}{\mathbf{g}(\mathbf{x})}(\mathbf{K}\mathbf{e} + \dot{\mathbf{x}}_d) \qquad (13.23)$$

$$= \begin{bmatrix} \cos\theta & \sin\theta \\ -\frac{\sin\theta}{L} & \frac{\cos\theta}{L} \end{bmatrix} \left(\begin{bmatrix} k_1 & 0 \\ 0 & k_2 \end{bmatrix} \begin{bmatrix} x_d - x \\ y_d - y \end{bmatrix} + \begin{bmatrix} \dot{x}_d \\ \dot{y}_d \end{bmatrix} \right) \qquad (13.24)$$

The controller can be realized through simulations, and the results are given in Figure 13.7-13.8. The state as well as control trajectories indicate that the robot starting at $\mathbf{x} = [0.5, 0.5]^T$ is making a circular trajectory of 1 m radius centered at the origin, with an angular velocity of 0.1 rad/s and linear velocity of 0.1 m/s.

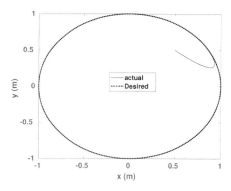

FIGURE 13.7: Feedback linearization: x-y positions.

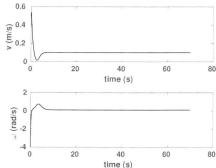

FIGURE 13.8: Feedback linearization: Control inputs.

```
MATLAB CODE
-----------
clear all;
x=0.5; y=0.5; v=0; omega=0; L=0.1; n=7000;
T=.01;t=0;theta=0;k1=1;k2=1;
for i=1:n
    t=t+T; x_d=cos(.1*t); y_d=sin(.1*t);
    xddot=-0.1*sin(.1*t); yddot= 0.1*cos(.1*t);
    xdot=v*cos(theta)-L*omega*sin(theta);
    ydot=v*sin(theta)+L*omega*cos(theta);
    x=x+T*xdot; y=y+T*ydot; ex=x-x_d; ey=y-y_d;
    v=-(k1)*cos(theta)*ex+(xddot)*cos(theta)-
    (k2)*sin(theta)*ey+(yddot)*sin(theta);
    omega=(k1/L)*sin(theta)*ex-(xddot/L)*sin(theta)
    -(k2/L)*cos(theta)*ey+(yddot/L)*cos(theta);
    theta=theta+omega*T; x_save(i)=x; y_save(i)=y;
xd_save(i)=x_d; yd_save(i)=y_d;v_save(i)=v; omega_save(i)=
    ↪ omega;
end
    time=T*(1:n);
 figure
 plot(x_save,y_save); hold on
 plot(xd_save,yd_save);
 figure
 subplot(2,1,1); plot(time(1:n),v_save(1:n));
 subplot(2,1,2); plot(time(1:n),omega_save);
```

13.6.2 Backstepping

Backstepping is another control technique widely used in nonlinear systems [191, 393], in which the controller is having a recursive structure. Unlike feedback linearization, in back stepping approach, it is not essential to make the closed loop dynamics linear. Hence, it doesn't require the cancellation of all useful nonlinearities in the system. In this method, by considering the state variables as the virtual control inputs, stabilizing control laws are derived for each intermediate subsystems in a progressive fashion. And, this process continues until the external control input is reached. It is suited for systems with a strict feedback form given as follows:

$$\dot{x}_1 = f_1(x_1) + g_1(x_1)x_2 \quad (13.25)$$
$$\dot{x}_2 = f_2(x_1, x_2) + g_2(x_1, x_2)x_3 \quad (13.26)$$
$$\vdots \quad (13.27)$$
$$\dot{x}_n = f_n(x_1, x_2 \ldots x_n) + g_2(x_1, x_2 \ldots x_n)u \quad (13.28)$$

where $x_i \in \Re^m$, $i = 1\ldots n$, and $u \in \Re^m$ denote the states and the control inputs of the system; $f_i \in \Re^m X m$, $i = 1\ldots n$ are nonlinear functions.

In this case, by considering x_2 as the virtual control input for the subsystem given by (13.25), a stabilizing control law can be derived to ensure that x_1 tracks x_{1d}. This virtual control law acts as the reference input, x_{2d}, for the second subsystem, where the virtual control law is derived for x_3, to ensure that x_2 tracks x_{2d}. This process continues in a recursive manner until a stabilizing control law is derived for the actual external control input, u.

Example 13.5. Consider a two-wheel differential-drive mobile robot with the dynamic model given by (13.7). Design and implement a backstepping-based stable controller for the robot starting from $[0, 0.5]^T$ of the reference frame to track a lemniscates shape trajectory about the point $[0.5, 0]^T$, with a lobe length of 0.5 m, and a desired linear velocity (v) of 0.1 m/s.

Solution 13.5. *For the lemniscates shaped desired trajectories with the given specifications, the reference inputs can be obtained as*
$\mathbf{x}_{1d} = [0.5 + 0.5\sin(\pi t/50), 0.5\sin(2\pi t/50)]^T$.

Consider $\mathbf{x}_1 = \mathbf{x}$, *and* $\mathbf{x}_2 = \dot{\mathbf{x}}$. *The dynamic model given by (13.7) can be expressed in strict feedback form given by (13.25)-(13.28) as*

$$\dot{\mathbf{x}}_1 = \mathbf{f}_1(x_1) + \mathbf{g}_1(\mathbf{x}_1)\mathbf{x}_2 \quad (13.29)$$
$$\dot{\mathbf{x}}_2 = \mathbf{f}_2(\mathbf{x}_1, \mathbf{x}_2) + \mathbf{g}_2(\mathbf{x}_1, \mathbf{x}_2)\mathbf{u} \quad (13.30)$$

where $\mathbf{f}_1 = \mathbf{0}_{2\times 2}$, $\mathbf{g}_1 = I_{2\times 2}$,
$\mathbf{f}_2 = \begin{pmatrix} -v\omega\sin\theta - L\omega^2\cos\theta \\ v\omega\cos\theta - L\omega^2\sin\theta \end{pmatrix}$, $\mathbf{g}_2 = \begin{pmatrix} \frac{1}{m}\cos\theta & -\frac{L}{I}\sin\theta \\ \frac{1}{m}\sin\theta & \frac{L}{I}\cos\theta \end{pmatrix}$;
and $\mathbf{u} = \begin{pmatrix} F \\ \tau \end{pmatrix}$

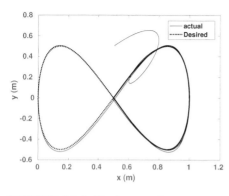
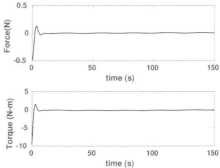

FIGURE 13.9: Back stepping: x-y positions.

FIGURE 13.10: Backstepping: Control inputs.

For the first subsystem given by (13.29), considering $\mathbf{x}_2$ as the virtual control input, the tracking error can be defined as, $\mathbf{e}_1 = \mathbf{x}_1 - \mathbf{x}_{1d}$, the stabilizing control law is designed as $x_2 = -k_1 \mathbf{e} + \dot{\mathbf{x}}_{1d}$. This virtual control law acts as the reference input, x_{2d}, for the second subsystem.

Hence, the control input can be designed as

$$\mathbf{u} = \frac{1}{g_2(\mathbf{x}_1, \mathbf{x}_2)}(-\mathbf{f}_2(\mathbf{x}_1, \mathbf{x}_2) - k_1 \dot{\mathbf{e}} - k_2 \mathbf{e}) \qquad (13.31)$$

The results are given in Figure 13.9-13.10.

13.6.3 Sliding Mode Control

Sliding mode control is a robust nonlinear technique [393] widely used in robotic applications.

Consider a single input nonlinear system,

$$\dot{\mathbf{x}} = f(\mathbf{x}, u, t), \qquad (13.32)$$

in which the structure of the system varies according to the switching logic [394],

$$u = \begin{cases} u^+(t), & \text{if, } S(x) > 0, \\ u^-(t), & \text{if, } S(x) < 0 \end{cases} \qquad (13.33)$$

where $\mathbf{x} \in \Re^n$ is the state vector, and u is the control input; $S(x)$ represents the switching surface, and the motion along $\mathbf{S}(x) = 0$ is termed as sliding mode. The system trajectories can be directed to $S(x) = 0$, and it can be retained there for subsequent time, if the following conditions hold:

$$\lim_{S \to -0} \dot{S} > 0; \quad \lim_{S \to +0} \dot{S} < 0 \qquad (13.34)$$

Control Approaches

so that $S\dot{S} \leq 0$, ie. the distance to the switching surface, and $\dot{S}$ should have opposite signs [394]. Hence, the solution of the equations

$$\dot{x} = f(\mathbf{x}, u^+, t),\ S > 0;\ \text{and}\ \dot{x} = f(\mathbf{x}, u^-, t),\ S < 0$$

approach $S(x) = 0$ in finite time [395]. The motion of the trajectories along the sliding surface is the average of the dynamics on both sides of the switching surface.

For an nth order tracking problem in $\mathbf{x}$, the equation for sliding surface can be generalized as

$$S(t) = \left(\frac{d}{dt} + \lambda\right)^{n-1} \tilde{x} \qquad (13.35)$$

where $\tilde{x} = \mathbf{x} - \mathbf{x}_d$, $\tilde{x} \in \Re^n$, and $\mathbf{x}_d$ is the reference state vector; and λ is a positive constant. From (13.35), one can find that fulfilling the control objective is equivalent to retaining the system trajectory on the sliding surface; and the nth order tracking problem is converted to a first order stabilization problem in $S(t)$ [393].

This can be achieved by designing the control law such that following condition holds [394]

$$\frac{1}{2}\frac{d}{dt}S(t)^2 \leq -\eta|S(t)| \quad \text{or} \quad S(t)\dot{S}(t) \leq -\eta|S(t)| \qquad (13.36)$$

This is known as reaching condition. The time taken by the trajectories to reach the sliding surface is known as reaching time, and this phase is termed as reaching phase. The motion of the trajectories along the sliding surface is termed as sliding mode.

Consider a Lyapunov candidate

$$V = \frac{1}{2}S(t)^2 \qquad (13.37)$$

Taking the derivative,

$$\dot{V} = S(t)\dot{S}(t)$$

If the control law is designed such that the following sliding condition is satisfied,

$$S(t)\dot{S}(t) \leq -\eta|S(t)| \qquad (13.38)$$

then, $\quad \dot{V} \leq -\eta|S(t)|$

Since $\dot{V}$ is negative definite and the system is asymptotically stable. The sliding condition ensures the asymptotic stability of the system. That is, the system trajectories can converge to the sliding surface from any initial condition asymptotically; and the reaching time (t_r) can be obtained as

$$t_r \leq |S(t=0)|/\eta \qquad (13.39)$$

Chattering is one of disadvantage of SMC [393, 394]. This is the condition in which high frequency oscillations appear in the control signal due to the imperfections in the associated control switching, owing to disturbances/ model uncertainties. Numerous chattering reduction techniques are available in literature, viz. fuzzy sliding mode technique, boundary layer technique and hyperbolic tangent function approach. Different variants of SMC are available in literature. A few of them are detailed in the following subsections.

13.6.4 Conventional SMC

Consider the nonlinear system,

$$\ddot{x} = \mathbf{f}(\mathbf{x}) + \mathbf{g}(\mathbf{x})\mathbf{u}$$

where $\mathbf{x} \in \Re^n$, $\mathbf{u} \in \Re^m$ represent the state and control input vectors respectively.

The tracking error can be defined as

$$\tilde{x} = x - x_d$$

For the given system, the conventional type sliding surface [393] can be defined in terms of the tracking error as,

$$S = \dot{\tilde{x}} + \lambda \tilde{x}$$

To satisfy the sliding condition, the reaching law can be chosen as

$$\dot{S} = -K sign(S)$$

and the control input can be designed as

$$\mathbf{u} = \frac{1}{\mathbf{g}(\mathbf{x})}[-\mathbf{f}(\mathbf{x}) + \ddot{\mathbf{x}}_\mathbf{d} - \lambda \dot{\tilde{\mathbf{x}}} - \mathbf{K sign(S)}] \tag{13.40}$$

In this case, the finite time convergence cannot be theoretically proved. Moreover, the chattering phenomenon is also prevalent in this type of design. Finite-time convergent SMC strategies are detailed in next sections.

Example 13.6. For a two-wheel differential-drive mobile robot with the dynamic model given by (13.7), design a sliding mode controller for the robot based on conventional SMC technique, to track a circular reference trajectory of radius 1 m centered at the origin, with a velocity of 0.2 m/s.

Solution 13.6. *Based on the desired specifications, the reference trajectories can be defined as* $\mathbf{x}_d = [\cos 0.1t, \sin 0.1t]^T$ *and* $\dot{\mathbf{x}}_d = [-0.1 \sin 0.1t, 0.1 \cos 0.1t]^T$.

The tracking error is defined as $\mathbf{e} = \mathbf{x}_d - \mathbf{x}$. *The sliding surface is given by*

$$\mathbf{S} = \dot{\tilde{\mathbf{x}}} + \lambda \tilde{\mathbf{x}}$$

Control Approaches

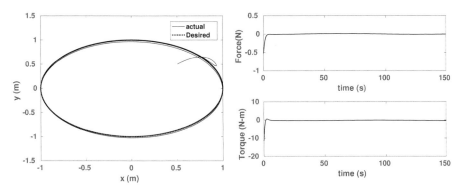

FIGURE 13.11: Conventional SMC: x-y positions.

FIGURE 13.12: Conventional SMC: Control inputs.

The control input is obtained as

$$\mathbf{u} = \frac{1}{\mathbf{g}(\mathbf{x})}[-\mathbf{f}(\mathbf{x}) + \ddot{\mathbf{x}}_d - \lambda \dot{\tilde{\mathbf{x}}} - \mathbf{K}\text{sign}(\mathbf{S})]$$

where

$$\mathbf{f}(\mathbf{x}) = \begin{pmatrix} -v\omega\sin\theta - L\omega^2\cos\theta \\ v\omega\cos\theta - L\omega^2\sin\theta \end{pmatrix}, \mathbf{g}(\mathbf{x}) = \begin{pmatrix} \frac{1}{m}\cos\theta & -\frac{L}{I}\sin\theta \\ \frac{1}{m}\sin\theta & \frac{L}{I}\cos\theta \end{pmatrix} \quad (13.41)$$

and $\mathbf{u} = \begin{pmatrix} F \\ \tau \end{pmatrix}$ The controller gains, $\mathbf{K} = diag\{1,1\}$, and $\lambda = 1$ are tuned using trial and error method. To reduce chattering inherent in SMC, the signum function in the control law can be replaced by a hyperbolic tangent function. The results are given in Figure 13.11-13.12

13.6.5 Terminal SMC

In a terminal sliding mode control [396, 397], nonlinear sliding surfaces are employed, in which finite time convergence can be guaranteed.

The sliding surface is defined as

$$S = \dot{\tilde{x}} + \lambda \tilde{x}^{\frac{p}{q}}, \quad q > p > 0 \quad (13.42)$$

The settling time is given by

$$T_{tsm} = \frac{q}{\lambda(q-p)}|\tilde{x}(0)|^{\frac{q-p}{q}}$$

The control input is given by

$$u = \frac{1}{\mathbf{g}(\mathbf{x})}[-\mathbf{f}(\mathbf{x}) + \ddot{x}_d - \left(\frac{\lambda p}{q}\right)\tilde{x}^{\frac{(p-q)}{q}}\dot{\tilde{x}} - K sign(S)] \quad (13.43)$$

Even though, finite time convergence can be guaranteed for TSMC, when $\tilde{x} = 0$, singularity will appear inside the control law, since $q > p$. Nonsingular type terminal sliding surfaces can be utilized to avoid this situation.

13.6.6 Nonsingular TSMC (NTSMC)

In NTSMC [398–400], the nonlinear sliding surface and the parameters are chosen such that for any given operating condition, the singularity won't appear in the control law. For the sliding surface given by

$$S = \tilde{x} + \lambda |\dot{\tilde{x}}|^\alpha sign(\dot{\tilde{x}}), \ \lambda > 0, \ 1 < \alpha < 2, \quad (13.44)$$

the control law can be obtained as

$$u = \frac{1}{\mathbf{g(x)}}[-\mathbf{f(x)} + \ddot{x}_d - (1/\lambda\alpha)|\dot{\tilde{x}}|^{(2-\alpha)} sign(\dot{\tilde{x}}) - K sign(S)] \quad (13.45)$$

Since $1 < \alpha < 2$, from the control law, it is evident that the controller is free from singularity issue, even if the tracking error/ error derivative goes to zero.

The settling time is given by

$$T_{ntsm} = \frac{\lambda^{\frac{1}{\alpha}} |\tilde{x}(0)|^{1-1/\alpha}}{1 - 1/\alpha}$$

Speed of convergence is also equally significant for a robust controller. For faster convergence, fast TSMC techniques are available in literature.

Example 13.7. Design a nonsingular terminal sliding mode controller to drive a mobile robot through a sinusoidal trajectory with a velocity of .2 m/s using the dynamic model given by (13.7).

Solution 13.7. *Based on the desired specifications, the reference trajectories can be defined as* $\mathbf{y}_d = \sin(0.1 x_d)$

The tracking error is defined as $\mathbf{e} = \mathbf{x}_d - \mathbf{x}$. *The sliding surface is given by*

$$S = \tilde{x} + \lambda |\dot{\tilde{x}}|^\alpha sign(\dot{\tilde{x}})$$

Based on the dynamic model given by (13.7), the control input is obtained as

$$u = \frac{1}{\mathbf{g(x)}}[-\mathbf{f(x)} + \ddot{x}_d - (1/\lambda\alpha)|\dot{\tilde{x}}|^{(2-\alpha)} sign(\dot{\tilde{x}}) - K sign(S)]$$

The parameters can be chosen as follows: $\mathbf{K} = diag\{1, 1\}$, $\alpha = 0.5$ *and* $\lambda = 1$. *To reduce chattering inherent in SMC, the signum function in the control law can be replaced by a hyperbolic tangent function. The trajectories are shown in Figure 13.13-13.14.*

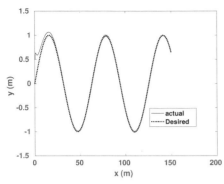

FIGURE 13.13: NTSMC: x-y positions.

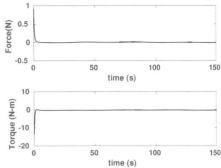

FIGURE 13.14: NTSMC: Control inputs.

13.6.7 Fast Nonsingular TSMC (FNTSMC)

In FNTSMC [399, 400], a sliding surface is designed such that it can ensure fast and finite time convergence, avoiding singularity, and is given by

$$S = \tilde{x} + \lambda_1 |\tilde{x}|^{\gamma_1} sign(\tilde{x}) + \lambda_2 |\dot{\tilde{x}}|^{\gamma_2} sign(\dot{\tilde{x}}) \tag{13.46}$$

where $1 < \gamma_1$, $\gamma_2 < 2$, λ_1, $\lambda_2 > 0$.

The control law is given by

$$u = \frac{1}{\mathbf{g}(\mathbf{x})}[-\mathbf{f}(\mathbf{x}) - \frac{1}{\lambda_2 \gamma_2}|\dot{\tilde{x}}|^{2-\gamma_2} sign(\dot{\tilde{x}})(1 + \lambda_1 \gamma_1 |\tilde{x}|^{\gamma_1 - 1}) - K sign(S)] \tag{13.47}$$

The settling time can be computed as

$$T_{fntsm} = \frac{\eta_2 |\tilde{x}(0)|^{1-1/\gamma_2}}{\lambda_1(\gamma_2 - 1)} \cdot f(\frac{1}{\gamma_2}, \frac{\gamma_2 - 1}{(\gamma_1 - 1)\gamma_2}; 1 + \frac{\gamma_2 - 1}{(\gamma_1 - 1)\gamma_2}; -\lambda_1 |\tilde{x}(0)|^{\gamma_1 - 1})$$

$f(a, b; c; x)$ represents a convergent Gauss' hypergeometric function [399]. From this, one can easily prove that $T_{fntsm} < T_{ntsm}$.

Example 13.8. Design and implement a stable controller based on fast nonsingular terminal SMC technique for the robot starting from $[0, 0.5]^T$ of the reference frame to track a lemniscates shape trajectory about the point $[0.5, 0]^T$, with a lobe length of 0.5 m, and a desired linear velocity (v) of 0.1 m/s.

Solution 13.8. *For the lemniscates shaped desired trajectories with the given specifications, the reference inputs can be obtained as*

$\mathbf{x}_{1d} = [0.5 + 0.5 \sin(\pi t/50), 0.5 \sin(2\pi t/50)]^T$. *Based on the dynamic model given by* (13.7), *the control input is obtained as*

$$u = \frac{1}{\mathbf{g}(\mathbf{x})}[-\mathbf{f}(\mathbf{x}) - \frac{1}{\lambda_2 \gamma_2}|\dot{\tilde{x}}|^{2-\gamma_2} sign(\dot{\tilde{x}})(1 + \lambda_1 \gamma_1 |\tilde{x}|^{\gamma_1 - 1}) - K sign(S)] \tag{13.48}$$

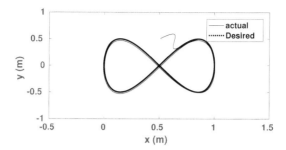

FIGURE 13.15: FNTSMC: x-y positions.

where $\mathbf{f}(\mathbf{x})$ and $\mathbf{g}(\mathbf{x})$ are defined by (13.41).

The parameters can be chosen as follows: $\lambda_1 = \lambda_2 = diag\{0.8, 0.8\}$, $\gamma_1 = 1.006$, $\gamma_2 = 1.003$. A hyperbolic tangent function-based reaching law is also used to mitigate the chattering. The results are shown in Figure 13.15.

13.6.8 Fractional Order SMC (FOSMC)

The dynamics of a system can be represented more precisely in fractional order mode. This is the main motivation for the development of FOSMC [401–403]. In this approach, the system dynamics is forced to follow a fractional order sliding dynamics.

Consider a fractional order sliding surface given by

$$S = \lambda \tilde{x} + D^{\alpha-1} \dot{\tilde{x}} \tag{13.49}$$

where $D^{\alpha-1}$ represents the fractional differential operator, and α is a fraction.

The control input can be computed as

$$u = \frac{1}{\mathbf{g}(\mathbf{x})}[(D^{1-\alpha}(\lambda \dot{\tilde{x}})) - \mathbf{f}(\mathbf{x}) + \ddot{x}_d - K sign(S)] \tag{13.50}$$

Though, it can add robustness to the system, it may increase the complexity of the design. Chattering is found to be slightly less with this architecture. The fractional derivative can be computed using any one of the following three methods: Grunwald-Letnikov approximation/ Riemann-Liouville approximation / Caputo approximation.

Grunwald-Letnikov Approximation

The αth order fractional derivative of a function f(x) can be computed as [404]

$$D^\alpha f(x) = Lim_{\Delta t \to 0} \frac{1}{\Delta t^\alpha} \sum_0^{(x-\beta)/\Delta t} (-1)^j \begin{bmatrix} \alpha \\ j \end{bmatrix} f(x - j\Delta t) \tag{13.51}$$

in which $\begin{bmatrix} \alpha \\ j \end{bmatrix}$ represents the binomial coefficient, β is the the initial value, and Δt represents the sampling time.

Riemann-Liouville Approximation

Using Riemann-Liouville Approximation, the fractional derivative approximation for a function f(x) is given by [404]

$$D^\alpha f(x) = \frac{1}{\Gamma(n-\alpha)} \frac{d^n}{dx^n} \left(\int_\beta^x (x-t)^{(n-\alpha-1)} f(t) dt \right) \quad (13.52)$$

where Γ is the Gama function, β is the initial value, and n is the nearest integer, which is greater than α.

Caputo Approximation

The Caputo approximation for the αth order fractional derivative of a function f(x) is given by [404]

$$D^\alpha f(x) = \frac{1}{\Gamma(n-\alpha)} \left(\int_\beta^x (x-t)^{(n-\alpha-1)} f^{(n)}(t) dt \right) \quad (13.53)$$

13.6.9 Higher Order SMC (HOSMC)

Contrary to the above mentioned approaches, A. Levant [405, 406] has introduced a new technique called higher order SMC to reduce chattering and to improve the control precision. In HOSMC, the system is enforced to follow higher order sliding dynamics, ie., in an nth order HOSMC, the controller is designed so as to satisfy the following conditions:

$$S = \dot{S} = \ddot{S} = \ldots S^{n-1} = 0$$

It gives a sliding precision of order n.

For example, in a second order sliding mode control, reaching laws are designed so as to enforce second order sliding modes. ie. $S = \dot{S} = 0$.

A few of the popularly used reaching law structures are listed below:

$$u_R = -K_1 sgn(S) - K_2 sgn(\dot{S})$$

$$u_R = -K_1 sgn[\dot{S} - K_2 |S|^{1/2} sgn(S)]$$

$$u_R = -K_1 |S|^{1/2} sgn(S) + v$$
$$\dot{v} = -K_2 sgn(S) \quad (13.54)$$

where (13.54) is termed as super-twisting structure. Studies [407, 408] show that the finite time convergence and reduced chattering are the major advantages of HOSM. Unlike other SMC approaches, HOSM doesn't need the relative degree of the system to be strictly equal to one. But in a recent technical note [409], Utkin has claimed that all such potential benefits of HOSM are system dependent. The author has pointed out that the performance of an HOSM depends upon the disturbance level, chosen modeling approach, etc.

Example 13.9. Design a kinematic controller for a mobile robot based on super twisting SMC technique to track a circular reference trajectory of radius 1 m.

Solution 13.9. *The kinematic model of the robot can be defined as* $\dot{\mathbf{x}} = \mathbf{g}(\mathbf{x})\mathbf{u}$, *where* $\mathbf{g}(\mathbf{x}) = \begin{pmatrix} \cos\theta & -L\sin\theta \\ \sin\theta & L\cos\theta \end{pmatrix}$. $\mathbf{g}(\mathbf{x})$ *is nonsingular, since* $|\mathbf{g}(\mathbf{x})| \neq 0$. *For a circular reference trajectory,* $\mathbf{x}_d = [\cos 0.1t, \sin 0.1t]^T$. *For the super twisting controller, the sliding surface can be defined as* $\mathbf{S} = \tilde{x} = \mathbf{x} - \mathbf{x}_d$, *where* $\tilde{x}$ *is the tracking error. The control law is given by*

$$\mathbf{u}_R = \frac{1}{\mathbf{g}(x)}(-K_1|\mathbf{S}|^{1/2}\mathrm{sgn}(\mathbf{S}) + \mathbf{v} + \dot{\mathbf{x}}_\mathbf{d})$$
$$\dot{\mathbf{v}} = -K_2\mathbf{sgn}(\mathbf{S}) \qquad (13.55)$$

The values of $K_1 = diag\{1,1\}$ *and* $K_2 = diag\{1,1\}$ *are chosen heuristically. In this case, the control law is such that* $S = \dot{S} = 0$. *The state trajectories given by Figure 13.16 indicate that the sliding precision is more for super twisting control.*

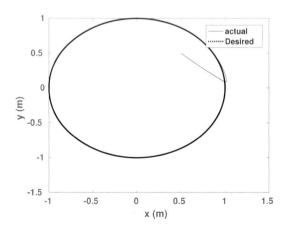

FIGURE 13.16: Super twisting SMC: x-y positions.

13.7 Summary

This chapter gives an overview of the modeling and control of differential-drive wheeled mobile robots. We have detailed the kinematic as well as the dynamic modeling of the system. The coordinate frames, the robot attitude, and the frame transformations are also presented with examples. The most commonly used nonlinear control techniques such as feedback linearization, backstepping, and sliding mode control with its different variants are also discussed in detail. The numerical examples are solved in each section, which can aid the readers to acquire a thorough understanding of the modeling and control design aspects of a mobile robotic system.

14
Multi-robot Formation

14.1 Introduction

The concept of cooperative infrastructure has received considerable attention among the robotic research community in recent years due to the diverse applications including surveillance, search, rescue, security, survey, farm aerial spraying, and disease detection and cooperative transportation. Compared to single robotic system, multi-robotic platforms have numerous advantages such as improved mission efficiency in terms of time and quality, high flexibility, adaptability, accuracy, and energy efficiency. However, the success of these types of missions greatly depends upon the distributed control module, and its efficacy to overcome the detrimental effects due to unpredictable disturbances/ modeling uncertainties, and the faults in the system. This chapter addresses the tracking control problem in a a multi-robotic system (MRS), and the various implementation issues associated with it. The formation control architectures can be primarily categorized as follows:

- **Leader-Follower-Based Approach.** In a leader-follower-based approach [410], [411], one or more agents are assigned as leaders and the rest of them as followers, as shown in Figure 14.1. The follower will track the leader with a desired offset. The desired trajectory of motion is determined by

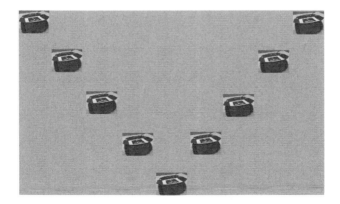

FIGURE 14.1: Multi-robot formation.

the leader. The main advantage is that it is easy to add new members to the formation. Though the approach is simple in its architecture, it has the inherent disadvantage of not having any explicit feedback from the followers to the leader.

- **Virtual Leader-Based Approach.** In a virtual-leader-based approach [412], [413], the formation as a whole acts as a rigid structure. In this architecture, a virtual point is defined in the rigid formation structure termed as virtual leader, and the desired trajectories of each of the agents are defined with respect to the virtual leader. It is difficult to do formation reconfiguration with this architecture. It is usually a centralized approach, hence, the communication/ computational cost will be more.

- **Behavior-Based Approach.** In a behavior-based approach [414], [415], the agents will try to mimic the swarming behaviors, such as flock joining, path following, collision avoidance, obstacle avoidance, reconfiguration, etc. There is no priority in particular for any of the agents. The control input is generally the weighted average of the inputs corresponding to the assigned behaviors. The weightage depends upon the mission objective.

Among all these approaches, the leader-follower-based approach is widely studied due to its simplicity in the architecture and the broad range of applications. A robust guidance and control module plays a crucial role in retaining the stability of an MRS. A brief survey on the multi-agent formation control has been presented in [416], where the author has detailed the various challenges in the control and co-ordination of formation. system. In an MRS, speed of convergence, fault tolerance capability, robustness to perturbations, adaptability to varying operating conditions, etc., are some of the major performance measures to be considered in validating the efficacy of a control module. In a generalized multi-robot formation problem, the major objectives can be categorized as follows:

- Formation maintenance/ shape formation
- Formation reconfiguration
- Trajectory following
- Obstacle avoidance
- Fault detection and isolation
- Communication consensus-based formation

Owing to the simplicity in architecture and the immense scope in real time applications, the leader-follower-based formation approach has been widely explored in literature.

14.2 Path Planning Schemes

In MRS, the choice of the path planning scheme/ guidance scheme depends upon the formation objectives.

Geometric Approach

For shape generation, path planning based on formation geometry has been used in literature. In [391], two types of approaches, viz. separation-separation and separation-bearing are used in a leader-follower-based framework. In separation-separation approach, the formation objective is to maintain and control the desired relative position between the robots. Whereas, in separation-bearing, the control objective is to maintain the relative separation as well as orientation (bearing angle) between the robots. The schematic of a two-robot system in a leader-follower-based formation is shown in fig. 14.2. The kinematics of the two robot system is given by

$$\dot{z}_{ij} = G_1(z_{ij}, \beta_{ij})u_j + F_1(z_{ij})u_i, \quad \dot{\beta}_{ij} = \omega_i - \omega_j \quad (14.1)$$

where $z_{ij} = [l_{ij}, \psi_{ij}]^T$ is the system output, $\beta_{ij} = \theta_i - \theta_j$ is the relative orientation, $u_i = [v_i, \omega_i]$ is the input to the leader and $u_j = [v_j, \omega_j]$ is the input to j^{th} follower.

$$G_1 = \begin{bmatrix} cos\gamma & dsin\gamma \\ -sin\gamma/l & dcos\gamma/l \end{bmatrix}.$$

$$F_1 = \begin{bmatrix} -cos\psi_{ij} & 0 \\ sin\psi_{ij}/l_{ij} & -1 \end{bmatrix}.$$

The formation can be maintained by controlling l_{ij} ans ψ_{ij}.

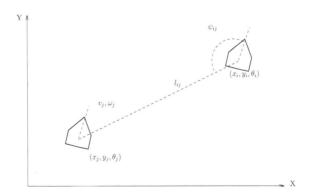

FIGURE 14.2: Two-robot system in leader-follower formation.

Example 14.1. Consider a two-robot formation in leader-follower-based framework. Develop a feedback linearization-based controller for the trailing robots to maintain a desired distance separation of 1 m with respect to the leader.

Solution 14.1. *Based on the kinematic model given by (14.1), and using the feedback linearization technique depicted in Chapter 13, the control law can be designed for the follower robots as*

$$u_j = G_1^{-1}(p_1 - F_1 u_i) \quad (14.2)$$

$$p_1 = \begin{bmatrix} k_1(l_{ij}^d - l_{ij}) \\ k_2(\psi_{ij}^d - \psi_{ij}) \end{bmatrix}.$$

Therefore

$$v_j = cos\gamma k_1(l_{ij}^d - l_{ij}) + cos\gamma cos\psi_{ij} v_i \\ - l_{ij} sin\gamma k_2(\psi_{ij}^d - \psi_{ij}) + sin\gamma sin\psi_{ij} v_i - l sin\gamma \omega_i \quad (14.3)$$

$$\omega_j = \frac{sin\gamma}{d} k_1(l_{ij}^d - l_{ij}) + \frac{cos\psi_{ij} v_i sin\gamma}{d} + \frac{l_{ij} k_2(\psi_{ij}^d - \psi_{ij})}{d} \\ \frac{-cos\gamma sin\psi v_i}{d} + \frac{l_{ij} cos\gamma \omega_i}{d} \quad (14.4)$$

$$\gamma_{ij} = \beta_{ij} + \psi_{ij} \quad (14.5)$$

$$\beta_{ij} = \theta_i + \theta_j \quad (14.6)$$

$$l_{ij} = \sqrt{(x_i - (x_j + d cos\theta_j))^2 + (y_i - (y_j + d sin\theta_j))^2} \quad (14.7)$$

$$\psi_{ij} = 180 - (\theta_i - tan^{-1} \frac{(y_i - y_j - d sin\theta_j)}{(x_i - x_j - d cos\theta_j)} \quad (14.8)$$

For a triangular formation with the given specifications, $\psi_d = 120°$ for the left follower and $\psi_d = 240°$ for the right follower, $l_{ij}^d = 1m$ and $d = 0.25m$.

Example 14.2. Design and develop a sliding mode-based controller for a three-robot system in triangular formation. Consider a desired distance separation of 2 m between each of the robots.

Solution 14.2. *The sliding surface can be defined as*

$$S = e = x - x_d \quad (14.9)$$

where $x = [l_{ij}, \psi_{ij}]^T$. Using the sliding mode technique depicted in chapter 13, the control law can be designed for the follower robots as

$$u_j = G_1^{-1}(K sign(s) - F_1 u_i) \tag{14.10}$$

where $K = diag\{k_1, k_2\}$. Therefore

$$v_j = cos\gamma k_1 sign(l_{ij}^d - l_{ij}) + cos\gamma cos\psi_{ij} v_i$$
$$- l_{ij} sin\gamma k_2 sign(\psi_{ij}^d - \psi_{ij}) + sin\gamma sin\psi_{ij} v_i - l sin\gamma \omega_i \tag{14.11}$$

$$\omega_j = \frac{sin\gamma}{d} k_1 sign(l_{ij}^d - l_{ij}) + \frac{cos\psi_{ij} v_i sin\gamma}{d} + \frac{l_{ij} k_2 sign(\psi_{ij}^d - \psi_{ij})}{d}$$
$$\frac{-cos\gamma sin\psi v_i}{d} + \frac{l_{ij} cos\gamma \omega_i}{d} \tag{14.12}$$

But, it is difficult to realize obstacle avoidance, along with target seeking using geometric approaches. Moreover, the geometric framework will become complex, if the number of robots is more. For realizing these schemes, agents need to maintain the communication connectivity through out.

Potential Function-based Approach

The APF-based method proposed by Khatib [417] has already been proven as a computationally efficient and a simple path planning scheme for obstacle avoidance, target seeking and collision avoidance in robotic systems. In this approach, the trajectory is determined by the computed net potential force with respect to the obstacles, target, as well as with the other agents in the formation. Depending upon the formation objectives, the potential function may have attractive and repulsive parts, which are functions of the relative positions/ orientations. A potential function is a real valued differentiable energy function, and its gradient is the force. In robot motion planning, the robot velocities can be enforced to follow this gradient potential, where the robot follows the path of the negative gradient of the potential, so that the motion ceases, when the gradient reaches zero. In general, the potential function is attractive/ repulsive in nature, where the agents are attracted to the goal/ far distant agents for making the formation, due to the attractive part, and are repelled away from the obstacles/ neighboring agents for collision avoidance, due to the repulsive parts. For goal oriented motion of agents, avoiding the obstacles, the potential function [418] can be expressed as

$$\mathcal{P} = \mathcal{P}_{goal} + \mathcal{P}_{obst} \tag{14.13}$$

where $\mathcal{P}_{goal}$ and $\mathcal{P}_{obst}$ represent the attractive potential toward the goal, and the repulsive potential to avoid the obstacles respectively. The widely used expressions for these potentials are given below: [418, 419]

$$P_{goal} = \frac{1}{2}\gamma(x - x_{goal})^2 \tag{14.14}$$

$$P_{obst} = \begin{cases} \frac{1}{2}\beta\left(\frac{1}{x} - \frac{1}{\rho_0}\right)^2, & \text{if } x \leq \rho_0 \\ 0, & \text{if } x > \rho_0. \end{cases} \tag{14.15}$$

where γ and β represent the gains, and ρ_0 represents the minimum distance threshold to be maintained with respect to the obstacle. The force can be computed using the gradient of these potential given by

$$\mathcal{F} = -\nabla \mathcal{P} \tag{14.16}$$

The artificial potential function-based kinematic model for target tracking proposed by Gazi [420–422] has been reproduced here for the completeness of the discussion. This section addresses the path planning problem of a group of agents in formation in an $n-$ dimensional Euclidean space. The motion dynamics of each of the swarm members can be represented as

$$\dot{\mathbf{x}}^i = \sum_{j=1, j\neq i}^{N} \mathbf{g}(x^i - x^j), i = 1....N \tag{14.17}$$

where $\mathbf{x}^i \in \Re^n$ is the position vector of the ith agent, and $\mathbf{g} : \Re^n \to \Re^n$ is the function of attraction/repulsion between the swarm members.

The function $\mathbf{g}(.)$ can be represented by

$$\begin{aligned} g(y) &= -y[g_a(||y||) - g_r(||y||)] \\ &= -y\left[a - b\exp\left(-\frac{||y||^2}{c}\right)\right] \end{aligned} \tag{14.18}$$

where a, b, and c are positive constants, $||y|| = \sqrt{y^T y}$ and $y \in \Re^n$.

Equation (14.17) can be represented also by

$$\dot{\mathbf{x}}^i = -\nabla_{x^i} \mathbf{J}(x), i = 1, 2...N, \tag{14.19}$$

where $\mathbf{J}$ is the potential function determining the inter agent interaction and it has the following form

$$\mathbf{J}(x) = \sum_{i=1}^{N-1} \sum_{j=i+1}^{N} \mathbf{J}_{ar}^{ij}(||x^i - x^j||) \tag{14.20}$$

where $\mathbf{J}_{ar}^{ij}(||x^i - x^j||)$ represents the potential between the agents in the group. With such a form, each agent will be forced to maintain the desired distance d_{ij} with respect to its neighbors. The potential between i^{th} and j^{th} agent, i.e., $\mathbf{J}_{ar}^{ij}(||x^i - x^j||)$, can be different for different pairs and it will satisfy the following conditions [420].

Condition 14.1. *The potential, $\mathbf{J}_{ar}^{ij}(||x^i - x^j||)$, is symmetric and it satisfies the following condition,*

$$\nabla_{x^i} \mathbf{J}_{ar}^{ij}(||x^i - x^j||) = -\nabla_{x^j} \mathbf{J}_{ar}^{ij}(||x^i - x^j||) \qquad (14.21)$$

Condition 14.2. *[423] There exists corresponding functions $\mathbf{J}_a^{ij} : \Re^+ \to \Re$ and $\mathbf{J}_r^{ij} : \Re^+ \to \Re$ such that $\mathbf{J}^{ij}(.) = \mathbf{J}_a^{ij}(.) - \mathbf{J}_r^{ij}(.)$, $\nabla_y \mathbf{J}_a(||y||) = y\mathbf{g}_a^{ij}(||y||)$ and $\nabla_y \mathbf{J}_r(||y||) = y\mathbf{g}_r^{ij}(||y||)$, where $\mathbf{J}_a^{ij}(||y||)$, the attractive potential, dominates on long distances, and $\mathbf{J}_r^{ij}(||y||)$, i.e., the repulsive potential, dominates on short distances. Then, the motion of agents toward each other is restricted along the combined gradient of these potentials.*

Condition 14.3. *For any desired formation, which is uniquely defined by the formation constraints $||x^i - x^j|| = d_{ij}, 1 \leq i, j \leq N$, the potential function is chosen such that the unique minimum of $\mathbf{J}_{ar}^{ij}(||x^i - x^j||)$ occurs at d_{ij} and $\nabla_{x^i} \mathbf{J}_{ar}^{ij}(||x^i - x^j||) = \mathbf{0}$, when $||x^i - x^j|| = d_{ij}$, i.e. $\mathbf{g}_a^{ij}(d_{ij}) = \mathbf{g}_r^{ij}(d_{ij})$.*

Note that, for the attraction/repulsion function,

$$J_a(y) = \frac{a}{2}||y||^2,$$

$$J_r(y) = -\frac{bc}{2}\exp\left(-\frac{||y||^2}{c}\right)$$

Assuming that the motion of the agents is given by (14.17), it can be shown that if $\mathbf{J}(x)$ is bounded from below, i.e., $\mathbf{J}(x) > a$ for some finite $a \in \Re$, then, for any initial condition $x(0) \in \Re^{nN}$, as $t \to \infty$, $x(t) \to \sigma_e$, where $\sigma_e = \{x : \dot{x} = 0\}$ and $x^T = [x^{1T}....x^{NT}]$. It is inferred that, given the initial positions of the individuals $x^i(0)$, $i = 1, 2..N$, the final configuration to which the individuals will converge is unique. Since $J(x)$, i.e., the formation function, has unique minimum at the desired formation, the desired formation will be asymptotically achieved. Once the desired formation is attained, the potential gradient becomes zero, and the velocities of all the agents become equal. In order to show the nature of function, the attraction/repulsion function for a distance separation of 5 km is given in Figure 14.3. From this plot, one can find that the gradient potential becomes zero at the respective distance separations in both the directions.

But APF has the inherent drawback of local minima, if the parameters are not chosen properly. In some of the related works, the region of operation is constrained to avoid this. In [424], Jin and Pradeep have introduced a potential field based on harmonic function to overcome the local minima problem. But this type of harmonic potential function will work well in the case of non-point robots. In [425], a fuzzy inferencing-based adaptive potential function has been developed for obstacle avoidance and navigation of mobile robots in a cluttered environment. Apart from the fact that this approach is constrained to obstacle avoidance of a single robot, it also needs complex sensing requirements.

Another method to avoid the local minima problem is to introduce separate local and global path planners for navigation [426]. In this scenario,

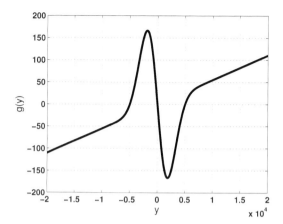

FIGURE 14.3: Attraction/repulsion function.

whenever the local path planner met with a local minima, a global path planner comes into action, and it does the further path planning to reach the goal, avoiding the local minima. But the global path planner will work only in a predefined environment. Hence, this strategy cannot be extended to complex systems working in a cluttered environment. Another approach is to place virtual obstacles at local minima points [427], so that the obstacle avoidance module will take care of them. But the accuracy of this depends upon the choice of parameters of the residual signals utilized for placing the virtual obstacles. Hence, this may not work well for all operating conditions. Gazi and Passino [428] have discussed a class of attraction/repulsion potential functions for stable swarm aggregations. Similar to this technique, Izzo and Pettazi [429] have developed a path planning technique for a system of spacecrafts in a behavior-based framework with limited sensing, where the individual responses are co-ordinated to achieve a common task.

Example 14.3. Design and implement an APF-based path planner for a simple multi-agent system consisting of three members, to form a triangular formation, with an inter-agent distance separation of 5 m.

Solution 14.3. *Consider the dynamics of the ith agent as*

$$\dot{\mathbf{x}}^i = u^i, i = 1, 2, 3, \qquad (14.22)$$

The attractive and repulsive potentials for i^{th} can be defined as

$$J_a(x^i - x^j) = \frac{a}{2}\|x^i - x^j\|^2,$$

$$J_r(x^i - x^j) = -\frac{bc}{2}exp\left(-\frac{\|x^i - x^j\|^2}{c}\right)$$

Path Planning Schemes

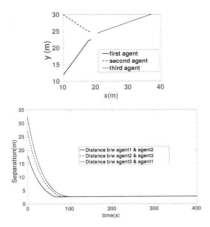

FIGURE 14.4: Path of the agents (left); distance separation between the agents (right).

where the parameters of the potential function can be chosen as $a=0.00551819$, $b=0.15$, and $c=0.302793$. The path planner can be designed such that the agent velocities need to follow the gradient of the respective net potential. [423]

$$J_x = \sum_{i=1}^{N-1} \sum_{j=i+1}^{N} (J_a(x^i - x^j) + J_r(x^i - x^j)) \tag{14.23}$$

$$u^i = -\nabla_{x^i} \mathbf{J}(x) \tag{14.24}$$

The gradient can be computed from the net potential, and the agents can be commanded to follow these velocities. The results showing the path of the agents and the respective distance separations are shown in Fig 14.4.

```
MATLAB CODE FOR APF-BASED SWARM AGGREGATION
-------------------------------------------
clear all;
a=.01;b=20;c=1;n=40000;n1=7000; m=1;u0=25;beta=10;
xdot(:,1)=[0;0;0]; ydot(:,1)=[0;0;0];t=0; T=.01;
x(:,1)=10*[1 ;1;3.7]; y(:,1)=10*[1.2;3;3];
for k=1:n
    t=t+T;
for i=1:3
      Jx=0; Jy=0;
for j=1:3
dist=sqrt((x(i)-x(j))^2+(y(i)-y(j))^2);
Jx=Jx+(x(i)-x(j))*(a-b*exp(-(dist^2/c)));
Jy=Jy+(y(i)-y(j))*(a-b*exp(-(dist^2/c)));
end
u2=-Jx; u3=-Jy;xdot(i)=u2; ydot(i)=u3;x(i)=x(i)+T*xdot(i);
y(i)=y(i)+T*ydot(i);u_save(i,k)=u2;u1_save(i,k)=u3;
```

```
x_save(i,k)=x(i);  y_save(i,k)=y(i);
end
xbar(k)=(x_save(1,k)+x_save(2,k)+x_save(3,k))/3;
 ybar(k)=(y_save(1,k)+y_save(2,k)+y_save(3,k))/3;
 center(k)=sqrt(xbar(k)^2+ybar(k)^2);
 d1(k)=sqrt((x_save(1,k)-x_save(2,k))^2
 +(y_save(1,k)-y_save(2,k))^2);
 d2(k)=sqrt((x_save(2,k)-x_save(3,k))^2
 +(y_save(2,k)-y_save(3,k))^2);
 d3(k)=sqrt((x_save(3,k)-x_save(1,k))^2
 +(y_save(3,k)-y_save(1,k))^2);
end
time = T*(1:n);
 figure
 plot(x_save(1,m:n),y_save(1,m:n))   hold on
 plot(x_save(2,m:n),y_save(2,m:n))   hold on
 plot(x_save(3,m:n),y_save(3,m:n))
 xlabel('x(m)');   ylabel('y (m)')
 figure
 plot(time(1:n),d1(1:n))   hold on
 plot(time(1:n),d2(1:n))   hold on
 plot(time(1:n),d3(1:n))
 xlabel('time(s)');   ylabel('Separation(m)')
```

Example 14.4. Consider a simple multi-agent system with three members. Design an APF-based path planner for the system to navigate to the target point at the location (2,2), keeping a triangular formation.

Solution 14.4. *For an MAS moving to any goal position, the attractive part of the potential function will have an additional term given by (16.27), to get the system attracted to the goal position [430] .*

$$\mathcal{P}_{goal} = -\frac{1}{2}\gamma(x^i - x_{goal})^2 \tag{14.25}$$

The attractive and repulsive potentials for i^{th} agent can be defined as

$$J_a(x^i - x^j) = \frac{a}{2}\|x^i - x^j\|^2 - \frac{1}{2}\gamma(x^i - x_{goal})^2,$$
$$J_r(x^i - x^j) = -\frac{bc}{2}exp\left(-\frac{\|x^i - x^j\|^2}{c}\right)$$

The parameters can be chosen as $\gamma = 0.5$, a=0.00551819, b=0.15, and c=0.302793. For a multi-agent system with the dynamics of each agent defined by (14.22), the control input can be obtained by computing the gradient of the net potential.

$$J_x = \sum_{i=1}^{N-1}\sum_{j=i+1}^{N}(J_a(x^i - x^j) + J_r(x^i - x^j)) \tag{14.26}$$

$$u^i = -\nabla_{x^i}\mathbf{J}(x) \tag{14.27}$$

Path Planning Schemes

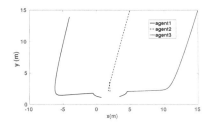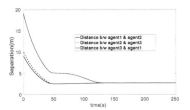

FIGURE 14.5: Path of the agents (left); distance separation between the agents (right).

The results given by Figure 14.5 show that the agents are making the triangular formation with the desired separation, and the center of the swarm is at the desired final position $[2, 2]^T$.

```
MATLAB CODE FOR APF-BASED TARGET TRACKING
------------------------------------------
clear all;
a=.01; b=20; c=1; n=25000; n1=7000; m=1; u0=25;
beta=10; xdot(:,1)=[0;0;0]; ydot(:,1)=[0;0;0];
x(:,1)=5*[-.8;1;3]; y(:,1)=5*[2.8;3;3];
    t=0;  T=.01;   gama=0.8;  x_g=2;  y_g=2;
for k=1:n
    t=t+T;
for i=1:3
Jx=0;  Jy=0; c_n_x=0; c_n_y=0;
for j=1:3
c_n_x=c_n_x+x(j); c_n_y=c_n_y+y(j);
dist=sqrt((x(i)-x(j))^2+(y(i)-y(j))^2);
Jx=Jx+(x(i)-x(j))*(a-b*exp(-(dist^2/c)));
Jy=Jy+(y(i)-y(j))*(a-b*exp(-(dist^2/c)));
end
Jx=Jx+gama*((c_n_x/3)-x_g);
Jy=Jy+gama*((c_n_y/3)-y_g);
u2=-Jx;  u3=-Jy; xdot(i)=u2; ydot(i)=u3;
x(i)=x(i)+T*xdot(i); y(i)=y(i)+T*ydot(i); u_save(i,k)=u2;
u1_save(i,k)=u3; x_save(i,k)=x(i); y_save(i,k)=y(i);
end
xbar(k)=(x_save(1,k)+x_save(2,k)+x_save(3,k))/3;
ybar(k)=(y_save(1,k)+y_save(2,k)+y_save(3,k))/3;
center(k)=sqrt(xbar(k)^2+ybar(k)^2);
d1(k)=sqrt((x_save(1,k)-x_save(2,k))^2 +(y_save(1,k)-
    ↪ y_save(2,k))^2);
d2(k)=sqrt((x_save(2,k)-x_save(3,k))^2 +(y_save(2,k)-
    ↪ y_save(3,k))^2);
d3(k)=sqrt((x_save(3,k)-x_save(1,k))^2 +(y_save(3,k)-
    ↪ y_save(1,k))^2);
```

```
end
 time = T*(1:n);
 figure
 plot (x_save(1,m:n),y_save(1,m:n))     hold on
 plot(x_save(2,m:n),y_save(2,m:n))      hold on
 plot(x_save(3,m:n),y_save(3,m:n))
 figure
 plot(time(1:n),d1(1:n))    hold on
 plot(time(1:n),d2(1:n))    hold on
 plot(time(1:n),d3(1:n))
```

14.3 Multi-Agent Formation Control

The APF-based schemes given in the previous section do not specify how swarming can be achieved in engineering applications, with given agent dynamics. To achieve the coordinated, smooth motion of agents, with general fully actuated vehicle dynamics, even in the presence of disturbances and uncertainties, robust control techniques can be utilized. The potential field based guidance scheme promises a collision-free navigation and formation in an optimal time period. Now, the aim is to design a control law, for individual agents with dynamic uncertainties, such that (14.19) is satisfied. In short, the control inputs are to be designed such that the relative velocity of each agent is enforced along the negative gradient of the respective net potential. Owing to the simplicity and flexibility in the architecture, we have chosen the leader-follower-based formation approach in this section. In a leader-follower-based approach, the leader is free to navigate, and the followers have to track the leader, maintaining a desired offset w. r. t each other as well as with the leader. Once they are in formation, the followers (deputies) track the natural dynamics of the leader.

The tracking control schemes can be designed for the deputies to ensure the same, amid the unforeseen disturbances acting on the system. This is accomplished by using an SMC approach, which is one of the most powerful and robust nonlinear control techniques, that enables separation of the overall system motion into independent partial components of lower dimensions. Owing to the robustness features and finite time convergence property, as discussed in the preceding chapters, NTSMC is equally advisable in multi-robot system also. The speed of tracking error convergence is extremely critical in the event of system failure and reconfiguration, as well as with other changes in operating conditions. To ensure a faster convergence, and to overcome the singularity issue inherent in conventional TSMC, a fast nonsingular type of nonlinear sliding surface is highly preferable in this case. The speed of convergence and the

Multi-Agent Formation Control 519

singularity problem inherent in conventional TSMC have been addressed in numerous literature [399, 431–433].

Consider a group of agents with one leader and N number of followers. The motion dynamics of the ith follower w. r. t leader can be represented as,

$$\dot{\mathbf{x}}^i - \dot{\mathbf{x}}^\ell = -\nabla_{\mathbf{x}^i} J(\mathbf{x}), i = 1....N \qquad (14.28)$$

where $\mathbf{x}^i \in \Re^n$ represents the position vector of ith follower; $\mathbf{x}^\ell \in \Re^n$ is the position vector of the leader, and J is the net potential function [418] determining the interaction between the agents and it has the following form:

$$J(\mathbf{x}) = \sum_{i=1}^{N-1} \sum_{j=i+1}^{N} J_a^{ij}(\|\mathbf{x}^i - \mathbf{x}^j\|) - J_r^{ij}(\|\mathbf{x}^i - \mathbf{x}^j\|)$$
$$+ \sum_{i=1}^{N} J_a^{i\ell}(\|\mathbf{x}^i - \mathbf{x}^\ell\|) - J_r^{i\ell}(\|\mathbf{x}^i - \mathbf{x}^\ell\|) \qquad (14.29)$$

The attractive potential $J_a^{\{\cdot\}} : \Re^+ \to \Re$ dominates on long distances, whereas the repulsive potential $J_r^{\{\cdot\}} : \Re^+ \to \Re$ dominates on short distances. The motion of each follower is restricted along the combined gradient of these potentials.

Remark 14.1. The APF parameters are chosen such that the unique global minimum of $J(\mathbf{x}^i)$ occurs when $\|\mathbf{x}^i - \mathbf{x}^j\| = d^{ij}$, and $\nabla J_a^{ij}(d^{ij}) = \nabla J_r^{ij}(d^{ij})$. Hence, the desired formation can be asymptotically achieved [434]. Different APF parameters can be chosen for different inter-agent distances. The choice of the initial position determines the unique final position to which each agent converges. Once they are in formation, the net gradient potential becomes zero and each follower reaches a consensus with the leader.

Lemma 14.1. *[398]* For the system given by $\dot{\mathbf{x}} = \mathbf{f}(\mathbf{x})$, $f(0) = 0$, $\mathbf{x}(0) = \mathbf{x}^0$, $\mathbf{x} \in U_0 \subset \Re^n$, suppose that there exists a positive definite continuous function, $V(\mathbf{x})$, in the neighborhood of the origin, real numbers $k > 0$, $\eta \in (0,1)$, so that $\dot{V}(\mathbf{x}) + kV^\eta(\mathbf{x}) \leq 0$, then the system is finite time stable, and the settling time T is such that $T \leq \frac{V^{1-\eta}(\mathbf{x}(0))}{k(1-\eta)}$.

14.3.1 Fast Adaptive Gain NTSMC

In order to achieve the control objective, a novel formation error has been defined [435] in terms of position measurements alone, and it is given by

$$e^i = \mathbf{x}^i - \mathbf{x}^\ell + \underbrace{\int_0^t \nabla_{\mathbf{x}^i} J(\mathbf{x}) dt}_{-\mathbf{x}_d^i} \qquad (14.30)$$

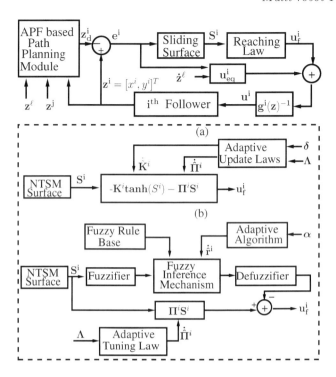

FIGURE 14.6: Block schematic of the proposed scheme; (a)-(b) adaptive reaching laws.

In order to improve the speed of convergence, a fast nonsingular terminal sliding surface [399], [432], [433] has been employed to design the tracking controller. The block schematic of the proposed scheme is shown in Figure 14.6. The proposed sliding surface utilizes only the relative position error. For the ith member, it can be defined as

$$\mathbf{S}^i = \mathbf{e}^i + \int_0^t \left(\boldsymbol{\lambda}_1^i |\mathbf{e}^i|^{1+\frac{1}{\beta}} \mathbf{sign}(\mathbf{e}^i) + \boldsymbol{\lambda}_2^i |\mathbf{e}^i|^{1-\frac{1}{\beta}} \mathbf{sign}(\mathbf{e}^i) \right) dt \qquad (14.31)$$

where $\mathbf{S}^i = [S_x^i, S_y^i]^T$, $\mathbf{e}^i = [e_x^i, e_y^i]^T$, $|\mathbf{e}^i|^{\{\cdot\}} = diag\{|e_x^i|^{\{\cdot\}}, |e_y^i|^{\{\cdot\}}\}$,

$\mathbf{sign}(\mathbf{e}^i) = [sign(e_x^i), sign(e_y^i)]^T$, $\boldsymbol{\lambda}_1^i = diag\{\lambda_{1x}^i, \lambda_{1y}^i\}$, $\boldsymbol{\lambda}_2^i = diag\{\lambda_{2x}^i, \lambda_{2y}^i\}$, $\lambda_{1x,1y}^i, \lambda_{2x,2y}^i > 0$, $\forall i = 1\ldots N$; and $\beta > 1$. To further improve the speed of convergence, a fast reaching law given by,

$$\dot{\mathbf{S}}^i = -\mathbf{K}^i \mathbf{sign}(\mathbf{S}^i) - \mathbf{\Pi}^i \mathbf{S}^i, \qquad (14.32)$$

has been used.

Multi-Agent Formation Control

Remark 14.2. To reduce the chattering effect, **sign** function in the reaching law can be replaced by **tanh** function. The time of convergence for the tracking error, $e_{x,y}^i$, to reach the origin is given by

$$t_{x,y}^i = \frac{\beta}{\lambda_{1x,1y}^i \lambda_{2x,2y}^i} tan^{-1}\left(\sqrt{\frac{\lambda_{1x,1y}^i}{\lambda_{2x,2y}^i}} |e_{x,y}^i(0)|^{\frac{1}{\beta}}\right)$$

The control law can be obtained as

$$\mathbf{u}^i = \mathbf{g}^i(\mathbf{x})^{-1}\left(-\nabla_{\mathbf{x}^i}J(\mathbf{x}) + \dot{\mathbf{x}}^\ell - \lambda_1^i|\mathbf{e}^i|^{1+\frac{1}{\beta}}\mathbf{sign}(\mathbf{e}^i)\right.$$
$$\left.-\lambda_2^i|\mathbf{e}^i|^{1-\frac{1}{\beta}}\mathbf{sign}(\mathbf{e}^i) - \mathbf{K}^i\mathbf{tanh}(\mu\mathbf{S}^i) - \mathbf{\Pi}^i\mathbf{S}^i\right) \quad (14.33)$$

where $\mathbf{g}^i(\mathbf{x}) = \begin{pmatrix} cos(\theta^i) & -Lsin(\theta^i) \\ sin(\theta^i) & Lcos(\theta^i) \end{pmatrix}$, $\dot{\mathbf{x}}^\ell = [\dot{x}^\ell, \dot{y}^\ell]^T$; $\mathbf{K}^i = diag\{K_x^i, K_y^i\}$, and $\mathbf{\Pi}^i = diag\{\Pi_x^i, \Pi_y^i\}$, represent the controller gains, $\Pi_{x,y}^i, K_{x,y}^i > 0, \forall i$; $\mathbf{tanh}(\mu\mathbf{S}^i) = [tanh(\mu S_x^i), tanh(\mu S_y^i)]^T$.

Since $det(\mathbf{g}^i(\mathbf{x})) \neq 0$, $\mathbf{g}^i(\mathbf{x})$ is nonsingular and invertible.

Assumption 14.1. But in practical applications, disturbances will be acting on the system. The lumped uncertainty in the system, say $\mathbf{d}^i$, is assumed to be bounded. Hence the motion dynamics can be modified as

$$\dot{\mathbf{x}}^i = \mathbf{g}^i(\mathbf{x})\mathbf{u}^i + \mathbf{d}^i \quad (14.34)$$

Differentiating (14.31) w. r. t. time, and by successively substituting for $\dot{\mathbf{e}}$, $\dot{\mathbf{x}}^i$, and $\mathbf{u}^i$ using (14.30), (14.34) and (14.33) respectively yields,

$$\dot{\mathbf{S}}^i = -\mathbf{K}^i\mathbf{tanh}(\mu\mathbf{S}^i) - \mathbf{\Pi}^i\mathbf{S}^i + \mathbf{d}^i \quad (14.35)$$

Theorem 14.1. *[435]* For the given system, with a nonsingular terminal sliding surface as given in (14.31), the control law defined by (14.33) can make the system finite time stable, and the system error will converge to origin in finite time without any singularity, if the controller gains are updated using the following adaptive laws:

$$\dot{\hat{\mathbf{K}}}^i = \frac{1}{\delta}\mathbf{tanh}(\mu\mathbf{S}^i)\mathbf{S}^{iT}$$
$$\dot{\hat{\mathbf{\Pi}}}^i = \Lambda^{-1}\mathbf{S}^i\mathbf{S}^{iT}, \quad \alpha, \Lambda > 0.$$

Proof. Consider the Lyapunov function as

$$V^i = \frac{1}{2}\mathbf{S}^{iT}\mathbf{S}^i \quad (14.36)$$

Taking the derivative, and substituting for $\dot{\mathbf{S}}^i$ from (14.35),

$$\dot{V}^i = \mathbf{S}^{iT}(-\mathbf{K}^i\mathbf{tanh}(\mu\mathbf{S}^i) - \mathbf{\Pi}^i\mathbf{S}^i + \mathbf{d}^i) \quad (14.37)$$

It has been assumed that, for any given operating condition, there exists finite, nominal values for the gains, $\mathbf{K}^{i*}$ and $\mathbf{\Pi}^{i*}$ respectively, such that the sliding condition can be satisfied.

$$\text{Let, } \tilde{\mathbf{K}}^i = \hat{\mathbf{K}}^i - \mathbf{K}^{i*}; \tilde{\mathbf{\Pi}}^i = \hat{\mathbf{\Pi}}^i - \mathbf{\Pi}^{i*}, \tag{14.38}$$

where $\hat{\mathbf{K}}^i$ and $\hat{\mathbf{\Pi}}^i$ are the estimated values of the respective gains; and $\tilde{\mathbf{K}}^i$ and $\tilde{\mathbf{\Pi}}^i$ are the adaptation errors.

The sliding condition ($S\dot{S} \leq 0$) can be satisfied, if the gains are chosen such that

$$\mathbf{K}^i \geq \mathbf{d}^i (\tanh(\mu \mathbf{S}^i))^{-1} + \boldsymbol{\eta} \tag{14.39}$$

$$\mathbf{\Pi}^i \geq -\eta \tanh(\mu \mathbf{S}^i)(\mathbf{S}^i)^{-1} + (\mathbf{S}^{iT})^{-1}[\mathbf{S}^T\mathbf{S}$$
$$+ n\lambda_{max}(\delta\tilde{\mathbf{K}}^i\tilde{\mathbf{K}}^i) + n\lambda_{max}(\Lambda\tilde{\mathbf{\Pi}}^i\tilde{\mathbf{\Pi}}^i)]^\sigma (\mathbf{S}^i)^{-1} \tag{14.40}$$

where $\lambda_{max}(.)$ represents the maximum eigen value; δ, $\Lambda > 0$; n is the order of $\tilde{\mathbf{K}}$; $\sigma \in (0,1)$; and $\boldsymbol{\eta}$ is a diagonal positive definite matrix. The involvement of disturbance term makes it difficult to compute the gains using (14.39) and (14.40). This is the motivation to develop adaptive update laws for tuning these parameters.

Consider the modified Lyapunov function

$$V(\mathbf{S}, \tilde{\mathbf{K}}, \tilde{\mathbf{\Pi}}) = \frac{1}{2}\mathbf{S}^{iT}\mathbf{S}^i + tr\{\delta\tilde{\mathbf{K}}^i\tilde{\mathbf{K}}^i\} + tr\{\Lambda\tilde{\mathbf{\Pi}}^i\tilde{\mathbf{\Pi}}^i\} \tag{14.41}$$

where $tr\{.\}$ represents the trace of a matrix.

Take the derivative, and substitute for $\dot{\mathbf{S}}^i$

$$\dot{V}(\mathbf{S}, \tilde{K}, \tilde{\Pi}) = \mathbf{S}^{iT}\dot{\mathbf{S}}^i + tr\{\delta\tilde{\mathbf{K}}^i\dot{\hat{\mathbf{K}}}^i\} + tr\{\Lambda\tilde{\mathbf{\Pi}}^i\dot{\hat{\mathbf{\Pi}}}^i\}$$
$$= \mathbf{S}^{iT}(-\hat{\mathbf{K}}^i\tanh(\mu\mathbf{S}^i) - \hat{\mathbf{\Pi}}^i\mathbf{S}^i + \mathbf{d}^i)$$
$$+ tr\{\delta\tilde{\mathbf{K}}^i\dot{\hat{\mathbf{K}}}^i\} + tr\{\Lambda\tilde{\mathbf{\Pi}}^i\dot{\hat{\mathbf{\Pi}}}^i\}$$
$$= \mathbf{S}^{iT}(-\hat{\mathbf{K}}^i\tanh(\mu\mathbf{S}^i) - \hat{\mathbf{\Pi}}^i\mathbf{S}^i + \mathbf{K}^{i*}\tanh(\mu\mathbf{S}^i)$$
$$- \mathbf{K}^{i*}\tanh(\mu\mathbf{S}^i) + \mathbf{\Pi}^{i*}\mathbf{S}^i - \mathbf{\Pi}^{i*}\mathbf{S}^i + \mathbf{d}^i)$$
$$+ tr\{\delta\tilde{\mathbf{K}}^i\dot{\hat{\mathbf{K}}}^i\} + tr\{\Lambda\tilde{\mathbf{\Pi}}^i\dot{\hat{\mathbf{\Pi}}}^i\}$$
$$= \mathbf{S}^{iT}(-\tilde{\mathbf{K}}^i\tanh(\mu\mathbf{S}^i) - \tilde{\mathbf{\Pi}}^i\mathbf{S}^i - \mathbf{K}^{i*}\tanh(\mu\mathbf{S}^i)$$
$$- \mathbf{\Pi}^{i*}\mathbf{S}^i + \mathbf{d}^i) + tr\{\delta\tilde{\mathbf{K}}^i\dot{\hat{\mathbf{K}}}^i\} + tr\{\Lambda\tilde{\mathbf{\Pi}}^i\dot{\hat{\mathbf{\Pi}}}^i\}$$

The nominal values of the gains are such that the conditions given by (14.39)-(14.40) have to be satisfied.

Using these conditions, one can find that

$$\dot{V}(\mathbf{S},\tilde{\mathbf{K}},\tilde{\mathbf{\Pi}}) \leq \mathbf{S}^{i^T}\left(-\tilde{\mathbf{K}}^i\tanh(\mu\mathbf{S}^i) - \tilde{\mathbf{\Pi}}^i\mathbf{S}^i\right) - (\mathbf{S}^T\mathbf{S} + n\lambda_{max}(\delta\tilde{\mathbf{K}}^i\tilde{\mathbf{K}}^i)$$
$$+ n\lambda_{max}(\Lambda\tilde{\mathbf{\Pi}}^i\tilde{\mathbf{\Pi}}^i))^\sigma + tr\{\delta\tilde{\mathbf{K}}^i\dot{\hat{\mathbf{K}}}^i\} + tr\{\Lambda\tilde{\mathbf{\Pi}}^i\dot{\hat{\mathbf{\Pi}}}^i\}$$

From the properties of trace,

$$\dot{V}(\mathbf{S},\tilde{\mathbf{K}},\tilde{\mathbf{\Pi}}) \leq -(\mathbf{S}^T\mathbf{S} + n\lambda_{max}(\delta\tilde{\mathbf{K}}^i\tilde{\mathbf{K}}^i) + n\lambda_{max}(\Lambda\tilde{\mathbf{\Pi}}^i\tilde{\mathbf{\Pi}}^i))^\sigma$$
$$+ tr\{\delta\tilde{\mathbf{K}}^i\dot{\hat{\mathbf{K}}}^i - \tilde{\mathbf{K}}^i\tanh(\mu\mathbf{S}^i)\mathbf{S}^{i^T}\} + tr\{\Lambda\tilde{\mathbf{\Pi}}^i\dot{\hat{\mathbf{\Pi}}}^i - \tilde{\mathbf{\Pi}}^i\mathbf{S}^i\mathbf{S}^{i^T}\}$$

If,

$$\dot{\hat{\mathbf{K}}}^i = \frac{1}{\delta}\tanh(\mu\mathbf{S}^i)\mathbf{S}^{i^T} \tag{14.42}$$
$$\dot{\hat{\mathbf{\Pi}}}^i = \Lambda^{-1}\mathbf{S}^i\mathbf{S}^{i^T}$$

then,

$$\dot{V}(\mathbf{S},\tilde{\mathbf{K}},\tilde{\mathbf{\Pi}}) \leq -(\mathbf{S}^T\mathbf{S} + n\lambda_{max}(\delta\tilde{\mathbf{K}}^i\tilde{\mathbf{K}}^i) + n\lambda_{max}(\Lambda\tilde{\mathbf{\Pi}}^i\tilde{\mathbf{\Pi}}^i))^\sigma \tag{14.43}$$

$$tr\{\delta\tilde{\mathbf{K}}^i\tilde{\mathbf{K}}^i\} + tr\{\Lambda\tilde{\mathbf{\Pi}}^i\tilde{\mathbf{\Pi}}^i\}\} \leq n(\lambda_{max}(\delta\tilde{\mathbf{K}}^i\tilde{\mathbf{K}}^i) + \lambda_{max}(\Lambda\tilde{\mathbf{\Pi}}^i\tilde{\mathbf{\Pi}}^i))$$

Hence,

$$V(\mathbf{S},\tilde{\mathbf{K}},\tilde{\mathbf{\Pi}}) \leq \frac{1}{2}(\mathbf{S}^T\mathbf{S}) + n(\lambda_{max}(\delta\tilde{\mathbf{K}}^i\tilde{\mathbf{K}}^i) + \lambda_{max}(\Lambda\tilde{\mathbf{\Pi}}^i\tilde{\mathbf{\Pi}}^i))$$
$$V(\mathbf{S},\tilde{\mathbf{K}},\tilde{\mathbf{\Pi}}) \leq \mathbf{S}^T\mathbf{S} + n(\lambda_{max}(\delta\tilde{\mathbf{K}}^i\tilde{\mathbf{K}}^i) + \lambda_{max}(\Lambda\tilde{\mathbf{\Pi}}^i\tilde{\mathbf{\Pi}}^i)) \tag{14.44}$$

Using (14.43) and (14.44), one can find that

$$\dot{V}(\mathbf{S},\tilde{\mathbf{K}},\tilde{\mathbf{\Pi}}) \leq -V^\sigma(\mathbf{S},\tilde{\mathbf{K}},\tilde{\mathbf{\Pi}})$$

Since $0 < \sigma < 1$, as per Lemma 14.1 for finite stability, the system is finite time stable.

Hence, one can conclude that the system error will converge to origin in finite time without any singularity, and the settling time is given by

$$T \leq \frac{V^{1-\sigma}(\mathbf{S},\tilde{\mathbf{K}},\tilde{\mathbf{\Pi}})}{(1-\sigma)} \tag{14.45}$$

□

From (14.42), it is clear that unlike conventional approaches, the knowledge of the disturbance bound is not needed for tuning the controller gains.

14.3.2 Fast Adaptive Fuzzy NTSMC (FAFNTSMC)

As an alternative to reduce chattering in fast NTSMC, a fuzzy inference mechanism [435–437] has been employed to emulate the discontinuous term in the fast reaching law defined in the previous Subsection. Here, the discontinuous term, $K^i sign(S^i)$, is being replaced by a fuzzy inference mechanism. The input to the fuzzy system is the sliding surface S^i, and the output is fuzzy reaching control law represented by u_d^i. The fuzzy sets for S^i include positive, zero and negative represented by P, Z and N respectively, whereas for u_d^i, they are positive effort, zero effort, and negative effort represented by PE, ZE and NE respectively. The corresponding membership functions are given in Figure 14.7. Taking into account the computational simplicity as well as the feasibility factors, we have chosen the triangular type input membership functions, singleton type output membership functions, and the center of gravity-based defuzzification method. The parameters of the input membership functions, S_a and S_b, are chosen heuristically considering the control constraints as well as the stability requirements. Based on the reaching law, $K^i sign(S^i)$, the rule base can be developed as follows:

- If S^i is P, then u_d^i is PE.
- If S^i is Z, then u_d^i is ZE.
- If S^i is N, then u_d^i is NE.

Using the COG-based defuzzification, the fuzzy inferencing-based reaching law [437], [436] can be defined as

$$u_d^i = \sum_{j=1}^{3} w_j^i r_j^i / \sum_{j=1}^{3} w_j \qquad (14.46)$$

where w_1^i, w_2^i, w_3^i are the firing strengths of the rules; $0 \leq w_1^i, w_2^i, w_3^i \leq 1$, and $w_1^i + w_2^i + w_3^i = 1$. The centers of the output membership functions are $r_1^i = r^i$, $r_2^i = 0$, and $r_3^i = -r^i$ respectively, where r^i is the width of the output membership function. To cope with the uncertainties in the system, an adaptive tuning algorithm can be utilized for updating r^i.

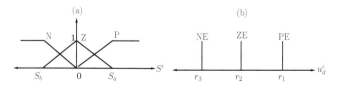

FIGURE 14.7: (a) Input membership function; (b) output membership function.

Multi-Agent Formation Control

Remark 14.3. Computing the firing strengths and the control input, u_d^i, for all possible values of $\mathbf{S}^i$, one can find that $sign(S_j^i) = sign((w_1 - w_3)_j^i)$, where S_j^i and $(w_1 - w_3)_j^i$ are the corresponding elements of $\mathbf{S}^i$ and $(\mathbf{w_1} - \mathbf{w_3})^i$ respectively. $S^i(w_1 - w_3)^i = |S^i||(w_1 - w_3)^i| \geq 0$, and $u_d^i = r^i(w_1 - w_3)^i$. The fast adaptive fuzzy reaching law can be obtained as follows:

$$\mathbf{u}_f^i = -\mathbf{r}^i(\mathbf{w_1} - \mathbf{w_3})^i - \mathbf{\Pi}^i \mathbf{S}^i \qquad (14.47)$$

The final FAFNTSMC law can be defined as

$$\mathbf{u}^i = \mathbf{g}^i(\mathbf{x})^{-1}\left(-\nabla_{\mathbf{x}^i} J(\mathbf{x}) - \lambda_1^i |\mathbf{e^i}|^{1+\frac{1}{\beta}} \text{sign}(\mathbf{e}^i)\right. \qquad (14.48)$$
$$\left. -\lambda_2^i |\mathbf{e^i}|^{1-\frac{1}{\beta}} \text{sign}(\mathbf{e}^i) + \dot{\mathbf{x}}^\ell - \mathbf{r}^i(\mathbf{w_1} - \mathbf{w_3})^i - \mathbf{\Pi}^i \mathbf{S}^i\right)$$

where $\mathbf{r}^i \in \Re^{n \times n}$ is the fuzzy parameter matrix; and $(\mathbf{w_1} - \mathbf{w_3})^i \in \Re^n$ is the weight matrix.

Theorem 14.2. For the given MRS, the control law given by (14.48) can make the system finite time stable, and the system error will converge to origin in finite time without any singularity, if the parameters of the fuzzy-based fast reaching law, given by (14.47), are updated using the following adaptive tuning laws:

$$\dot{\hat{\mathbf{r}}}^i = \frac{1}{\alpha}(\mathbf{w_1} - \mathbf{w_3})^i \mathbf{S}^{iT}$$
$$\dot{\hat{\mathbf{\Pi}}}^i = \Lambda^{-1} \mathbf{S}^i \mathbf{S}^{iT} \quad \alpha, \Lambda > 0$$

Proof. In this case,

$$\dot{\mathbf{S}}^i = -\mathbf{r}^i(\mathbf{w_1} - \mathbf{w_3})^i - \mathbf{\Pi}^i \mathbf{S}^i + \mathbf{d}^i \qquad (14.49)$$

As in the proof for Theorem 14.1, taking the derivative of the Lyapunov candidate given by (14.36), and substituting for $\dot{\mathbf{S}}^i$,

$$\dot{V}^i = \mathbf{S}^{iT}(-\mathbf{r}^i(\mathbf{w_1} - \mathbf{w_3})^i - \mathbf{\Pi}^i \mathbf{S}^i + \mathbf{d}^i) \qquad (14.50)$$

It has been assumed that, for any given operating condition, there exists finite, nominal values for the gains, $\mathbf{r}^{i*}$ and $\mathbf{\Pi}^{i*}$ respectively, such that the sliding condition can be satisfied.

$$\text{Let,} \quad \tilde{\mathbf{r}}^i = \hat{\mathbf{r}}^i - \mathbf{r}^{i*}; \quad \tilde{\mathbf{\Pi}}^i = \hat{\mathbf{\Pi}}^i - \mathbf{\Pi}^{i*}, \qquad (14.51)$$

where $\hat{\mathbf{r}}^i$ and $\hat{\mathbf{\Pi}}^i$ are the estimated values of the respective gains; and $\tilde{\mathbf{r}}^i$ and $\tilde{\mathbf{\Pi}}^i$ are the adaptation errors. The sliding condition ($S\dot{S} \leq 0$) can be satisfied, if the gains are chosen such that

$$\mathbf{r}^i \geq \mathbf{d}^i(\mathbf{w_1} - \mathbf{w_3})^{i^{-1}} + \eta \qquad (14.52)$$

$$\Pi^i \geq -\eta(\mathbf{w_1} - \mathbf{w_3})^{\mathbf{i}}(\mathbf{S}^i)^{-1} + (\mathbf{S}^{iT})^{-1}(\mathbf{S}^T\mathbf{S} + n\lambda_{max}(\delta\tilde{\mathbf{r}}^i\tilde{\mathbf{r}}^i)$$
$$+ n\lambda_{max}(\Lambda\tilde{\mathbf{\Pi}}^i\tilde{\mathbf{\Pi}}^i))^\sigma(\mathbf{S}^i)^{-1} \tag{14.53}$$

where $\lambda_{max}(.)$ represents the maximum eigen value; $\delta, \Lambda > 0$; n is the order of $\tilde{\mathbf{r}}$ and $\sigma \in (0,1)$; and $\boldsymbol{\eta}$ is a diagonal positive definite matrix. The involvement of disturbance term makes it difficult to compute the gains using (14.52) and (14.53). This is the motivation to develop adaptive update laws for tuning these parameters. The rest of the proof proceeds in the same way as that of Theorem 14.1, except that the terms $\mathbf{K}^i$ and $\tanh(\mu \mathbf{S}^i)$ are replaced by $\mathbf{r}^i$ and $(\mathbf{w_1} - \mathbf{w_3})^i$ respectively. Finally, when the parametric update laws are chosen as follows:

$$\dot{\hat{\mathbf{r}}}^i = \frac{1}{\alpha}(\mathbf{w_1} - \mathbf{w_3})^i \mathbf{S}^{iT}$$
$$\dot{\hat{\mathbf{\Pi}}}^i = \Lambda^{-1}\mathbf{S}^i\mathbf{S}^{iT}, \quad \alpha, \Lambda > 0 \tag{14.54}$$

it can be found that

$$\dot{V}(\mathbf{S}, \tilde{\mathbf{r}}, \tilde{\mathbf{\Pi}}) \leq -(\mathbf{S}^T\mathbf{S} + n\lambda_{max}(\delta\tilde{\mathbf{r}}^i\tilde{\mathbf{r}}^i) + n\lambda_{max}(\Lambda\tilde{\mathbf{\Pi}}^i\tilde{\mathbf{\Pi}}^i))^\sigma \tag{14.55}$$

As in the proof for Theorem 14.1,

$$tr\{\delta\tilde{\mathbf{r}}^i\tilde{\mathbf{r}}^i\} + tr\{\Lambda\tilde{\mathbf{\Pi}}^i\tilde{\mathbf{\Pi}}^i\}\} \leq n(\lambda_{max}(\delta\tilde{\mathbf{r}}^i\tilde{\mathbf{r}}^i) + \lambda_{max}(\Lambda\tilde{\mathbf{\Pi}}^i\tilde{\mathbf{\Pi}}^i))$$

Hence,

$$V(\mathbf{S}, \tilde{\mathbf{r}}, \tilde{\mathbf{\Pi}}) \leq \frac{1}{2}(\mathbf{S}^T\mathbf{S}) + n(\lambda_{max}(\delta\tilde{\mathbf{r}}^i\tilde{\mathbf{r}}^i) + \lambda_{max}(\Lambda\tilde{\mathbf{\Pi}}^i\tilde{\mathbf{\Pi}}^i))$$
$$V(\mathbf{S}, \tilde{\mathbf{r}}, \tilde{\mathbf{\Pi}}) \leq \mathbf{S}^T\mathbf{S} + n(\lambda_{max}(\delta\tilde{\mathbf{r}}^i\tilde{\mathbf{r}}^i) + \lambda_{max}(\Lambda\tilde{\mathbf{\Pi}}^i\tilde{\mathbf{\Pi}}^i)) \tag{14.56}$$

Using (14.55) and (14.56), it can be found that,

$$\dot{V}(\mathbf{S}, \tilde{\mathbf{r}}, \tilde{\mathbf{\Pi}}) \leq -V^\sigma(\mathbf{S}, \tilde{\mathbf{r}}, \tilde{\mathbf{\Pi}})$$

Since $0 < \sigma < 1$, as per Lemma 14.1 for finite stability, the system is finite time stable. This completes the proof. □

In a two-wheel differentially driven robotic system, even a single critical actuator fault can lead to complete system failure. Hence, in an MRS, the relevance is more for the control scheme, which is robust enough to retain the stability of formation, even if the formation gets reconfigured owing to the isolation of one or more critically faulty agents. The delay introduced by the fault detection and isolation (FDI) module also affects the system stability. In the case of all such detrimental, unforeseen situations, the speed at which the system adapts to them is equally critical. In addition to this, chattering is another issue to be addressed, while designing a controller based on TSMC scheme.

14.3.3 Fault Detection, Isolation and Collision Avoidance Scheme

In this section, we have considered the case of critical actuator faults in mobile agents, that can lead to complete agent failure, viz., short circuit faults, critical rotor/stator faults, mechanical failures, insulation failure, etc., in the actuators, such that there is reasonably a big difference between the measured and the computed control inputs. The block schematic of the proposed fault tolerant FAGNTSMC scheme is shown in Figure 14.8. To isolate the agents in case of any critical system fault, and to facilitate the formation reconfiguration avoiding the possible collision between the healthy robots and the faulty ones, the APF has been modified in terms of fault parameters. The fault parameters are estimated using a residual-based synchronous fault detection strategy. In the fault detection and isolation (FDI) [435] scheme given in [438], each agent has to compute the residual signal for all the other agents at every instant. It includes the computation of the velocity of each agent from the noisy position data using conventional differentiation as well as the control inputs of all other agents. This will increase the computational complexity, and can make the system slow. Moreover, in a distributed fault detection system, the faulty agent has to be identified and isolated at the same instant by all the other agents. Otherwise, the asynchronous recovery behavior may lead to system instabilities. To avoid this, we have devised the FDI protocol in such a way that each member need to compute only its own residual signal and a fault flag has to be set/ reset based on this residual value. The value of the fault flag has to be communicated with all the other members. The fault flag, f^i, at every sampling instant can be computed as follows:

$$f^i = \begin{cases} 1, & \text{if } \|\mathbf{u}^i - \mathbf{u}_m^i\| > (D + \xi \|\mathbf{u}^i\|) \\ 0, & \text{otherwise.} \end{cases} \quad (14.57)$$

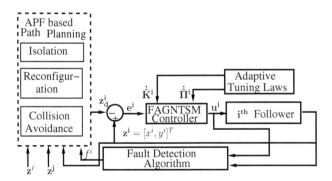

FIGURE 14.8: Fault tolerant formation control using fast adaptive gain NTSMC scheme.

where $\mathbf{u}^i$ is the control input computed using (14.33), $\mathbf{u}^i_m$ is the measured input, D is the bound on the lumped uncertainty, and ξ is a constant. To ensure the criticality of the fault, ξ can be set to a relatively high value. Minor faults can be treated as disturbances, so that the FAGNTSMC can handle them using adaptation. To isolate the jth faulty agent and to facilitate the formation reconfiguration avoiding the collision between healthy robot and the isolated faulty ones, the gradient potential to be computed by the ith member is modified as

$$\nabla_{\mathbf{z}^i} J(\mathbf{z}) = \sum_{j=1, j \neq i}^{N} [\nabla_{\mathbf{z}^i} J^{ij}(\|\mathbf{z}^i - \mathbf{z}^j\|)](1 - f^j) + f^j \nabla_{\mathbf{z}^i} F^i$$
$$+ \nabla_{\mathbf{z}^i} J^{il}(\|\mathbf{z}^i - \mathbf{z}^l\|) \quad (14.58)$$

where F represents the collision avoidance potential between the healthy robots and the faulty ones. The structure of this term has been chosen as similar to that of the repulsive potentials employed in obstacle avoidance-based motion planning [419], and it is given by

$$F^i = \begin{cases} \frac{1}{2}\gamma \left(\frac{1}{\|\mathbf{z}^i - \mathbf{z}^j\|} - \frac{1}{d_f} \right)^2, & \text{if } \|\mathbf{z}^i - \mathbf{z}^j\| < d_f \\ 0, & \text{otherwise.} \end{cases} \quad (14.59)$$

where d_f is the distance of influence of the faulty robot, in other words, the minimal distance to be maintained by the healthy robot with respect to the faulty one. γ is a positive constant, which indicates the maximum applicable potential. F^i is a continuous differentiable function and its structure is such that, for the ith agent, when the distance with respect to the jth faulty agent falls below d_f, it gets repelled away, as we are enforcing the velocity of the agent to follow the negative gradient of the respective net potential. This potential will be triggered only when the fault flag f^j is being set and communicated with the neighboring agents. It has been assumed that there are no communication faults in the system.

Example 14.5. Consider a system of five robots with one leader and the rest of the followers making a square shaped formation, with an edge size of 1m. Design and implement an APF-based path planner and an FAGNTSMC-based control scheme for the robots to follow a circular trajectory.

Solution 14.5. *The path planning scheme can be developed based on the relative motion dynamics of the ith follower w. r. t leader, given by*

$$\dot{\mathbf{x}}^{\mathbf{i}} - \dot{\mathbf{x}}^{\ell} = -\nabla_{\mathbf{x}^i} J(\mathbf{x}), i = 1....N \quad (14.60)$$

where $\mathbf{x}^{\mathbf{i}} \in \Re^n$ represents the position vector of ith follower, and $\mathbf{x}^{\ell} \in \Re^n$ is the position vector of the leader,

$$J(\mathbf{x}) = \sum_{i=1}^{N-1} \sum_{j=i+1}^{N} J^{ij}_a(\|\mathbf{x}^i - \mathbf{x}^j\|) - J^{ij}_r(\|\mathbf{x}^i - \mathbf{x}^j\|)$$

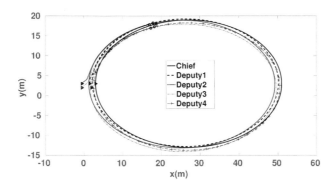

FIGURE 14.9: Path of the robots in formation.

$$+ \sum_{i=1}^{N} J_a^{i\ell}(\|\mathbf{x}^i - \mathbf{x}^\ell\|) - J_r^{i\ell}(\|\mathbf{x}^i - \mathbf{x}^\ell\|) \qquad (14.61)$$

$$J_a(x^i - x^j) = \frac{a}{2}\|x^i - x^j\|^2,$$
$$J_r(x^i - x^j) = -\frac{bc}{2}exp\left(-\frac{\|x^i - x^j\|^2}{c}\right)$$

The simulation parameters are chosen as: $\alpha = 10$, $L = 0.21$ m, $\boldsymbol{\lambda}_1 = diag\{0.1, 0.1\}$, $\boldsymbol{\lambda}_2 = diag\{0.1, 0.1\}$, $\beta = 3$, $\delta = 1000$, $\Lambda = 50000$, $S_a = 1$, $S_b = -1$. The values of the potential function parameters are chosen based on the desired inter-agent distances. Based on the motion dynamics of the non-holonomic robots given by (14.34), the control law for the ith follower can be obtained as

$$\mathbf{u}^i = \mathbf{g}^i(\mathbf{x})^{-1}\left(-\boldsymbol{\nabla}_{\mathbf{x}^i}J(\mathbf{x}) + \dot{\mathbf{x}}^\ell - \boldsymbol{\lambda}_1^i|\mathbf{e}^i|^{1+\frac{1}{\beta}}\mathbf{sign}(\mathbf{e}^i)\right.$$
$$\left.-\boldsymbol{\lambda}_2^i|\mathbf{e}^i|^{1-\frac{1}{\beta}}\mathbf{sign}(\mathbf{e}^i) - \mathbf{K}^i\mathbf{tanh}(\mu\mathbf{S}^i) - \mathbf{\Pi}^i\mathbf{S}^i\right) \qquad (14.62)$$

where $\mathbf{g}^i(\mathbf{x}) = \begin{pmatrix} cos(\theta^i) & -Lsin(\theta^i) \\ sin(\theta^i) & Lcos(\theta^i) \end{pmatrix}$, $\dot{\mathbf{x}}^\ell = [\dot{x}^\ell, \dot{y}^\ell]^T$; $\mathbf{K}^i = diag\{K_x^i, K_y^i\}$, and $\mathbf{\Pi}^i = diag\{\Pi_x^i, \Pi_y^i\}$, represent the controller gains, $\Pi_{x,y}^i, K_{x,y}^i > 0$, $\forall i$; $\mathbf{tanh}(\mu\mathbf{S^i}) = [tanh(\mu S_x^i), tanh(\mu S_y^i)]^T$.

Since $det\left(\mathbf{g}^i(\mathbf{x})\right) \neq 0$, $\mathbf{g}^i(\mathbf{x})$ is nonsingular and invertible. In the simulations, the robots are indexed as follows: Leader→ 1, Deputy 1 → 2, Deputy 2 → 3, Deputy 3 → 4, Deputy 4 → 5. The separation distances are defined according to these index representations. The results given by Figure 14.9–14.11 indicate that the required formation is achieved within 5s.

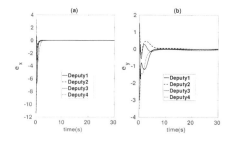
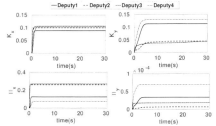

FIGURE 14.10: Tracking error for the follower robots.

FIGURE 14.11: Tuning response of the gains.

14.4 Experiments

The real time experimentations are done using *FireBird VI* and *Pioneer P3 − DX* robotic platforms. Both are two-wheel differentially driven robots with a caster wheel support. FireBird VI robot is equipped with an onboard computer with atom processor, running on Ubuntu 12.04 platforms, sonars, position encoders *etc.*, weighing around 7 kg. All the peripherals of the robot are interfaced with the main microcontroller LPC1769 ARM cortex-M3 over UART, I2C and SPI bus. The Pioneer P3-DX robot is also equipped with an onboard PC running on Ubuntu platform, and other sensors as in FireBird VI, weighing around 16 kg.

The robots are communicating through a local WIFI network, with which static IPs are generated and assigned to the robots. The *Player* 3.0.2 robot device server is running on all the platforms, which facilitates ultimate control over all sensors and actuators. A static overhead low cost web cam has been utilized to find the initial relative positions and orientations of the robots. The camera is connected to an external PC, which will communicate with the robots through WIFI. Different patterns of black and white markers are fixed on the top of each robot, with which the platforms can be identified. The markers are detected using $ARUCO$, an open source augmented virtual reality library; and the initial relative positions and orientations can be computed using the marker IDs. Once the robot starts moving, relative positions and orientations are computed with the help of odometry data. The marker ID based estimated data will serve as initial offset to the odometry readings.

The values of the experimental parameters are chosen as: $S_a = 1$, $S_b = -1$, $\alpha = 10$, $L = 0.21\ m$, $\lambda_1 = diag\{0.1, 0.1\}$, $\lambda_2 = diag\{0.1, 0.1\}$, $\beta = 3$, $\delta = 1000$, $\Lambda = 50000$, $a = 0.00551819$, $b = 0.15$, and $c = 0.302793$. The APF parameters are chosen based on the desired formation separation.

We have considered a system of three robots making a triangular formation, with one leader and two followers. The control graph is such that the $deputy - 1$ has to follow only the chief, whereas the $deputy - 2$ has to follow

Experiments 531

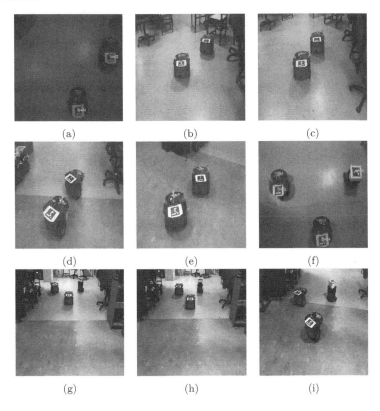

FIGURE 14.12: Taped video images of robots in formation with: (a)-(e) two similar robots; (f)-(i) three dissimilar robots; (a) and (f) images from the overhead camera with the markers detected.

both the *chief* as well as *deputy* − 1 simultaneously, maintaining the respective desired offsets. The control graph determines the extent of computation of the gradient potentials as well as the control inputs. In this case, one of the Pioneer $P3 - DX$ robots has been assigned as the chief and the other one as $deputy-2$, with a FireBird VI robot as the $deputy-1$. Here, the chief is free to navigate, and it decides the formation trajectory, deputies have to follow the chief with a desired distance separation of 0.7 m with respect to chief as well as between themselves. Since it is APF-based path planning, the final configuration to which the deputies converge, depends on the initial positions. Once the robots are in formation, the deputies are lying along the circumference of 0.7 m circle centered around the chief, maintaining the desired separation of 0.9 m between themselves. Taped video images of robots in formation are shown in Figure 14.12. The results given by Figure 14.13 − 14.15, show that the the tracking errors are converging rapidly to the origin, and the desired

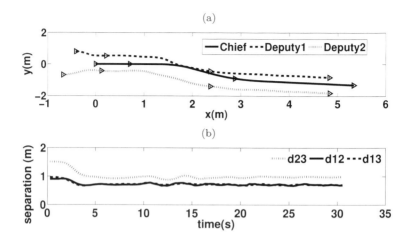

FIGURE 14.13: Experimental results with FAGNTSMC (three robots): (a) Position trajectories, (b) Separation distances.

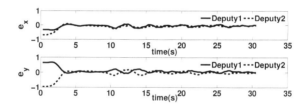

FIGURE 14.14: Experimental results with FAGNTSMC (three robots): Tracking error.

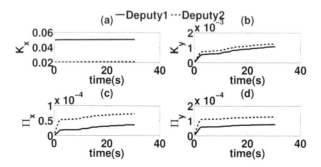

FIGURE 14.15: Experimental results with FAGNTSMC (three robots): Tuning response of the gains.

Experiments

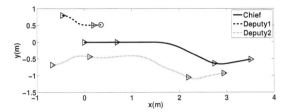

FIGURE 14.16: Experimental results with FAGNTSMC (fault case): Path of the robots.

formation is achieved within $5s$, even if the dissimilar platforms are used. The taped video images of the formation are shown in Figure 14.12, with Figure 14.12 (a) and (f) showing the images captured by the overhead camera, with the detected markers [435].

For demonstrating the robustness and fault tolerance capability of the proposed scheme, the experiments are carried out using three robots, in which one of the agents goes to a complete failure on runtime. FDI algorithm depicted in Section $IV\ B$ is utilized to identify and isolate the faulty member. In order to study the effect of delay between fault occurrence and isolation on the system performance, a delay of almost $3s$ has been induced in the FDI module. The parameters of the residual signal are chosen as, $D = 0.2$ and $\xi = 0.7$ respectively. The deliberate choice of the high value for ξ is to induce the necessary delay in the FDI module, as well as to ensure the criticality of the fault. The fast and finite sliding surface, together with an adaptive, fast reaching law can save the system from possible instability, which may occur due to this delay. The faster convergence along with gain adaptation, can empower the system to adapt to such a situation upto a certain extent. In this case, the *deputy* -1 goes to complete failure after $10s$, and the *deputy* -2, which is supposed to follow *chief* as well as *deputy* -1, reconfigures itself, maintaining the desired formation separation. Based on the FDI output, the APF-based

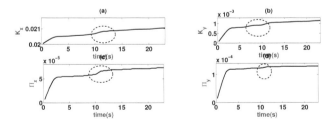

FIGURE 14.17: Experimental results with FAGNTSMC (fault case): Tuning response of the gains for Deputy 2.

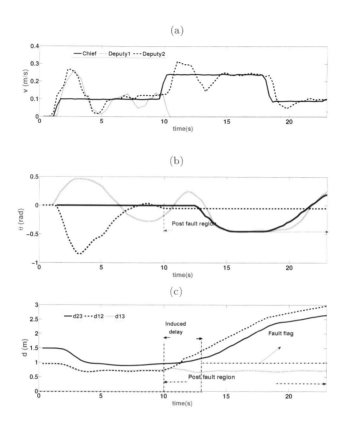

FIGURE 14.18: Experimental results with FAGNTSMC (fault case): (a) Velocity, (b) Turn angle, (c) Separation distances.

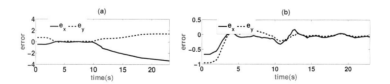

FIGURE 14.19: Experimental results with FAGNTSMC (fault case): Tracking errors (a) Deputy 1, (b) Deputy 2.

path planning module will switch it to a two-robot formation. The path of the robots are shown in Figure 14.16, where the circle shows the system failure point, and the positions of the agents are represented by colored arrows. The deflections observed at around $10s$ (region shown by ellipse) in all the results are due to the formation switching owing to agent failure. From Figure 14.18. c, it is evident that even though the fault occurs at around $10s$, the fault flag is being set only after $13s$. The results given by Figure 14.19 indicate that the error trajectories are converging again to the origin rapidly after the faulty member is isolated. The controller gains settle down to their respective optimal values far before the occurrence of the fault. Once the failed node is identified and isolated, the formation reconfigures, and the system adapts to this by re-updating the controller gains. This is evident from the tuning drifts of these parameters shown in Figure 14.17. The **Youtube link** for the videos recorded for all the experiments is given in [439].

14.5 Summary

This chapter gives an overview of different control and coordination schemes for the formation control of multi-robot systems. We have detailed the geometric -based and artificial potential function-based path planning schemes with examples. The fault adaptive nonsingular TSMC technique for the formation control of multi-robot systems is also demonstrated with real-time experimental results. The numerical examples are solved in each section, which can aid the readers to acquire a through understanding of various control and coordination schemes in the formation control of multi-robotic systems.

15

Event Triggered Multi-Robot Consensus

15.1 Introduction to Event Triggered Control

Resource optimality is a significant design constraint in multi-robotic systems. The major operation costs are for the resource requirements associated with communication and sensing. In the last chapter, we have discussed different types of control and co-ordination schemes for multi-robot systems. But, they are all time triggered control protocols, i.e., control is triggered at all sampling instants as shown in Figure 15.1(a), where resource requirements are not optimal.

Event triggered control (ETC) is an effective technique to improve the resource saving capability of the system. In this design, the control input is updated, only when it is required. i.e., the control is triggered only when an event occurs, as shown in Figure 15.1(b). Event triggered control designs consist of two steps: first is the development of a controller to compute the control inputs and the second one includes the design of triggering condition, which decides when the control needs to be updated again. But ETC is based on the assumption that all the parameteric measurements/ estimates are available at all sampling instants. To optimize the sensing requirements, self triggered control (STC) techniques can be used, where the sensing is also event triggered as given in Figure 15.2. In this case, the event instants are pre-computed based on a predictive model or previous instant measurements. Hence, the accuracy depends upon the precision of the estimation scheme. This chapter aims to

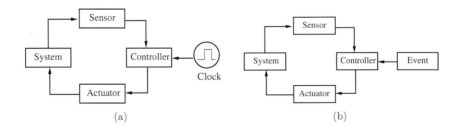

FIGURE 15.1: (a) Time triggered control, (b) Event triggered control.

537

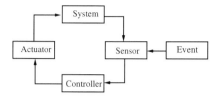

FIGURE 15.2: Self triggered control.

give an overview of the implementation of event triggered control scheme in the consensus-based formation of multi-robot systems.

Example 15.1. Consider a simple linear system with the dynamics given by

$$\dot{x} = Ax + Bu, \tag{15.1}$$

where $A = \begin{bmatrix} 1 & -1 \\ 1 & 1 \end{bmatrix}$ and $B = \begin{bmatrix} 0 & 1 \end{bmatrix}^T$.

Design and implement an event triggering based state feedback controller for the system.

Solution 15.1. Let $x(t_k)$ be the state measurements at the event instants, and $t \in (t_k, t_{k+1})$ be the inter-event time instant. Let the measurement error be defined as

$$e = x(t) - x(t_k)$$

The event triggering based state feedback control law can be designed as

$$u = -kx(t_k) \tag{15.2}$$

Consider a Lyapunov function

$$V = \frac{1}{2}x(t)^T x(t) \tag{15.3}$$

Taking the derivative and substituting

$$\begin{aligned} \dot{V} &= x(t)^T \dot{x}(t) \\ &= x(t)^T (Ax(t) - Bkx(t_k)) \\ &= x(t)^T (Ax(t) - Bk(x_t) - Bke) \\ &= x(t)^T (A - Bk)x(t) - x(t)^T Bke \\ &\leq -A'\|x(t)\|^2 - Bk\|x(t)\|\|e\| \end{aligned} \tag{15.4}$$

where $A' = Bk - A$.

Event Triggered Consensus

If the triggering condition is chosen as

$$e <= \sigma\|x(t)\|, \quad \sigma > 0 \tag{15.5}$$

and the controller gain, k, is chosen such that $k \geq B^{-1}A$,

$$\dot{V} \leq -(A' + Bk\sigma)\|x(t)\|^2 \tag{15.6}$$

Since $\dot{V}$ is negative definite, hence, one can conclude that with the triggering condition given by (15.5), the state feedback controller, given by (15.2) can make the system asymptotically stable.

```
MATLAB CODE FOR EVENT TRIGGERED CONTROL
---------------------------------------
clear all;
A=[1,-1;1,1];B=[0;1];x=[0;0]; x_pr=[0;0];
t=0,n=1000;k=[0.8, 0;0,0.8];
for i=1:n
xdot=Ax+Bu;x=x+xdot*T;
e_nor=sqrt((x-x_pr)'*(x-x_pr));x_nor=sqrt(x'*x); x_pr=x;
if(e_nor>=sigma*x_nor)
u=-k*x;
end
x_s(:,i)=x;u_s(:,i)=u;
end
time=T*(1:n);
figure;plot(time,x_s)
figure;plot(time,u_s)
```

Zeno phenomenon [440], i.e., infinite number of switchings / triggers happening in a finite time period, is a major drawback in an event triggered approach. The event triggered strategy should be designed in such a way that the zeno phenomenon should not occur.

15.2 Event Triggered Consensus

Communication cost is also a significant constraint in multi-robot formation control. In the preceding chapter, the communication consensus has not been considered. In this chapter, we are introducing a resource optimal multi-robot formation control scheme through consensus-based path planning and event triggered control technique. The event triggering technique can be extended to multi-agent system with a consensus-based framework, so that the control input to each agent, for the formation keeping, need to be triggered at its own event instants/ neighboring agents.

Numerous event triggering-based multi-agent consensus algorithms are available in literature [441–444]. In most of these works, only the asymptotic convergence is proved, where the finite time consensus cannot be guaranteed. Finite time convergence capability is extremely critical in a multi-agent system (MAS). The performance of the control module depends upon it. In [445], a nonlinear finite time event triggered consensus protocol has been developed for both leader-following as well as leaderless cases. But it does not take into account the practical disturbances existing in the system. The disturbance-rejection capability is also equally significant in the design of consensus-based MAS protocols.

In [446], an H_∞-based control approach has been utilized for the consensus of MAS with energy-bounded disturbances, in which an accurate model is required to implement it successfully. Apart from this, finite time convergence is also not guaranteed in this case. In [447], an ISMC-based strategy has been considered to address the consensus problem of MAS with bounded disturbances. Similarly, in [448], Rao and Ghose have developed a sliding mode-based algorithm for the leaderless consensus of UAVs. But, in both these works, an event triggering based resource saving criterion addressed in the preceding papers is not considered.

Sliding mode-based event triggered as well as self-triggered strategies for generalized linear and nonlinear systems are depicted in [440] and [449] respectively, in which a linear measurement error is defined. Though these approaches are very robust, it is difficult to obtain the triggering condition for the finite time consensus of MAS with a nonlinear consensus protocol using these techniques. Moreover, in [440], it has been assumed that the nonlinear function involved in the system dynamics has a unique equilibrium point, and it is locally Lipschitz continuous. This cannot always be guaranteed in a complex MAS.

Another issue is that most of the event triggered consensus protocols depicted in literature are restricted to simulation based validations. The real time implementation issues are not being addressed in these works.

The objectives of the chapter are as follows:

(i) To design a resource optimal consensus protocol for a multi-agent system with bounded disturbances, in a leader-follower-based framework.

(ii) To develop a robust event triggered consensus protocol based on ISMC technique to deal with the bounded disturbances.

(iii) The control protocol should be designed to ensure a faster convergence.

(iv) The triggering condition should be designed so as to guarantee the finite time sliding mode stability as well as finite time consensus.

(v) Theoretically validate the zeno-free behavior of proposed ETC scheme.

Event Triggered Consensus

(vi) The consensus protocol should be designed in such a way that the desired relative state deviation between the agents can be achieved with a directed graph topology.

(vii) To implement the proposed strategies in nonholonomic robots and to validate the robustness.

In this chapter, a novel integral sliding mode-based event triggering scheme has been designed for the consensus-based tracking control of multi-robotic systems with modeling uncertainties / disturbances. A novel measurement error has been defined, and the event triggering based robust consensus algorithm has been derived based on this. A fast reaching law has been utilized in the consensus protocol to improve the speed of convergence. The finite time consensus convergence and the finite time sliding mode stability have been proved for the proposed design using Lyapunov- based analysis. A lower bound for the inter-execution time has been derived to ensure that the zeno behavior does not exist. The event triggered consensus protocol is designed for the MAS in a leader-follower-based framework such that the desired relative state deviation between the agents can be achieved with a directed graph topology. The implementation issues are studied through real time experimentations.

15.2.1 Preliminaries

Consensus algorithms are the rules for inter-agent interactions, so that all the agents will reach an agreement regarding the quantity of interest, once they are in consensus, and the information states of all the agents will converge to a common value. Consider a multi-agent system (MAS) comprising N members. The communication topology among the agents is represented by a directed graph $\mathcal{G} = \{\mathcal{V}, \mathcal{E}, \mathcal{A}\}$, in which the set of nodes is denoted by $\mathcal{V} = \{\mathcal{V}_1, \mathcal{V}_2, \mathcal{V}_3 \ldots \mathcal{V}_N\}$, the set of edges by $\mathcal{E} \subseteq \mathcal{V} \times \mathcal{V}$. Let $A = [a_{ij}]$ represents the weighted adjacency matrix, in which $a_{ij} = 1$, if $(\mathcal{V}_i, \mathcal{V}_j) \in \mathcal{E}$, otherwise $a_{ij} = 0$. For $\mathcal{V}_i$, the neighbor set is given by $\mathcal{N}_i = \{\mathcal{V}_j : (\mathcal{V}_i, \mathcal{V}_j) \in \mathcal{E}\}$. The Laplacian is represented by $\mathcal{L} = \mathcal{D} - \mathcal{A}$, where $\mathcal{D} = \text{diag}\{d_1, d_2, \ldots d_n\}$ is the degree matrix, and $d_i = \sum_{j=1}^{n} a_{ij}$.

For a leader-follower-based MAS, with one leader and N followers, the leader is represented by node $\mathcal{V}_0$, and followers by $\{\mathcal{V}_1, \mathcal{V}_2, \mathcal{V}_3 \ldots \mathcal{V}_N\}$. The leader-follower connection weight matrix is defined as $\mathcal{B} = \text{diag}\{b_1, b_2, \ldots b_n\}$, $b_i > 0$, if the leader is connected to i^{th} follower, otherwise $b_i = 0$.

For a system of n agents with $x_i(t)$ as the information state [450] for the ith agent, the most commonly used consensus protocol is given by

$$\dot{x}_i(t) = a_{ij}(x_i(t) - x_j(t)), \quad i = 1 \ldots n \quad (15.7)$$
$$= -\mathcal{L}x \quad (15.8)$$

where $x = [x_1, x_2 \ldots x_n]$.

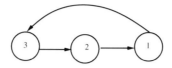

FIGURE 15.3: Graph.

Example 15.2. Consider a system of three agents communicating based on a directed graph given in Figure 15.3. Develop a consensus protocol for the MAS.

Solution 15.2. The consensus protocol is given by

$$\dot{x}_i(t) = a_{ij} * (x_i(t) - x_j(t)), \quad i = 1\ldots3 \tag{15.9}$$
$$= -\mathcal{L}x \tag{15.10}$$

where $x = [x_1, x_2, x_3]^T$.

From the graph, the adjacency matrix can be obtained as $A = \begin{bmatrix} 0 & 1 & 0 \\ 0 & 0 & 1 \\ 1 & 0 & 0 \end{bmatrix}$

and $D = \begin{bmatrix} 1 & 0 & 0 \\ 0 & 1 & 0 \\ 0 & 0 & 1 \end{bmatrix}$.

Laplacian can be computed as

$$\mathcal{L} = \begin{bmatrix} 1 & -1 & 0 \\ 0 & 1 & -1 \\ -1 & 0 & 1 \end{bmatrix} \tag{15.11}$$

For a leader-follower-based framework, the dynamics of the MAS can be defined as

$$\dot{\mathbf{x}_i}(\mathbf{t}) = \mathbf{u}_i(t) + \mathbf{d}_i(t), \quad i = 1\ldots N \tag{15.12}$$
$$\dot{\mathbf{x}_0}(\mathbf{t}) = \mathbf{u}_0(t)$$

where $\mathbf{x}_i(t) \in \Re^n$, $\mathbf{u}_i(t) \in \Re^n$ and $\mathbf{d}_i(t) \in \Re^n$ represent the position vector, the control input, and the bounded disturbance input for the ith follower respectively; $\mathbf{x}_0(t) \in \Re^n$ and $\mathbf{u}_0(t) \in \Re^n$ are the position vector and control input to the leader respectively. Our objective is to develop a finite time event triggering based control law for achieving consensus in a multi-agent system with a leader-follower-based framework, in spite of external bounded disturbances.

Let $\tilde{\mathbf{x}_i}(\mathbf{t}) = \mathbf{x}_i(t) - \mathbf{x}_0(t) + \boldsymbol{\delta}_i$, and $\tilde{\mathbf{u}_i}(\mathbf{t}) = \mathbf{u}_i(t) - \mathbf{u}_0(t)$ \tag{15.13}

where δ_i is the desired state deviation.

Event Triggered Consensus

The relative dynamics can be defined as

$$\dot{\tilde{\mathbf{x}}}_\mathbf{i}(\mathbf{t}) = \tilde{\mathbf{u}}_\mathbf{i}(\mathbf{t}) + \mathbf{d}_i \qquad (15.14)$$

Lemma 15.1. *[445]* In the absence of any disturbances, the system can achieve the consensus tracking in finite time, if the following protocol is chosen.

$$\tilde{u}_i(t) = \chi_i^\eta(t)$$

where,

$$\begin{aligned}\chi_i(t) &= -\frac{\mu_i}{n_i+1} \sum_{j \in \mathcal{N}_i} a_{ij}[(\mathbf{x}_i(t) + \mathbf{x}_0(t) + \boldsymbol{\delta}_i) \\ &\quad - (\mathbf{x}_j(t) + \mathbf{x}_0(t) + \boldsymbol{\delta}_j)] + b_i(\mathbf{x}_i(t) - \mathbf{x}_0(t) + \boldsymbol{\delta}_i) \\ &= -\frac{\mu_i}{n_i+1} \sum_{j \in \mathcal{N}_i} a_{ij}(\tilde{\mathbf{x}}_\mathbf{i}(t) - \tilde{\mathbf{x}}_j(t)) + b_i(\tilde{\mathbf{x}}_i(t)) \end{aligned} \qquad (15.15)$$

where $\mu_i > 0$, and $\eta \in (0.5, 1)$ is strictly the ratio of positive odd numbers; and $1 \leq n_i \leq N$ is the number of neighboring agents for the ith member.

Lemma 15.2. *[445], [451]* For the system given by $\dot{\mathbf{x}} = \mathbf{f}(\mathbf{x})$, $f(0) = 0$, $\mathbf{x}(0) = \mathbf{x}^0$, $\mathbf{x} \in U_0 \subset \Re^n$, suppose that there exists a positive definite continuous function, $V(\mathbf{x})$, in the neighborhood of the origin, real numbers $k > 0$, $\eta \in (0,1)$, so that $\dot{V}(\mathbf{x}) + kV^\eta(\mathbf{x}) \leq 0$, then $V(\mathbf{x})$ approaches zero in finite time, and the settling time T is such that $T \leq \frac{V^{1-\eta}(\mathbf{x}(0))}{k(1-\eta)}$.

Lemma 15.3. *[451] − [452]* For the system given by $\dot{\mathbf{x}} = \mathbf{f}(\mathbf{x})$, $f(0) = 0$, $\mathbf{x} \in U_0 \subset \Re^n$, suppose that there exists a positive definite continuous function, $V(\mathbf{x})$, in the neighborhood of the origin, and the real numbers $C_1, C_2 > 0$, $\eta \in (0,1)$, so that $\dot{V}(\mathbf{x}) + C_1 V^\eta(\mathbf{x}) + C_2 V(\mathbf{x}) \leq 0$, then $V(\mathbf{x})$ approaches zero in finite time, and the settling time T is such that $T \leq \frac{1}{C_2(1-\eta)} \ln\left[\frac{C_2 V^{1-\eta}(\mathbf{x}(0)) + C_1}{C_1}\right]$.

Lemma 15.4. *[445], [453] − [454]* For any leader-follower-based MAS, with the communication topology as defined in Section 15.2.1, if the graph contains a directed spanning tree, then all the eigen values of $\mathcal{L} + \mathcal{B}$ have positive real parts.

Lemma 15.5. *[445], [453]* For $x_i \in \Re^n$, $i = 1 \ldots n$, $\alpha \in (0,1]$, then $\left(\sum_{i=1}^n |x_i|\right)^\alpha \leq \sum_{i=1}^n |x_i|^\alpha \leq n^{1-\alpha}\left(\sum_{i=1}^n |x_i|\right)^\alpha$, and for $|\alpha| \in (0,1)$, $\|x_i^\alpha\| \leq n^{1-\alpha}\|x_i\|^\alpha$.

Assumption 15.1. For the MAS, with the agent dynamics given by (15.14), the the unmodelled dynamics/ disturbance has been lumped together to form $\mathbf{d}_i$, and it is assumed to be bounded, i.e., $\|\mathbf{d}_i\| < D_i$, where D_i is the bound on it.

15.2.2 Sliding Mode-Based Finite Time Consensus

Considering the disturbances in the system, the consensus protocol can be modified employing sliding mode-based approach. The integral type sliding surface is defined as

$$\mathbf{S}_i(t) = \tilde{\mathbf{x}}_i(t) - \tilde{\mathbf{x}}_i(0) - \int_0^t \chi_i^\eta(t)dt, \quad i = 1\ldots n \quad (15.16)$$

where $\mathbf{S}_i(t) = [s_1(t), s_2(t), \ldots s_n(t)]^T$. Once sliding mode occurs, $s_i(t) = 0$, $\dot{s}_i(t) = 0$

$$\dot{\tilde{\mathbf{x}}}_i(t) = \chi_i^\eta(t) \quad (15.17)$$

For faster convergence, we have chosen a fast reaching law [451], [455] given by,

$$\dot{\mathbf{S}}_i(t) = -\mathbf{K}_1 \mathbf{sign}(\mathbf{S}_i(t)) - \mathbf{K}_2 \mathbf{S}_i(t) \quad (15.18)$$

where $\mathbf{K}_1 = \mathrm{diag}\{k_{11}, k_{12}\ldots k_{1n}\}$, $\mathbf{K}_2 = \mathrm{diag}\{k_{21}, k_{22}\ldots k_{2n}\}$, $k_{ij} > 0$, $\forall i, j$, and $\mathbf{sign}(\mathbf{S}_i(t)) = [\mathrm{sign}(s_1(t)), \mathrm{sign}(s_2(t)), \ldots \mathrm{sign}(s_n(t))]^T$. The sliding mode-based consensus protocol can be defined as,

$$\tilde{\mathbf{u}}_i(t) = \chi_i^\eta(t) - \mathbf{K}_1 \mathbf{sign}(\mathbf{S}_i(t)) - \mathbf{K}_2 \mathbf{S}_i(t) \quad (15.19)$$

The design of event triggering based control algorithm is detailed in the next Section.

15.3 Event Triggered Sliding Mode-based Consensus Algorithm

The integral sliding mode-based event triggered consensus protocol can be defined as

$$\tilde{\mathbf{u}}_i(t) = \chi_i^\eta(t_k^i) - \mathbf{K}_1 \mathbf{sign}(\mathbf{S}_i(t_k^i)) - \mathbf{K}_2 \mathbf{S}_i(t_k^i) \quad (15.20)$$

where $t \in [t_k^i, t_{k+1}^i)$, and t_k^i is the triggering time.

We have defined a novel measurement error, suited for the system, and it is given by

$$\begin{aligned}\mathbf{e}_i(t) = &\chi_i^\eta(t_k^i) - \mathbf{K}_1 \mathbf{sign}(\mathbf{S}_i(t_k^i)) - \mathbf{K}_2 \mathbf{S}_i(t_k^i) \\ &- (\chi_i^\eta(t) - \mathbf{K}_1 \mathbf{sign}(\mathbf{S}_i(t)) - \mathbf{K}_2 \mathbf{S}_i(t))\end{aligned} \quad (15.21)$$

The triggering scheme is designed such that control will be triggered for each agent at its own event time only. The control input remains piecewise constant during the inter execution time. The block schematic of the proposed scheme is shown in Fig. 15.4.

Event Triggered Sliding Mode-based Consensus Algorithm

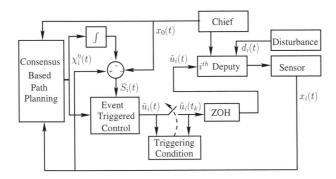

FIGURE 15.4: Proposed event triggering scheme.

Theorem 15.1. *[450]* Consider the leader-follower-based MAS with the dynamics as given in (15.14), the event triggered control protocol defined by (15.20) can make the system finite time stable, irrespective of the disturbances, if the triggering function is chosen as

$$f_i(t) = \|\mathbf{e}_i(t)\| - \rho_i \tag{15.22}$$

where, $\rho_i < \lambda_{min}(\mathbf{K}_1) - D_i$, $\lambda_{min}(\mathbf{K}_1) > D_i$.

Proof. consider the Lyapunov function as

$$\mathbf{V}_1(t) = \frac{1}{2}\mathbf{S}_i^T(t)\mathbf{S}_i(t) \tag{15.23}$$

$$\dot{\mathbf{V}}_1(t) = \mathbf{S}_i^T(t)\dot{\mathbf{S}}_i(t)$$

$$\dot{\mathbf{S}}_i(t) = \dot{\tilde{\mathbf{x}}}_i(t) - \chi_i^\eta(t)$$

$$\dot{\mathbf{S}}_i(t) = \chi_i^\eta(t_k^i) - \mathbf{K}_1\mathbf{sign}(\mathbf{S}_i(t_k^i)) - \mathbf{K}_2\mathbf{S}_i(t_k^i)$$
$$\quad + \mathbf{d}_i(t) - \chi_i^\eta(t) \tag{15.24}$$
$$= \mathbf{e}_i(t) - \mathbf{K}_1\mathbf{sign}(\mathbf{S}_i(t)) - \mathbf{K}_2\mathbf{S}_i(t) + \mathbf{d}_i(t)$$

Hence,

$$\dot{V}_1(t) = \mathbf{S}_i^T(t)[\mathbf{e}_i(t) - \mathbf{K}_1\mathbf{sign}(\mathbf{S}_i(t)) - \mathbf{K}_2\mathbf{S}_i(t) + \mathbf{d}_i(t)]$$
$$\leq \|\mathbf{S}_i(t)\|\|\mathbf{e}_i(t)\| - \lambda_{min}(\mathbf{K}_1)\mathbf{S}_i^T(t)\mathbf{sign}(\mathbf{S}_i(t))$$
$$\quad - \lambda_{min}(\mathbf{K}_2)\mathbf{S}_i^T(t)\mathbf{S}_i(t) + \|\mathbf{S}_i(t)\|D_i$$
$$\leq \|\mathbf{S}_i(t)\|\|\mathbf{e}_i(t)\| - \lambda_{min}(\mathbf{K}_1)\|\mathbf{S}_i(t)\|$$
$$\quad - \lambda_{min}(\mathbf{K}_2)\|\mathbf{S}_i(t)\|^2 + \|\mathbf{S}_i(t)\|D_i \tag{15.25}$$

If, $\|\mathbf{e}_i(t)\| \leq \rho_i$, where $\rho_i < \lambda_{min}(\mathbf{K}_1) - D_i$

$$\dot{V}_1(t) \leq -(\lambda_{min}(\mathbf{K}_1) - D_i - \rho_i)\|\mathbf{S}_i(t)\| - \lambda_{min}(\mathbf{K}_2)\|\mathbf{S}_i(t)\|^2$$

$\lambda_{min}(.)$ represents the smallest eigen value. Choosing $\lambda_{min}(\mathbf{K}_1) > D_i + \rho_i$, we can apply Lemma 15.3.

$$\dot{V}_1(t) \leq -C_1 V_1(t)^{\frac{1}{2}} - C_2 V_1(t), \tag{15.26}$$

$$\text{where,} \quad C_1 = \sqrt{2}(\lambda_{min}(\mathbf{K}_1) - D_i - \rho_i)$$
$$C_2 = \lambda_{min}(\mathbf{K}_2), \quad C_1, C_2 > 0$$

The reaching time can be computed as

$$T_i \leq \frac{2}{C_2} ln\left[\frac{C_2\sqrt{V_1(0)} + C_1)}{C_1}\right] \tag{15.27}$$

Hence, we can conclude that the sliding mode-based event triggered control law defined by (15.20), can make the system finite time stable, if $f(t) < 0$. The control is triggered, whenever $f(t) \geq 0$. □

Theorem 15.2. *[450]* Consider the leader-follower-based MAS given by (15.14), with the communication topology represented by connected graph $\mathcal{G}$. If the event triggered consensus protocol and the triggering function are defined by (15.20) and (15.22) respectively, once the sliding mode is reached, for any initial condition, the consensus can be reached in finite time, and the settling time satisfies the following condition,

$$T_2 \leq \frac{2V^{\frac{1-\eta}{2}}(\tilde{\mathbf{x}}(0))}{q(1-\eta)} \tag{15.28}$$

where, $q = \lambda_{min}(\xi \otimes I_n)\lambda_{min}((\mathcal{L} + \mathcal{B}) \otimes I_n)2^{\frac{1+\eta}{2}}$, and $\xi = diag\{(\mu_1/n_1 + 1)^\eta, (\mu_2/n_2 + 1)^\eta \ldots (\mu_N/n_N + 1)^\eta\}$.

Proof. Let

$$\mathbf{P}_i(t) = \sum_{j \in \mathcal{N}_i} a_{ij}(\tilde{\mathbf{x}}_\mathbf{i}(t) - \tilde{\mathbf{x}}_\mathbf{j}(t)) + b_i(\tilde{\mathbf{x}}_\mathbf{i}(t)) \tag{15.29}$$

Consider a Lyapunov function as

$$V_2(t) = (1/2)\tilde{\mathbf{x}}^T(t)((\mathcal{L} + \mathcal{B}) \otimes I_n)^T((\mathcal{L} + \mathcal{B}) \otimes I_n)\tilde{\mathbf{x}}(t)$$
$$= (1/2)\mathbf{P}^T\mathbf{P} \tag{15.30}$$

where $\mathbf{P} = [\mathbf{P}_1, \mathbf{P}_2 \ldots \mathbf{P}_n]^T$. Substituting using (15.29), and taking the derivative,

$$\dot{V}_2(t) = \mathbf{P}^T((\mathcal{L} + \mathcal{B}) \otimes I_n)\dot{\tilde{\mathbf{x}}}(t) \tag{15.31}$$

Event Triggered Sliding Mode-based Consensus Algorithm 547

When the sliding mode is reached, *i.e.*, $S_i(t) = 0$, from (15.17), we can find that $\dot{\tilde{\mathbf{x}}}(t) = \boldsymbol{\chi}^\eta(t)$. Since η is strictly the ratio of positive odd numbers, $\text{sign}(.) = \text{sign}((.)^\eta)$. Using (15.15) and (15.29),

$$\dot{V}_2(t) = \mathbf{P}^T((\mathcal{L}+\mathcal{B})\otimes I_n)\boldsymbol{\chi}^\eta(t) \quad (15.32)$$
$$\leq -\lambda_{min}(\xi\otimes I_n)\lambda_{min}((\mathcal{L}+\mathcal{B})\otimes I_n)\|\mathbf{P}\|^{1+\eta}$$
$$\leq -\lambda_{min}(\xi\otimes I_n)\lambda_{min}((\mathcal{L}+\mathcal{B})\otimes I_n)(\mathbf{P}^T\mathbf{P})^{(1+\eta)/2}$$
$$\leq -\lambda_{min}(\xi\otimes I_n)\lambda_{min}((\mathcal{L}+\mathcal{B})\otimes I_n)(2V_2(t))^{\frac{1+\eta}{2}}$$
$$\leq -qV_2(t)^{\frac{1+\eta}{2}} \quad (15.33)$$

Using Lemma 15.2, we can conclude that $V_2(t) \to 0$ in finite time, and the settling time is given by

$$T_2 \leq \frac{2V_2(\tilde{\mathbf{x}}(0))^{\frac{1-\eta}{2}}}{q(1-\eta)}, \quad (15.34)$$

where, $q = \lambda_{min}(\xi\otimes I_n)\lambda_{min}((\mathcal{L}+\mathcal{B})\otimes I_n)2^{\frac{1+\eta}{2}}$ & $\tilde{\mathbf{x}}(0) = \tilde{\mathbf{x}}^0$.

Then, $V_2(t) = 0, \forall t \geq T_2$, and $\mathbf{x}_i(t) = \mathbf{x}_0(t) - \boldsymbol{\delta}_i, \forall i, t \geq T_2$.

Hence, finite time consensus can be reached. □

Theorem 15.3. *[450]* Consider the leader-follower-based MAS given by (15.14), with the communication topology represented by the connected graph $\mathcal{G}$, and the event triggered consensus protocol defined by (15.20). If the triggering function is defined by (15.22), then, the inter-event time is lower bounded by

$$T_i \geq \frac{\rho_i}{\gamma_i}, \text{ where,} \quad (15.35)$$
$$\gamma_i = 2^{\frac{2\eta-1}{2}}\eta n^{3-2\eta}\|\Omega\otimes I_n\|^{2\eta}\|((\mathcal{L}+\mathcal{B})\otimes I_n)\|V_2(0)^{\frac{2\eta-1}{2}}$$
$$+ (\|\mathbf{K}_1\|\beta n + \|\mathbf{K}_2\|)(\|\boldsymbol{\chi}_i^\eta(t_k^i)\| + \|\mathbf{K}_1\text{sign}(\mathbf{S}_i(t_k^i))\|$$
$$+ \|\mathbf{K}_2\mathbf{S}_i(t_k^i)\| + D_i + 2^{\frac{\eta}{2}}n^{1-\eta}\|\Omega\otimes I_n\|^\eta V_2(0)^{\frac{\eta}{2}})$$

so that the zeno behavior can be avoided, where $\Omega = \text{diag}\{\mu_1/(n_1+1), \mu_2/(n_2+1)\ldots\mu_N/(n_N+1)\}$

Proof. Let T_i be the inter-event time, *i.e.*, the time required for the error to grow from zero to ρ_i. At $t = t_k^i$, the control is updated, hence the error becomes zero, *i.e.*, $e(t_k^i) = 0$. During the inter-event interval, $\|e(t)\| \leq \rho_i$, so that the consensus condition holds. Employing Lemma 15.5, we can find $\|\dot{\boldsymbol{\chi}}_i(t)\| \leq \|\dot{\boldsymbol{\chi}}(t)\| \leq \|\Omega\otimes I_n\|\|(\mathcal{L}+\mathcal{B})\otimes I_n\|\|\boldsymbol{\chi}^\eta(t)\| \leq \|\Omega\otimes I_n\|\|(\mathcal{L}+\mathcal{B})\otimes I_n\|n^{1-\eta}\|\boldsymbol{\chi}(t)\|^\eta$, and $\|\boldsymbol{\chi}_i^{\eta-1}(t)\| \leq \|\boldsymbol{\chi}^{\eta-1}(t)\| \leq n^{2-\eta}\|\boldsymbol{\chi}(t)\|^{\eta-1}$. In practical scenario, the system trajectory may deviate from the ideal sliding manifold. But it will remain bounded, which depends on ρ_i.

$$\frac{d}{dt}\|\mathbf{e}_i(t)\| \leq \|\frac{d}{dt}\mathbf{e}_i(t)\|$$
$$\leq \|\frac{d}{dt}[\boldsymbol{\chi}_i^\eta(t) - \mathbf{K}_1\mathbf{sign}(\mathbf{S}_i(t)) - \mathbf{K}_2\mathbf{S}_i(t)]\|$$

For obtaining the derivative of a **sign** function, we can approximate it using **tanh** function [456], *i.e.*, $\mathbf{sign}(\mathbf{S}_i(t)) \approx \mathbf{tanh}(\beta\mathbf{S}_i(t))$, where $\beta \gg 1$. Using Lemma 15.5 yields,

$$\frac{d}{dt}\|\mathbf{e}_i(t)\| \leq \|\frac{d}{dt}\boldsymbol{\chi}_i^\eta(t)\| + \|\frac{d}{dt}\mathbf{K}_1\mathbf{tanh}(\beta\mathbf{S}_i(t))\|$$
$$+ \|\frac{d}{dt}\mathbf{K}_2\mathbf{S}_i(t)\|$$
$$\leq \eta n^{3-2\eta}\|\boldsymbol{\chi}(t)\|^{\eta-1}\|\Omega \otimes I_n\|\|(\mathcal{L} + \mathcal{B}) \otimes I_n\|$$
$$\|\boldsymbol{\chi}(t)\|^\eta + \|\mathbf{K}_1\|\|[\mathbf{1} - \mathbf{tanh}^2(\beta\mathbf{S}_i(t))]\beta\dot{\mathbf{S}}_i(t)\|$$
$$+ \|\mathbf{K}_2\|\|\dot{\mathbf{S}}_i(t)\|$$

$\|\mathbf{1}_{n\times n} - \mathbf{tanh}^2(\beta\mathbf{S}_i(t))\| \leq \|\mathbf{1}_{n\times n}\| = n$, and using (15.24), we can find

$$\frac{d}{dt}\|\mathbf{e}_i(t)\| \leq \eta n^{3-2\eta}\|\Omega \otimes I_n\|\|((\mathcal{L} + \mathcal{B}) \otimes I_n)\|\|\boldsymbol{\chi}(t)\|^{2\eta-1}+$$
$$(\|\mathbf{K}_1\|\beta n + \|\mathbf{K}_2\|)(\|\boldsymbol{\chi}_i^\eta(t_k^i)\| + \|\mathbf{K}_1\mathbf{sign}(\mathbf{S}_i(t_k^i))\|$$
$$+ \|\mathbf{K}_2\mathbf{S}_i(t_k^i)\| + D_i + \|\boldsymbol{\chi}_i^\eta(t)\|)$$

From the proof for Theorem 15.2, we can find that

$$\|\boldsymbol{\chi}_i(t)\| \leq \|\boldsymbol{\chi}(t)\| \leq \|\Omega \otimes I_n\|\|\mathbf{P}(t)\|$$
$$\leq \sqrt{2}\|\Omega \otimes I_n\|\sqrt{V_2(t)} \leq \sqrt{2}\|\Omega \otimes I_n\|\sqrt{V_2(0)}$$

$$\frac{d}{dt}\|\mathbf{e}_i(t)\| \leq 2^{\frac{2\eta-1}{2}}\eta n^{3-2\eta}\|\Omega \otimes I_n\|^{2\eta}\|((\mathcal{L} + \mathcal{B}) \otimes I_n)\|V_2(0)^{\frac{2\eta-1}{2}}$$
$$+ (\|\mathbf{K}_1\|\beta n + \|\mathbf{K}_2\|)(\|\boldsymbol{\chi}_i^\eta(t_k^i)\| + \|\mathbf{K}_1\mathbf{sign}(\mathbf{S}_i(t_k^i))\|$$
$$+ \|\mathbf{K}_2\mathbf{S}_i(t_k^i)\| + D_i + 2^{\frac{\eta}{2}}n^{1-\eta}\|\Omega \otimes I_n\|^\eta V_2(0)^{\frac{\eta}{2}})$$

Solving this for t, with the initial condition, $\|\mathbf{e}_i(t)\| = 0$,

$$\|\mathbf{e}_i(t)\| \leq (t - t_k^i)\left[2^{\frac{2\eta-1}{2}}\eta n^{3-2\eta}\|\Omega \otimes I_n\|^{2\eta}\|((\mathcal{L} + \mathcal{B}) \otimes I_n)\|\right.$$
$$V_2(0)^{\frac{2\eta-1}{2}} + (\|\mathbf{K}_1\|\beta n + \|\mathbf{K}_2\|)(\|\boldsymbol{\chi}_i^\eta(t_k^i)\|$$
$$+ \|\mathbf{K}_1\mathbf{sign}(\mathbf{S}_i(t_k^i))\| + \|\mathbf{K}_2\mathbf{S}_i(t_k^i)\| + D_i$$
$$\left.+ 2^{\frac{\eta}{2}}n^{1-\eta}\|\Omega \otimes I_n\|^\eta V_2(0)^{\frac{\eta}{2}})\right]$$

Event Triggered Sliding Mode-based Consensus Algorithm 549

The event is triggered, when $f(t) \geq 0$, i.e., $\rho_i \leq \|\mathbf{e}_i(t)\|$. Moreover, $(t - t_k^i) \leq T_i$. Hence,

$$\rho_i \leq \|\mathbf{e}_i(t)\| \leq T_i \left[2^{\frac{2\eta-1}{2}} \eta n^{3-2\eta} \|\Omega \otimes I_n\|^{2\eta} \|((\mathcal{L} + \mathcal{B}) \otimes I_n)\| \right.$$
$$V_2(0)^{\frac{2\eta-1}{2}} + (\|\mathbf{K}_1\|\beta n + \|\mathbf{K}_2\|)(\|\chi_i^\eta(t_k^i)\|$$
$$+ \|\mathbf{K}_1 \mathbf{sign}(\mathbf{S}_i(t_k^i))\| + \|\mathbf{K}_2 \mathbf{S}_i(t_k^i)\| + D_i$$
$$\left. + 2^{\frac{\eta}{2}} n^{1-\eta} \|\Omega \otimes I_n\|^\eta V_2(0)^{\frac{\eta}{2}} \right)\right]$$

where $1 > \eta > (1/2)$. Rearranging this equation yields the lower bound for the inter-execution time given by (15.35). From this, we can ensure that it is strictly a positive value. □

Remark 15.1. By replacing the **sign** function by **tanh** function, the chattering inherent in the sliding mode control can be reduced [423].

15.3.1 Consensus-based Tracking Control of Nonholonomic Multi-robot Systems

The ISMC-based event triggered control algorithm can be validated using nonholonomic robots. The kinematic model of the nonholonomic robot given in (13.6) in Chapter 13 can be used here. Considering the kinematic disturbances (d_i^x, d_i^y), acting on the system, (13.6) can be redefined as

$$\dot{x}_i = u_i^x - \ell \omega_i \sin(\theta_i) + d_i^x$$
$$\dot{y}_i = u_i^y + \ell \omega_i \cos(\theta_i) + d_i^y \quad (15.36)$$

where $u_i^x = v_i \cos(\theta_i)$, and $u_i^y = v_i \sin(\theta_i)$. (15.36) can be approximated as a single integrator model given by,

$$\dot{\mathbf{x}}_i = \mathbf{u}_i + \mathbf{d}_i \quad (15.37)$$

where the unmodelled dynamics and the external disturbances are represented as lumped uncertainty, given by $\mathbf{d}_i$; and $\mathbf{u}_i = [u_i^x, u_i^y]^T$ is the control input, which can be computed using (15.20). The dynamics looks similar to that of (15.12).

Example 15.3. Consider a multi-agent system of four agents with one leader and three followers making a rectangular formation with the agent dynamics given by (15.12), and the relative state deviations of the followers w.r.t leader are as follows: $\boldsymbol{\delta}_1 = (1, -1)$, $\boldsymbol{\delta}_2 = (1, 1)$ and $\boldsymbol{\delta}_3 = (2, 0)$.

Design and implement SMC-based event triggered consensus protocol for the trailing agents, with a and the communication topology given by Figure 15.5. Consider that the leader is free to navigate, and the deputies will reach a consensus tracking the leader, and there is no feedback from the followers to the chief.

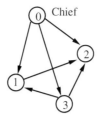

FIGURE 15.5: Communication topology.

Solution 15.3. *Based on the graph topology, the Laplacian and the leader-follower connection weight matrix can be obtained as*

$$\mathcal{L} = \begin{pmatrix} 1 & 0 & -1 \\ -1 & 2 & -1 \\ 0 & 0 & 0 \end{pmatrix}, \quad \text{and} \quad \mathcal{B} = \begin{pmatrix} 1 & 0 & 0 \\ 0 & 1 & 0 \\ 0 & 0 & 1 \end{pmatrix} \quad \text{respectively.}$$

With the agent dynamics given by (15.12), the ISMC-based event triggered control protocol can be obtained as

$$\tilde{\mathbf{u}}_i(t) = \boldsymbol{\chi}_i^\eta(t_k^i) - \mathbf{K}_1 \text{sign}(\mathbf{S}_i(t_k^i)) - \mathbf{K}_2 \mathbf{S}_i(t_k^i)$$

The values of the simulation parameters are as follows: $\rho_i = 0.05$, $\forall i$, $\mathbf{K}_1 = diag\{0.25, 0.25, 0.25\}$, $\mathbf{K}_2 = diag\{0.0008, 0.0008, 0.0008\}$, $\boldsymbol{\delta}_1 = (1, -1)$, $\boldsymbol{\delta}_2 = (1, 1)$, $\boldsymbol{\delta}_3 = (2, 0)$, $D_i = 0.195$ *and* $\eta = 5/7$. *The control is triggered at the event instants only, where the triggering function is given by*

$$f_i(t) = \|\mathbf{e}_i(t)\| - \rho_i \tag{15.38}$$

and the measurement error, e, can be computed as

$$\begin{aligned} \mathbf{e}_i(t) = &\ \boldsymbol{\chi}_i^\eta(t_k^i) - \mathbf{K}_1 \text{sign}(\mathbf{S}_i(t_k^i)) - \mathbf{K}_2 \mathbf{S}_i(t_k^i) \\ &- (\boldsymbol{\chi}_i^\eta(t) - \mathbf{K}_1 \text{sign}(\mathbf{S}_i(t)) - \mathbf{K}_2 \mathbf{S}_i(t)) \end{aligned} \tag{15.39}$$

From the trajectories given by Fig. 15.6(a)-15.6(b), we can find that the deputies are reaching a consensus with respect to the chief in finite time with the desired state deviation. The event triggering instants and the corresponding measurement errors for the first 30 s shown in Fig. 15.6(c)-15.6(d), indicate that the condition for the lower bound for inter-execution time is well satisfied. Hence, zeno behavior can be avoided.

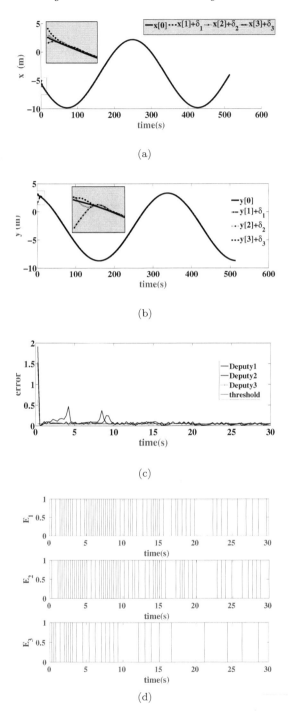

FIGURE 15.6: Example: (a) x position, (b) y position, (c) measurement error, (d) Triggering instants.

15.4 Experiments

For validating the event triggered based algorithm, we have used the same experimental set up as given in Section 14.4 in Chapter 14. For the experiments, we have used three robots, in which we have assigned FireBird VI robot as the chief and the two P3-DX robots as deputy 1 and deputy 2 respectively. The robots are communicating through a local WIFI network, within which static IPs are generated and assigned to the robots. The channel bandwidth is 20 MHz, and the maximum data rate is of 80 Mbps. The Player 3.0.2 robot device server is running on all the platforms, which facilitates ultimate control over all sensors and actuators. The sensor rate is configured to be 10 Hz. For the P3-DX robot, the accuracy of the odometry data lies in the range of $\pm$ 1.5 %, whereas, for the FireBird VI robot, it is of $\pm$ 6.2 %.

The experimental parameters are chosen as follows: $\ell = 0.21$ m, $\rho_i = 0.02$, $\forall i$, $\mathbf{K}_1 = \text{diag}\{0.05, 0.05, 0.05\}$, $\mathbf{K}_2 = \text{diag}\{0.0008, 0.0008, 0.0008\}$, $\boldsymbol{\delta}_1 = (0.7, -0.7)$, $\boldsymbol{\delta}_2 = (0.7, 0.7)$, and $\eta = 5/7$. From the path of the robots given by Fig. 15.9(a) and the position as well as velocity trajectories shown in Fig. 15.9(b)-15.9(d), it is clear that the deputies are reaching a consensus with the chief with the desired state deviation. We have tested with different velocities as well as turn maneuvers [450]. In order to confirm the robustness of the algorithm, the chief's trajectory is chosen such that the acceleration, deceleration, right turn as well as left turn maneuvers are included. The control input, $\|\mathbf{u}\|$ (velocity), trajectories are given in Fig. 15.9(d). c. The sensor noise is also predominant in the system. The results show that the controller

FIGURE 15.7: Taped video images of robots in formation.

Experiments

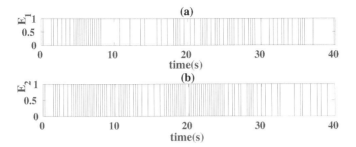

FIGURE 15.8: Experimental results: (a) and (b) Triggering instants for deputy 1 and deputy 2 respectively.

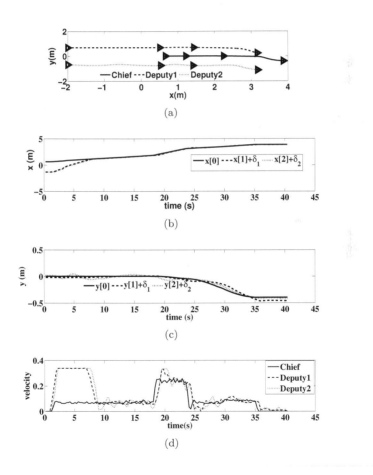

FIGURE 15.9: Experimental Results: Path of the agents in formation.

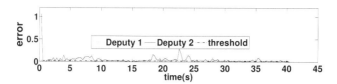

FIGURE 15.10: Experimental results: Measurement error for different agents.

is robust enough to deal with the modeling uncertainties as well as other disturbances.

The measurement error given in Fig. 15.10, indicates that the control is triggered whenever the triggering condition is violated. The threshold chosen based on the gains as well as the disturbance bound is also shown in the Fig. 15.10. From the triggering instants of the deputies given in Fig. 15.8, we can find that the control is triggered only at discrete event instants. Since there is a lower bound for the inter-execution time, the zeno phenomenon won't occur. To verify the same, we have computed the lower bound on the inter-execution time, using (15.35), at $t_k^i = 18.65$ s, where the inter-event interval attains the minimum value of 0.285 s, and it is found to be 0.0368 s. The taped video images of the formation are given in Figure 15.7.

15.5 Summary

In this chapter, we have discussed an event triggering based resource optimal control scheme. Through simulations as well as real-time experimentations, the consensus based resource-optimal formation control of a multi-agent system in a leader-follower framework, with bounded disturbances, has been explained in sufficient detail. Numerical examples are solved for a simple linear system as well as for a complex multi-agent system, so that the readers will get a complete outlook regarding the real implementation aspects of the event triggered control.

16

Vision-Based Tracking for a Human Following Mobile Robot

16.1 Visual Tracking: Introduction

Object tracking in a video sequence is an important problem in computer vision research with applications in areas like video surveillance, motion-based recognition, video indexing, human computer interaction, assistive robotics, and augmented reality. The field has witnessed an unprecedented advancement owing to the availability of high quality cameras and inexpensive computing power, commensurate with the development of ingenious techniques for image and video processing. An overview of visual tracking is provided in Figure 16.1. In spite of the advancements made in this field, visual tracking is still fraught with difficulties that can arise due to the following reasons.

16.1.1 Difficulties in Visual Tracking

- Abrupt object motion.
- Appearance pattern change including pose.
- Illumination variation.
- Non-rigid object structure.
- Partial/ full occlusion.
- Camera motion.

To develop a robust visual tracking system, these difficulties should be addressed properly along with some basic requirements of visual tracking. The basic requirements of visual tracking are described in the next subsection.

16.1.2 Required Features of Visual Tracking

- Robustness: The tracking algorithm should be able to follow the target object even in adverse condition.

FIGURE 16.1: Visual tracking at a glance.

- Adaptation: The tracking algorithm should be adaptive to the changes in the environment as well as to the changes in the target itself.

- Real-time processing: The algorithm should be computationally least intensive to enable it to be implementable in real-time. This is one of the primary concerns as the work is related to the navigation of the robot while tracking a target human.

The first step for visual object tracking is to have a description of the object to be tracked. Description can be anything like color, texture, shape or a template image of the object. The details of the feature descriptors commonly used for visual tracking are provided in the next subsection.

16.1.3 Feature Descriptors for Visual Tracking

In visual tracking, image feature selection plays a critical role. The most desirable property of a visual feature is its uniqueness so that the objects can be easily distinguished in the feature space. The commonly used visual features are as follows:

- *Color* :- The apparent color of an object is influenced primarily by two physical factors, (i) the spectral power distribution of the illuminant and (ii) the surface reflectance properties of the object. In image processing, the RGB color space is usually used to represent color. However, the RGB space is not a perceptually uniform color space. The L*u*v and L*a*b are perceptually uniform color spaces, while HSV is an approximately uniform color space.

Recent advances divide the color descriptors into two categories (i) histogram-based color descriptors, and (ii) SIFT-based color descriptors. In [457], the hue histogram is made robust by weighing each sample of the hue by its saturation. In [458], Gevers et al. used an rg-histogram descriptor, which is based on a normalized RGB color model. In [457], authors have introduced a concatenation of the hue histogram with the SIFT descriptor, which is scale-invariant and shift-invariant.

- *Texture* :- Texture is a measure of the intensity variation of a surface which quantifies properties such as smoothness and regularity [459,460]. A Gabor wavelet [461] is the commonly used texture feature. In recent years, the research is on investigating an image's local patterns for better detection and recognition. Ojala et al. [462] developed a very efficient texture descriptor Local Binary Patterns (LBP), which is defined as a gray-scale invariant texture measure. The important properties of the LBP operator is its tolerance against illumination changes and its computational simplicity.

- *Optical Flow* :- Optical flow is a dense field of displacement vectors which defines the translation of each pixel in a region. It is computed by assuming brightness constancy of corresponding pixels in consecutive frames [463]. Optical flow is commonly used as a feature in motion-based segmentation and tracking applications. Popular techniques for computing dense optical flow include methods by Horn and Schunck [1981] and Lucas and Kanade [464].

- *Gradient* :- The gradient features are mainly divided into two categories. The first category of gradient based methods is to use shape/contour to represent objects, such as the human body [465]. The second category is uses the statistical summarization of the gradients. In [466], Lowe introduced the well known SIFT descriptor for object recognition. Later, Bay et al. proposed SURF [467], which is a much faster scale and rotation invariant interest point descriptor. Dalal and Triggs [468] used the Histogram of Oriented Gradient (HOG) descriptor in training SVM classifier for pedestrian detection.

- *Spatio-temporal features* :- Local space-time features capture characteristic salient and motion patterns in video and provide relatively independent representation of events with respect to their spatio-temporal shifts and scales as well as background clutter and multiple motions in the scene [469].

Selecting a feature for an initial object model description is a crucial task, as the quality of the description directly translates to the quality of the tracking. Even after having a good object description available a priori, continual adaptation to appearance change is necessary to achieve robust tracking.

Handling appearance variations of a target object is an essential task of visual tracking. The appearance variations can be divided in two categories: (i) intrinsic and (ii) extrinsic. Pose variation and shape deformation of a target object are considered as the intrinsic appearance variations while the extrinsic variations are due to the changes resulting from different illumination, camera motion, and occlusion. These variations can only be handled with adaptive methods which are able to incrementally update their representations. Thus, there is a need for online algorithms that are able to learn continuously. A brief description of the online learning algorithms is presented in the following subsection.

This chapter aims at developing robust vision-based algorithms using point-based features, like SURF, to track a human under challenging conditions that include variation in illumination, pose change, full or partial occlusion, and abrupt camera motion. Since, the point-based methods use *tracking-by-detection* framework, the major problem lies in finding a sufficient number of descriptors in subsequent frames, as the target undergoes some of these variations. This chapter studies the problem of constructing an object model, which evolves over the time to deal with short-term changes, while maintaining stability on a longer term.

16.2 Human Tracking Algorithm using SURF Based Dynamic Object Model

A human-following robot needs to track a human walking in front of the robot using its on-board camera. The person may exhibit natural human motion including pose changes due to out-of-plane rotations. In this chapter, SURF is used as the visual feature, which is known to be robust to photometric and geometric distortions, and computationally efficient compared to other point features, like SIFT. The interest point based methods make use of object recognition for detecting a target in a given frame and are less influenced by abrupt object motion arising out of low frame rate or non-stationary camera. SURF-based tracking methods usually consist of two steps: (i) To represent the target or the reference model in terms of feature descriptors, and (ii) to infer the best location by computing the correspondences between the source frame and the target frame. The wrong correspondences are usually removed using RANSAC algorithm [470, 471].

The problem associated with interest-point based methods that use tracking-by-detection framework are as follows [472]:

- The number of matching points obtained vary significantly from one frame to another and may diminish over time, which leads to the failure of tracker in a long run.

- The computational complexity associated with computing SURF correspondence between a pair of images.

In order to obtain a healthy number of matching points necessary for object tracking, it is necessary to update the object model with time. In this chapter, a tracking algorithm is proposed that uses a dynamic object model description to detect a target in all subsequent frame, which consists of a set of SURF descriptor points, which evolves over the time. This object model derives its points from a *template pool* that helps in reinforcing the features, which occur more frequently compared to others. In this process, it aims to resolve the *stability-plasticity dilemma* in object tracking [472] without having to learn the actual motion model of the object [473, 474] or creating bag-of-words through clustering [471]. It is assumed that the human target is big enough so that a few descriptors could be extracted from it.

The problem of computational complexity associated with computing SURF correspondence between a pair of images is resolved by using temporal coherence in a video to skip consecutive frames and carry out matching only when a recognition step is warranted [473].

The proposed dynamic object model description is combined with the SURF-based mean-shift algorithm, which guarantees tracking of a non-rigid object with real-time computational power. The proposed approach uses the concept of dynamic object model description [475] but does not require frame-to-frame matching, therefore, can be implemented on a mobile robotic platforms. However, this approach is prone to drifting error, which might arise when background points get added to the pool. Therefore, a second approach is proposed in which background descriptors are prevented from getting added into an object template. In the second approach, a SURF-based human tracker is proposed, which searches the human in the expanded rectangular region around the last target location. The object model is updated over time by selecting new templates and projecting the points from previous templates using an affine transformation. A k-d tree [476] based classifier along with a Kalman filter predictor is used to differentiate between a case of pose change and a case of occlusion (partial/full). In this approach, a method is developed to detect and avoid tracking failure due to *out-of-plane rotations*, which is a very difficult problem to solve with point-based features. In order to deal with the case of full occlusion, a simple auto-regression predictor $AR(p)$ with $p = 5$ is learned for predicting the location of the target when no match points are obtained. The coefficients of the predictor are learned online using a *gradient-descent* algorithm [477].

16.2.1 Problem Definition

Consider a set of frames I_i, $i = 0, 1, 2, \cdots N$ of a video sequence, where a human identified by a user in the first frame is to be tracked over all the frames. The human is identified by a polygon P_0 drawn by selecting the points on the boundary of the human silhouette. Let $V(I) = \{(\mathbf{x}_1, \mathbf{v}_1), (\mathbf{x}_2, \mathbf{v}_2), \cdots, (\mathbf{x}_n, \mathbf{v}_n)\}$ be the set of SURF key-points of the image I, where $\mathbf{x}_i$ is a 2-dimensional key-point location of the 64-dimensional SURF descriptor $\mathbf{v}_i$. The SURF key-points lying within a polygon P_t is represented by the symbol $V_P(t)$ for any frame I_t. The corresponding rectangle bounding these points is represented by W_t.

The SURF correspondence between two images is the set of best matching key-points, which is defined as

$$\zeta(I_t \sim I_r) \triangleq \{(\mathbf{x}_1, \mathbf{v}_1)^t, \cdots\cdots, (\mathbf{x}_m, \mathbf{v}_m)^t, (\mathbf{x}_1, \mathbf{v}_1)^r, \cdots\cdots, (\mathbf{x}_m, \mathbf{v}_m)^r\}, \tag{16.1}$$

where I_t is the current image, and $I_r, r \in (0, t-1)$, is any other image encountered in the past. The set of SURF descriptors of the image I_t, which matches with those of the image I_r, lying within its own polygon P_r, is represented by $\psi_v(t)^r$. The corresponding key-point locations are denoted by $\psi_x(t)^r$. Mathematically, it can be written as

$$(\psi_x(t), \psi_v(t))^r = \{(\mathbf{x}_i, \mathbf{v}_i)^t | (\mathbf{x}_i, \mathbf{v}_i)^t \in \zeta(I_t \sim I_r) \wedge \mathbf{x}_i^r \in P_r, i = 1, 2, \ldots, p \leq m\}. \tag{16.2}$$

Hence, the set $(\psi_x(t), \psi_v(t))^r$ is a subset of $\zeta(I_t \sim I_r)$. The tracking polygon P_t, on an image I_t, is represented by $P_t = (B_t, \mathbf{c}_t)$, where $\mathbf{c}_t = (c_x, c_y)$ is the center of the polygon, and B_t is the set of key-point locations enclosed by the polygon defined as

$$B_0 = \{\mathbf{x}_i | (\mathbf{x}_i, \mathbf{v}_i) \in V_0 \wedge \mathbf{x}_i \in P_0\}. \tag{16.3}$$

Given I_0, $P_0(B_0, \mathbf{c}_0)$, and V_0, the task is to compute $P_t(B_t, \mathbf{c}_t)$ for all frames $t = 1, 2, \ldots N$.

The next subsection discusses the object model description that is used in the proposed tracking algorithm.

16.2.2 Object Model Description

The object model description consists of three different sets of SURF key-points and descriptors, as shown in Figure 16.2. The first set is obtained by finding the SURF correspondences between the consecutive images represented by $\zeta(I_t \sim I_{t-1})$. The set of points $(\psi_x(t), \psi_v(t))^{t-1}$, which is a subset of ζ, obtained above constitutes the set $M_1(t)$.

The second set is obtained by finding the SURF correspondences between the the first frame I_0 and the current frame I_t, represented by $\zeta(I_0 \sim I_t)$. The subset of key-points corresponding to the points within the polygon P_0 constitute the set $M_2(t) = (\psi_x(t), \psi_v(t))^0$, as explained in Section 16.2.1.

The third set is obtained by finding the correspondence between the SURF descriptors of a template pool with those of the current frame $V(I_t)$. Only those points of the current frame are retained in this set, which lie within the

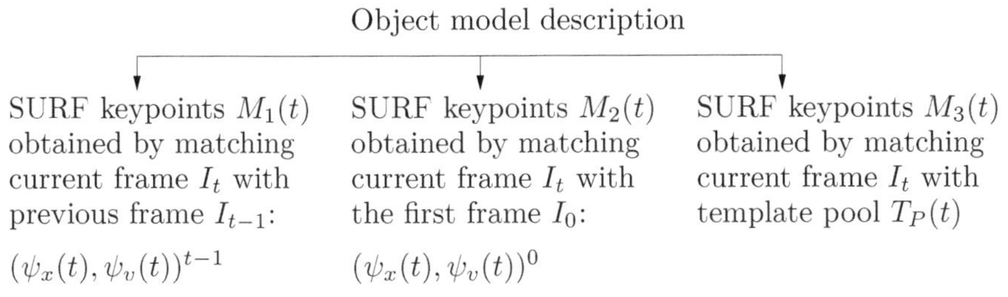

```
                    Object model description
┌─────────────────────────┬─────────────────────────┬─────────────────────────┐
```

SURF keypoints $M_1(t)$ obtained by matching current frame I_t with previous frame I_{t-1}:	SURF keypoints $M_2(t)$ obtained by matching current frame I_t with the first frame I_0:	SURF keypoints $M_3(t)$ obtained by matching current frame I_t with template pool $T_P(t)$
$(\psi_x(t), \psi_v(t))^{t-1}$	$(\psi_x(t), \psi_v(t))^0$	

FIGURE 16.2: Object model description.

scaled reference window W_t. This set of point is represented by $M_3(t)$. This window is centered at the center of the polygon formed by the key-points of $M_1(t)$. Hence, the object model is described as

$$O_m(t) = \{M_1(t) \cup M_2(t) \cup M_3(t)\}. \qquad (16.4)$$

In order to obtain a reliable object model, it is important to eliminate outliers while forming these sets of SURF descriptors. In this algorithm, RANSAC based on homography is used to avoid outliers in set $M_1(t)$. For the set $M_2(t)$, scaling [478] and k-nearest neighbor [472] are used to remove outliers. RANSAC cannot be used in this case, as one can expect a significant displacement of a target in these two images. For the third set $M_3(t)$, outliers are avoided by putting a threshold on the similarity between the descriptors of the template pool and the matching points obtained on the current frame I_t within the window W_t.

It should be noted that an object model $O_m(t)$ is a function of the current frame, therefore, it is dynamic in the nature. This model evolves over time, and hence, accommodates the variations that may arise due to change in poses or in motions. The set $M_1(t)$ provides high frequency temporal information about the model change that can occur between consecutive frames. The set $M_2(t)$ provides the stable information obtained from the first frame, which is expected to vary little over the frames. On the other hand, the set $M_3(t)$ obtained from the template pool of descriptors provides low frequency temporal information about the model variation over all previous frames I_{t-2} to I_0.

16.2.2.1 Maintaining a Template Pool of Descriptors

The template pool consists of a set of SURF descriptors, which is updated over the time. The number of descriptors in this set is limited to 5,000, which can be varied by the user. Mathematically, this set can be written as

$$T_p(t) = \{T_p(t-1) \cup \psi_v(t)^{t-1}\}. \qquad (16.5)$$

In order to keep a check on the size of the template pool, the oldest descriptors are pushed out by the same number of incoming new descriptors.

Algorithm 19 Human tracking algorithm using SURF-based dynamic object model

Input:	n video frames $I_0 \cdots, I_n$
	Polygon P_0 of human in the first frame
Output:	Polygon $P_1, \cdots, P_n$

Step 1: Initialization(for frame I_0):

- Initialize the reference target model with V_0 and push its descriptors ($\psi_v(0)$) into T_p

Step 2: For each new frame I_t do:

- Obtain the set $M_1(t)$ by matching current frame I_t with previous frame I_{t-1}: $(\psi_x(t), \psi_v(t))^{t-1}$ and update template pool $T_p(t)$

- Obtain the set $M_2(t)$ by matching current frame I_t with the first frame I_0: $(\psi_x(t), \psi_v(0))^0$

- Obtain the set $M_3(t)$ by matching current frame I_t with the template pool $T_p(t)$

Step 3:
if $O_M(T) >$ threshold **then**
  Fit polygon P_t on total matched points
else
  Occlusion detected
end if

if Occlusion detected **then**
  Predict target location
else
  Update $AR(p)$ predictor
end if

16.2.3 The Tracking Algorithm

The steps involved in the proposed method is given in Algorithm 19, and the method is described pictorially in the flowchart, as shown in Figure 16.3. The tracking algorithm consists of three modules, namely, target initialization, object recognition and template pool update, and prediction of target location in case of occlusion. These individual modules are explained in the following subsections.

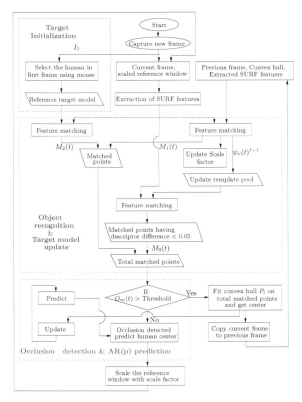

FIGURE 16.3: Block diagram of the human tracking algorithm using SURF-based dynamic object model.

16.2.3.1 Step 1: Target Initialization

Target is initially selected in the first frame I_0 by manually marking points on the boundary of the human silhouette. The resulting polygon is denoted by P_0. The SURF key-points lying within this polygon $V_P(0)$ is taken as the reference model, which is used for populating the set $M_2(t)$, as explained in Section 16.2.2. This set is bounded by the rectangle W_0 on the first frame.

16.2.3.2 Step 2: Object Recognition and Template Pool Update

In this module, the target is detected in the current frame I_t by finding all the SURF key-points matching with those present in the object model $O_m(t)$. The polygon P_t, bounding these matching points B_t, defines the target (object) in the current frame I_t. The set of matching descriptors obtained between two consecutive frames represented by the set $(\psi_x(t), \psi_v(t))^{t-1}$, is used to update the *template-pool*, as described in Section 16.2.2.1.

A part of this template pool is included to the object model description, as the set $M_3(t)$, as described above. This set, in turn, contributes toward defining the convex hull of the target in the current frame I_t. These matching points again become the part of the template pool for the next frame. Thus, the frequently matching descriptors survive in the pool for a longer period of time, as compared to those insignificant features, which appear once in a while during matching.

16.2.3.3 Step 3: Occlusion Detection, Target Window Prediction

In this algorithm, it is assumed that an occlusion may occur if the total number of matching points between the current frame $V(I_t)$ and the object model $O_m(t)$ is less than a user-defined threshold. The threshold is defined as the 20% of the points present in the object model. In case of occlusion, the center of the tracking polygon is predicted using a simple auto-regression model $AR(p)$, where $p = 5$. A search for the matching key-points with the object model is done within a rectangular region around this center. The dimension of the rectangular window is obtained after applying cumulative scaling on the original window W_0. If the number of matching points is less than the threshold, the target is still occluded.

If matching is found, then the tracking is reinitialized by creating a polygon around the match points. The parameters of the AR predictor are learned using a gradient-descent algorithm, whenever a target is detected on an image frame.

16.2.4 SURF-Based Mean-Shift Algorithm

Direct SURF correspondences between two consecutive frames cannot be used for tracking because of the inherent image noise, which makes many of the descriptors transient and will be thrown away when consecutive frames are matched [473]. Secondly, computing SURF correspondences between complete images to localize the target is computationally expensive as the number of descriptors available for a given image may be quite large. Hence, it is pertinent to carry out SURF matching only in a local region around the current location.

The best local search algorithm is mean-shift tracking, which is based on color histogram matching [479]. Mean-shift tracking requires a histogram of the object template, which is formed by creating a fixed number of clusters in the SURF feature space. SURF histograms have been used for object recognition [480, 481], and place recognition [482] using bag-of-words approach. The center of the new target window is computed by the mean-shift algorithm, and given by

$$\mathbf{z} = \frac{\sum_{i=1}^{n} w_i g\left(\left\|\frac{\mathbf{x} - \mathbf{x}_i}{h}\right\|^2\right) \mathbf{x}_i}{\sum_{i=1}^{n} w_i g\left(\left\|\frac{\mathbf{x} - \mathbf{x}_i}{h}\right\|^2\right)}, \qquad (16.6)$$

where $g(x) = -k'(x)$ is the derivative of the kernel profile and w_i is the weight associated with each key-point location $\mathbf{x}_i$ of the source window, which has a correspondence in the target window. The new center location depends on the number of correspondences n between the source and the target window. The SURF correspondences between windows are computed using the minimum distance criterion, and RANSAC [470] is used for removing outliers.

Histogram creation requires availability of sufficient number of SURF key-points for the object template, which may not be available if we start with a single template. The problem is partially solved by using *re-projection* method, where the source histogram is enriched by making homo-graphic projection of matching points from the target window to the source window at the end of each mean-shift convergence. The details of the algorithm are provided in [483]. The approach properly works when there is no significant change in the pose or shape, and sufficient matching SURF key-points are available between any two consecutive frames. However, in case of a severe change in the appearance of the target, the number of matching descriptors available may fall drastically causing the tracker to fail. Therefore, instead of using a single template, a template pool is used to create histogram. This template pool is updated by a modified object model description in each iteration to accommodate for the change in poses, and is explained in the following subsection.

16.2.5 Modified Object Model Description

The object model O_m consists of three different sets of SURF key-points and descriptors. The first set is obtained by the mean-shift tracker and is represented as $M_1(t)$. The mean-shift tracker gives the SURF correspondences between the object description $O_m(t-1)$ and the descriptors of the window W_t of image I_t as $\zeta(I_{t-1}^{O_m(t-1)} \sim I_t^{W_t})$. The matched key-points and descriptors of the $I_t^{W_t}$ (i.e. $V(I_t^{W_t})$) are directly added to the $M_1(t)$. Mathematically, it can be written as

$$V(I_t^{W_t}) = \{(\mathbf{x}_i, \mathbf{v}_i)^t | (\mathbf{x}_i, \mathbf{v}_i)^t \in \zeta(I_{t-1}^{O_d(t-1)} \sim I_t^{W_t}), i = 1, 2, \cdots, m\}. \quad (16.7)$$

The matched key-points of $O_m(t-1)$ are first projected on an image I_t by replacing the x_i^{t-1} by x_i^t and then key-points and descriptors of the $I_{t-1}^{O_d(t-1)}$ (i.e.$V(I_{t-1}^{O_d(t-1)}) \in \zeta(I_{t-1}^{O_d(t-1)} \sim I_t^{W_t})$) are added to the $M_1(t)$.

The second set is obtained by applying nearest neighbor approach on the set $M_1(t)$ and is represented as $M_2(t)$. To get the region, where nearest neighbor approach has to be applied, the mean-shift tracker window of images I_{t-1} and I_t are resized (increasing the window size to 120% of its original size) and the SURF correspondences between the two resized window is obtained as $\zeta(I_{t-1}^{W_{t-1}^R}, I_t^{W_t^R})$. The key-points lying in the region $W_{t-1}^R - W_{t-1}$ are obtained and a polygon P_R is formed by using their corresponding key-points on image

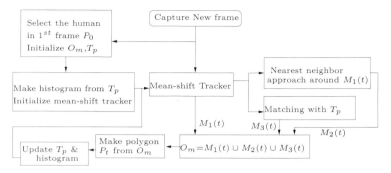

FIGURE 16.4: Block diagram of the proposed tracking algorithm.

I_t. All the key-points and descriptors of $I_t^{W_t^R}$ (i.e. $V(I_t^{W_t^R})$), which lie inside the polygon P_R and have a distance less than a user specified threshold with any point of the set $M_1(t)$, are added to set $M_2(t)$.

The third set is obtained by finding the correspondence between the SURF descriptors of the template pool with those of the $V(I_t^{W_t})$ i.e. $\zeta(T_p(t) \sim I_t^{W_t})$. Only those key-points and descriptors of the $\zeta(T_p(t) \sim I_t^{W_t})$ are retained in this set that have a distance less than a user specified threshold with any point of the set $(M_1(t) \cup M_2(t))$, and is represented by $M_3(t)$.

16.2.6 Modified Tracking Algorithm

The block diagram of the proposed algorithm is shown in Figure 16.4. The bounding rectangle of the polygon P_0 (drawn on the boundary of the human silhouette in the first frame) is used, as the initial window W_0 for the mean-shift tracker. The object model is initialized as $V_p(I_0)$ and the template pool is initialized with $\psi(V_p(I_0))$. The histogram for mean-shift tracker is made using the SURF descriptors of the template pool T_p. This histogram gets updated in each iteration, as the template pool is updated. The mean-shift tracker gives the best matched SURF correspondences between the object model (O_m) and the SURF descriptors of the window W_t of image I_t and constitutes the set $M_1(t)$. The object description is populated with set $M_2(t)$, which is made by using the nearest neighbor approach. The SURF matching of window W_t with template pool results in the set $M_3(t)$. The new object model is calculated, as per Equation (16.4). The template pool is updated in each iteration using the set $M_2(t)$. The convex hull of the points of the object description gives the desired polygon P_t.

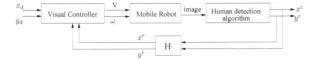

FIGURE 16.5: Schematic block diagram of a human following robot. Human detection and tracking algorithm locates the target to be followed by the robot. The location of the target in the image plane is used by the visual controller block to generate necessary motion commands for the mobile robot which allows it to follow the human. The depth value obtained from Kinect allows the robot to maintain a constant distance from the target.

16.3 Human Tracking Algorithm with the Detection of Pose Change due to Out-of-plane Rotations

In this section, a SURF-based human tracker is proposed which searches a human in the expanded rectangular region around the last target location. The object model is updated over time by selecting new templates and projecting points from previous templates using affine transformation. The proposed approach seeks to detect and avoid tracking failure due to out-of-plane rotations, which is a very difficult problem to solve with point-based features. The pose change due to out-of-plane rotations is confirmed by using the aspect ratio of the bounding region of points projected using affine transformation. Whenever the SURF-based tracker fails to detect the target human, a k-d tree based classifier is used to differentiate between a case of a pose change and an occlusion.

16.3.1 Problem Definition

Consider a set of frames I_i, $i = 0, 1, 2, \cdots N$ of a video sequence, where an object identified by a user in the first frame is to be tracked over all the frames. The object is identified by the user by selecting a rectangular region on the first frame. Let this rectangular region be denoted by W_0 corresponding to the first image I_0. Let $V(W_i) = \{(\mathbf{x}_1, \mathbf{v}_1, \omega_1), (\mathbf{x}_2, \mathbf{v}_2, \omega_2), \cdots, (\mathbf{x}_n, \mathbf{v}_n, \omega_n)\}$ be the set of SURF features of an image I_i within the window W_i, where $\mathbf{x}_i$ is the 2-dimensional key-point location of the 64-dimensional SURF descriptor $\mathbf{v}_i$, and ω_i is the weight assigned to the SURF descriptor $\mathbf{v}_i$. The initial weights are assigned so as to ensure that the descriptors survive at least for few frames. In the proposed approach, the initial weight for the descriptors are taken as 15. The set of key-point locations in a given window or a frame is denoted by V_x and corresponding set of descriptors is denoted by V_v. The

SURF correspondence between a source window W_s and a target window W_t is the set of best matching key-points and descriptors given by

$$V(W_s \sim W_t) \triangleq \{(\mathbf{x}_1, \mathbf{v}_1, \omega_1)^s, \cdots, (\mathbf{x}_m, \mathbf{v}_m, \omega_m)^s, (\mathbf{x}_1, \mathbf{v}_1, \omega_1)^t, \cdots, \\ (\mathbf{x}_m, \mathbf{v}_m, \omega_1)^t\}, \quad (16.8)$$

where superscript s represents source window and t represents target window.

The tracking window W is represented by $W = (\mathbf{c}, w, h)$ where $\mathbf{c} = (c_x, c_y)$ center of the window with width w and height h. Given I_0, W_0 and $V(W_0)$, the task is to compute the tracking window $W_i(\mathbf{c}_i, w_i, h_i)$ for all image frames $i = 1, 2, \cdots, N$.

16.3.2 Tracking Algorithm

The tracking problem could be described as follows. Consider a set of frames I_i, $i = 0, 1, 2, \cdots N$ of a video sequence, where an object identified by the user in the first frame is to be tracked by a mobile robot over all the frames. The block diagram of the proposed scheme is given in Figure 16.5. The object is identified by the user by selecting a rectangular region on the first frame. Let this rectangular region be denoted by W_0 corresponding to the first image I_0. Let $V(W_i) = \{(\mathbf{x}_1, \mathbf{v}_1, \omega_1), (\mathbf{x}_2, \mathbf{v}_2, \omega_2), \cdots, (\mathbf{x}_n, \mathbf{v}_n, \omega_n)\}$ be the set of SURF features of an image I_i within the window W_i, where $\mathbf{x}_i$ is the 2-dimensional key-point location of the 64-dimensional SURF descriptor $\mathbf{v}_i$ and ω_i is the weight assigned to the SURF descriptor $\mathbf{v}_i$. V_x is used to denote the set of key-point locations in a given window or a frame, and V_v is used to denote the corresponding set of descriptors. The tracking window W is represented by $W = (\mathbf{c}, w, h)$ where $\mathbf{c} = (c_x, c_y)$ center of the window with width w and height h. Given I_0, W_0 and $V(W_0)$, the task is to compute the tracking window $W_i(\mathbf{c}_i, w_i, h_i)$ for all image frames $i = 1, 2, \cdots, N$.

The proposed method for visual tracking of human using SURF is explained in the flowchart provided in Figure 16.6. It primarily consists of four parts: (1) Initialization, (2) Tracking, (3) Template Update, and (4) Error-recovery. In the *initialization* part, an object model is defined for the target, which is to be tracked in subsequent frames of the video. The initialization is carried out once in the first frame for a given video sequence. In the *tracking* module, a SURF-based tracker is used to locate the target in the next frame. A Kalman filter motion predictor is updated whenever the target is successfully detected by the SURF-based tracker. The *template update module*, selects a new template and updates the object model by projecting the points from previous template to current template. The *error recovery module* provides the way to deal with pose change and occlusion. All these modules are described in detail in the following subsections.

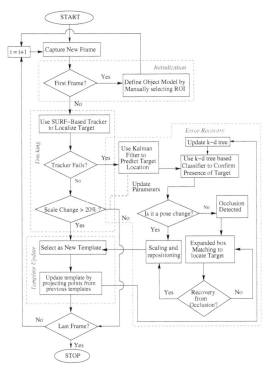

FIGURE 16.6: Flowchart of the proposed human tracking algorithm.

16.3.3 Template Initialization

In this module, the target human to be tracked is selected manually by drawing a rectangle box W_0 around it in the first frame I_0. The SURF descriptors present in this box, represented by $V(W_0)$ also include some descriptors from the background region. The inclusion of background descriptors into the object template are most hazardous, since they increase the probability of negative correspondence during the tracking phase. In order to remove the background SURF descriptors from the object template, an ellipse that fits inside the rectangle W_0 is drawn, as shown in Figure 16.7(b). The SURF descriptors present within this elliptical region (e_0) is represented as $V(e_0)$. The set $V(e_0)$ may still contain a few descriptors that belong to the background. In order to remove these remaining background descriptors from the object template, a region growing algorithm (flood-fill) is applied to the key-points that lie outside the elliptical region [484]. The flood-fill algorithm fills the adjacent cells having same pixel intensities with same color. In this way, the background region is segmented from the foreground, as shown in Figure 16.7(c). All the descriptors, which belong to this segmented background region (B_R) are now

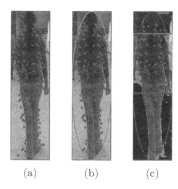

FIGURE 16.7: Object template selection - Set of SURF descriptors (a) in the selected rectangle region, (b) after fitting the ellipse in rectangle region, (c) after removing background inside the ellipse using region growing method.

removed from the object template. Mathematically, the object template (OT_0) can be defined as

$$OT_0 = \{(\mathbf{x}_i, \mathbf{v}_i, \omega_i) | (\mathbf{x}_i, \mathbf{v}_i, \omega_i) \in V(W_0) \wedge \mathbf{x}_i \in e_0 \wedge \mathbf{x}_i \notin B_R,$$
$$\omega_i = 15, i = 1, 2, \ldots, p \leq m\} \quad (16.9)$$

where m is the number of SURF descriptors present inside the window W_0. The object model (OM_0) is initialized with the object template. All the SURF descriptors of the object model are assigned an initial weight ω_i. The initial weight is selected so as to ensure that the descriptors survive at least for a few frames. In this work, the initial weight for a descriptor is chosen as 15. This step is repeated whenever a new template is selected. This is a crucial step that helps in avoiding tracker drift for a longer duration.

16.3.4 Tracking

In this module, the target is localized in the next frame using direct SURF matching. The SURF correspondences are computed between the current object model and the new frame within a bounded rectangular region around the last target location. The bounded region is 10% bigger than the last target window. The wrong matches are removed by using Random Sample Consensus (RANSAC) based on homography [470, 471]. The tracking is considered successful if the percentage match between the source and destination region is found to be greater than a user defined threshold (θ). The percentage match is defined as,

$$M_p = \frac{N_m}{N_s} \times 100 \quad (16.10)$$

where N_s is the number of SURF key-points present in the last selected template, and N_m is the number of matching key-points obtained through SURF correspondence. The match threshold (θ) used in this work is 20. If the tracking is successful, the weights of the descriptors of the object model is updated as

$$\omega_i(t+1) = \begin{cases} \omega_i(t) + 2 & \text{if match is found} \\ \omega_i(t) - 1 & \text{otherwise.} \end{cases} \quad (16.11)$$

The tracker window is scaled and repositioned, as described in the following section. The new tracker location is used to train the Kalman filter motion predictor.

16.3.4.1 Scaling and Re-positioning the Tracking Window

As the size of human being tracked may vary over time, the tracking window needs to be scaled and re-positioned in order to avoid spurious background descriptors being included into the template information. The bounding box around the human is divided into three parts. The torso region having maximum number of matching descriptors is used for computing the scaling factor, as explained in [478]. The re-positioning of the tracker window is done by obtaining the center of the points in the torso region and then shifting it by using the body ratio of the human.

16.3.5 Template Update Module

The object model used for tracking the target by SURF-based tracker needs to be updated in order to accommodate for the temporal changes that the target may undergo during its motion. The update of object model in this case includes two steps.

1. A new template is selected only when the current template is inadequate to detect the target in the next frame using SURF correspondences. This happens when substantial movement has accumulated over time rendering the current template unfit for detecting the target in the new frame. Therefore, a new template is selected whenever the KD-tree based classifier confirms the pose change case in the predicted tracker window. To avoid the frequent failure of the tracker, a new template is selected whenever the tracking window obtained from SURF-based tracker undergoes a scale change of more than 20% or the target recovered from occlusion.

2. Selecting a new template can detect the target over short-term, it might fail over longer run in the absence of stable descriptors, which matched frequently in previous frames. Hence, to maintain stability over long run, these stable features need to be incorporated into the new template so as to resolve the stability-versus-plasticity dilemma [472]. All the SURF descriptors, which have non-negative weight, along with their key-point

locations are added to the new template by using the affine transformation (AT). AT reflects the displacement between the new template and the last template. Although the human motion is non-affine in nature, the torso region under the assumption of upright human position can be considered as a rigid object and its motion can be considered, as an affine motion. The AT [485] model is given by

$$\begin{pmatrix} f_{x,k+1} \\ f_{y,k+1} \end{pmatrix} = \begin{pmatrix} a_0 & a_1 \\ a_3 & a_4 \end{pmatrix} \begin{pmatrix} f_{x,k} \\ f_{y,k} \end{pmatrix} + \begin{pmatrix} a_2 \\ a_5 \end{pmatrix} \quad (16.12)$$

The six parameters in the model is estimated using the least-square method for the set of points obtained in the torso region through SURF correspondence between the two templates.

Therefore, the updated object model can be defined as

$$OM_n = \{OT_n \cup (\mathbf{x}_i, \mathbf{v}_i, \omega_i) \mid (\mathbf{x}_i, \mathbf{v}_i, \omega_i) \in OT_{n-1}$$
$$\wedge \mathbf{x}_i \in g(V_x{}^{n-1}) \; \forall \; \omega_i \geq 0, i = 1, 2, \ldots, m\} \quad (16.13)$$

where m is the number of SURF-descriptors present in the OT_{n-1} and $g(.)$ represents the affine transformation. Note that a subset of the past features, which are occurring more frequently are projected onto the current template. This reduces the computational and memory requirement by avoiding linearly increasing number of descriptors in the object model with the increasing frames in a video. The computational and memory requirement of the proposed algorithm is much less compared to methods reported in [475, 486]. The projected descriptors with their key-point locations lead to better matches in subsequent frames.

16.3.6 Error Recovery Module

This module is executed whenever the SURF-based tracker fails to detect the human. The tracker fails to localize the target in a frame, when the number of matching points between the source and the target windows falls below a certain threshold. Such a case might arise under two conditions: (1) the target is present but its appearance has changed significantly from the current object template due to pose change such as *out-of-plane rotations*, and (2) the target is partially or fully occluded by other objects in the environment. Compared to other effects like variation in illumination or scaling, the pose change due to out-of-plane rotations lead to frequent tracking failures. In order to differentiate a case of pose change from that of an occlusion, a KD-tree based classifier is used is this algorithm.

16.3.6.1 KD-tree Classifier

In this algorithm, a KD-tree is used for feature matching or to classify a tracker window whether its belongs to the foreground or to the background.

Directly matching raw features extracted from the current tracker window with the older templates represents an reasonable similarity measurement and can be used for classification. The problem is that the direct feature matching via linear search can soon become intractable with the increasing number of templates to be processed and prevent its real-time implementation. In contrast, a tree structure is an efficient data structure for feature matching [487] and can provide online processing in classification due to its efficiency. In the proposed algorithm, a KD-tree is built over all the template features and a tree search is performed for the features extracted from the query tracker window. The idea of using KD-tree for feature matching is inspired by the following facts.

The KD-tree reduces the search complexity from linear to logarithmic and compares one dimension of the high-dimensional features each time, and thus, avoids the distance computations (the most time-consuming part in finding the correct nearest neighbor for a query feature with linear search). The second fact is that the distance ratio method [476] is an effective way for verifying putative matches. In distance ratio method, a correct match requires the ratio between the distances of the closest and second closest neighbor to the query feature to be below some given threshold. This work focus more on the applicability of the tree structure in identifying foreground descriptors by using distance ratio technique.

16.3.6.2 Construction of KD-Tree

The KD-tree is initially built with the descriptors of the first template and these descriptors are pushed into the feature pool D. Whenever a new template is selected, its descriptors are inserted to the feature pool D and the KD-tree is reconstructed over all the features in D. Hence, D always includes the features extracted from all the templates and the tree is updated every time a new template is selected. When the SURF-based tracker fails to converge due to unavailability of sufficient number of matching points, the descriptors of the tracker window, whose location is predicted by the Kalman filter is subjected to a KD-tree based classifier to distinguish the above two situations. Let Q be the set of SURF descriptors extracted from the current tracker window. For each descriptor in Q, it will go through a tree search in the current KD-tree and top two nearest neighbors are returned. Distance ratio is applied to determine whether the closest one is a good match or not. If it is a good match, then it is considered as a foreground descriptor, otherwise, it is considered as a background descriptor. If the number of foreground descriptors is more than 1.5 times of background descriptors in the tracker window, then it is considered a case of pose change, otherwise, it is considered a case of occlusion.

16.3.6.3 Dealing with Pose Change

Once the KD-tree based classifier confirms that the human is present in the tracker window predicted by the Kalman filter, it is necessary to figure out that

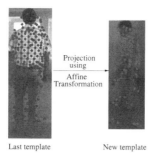

FIGURE 16.8: Detecting out-of-plane rotations by using the aspect ratio of points projected using affine transformation.

the pose change is due to in-plane-rotations or out-of-plane rotations. If the target undergoes a significant amount of out-of-plane rotation (side pose) and the object model is updated with this template, then a number of background descriptors will get include into the object model, and results in failure of the tracker. Therefore, the object model is only updated with the template, in which pose change is either due to in-plane rotations or due to small amount of out-of-plane rotations. The out-of-plane rotations can be checked by projecting the points of the previous template onto the current template through the affine transformation obtained between the two templates, as shown in Figure 16.8. As one can see, the projected points get concentrated on a line during significant out-of-plane rotations and this information could be utilized to avoid tracking failure.

To prove that the points get concentrated with increasing amount of out-of-plane rotations, thirty points are taken on the boundary of a circle in x-y plane, as shown in Figure 16.9 by red circles. In an image, the out-of-plane rotation is the rotation about y-axis and its rotation matrix is given by

$$R(\theta) = \begin{pmatrix} cos\theta & 0 & sin\theta \\ 0 & 1 & 0 \\ -sin\theta & 0 & cos\theta \end{pmatrix} \qquad (16.14)$$

In this experiment, a rotation of 0, 15, 30, 45, 60, 75 and 90 degrees is applied on these points and these points are plotted, as shown in Figure 16.9. The value of z-coordinate is taken as zero. One can see in Figure 16.9 that these points get concentrated with increasing amount of rotations and finally at 90 degree rotation, these points get concentrated on a straight line. This is pictorially demonstrated on dataset D6 in Figure 16.10.

16.3.6.4 Tracker Recovery from Full Occlusions

Once the occlusion is detected, the location of the tracker window is provided by a Kalman filter based predictor. The confirmation about occlusion is obtained from the KD-tree based classifier which identifies it as a background.

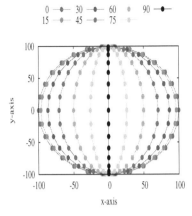

FIGURE 16.9: Thirty points are taken on the boundary of a circle in x-y plane. With increasing amount of out-of-plane rotation, these points get closer to each-other and finally merged in a straight line at 90° rotation.

FIGURE 16.10: Demonstration of projected points during out-of-plane rotation on dataset D6.

However, if the target is occluded for several consecutive frames, the motion predictor might drift from the actual trajectory of the target in absence of update over these frames. Therefore, in such cases, the SURF-based tracker is used to search around the location predicted by the Kalman filter with increasing size of the tracker window. In the worst case, a frame-to-frame matching between the current image and the past template image could be used to locate the target once it recovers from the occlusion. In most cases, a local search around predicted position is adequate to recover the target.

In order to implement the proposed visual human tracking algorithm on a mobile robot, a visual servo controller is designed in the next section which takes data from the tracking algorithm and gives motion commands to the mobile robot.

16.4 Human Tracking Algorithm Based on Optical Flow

Optical flow is used to track specific features (points) in image over the time, where the movements of the pixels (points) from one frame to another frame were converted into velocity vector. It is defined as follow

$$I(x, y, t) = I(x + dx, y + dy, t + dt), \quad (16.15)$$

where I is the intensity value of the image, and t is the displaced time for the intensity value. In this research work, Lucas-Kanade optical flow method is used [488]. The Lucas Kanade method assumes that the displacement of the image contents between two consecutive frames is small and approximately constant within a neighborhood of the point p under consideration. Thus, the optical flow equation can be assumed to hold for all pixels within a window centered at p. Therefore, the local image flow (velocity) vector (v_x, v_y) must satisfy

$$I_x(q_i)v_x + I_y(q_i)v_y = -I_t(q_i), \quad \text{where } i = 1, 2, ..., n; \quad (16.16)$$

$q_1, q_2, \ldots, q_n$ are the pixels inside the window; and $I_x(q_i), I_y(q_i), I_t(q_i)$ are the partial derivatives of the image I with respect to position x, y, and time t, evaluated at the point q_i, and at the current time. These equations can be written in matrix form $Av = b$, where

$$A = \begin{bmatrix} I_x(q_1) & I_y(q_1) \\ I_x(q_2) & I_y(q_2) \\ \vdots & \vdots \\ I_x(q_n) & I_y(q_n) \end{bmatrix}, \quad v = \begin{bmatrix} v_x \\ v_y \end{bmatrix}, \quad b = \begin{bmatrix} -I_t(q_1) \\ -I_t(q_2) \\ \vdots \\ -I_t(q_n) \end{bmatrix}. \quad (16.17)$$

The Lucas-Kanade method obtains the solution of this over-determined system by the least squares as

$$v = (A^T A)^{-1} A^T b. \quad (16.18)$$

Optical flow has been used in a large number of tracking algorithms, as in Kalal's TMD model [489], active appearance models (AAM) [490], and with SURF [491] etc. The optical flow tracker used in the proposed approach is similar to the SURF-based optical flow tracker described in [491]. However, the proposed approach does not compute the parameters of the warp matrix or update of the eigen basis representation of the model using incremental PCA

Human Tracking Algorithm Based on Optical Flow

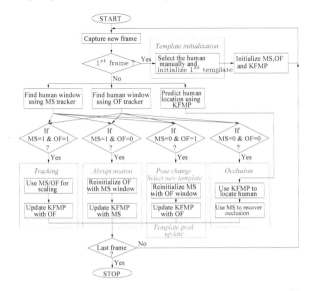

FIGURE 16.11: Flowchart of the proposed optical flow-based algorithm. Human is selected manually in first frame and tracked with optical flow and mean-shift tracker in subsequent frame.

algorithm [492], as has been done by the authors. The optical flow works under the assumption of temporal persistence and spatial coherence, and may fail in case of abrupt camera motion or low frame rate [493, 494]. In case of gradual pose change, optical flow may be used to localize the tracker window by estimating the positions of current SURF key-points in the next frame. The optical flow serves two purposes in the proposed approach. First, it helps in scaling the bounding box for the target being tracked. Secondly, it is used for selecting templates over time particularly in cases, where the target undergoes change in pose due to out-of-plane rotations.

The proposed method is explained in the next section that makes use of the above two trackers to yield robust tracking of human in a dynamic environment. The flowchart of the proposed scheme is shown in Figure 16.11 and the working of the proposed algorithm is illustrated in Figure 16.12.

16.4.1 The Template Pool and its Online Update

The template pool P is a collection of SURF descriptors obtained from the templates that are selected online. Each template is represented by a set of SURF descriptors available within the tracker window W_j. Mathematically, template pool can be defined as

$$P \triangleq \{V_v(W_j), j = 1, 2, \cdots, r\}, \tag{16.19}$$

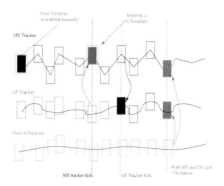

FIGURE 16.12: The working of the proposed algorithm. Two trackers and a motion predictor run in parallel. First tracker is a SURF-based mean-shift tracker (MS) while the other tracker is an flow tracker (OF). A Kalman Filter based motion predictor (KFMP) is used to learn the motion model of the human. MS and OF trackers correct each other whenever one of them fails to detect the target. When both of these trackers fail, it is considered as a case of occlusion where the motion predictor (KFMP) is used to localize the target.

where r is the number of templates in the pool. The template pool has an upper bound on its size and it is taken as 1,000 descriptors. Each descriptor is assigned an initial weight, whenever it is added to the template pool. The initial weights are selected so as to ensure that the descriptors survive at least for few frames. In the proposed approach, the initial weight for the descriptors are taken as 15. The weight is incremented by 2, whenever a match is found with this descriptor; otherwise, it is decremented by 1. Mathematically, it can be written as

$$\omega_i(t+1) = \begin{cases} \omega_i(t) + 2 \text{ if match is found} \\ \omega_i(t) - 1 \text{ else} \end{cases} \quad (16.20)$$

This is similar to weight update method provided in [495]. However, in the proposed approach, the weights are maintained only for the foreground model. The descriptors with negative weights are discarded, whenever new descriptors are added to the template pool. If the size of the template pool exceeds its upper bound size, then the descriptors with lowest weights are discarded in order to bring the template pool with-in its upper bound size.

16.4.1.1 Selection of New Templates

A new template is selected, whenever the mean-shift tracker fails to converge but the optical flow provides a sufficient number of matching points. This case arises when there is a pose change particularly due to an out-of-plane rotation, and the target is not occluded. In this case, the mean-shift tracking fails due to the absence of matching descriptors. However, the optical flow

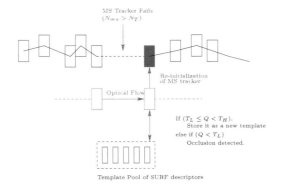

FIGURE 16.13: Using optical flow to re-initialize the mean-shift tracker. OF tracker is used for selecting templates. Mean-shift tracker fails when the number of matching correspondences falls below a threshold. This usually occurs during a pose change due to out-of-plane rotations. In this case, optical flow could be used to get an estimate of the tracker window location using the temporal coherence present in the video.

provides the estimates of the current key-point locations in the next frame, and the complete reinitialization process flow is illustrated by Figure 16.13. The basic assumptions of an optical flow algorithm, namely, brightness constancy, temporal persistence, and spatial coherence are satisfied in case of a pose change [494], and hence, OF tracker does not fail.

Every template obtained from the OF tracker may not be a right template to be included into the template pool P. Each template obtained from the OF tracker is checked for its quality. Therefore, SURF correspondence between the template provided by tracker window W_t and the template pool P is calculated, and is represented as $V(P \sim W_t)$. In this case, the SURF correspondence is obtained based on the Euclidean distance between the descriptors of the two sets. The set of descriptors of the current tracker window W_t that matches with the template pool is given by

$$\psi_v^P(W_t) = \{\mathbf{v}_i | \mathbf{v}_i \in W_t \wedge \mathbf{v}_i \in V(P \sim W_t), \quad i = 1, 2, \ldots, m\}, \quad (16.21)$$

where m is the number of matching descriptors found between the two sets.

Each template is checked for the following two aspects before including it into the template pool:

(i) Matching Quotient: It is the ratio of the SURF descriptors in the tracker window that matches with those in the template pool to the total number of descriptors present in this window, weighted by the distance between the centers of two consecutive windows. Mathematically, it is expressed as

$$M_p = \frac{N_m}{N_T} \times e^{-\lambda \|\mathbf{c}_t - \mathbf{c}_{t-1}\|^2}, \quad (16.22)$$

where N_m is the cardinality of set $\psi_v^P(W_t)$, which represents the number of descriptors of target window W_t that matches with those in the template pool; N_T is the cardinality of the set $V(W_t)$, which represents the total number of descriptors in the current tracker window W_t corresponding to the image I_t; $\mathbf{c}_t$ and $\mathbf{c}_{t-1}$ are the centers of current and previous tracker window, respectively.

If the M_p falls below a certain threshold (0.2), the template is considered an invalid one and should not be included into the template pool. This may indicate a case of partial or full occlusion, which is dealt with separately as explained in the later part of this section.

(ii) If the height of the bounding box of the tracker window changes more than a certain percentage (20%), it is considered an invalid template, which should not be included into the template pool. The height of the bounding box may change abruptly if trackers fail to find matching points for a part of the body for instance, the leg region.

16.4.2 Re-Initialization of Optical Flow Tracker

In the case of a sudden change in the motion of the target arising out of abrupt camera motion or drop in camera frame rate, the optical flow may fail to localize the tracker window due to the loss of temporal coherence [493]. Usually, the number of points available from optical flow go on decreasing over the time when started with the points from a single template. It is necessary to re-initialize the tracker, when the number of points available fall below a certain threshold. The re-initialization is also needed in cases, where the size of the OF tracker window suddenly changes due to the loss of points in one part of the body (e.g., leg region).

In such cases, the mean-shift (MS) tracker could be utilized to initialize the optical flow tracker. It should be noted that the proposed approach cannot be used to deal with cases where pose change, and sudden motion occurs simultaneously. Such cases are treated as occlusions, and treated differently as explained in the next section. The SURF key-point locations of the current mean-shift (MS) tracker are used as the input to the OF tracker, which now estimates the location of these points in the next frame.

16.4.3 Detection of Partial and Full Occlusion

In the case of partial occlusion of the target, it is still possible to obtain a few matching descriptors with the template pool. The partial occlusion is detected in two ways. Either the mean-shift tracker does not converge within a pre-specified number of iterations or the quality of the tracker window lies between a lower and an upper bound ($T_L \leq Q < T_H$). In case of full occlusion, the quality of of the tracker window falls below the threshold T_L. In this algorithm, the value of T_L and T_H are taken as 0.2 and 0.3, respectively.

Partial occlusions are easier to deal with as it is always possible to find a few matching points, which could be utilized to position the tracker window. Full occlusions are difficult to deal with. In the case of full occlusion, both mean-shift tracker and optical flow tracker fail due to the unavailability of matching feature points.

16.5 Visual Servo Controller

To design a visual servo controller for the mobile robot, equations are derived from the kinematic model of the robot and the pinhole model of the camera. Then, the problem is formulated using these equations. Let us consider a differentially driven mobile robot (P3-DX) carrying a fixed camera (Kinect). The task of the robot is to track a moving human on a 2-**D** plane (x-y plane). As shown in Figure 16.14, the camera is mounted along the heading direction of the mobile robot. A coordinate frame $\{R_c\}$ is attached to the optical center of the camera P_c. The x-axis being along the optical axis of the camera. z-axis is in the direction pointing out of the paper. Using Figure 16.14, the kinematic equations of a mobile robot can be obtained as follows.

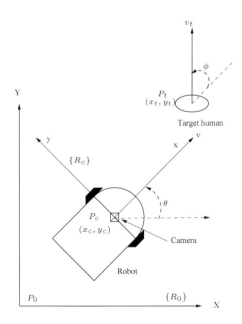

FIGURE 16.14: Mobile robot mounted with a fixed camera tracks a moving human on a 2-D plane.

16.5.1 Kinematic Model of the Mobile Robot

The posture of a mobile robot is presented by its position that is the middle point of the two driving wheels, and the heading direction θ. Figure 16.14 shows the position of the robot expressed in the X-Y coordinates. The posture vector of mobile robot is presented as $P_c = (x_c, y_c, \theta)^T$ with respect to initial frame $\{R_0\}$. The mobile robot has 2 degrees of freedom. It can move along the x-axis and rotate around the z-axis. The mobile robot's motion is controlled by the vector $\mathbf{q} = (v, \omega)^T$, where v is the linear velocity of the robot and ω is the angular velocity of the robot. Based on the assumption that the mobile robot moves on a 2-D plane without slips, the kinematic equation of a mobile robot can be written as

$$\dot{P}_c = \begin{bmatrix} \dot{x}_c \\ \dot{y}_c \\ \dot{\theta} \end{bmatrix} = \begin{bmatrix} v \cos \theta \\ v \sin \theta \\ \omega \end{bmatrix} = \begin{bmatrix} \cos \theta & 0 \\ \sin \theta & 0 \\ 0 & 1 \end{bmatrix} \mathbf{q} \qquad (16.23)$$

16.5.2 Pinhole Camera Model

The pinhole camera model describes the mathematical relationship between the coordinates of a 3-**D** point and its projection onto the image plane of an ideal pinhole camera, where the camera aperture is described as a point and no lenses are used to focus light [496]. Let $(^cx_t, {^c}y_t, h)$ is the position vector of the human center P_t with respect to the frame $\{R_c\}$, h is a known positive constant, (α_t, β_t) is the image coordinate of the point projected on the image plane, and f is the focal length of the camera. Using the pinhole model for the camera as shown in Figure 16.14, the following relationships are obtained:

$$\frac{^cx_t}{f} = \frac{^cy_t}{\alpha_t} = \frac{h}{\beta_t} \qquad (16.24)$$

16.5.3 Problem Formulation

Let the error posture between the target human and the robot be $P_e = (x_e, y_e, \theta_e)$. The error posture is a transformation of the target posture P_t from frame $\{R_0\}$ in a local coordinate frame $\{R_c\}$ with an origin of (x_c, y_c) and z-axis is the direction of θ_e.

$$\begin{aligned} P_e &= \begin{bmatrix} x_e \\ y_e \\ \theta_e \end{bmatrix} = \begin{bmatrix} ^cx_t \\ ^cy_t \\ \phi \end{bmatrix} \\ &= \begin{bmatrix} \cos \theta & -\sin \theta & 0 \\ \sin \theta & \cos \theta & 0 \\ 0 & 0 & 1 \end{bmatrix} \begin{bmatrix} x_t - x_c \\ y_t - y_c \\ \omega_t - \omega \end{bmatrix} \end{aligned} \qquad (16.25)$$

Visual Servo Controller

The equation of relative motion between the target human and the mobile robot can be calculated as follows [497]:

$$\dot{P}_e = \begin{bmatrix} ^c\dot{x}_t \\ ^c\dot{y}_t \\ \dot{\phi} \end{bmatrix} = \begin{bmatrix} -v + v_t \cos\phi + {}^c y_t.\omega \\ v_t \sin\phi - {}^c x_t.\omega \\ \omega_t - \omega \end{bmatrix}$$

$$= \begin{bmatrix} -v + v_{tx} + {}^c y_t.\omega \\ v_{ty} - {}^c x_t.\omega \\ \omega_t - \omega \end{bmatrix} \quad (16.26)$$

where v_t is the linear velocity of the target human, ϕ is the angle between moving directions of the target human and the mobile robot, ω_t is the angular velocity of the target human, and v_{tx} and v_{ty} are the target human velocity components, $v_{tx} = v_t \cos\phi$ and $v_{ty} = v_t \sin\phi$, respectively.

The objective of human tracking is to drive the mobile robot to keep the target human always in sight of the camera. This objective can be achieved if we keep the position of the human center P_t on the heading direction and within a short distance of the mobile robot, i.e.,

$$^c x_t = x_d \quad \text{and} \quad {}^c y_t = 0 \quad (16.27)$$

16.5.4 Visual Servo Control Design

For the objective of human tracking (16.27), the tracking errors, e_x and e_y can be defined as:

$$e = \begin{bmatrix} e_x \\ e_y \end{bmatrix} = \begin{bmatrix} ^c x_t - x_d \\ ^c y_t \end{bmatrix} \quad (16.28)$$

The system (16.26) can be rewritten in terms of tracking errors as:

$$\dot{e} = \begin{bmatrix} \dot{e}_x \\ \dot{e}_y \end{bmatrix} = \begin{bmatrix} -v + v_{tx} + {}^c y_t.\omega \\ v_{ty} - {}^c x_t.\omega \end{bmatrix} \quad (16.29)$$

or

$$\dot{e} = \begin{bmatrix} \dot{e}_x \\ \dot{e}_y \end{bmatrix} = \begin{bmatrix} -1 & ^c y_t \\ 0 & -^c x_t \end{bmatrix} \begin{bmatrix} v \\ \omega \end{bmatrix} + \begin{bmatrix} v_{tx} \\ v_{ty} \end{bmatrix} \quad (16.30)$$

In this work, the visual servo controller is designed using the approach of dynamic inversion. In dynamic inversion approach, the controller is synthesized such that the following stable linear error dynamics are satisfied

$$\dot{e} + K.e = 0, \quad (16.31)$$

where $K = \begin{bmatrix} k_1 & 0 \\ 0 & k_2 \end{bmatrix}$ is the gain matrix. Substituting the values of $\dot{e}$ and e, we get

$$\begin{bmatrix} -1 & {}^c y_t \\ 0 & -{}^c x_t \end{bmatrix} \begin{bmatrix} v \\ \omega \end{bmatrix} + \begin{bmatrix} v_{tx} \\ v_{ty} \end{bmatrix} + \begin{bmatrix} k_1 & 0 \\ 0 & k_2 \end{bmatrix} \begin{bmatrix} {}^c x_t - x_d \\ {}^c y_t \end{bmatrix} = 0 \quad (16.32)$$

After rearranging the equation, the desired control inputs are

$$v = v_{tx} + k_1({}^c x_t - x_d) + \frac{{}^c y_t(k_2 {}^c y_t + v_{ty})}{{}^c x_t} \quad (16.33)$$

$$\omega = \frac{(k_2 {}^c y_t + v_{ty})}{{}^c x_t}. \quad (16.34)$$

The values of $({}^c x_t, {}^c y_t)$ can be obtained using the pinhole camera equation (16.24) and the relative target velocities (v_{tx}, v_{ty}) are obtained using the Kalman filter. The Kalman filter provides the estimate of the velocities in image plane, which can be converted to real-world velocities using the interaction matrix as explained in [498].

16.5.5 Simulation Results

To illustrate the performance of the proposed control scheme, simulations were performed in which the mobile robot was required to track circular trajectories. The trajectories were generated by a path planner in the frame of the x-y coordinates. The units of x and y coordinates are in meter in all the experiments. In the simulations, for using the proposed control scheme, object trajectories were transformed into the robot local coordinate frame $\{R_c\}$ before generating control commands. Only later two trajectories are described below in the interest of the space limitation.

16.5.5.1 Example: Tracking an Object which Moves in a Circular Trajectory

Let the object move counterclockwise in a circle around $(0.7, 1)$ with respect to $\{R_0\}$ [499] and set the motion as follows:

$$x_t = 0.7 - 0.2\cos(t), \quad y_t = 1.0 - 0.2\sin(t) \quad (16.35)$$

Therefore, the velocity of moving object is described as follows:

$$\frac{dx_t}{dt} = 0.2\sin(t) = v_t \cos(\theta + \phi) \quad (16.36)$$

Experimental Results

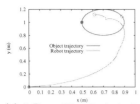
(a) 2-D position trajectories in the world coordinate frame

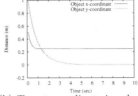
(b) Target coordinate in robot frame

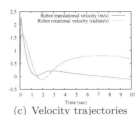
(c) Velocity trajectories

FIGURE 16.15: Simulation results for tracking a circular trajectory. The filled and empty circles show the starting and ending point of the trajectory respectively.

$$\frac{dy_t}{dt} = -0.2\cos(t) = v_t \sin(\theta + \phi) \tag{16.37}$$

The relative velocity between the object and the mobile robot can be derived as follows:

$$v_{tx} = v_t \cos(\phi) = -0.2\sin(\theta - t) \tag{16.38}$$
$$v_{ty} = v_t \sin(\phi) = -0.2\cos(\theta - t) \tag{16.39}$$

The robot initial pose is taken as $P_c(0) = (0, 0, 0.0)^T$. The desired tracking goal is set as $x_d = 0.25$ and $y_d = 0$. The values of k_1 and k_2 are taken as 3 and 1, respectively.

Figure 16.15 shows the simulation result for tracking a circular trajectory in the world coordinate ($\{R_0\}$ frame). As can be seen in this figure, the proposed controller results in a smooth trajectory. In order to maintain the desired distance x_d, the robot departs from the target when the target approaches to it. Figure 16.15(b) indicates that the tracking result achieves the tracking goal, i.e., $^c x_t = x_d = 0.25$ and $^c y_t = y_d = 0.0$, before four seconds. The velocities given to the robot are provided in Figure 16.15(c).

16.6 Experimental Results

16.6.1 Experimental Results for the Human Tracking Algorithm Based on SURF-based Dynamic Object Model

The experimental results are provided in two parts. In the first part, the performance of the proposed visual human tracking algorithm is evaluated while in the second part, the results obtained from the actual robot are analyzed.

16.6.2 Tracking Results

The performance of the proposed tracking algorithm is analyzed by testing it on a number of video datasets collected from different sources. These datasets exhibit different kinds of situations and different pose challenges. This is different from a person detection problem [500] or a video surveillance problem, where one needs to detect and track persons in the scene no matter how small they appear. The summary of various datasets used in this work are provided in Table 16.1. In total, six sets of videos are taken, out of which two are from a pedestrian dataset of ETH Zurich [501], two sets are from Youtube, and the last two are our own. The videos depict several challenging situations like variation in illumination, scaling, occlusion, camera motion, and change in pose with *out-of-plane rotations*.

The images in each video have a resolution of 640×480. The algorithm is implemented in C/C++ using OpenCV 2.0 library on a system with intel i7 processor running with 8 GB of RAM. The tracking results for these datasets are available on website [502] for the inspection and a few snapshots are shown in Figure 16.16. The templates generated by the proposed algorithm for all video datasets are shown in Figure 16.17. As one can see, the templates generated exhibit all kinds of variations as discussed earlier in the chapter. Online update of object model helps in tracking a person for a longer duration by avoiding tracking failures arising out of appearance changes that may occur over time.

It can be seen in Figure 16.18(a) that the maximum number of descriptors in the object model is less than 1,000, which is quite less as compared to the methods [486] that save a number of templates in the memory. Therefore, it can be concluded that the memory requirement of the proposed tracker is quite low, as compared to the existing methods. Computational efficiency of the proposed tracker can be seen in Figure 16.18(b). Average computation time for a video is around 150ms, i.e., 6 frame/second, which is comparable to

TABLE 16.1: Summary of attributes for datasets used for evaluating the performance of the proposed algorithm.

Dataset		Total no. of frames	Scaling (Upto)	No. of templates generated	Success Rate
ETH	D1	380	179%	25	98.15
	D2	251	214%	21	92.82
Youtube	D3	166	17%	9	80.12
	D4	266	87%	12	90.22
Own	D5	291	14%	19	94.15
	D6	979	51%	29	86.41

Experimental Results

FIGURE 16.16: Tracking results for different datasets. The green window is the main tracking window for the target being tracker while white window is the estimated target location obtained from a predictor.

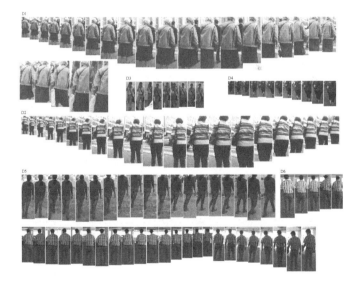

FIGURE 16.17: Templates generated by the algorithm for these datasets. As one can see, the templates contains different poses of the human with a varying degree of scaling and illumination.

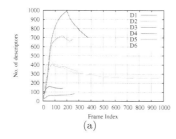 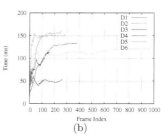

FIGURE 16.18: (a) Average number of descriptors and, (b) average computation time per frame for data sets mentioned in Table 16.1.

those reported by [474] and sufficient for tracking a human with mobile robot. The tracking results of real-time implementation of visual human tracking algorithm are available online [503].

In order to assess the utility of the proposed approach, its performance is compared with the well-known algorithms like mean-shift algorithm [479] and the SURF-based mean-shift algorithm (SBMS) [483].

Three measures are used, namely, percentage overlap, success rate, and computation time to quantify the efficacy a given tracking algorithm. In order to compute these values, the ground truths are labeled manually for all the video datasets. The percentage overlap between the tracker window (W_t) and its ground truth (W_g) is defined as

$$\% \; Overlap = \frac{area(W_g \cap W_t)}{area(W_g \cup W_t)} \times 100 \quad (16.40)$$

The overlap is measured as the percentage of area of the ground truth window that is common with the tracker window.

Similarly, the success rate [504] of the algorithm is defined as the ratio of number of frames where the target is correctly detected by the tracking algorithm (n) to the total number of frames (N). Mathematically, it is computed as follows:

$$Success \; Rate = \frac{n}{N} \times 100 \quad (16.41)$$

For a given frame, if the overlap between the tracker window and its ground truth exceeds 50%, the target detection is considered to be successful. The success rate for the videos is mentioned in Table 16.1. A tracking algorithm is considered to have a better performance if it has higher values for % overlap and success rate with lower value of computation time per image.

The performance comparison among the tracking algorithms in terms of the above quantitative measures is provided in Table 16.2. It is clear from the table that the success rate of the proposed algorithm is high for all the datasets, while mean-shift tracker tracks the human successfully only

Experimental Results

TABLE 16.2: Quantitative performance comparison of various tracking algorithms

Data-set	Algorithm	MS [479]	SBMS [483]	PA
D1	Success Rate	0	21.84	98.15
	Average % overlap	11.12	34.3	75.4
	Average time (ms)	121	682	134
D2	Success Rate	0	8.36	92.82
	Average % overlap	8.23	19.78	71.7
	Average time (ms)	121	565	179
D3	Success Rate	22.8	54.81	80.12
	Average % overlap	15.4	41.3	68.9
	Average time (ms)	120	486	101
D4	Success Rate	20.67	60.5	90.22
	Average % overlap	29.1	50.54	61.6
	Average time (ms)	121	423	51
D5	Success Rate	100	53.9	94.15
	Average % overlap	74.9	51.7	69.1
	Average time (ms)	121	748	160
D6	Success Rate	51.68	79.8	86.41
	Average % overlap	41.8	60.3	71.5
	Average time (ms)	124	771	131

for dataset D5, in which human color is different from the background color and human did not get occluded. However, the computation time of the proposed algorithm is greater than the mean-shift algorithm. As compared to the SURF-based mean-shift tracker, the proposed tracker outperforms in all the measures.

16.6.3 Human Following Robot

The schematic block diagram of a human following robot is shown in Figure 16.5. It consists of a P3-DX mobile robot from Adept Mobile robots [505] equipped with a camera, a human detection algorithm and a visual controller. The desired position of the human center in the robot frame is represented by the pair (x_d, y_d). V and ω are the translational and rotational velocities generated by the visual controller for the mobile robot so that it can reach the desired position. (x_c, y_c) is the human center position obtained by the human detection algorithm in the image frame. **H** represents the relationship between the variables defining the relative posture of the follower robot to the target human (x^r, y^r), and the image features. The experimental setup used for implementing the algorithms is shown in Figure 16.19. The Kinect is used as the visual sensor for detecting and tracking the human walking in front of the robot. The visual controller takes the human position in the robot frame as the input and with reference to the desired goal generates

FIGURE 16.19: The Experimental setup for a human following robot. This comprises a P3-DX mobile platform from Adept [505] and a Kinect. An external computer is used to run the tracking algorithm and the visual controller required for this task.

the velocity commands for the mobile robot. The mobile robot moves with the generated velocities to perform human following. The robot maintains a constant distance from the human target by using Kinect depth measurement. In order to analyze the performance of the robot, an experiment is performed where the subject walks in a corridor in front of the robot. Few snapshots of the person being tracked is shown in Figure 16.21 and the resulting trajectory of the robot is shown in Figure 16.20(a). The translational and rotational velocities of the robot is shown in Figures 16.20(b) and 16.20(c) respectively. As one can see, the robot trajectory is smooth and control velocities are well within the limits of the system. The average translational velocity of the robot is 0.4 m/s.

16.6.4 Discussion on Performance Comparison

To summarize, the advantages of the proposed approach are demonstrated by providing performance comparison with previous works only at individual module level, rather than for the full system. For instance, the performance of the vision-based tracking algorithm is compared with some of the earlier works as shown in Table 16.2. This improvement is achieved using lesser number of features compared to earlier works. Similarly, for the visual servoing module, an improved version of the controller used in [499] is proposed in this chapter which uses feedback linearization to overcome the problem of chattering present in sliding mode controller used in the previous work. Direct comparison for the overall system is difficult due to several factors such as, unavailability of common dataset, non-uniformity of sensors used, etc. A more rigorous comparison with uniform operating conditions will remain as a part of the future work.

Experimental Results 591

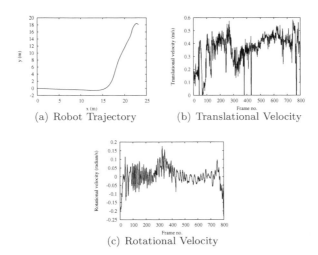

FIGURE 16.20: Robot motion trajectories in the actual experiment where a subject is tracked and followed in an outdoor environment. As one can see the velocity magnitudes are well within the physical limits of robot. The high frequency in the velocity component is due to sensor noise.

16.6.5 Experimental Evaluation of Human Tracking Algorithm Based on Optical Flow

In this experiment the proposed controller is combined with the human tracking algorithm based on optical flow. This experiment is performed in an outdoor environment where the target human walked in a corridor and was occluded twice by other humans. The trajectory generated by the robot to

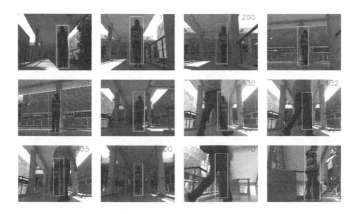

FIGURE 16.21: Snapshots of real-life experiment where the subject is walking in front of a robot in an outdoor environment.

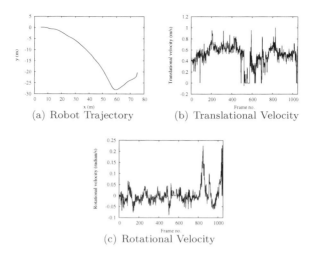

FIGURE 16.22: Experimental results for optical flow-based human tracking algorithm:

follow the target human using proposed visual servo controller is shown in Figure 16.22(a). It can be seen in this figure that the robot moves in a curve trajectory to follow the target human. The translational and rotational velocities of the robot are shown in Figure 16.22(b) and 16.22(c), respectively. The average translational velocity of the robot is 0.5 m/s. A few snapshots of the experiment are shown in Figure 16.23.

FIGURE 16.23: Snapshots of experiment-for optical flow-based human tracking algorithm.

16.7 Summary

In this chapter, the problems associated with a human following robot that uses a vision sensor for detecting and tracking a human subject has been discussed. The problem consists of two parts - first, detecting and tracking humans in a video sequence recorded from the on-board camera and secondly, designing a visual servo controller that generates necessary motion commands required for following the human target. The first part is fraught with several challenges, such as variation in shape, size, illumination, scale and pose, partial and full occlusion, pose-change due to out-of-plane rotations, etc. On the other hand, the visual servo controller has to deal with problems like inaccurate odometry information, field of view constraints, nonholonomic motion, and camera calibration errors. The proposed method uses a tracking-by-detection framework which uses point features for detection. The tracking algorithm uses a dynamic object model that evolves over time to accommodate for temporal changes while the stability is ensured by propagating stable features using affine motion model. The spread of the projected points is used for detecting pose change due to *out-of-plane rotations*, which is a novel contribution. A KD-tree based classifier is used to differentiate a case of full occlusion from that of partial occlusion and pose change. The visual servo controller uses a feedback-linearization based technique to ensure stability during motion. Experimental results are also provided to demonstrate the efficacy of the algorithms.

Exercises

Chapter 3

1. Consider the two link manipulator as given in Example 3.3. Actuate the manipulator with a random joint vector θ and note the tip position x in the Cartesian space. You can assume that you have training pair $\{x, y\}$. Train the KSOM network the way it is trained to learn a map in Example 3.2. After training, repeat the testing experiments as given in 3.3. Are you able to replicate the results? Why?

2. Consider the example of CRS Plus manipulator as given in 3.4. Integrate the camera model as given in [150]. Propose a KSOM network for the visual motor control. Train the network and track a straight line.

3. Consider the 7 DOF PowerCube manipulator without integrated with the camera. Learn the inverse kinematics with the standard KSOM network using 50,000 examples. Verify all the associated problems with the standard KSOM for learning the inverse kinematics of a redundant manipulator as given in Section 3.4.

Chapter 8

1. Consider following dynamics

$$\dot{\mathbf{x}} = \begin{bmatrix} 0 & 1 \\ 4 & 0 \end{bmatrix} \mathbf{x} + \begin{bmatrix} 0 \\ -1 \end{bmatrix} \mathbf{u} \tag{.42}$$

(i) Formulate the ARE for the above system.

(ii) Find the optimal controller for the above system.

(iii) Show the system is asymptotically stable with the optimal controller.

2. Consider any two DoF manipulator model with joint angles $\mathbf{q} = \begin{bmatrix} \theta_1 \\ \theta_2 \end{bmatrix}$ and τ is the joint torque:

$$\mathbf{B}(\mathbf{q})\ddot{\mathbf{q}} + \mathbf{C}(\dot{\mathbf{q}}, \mathbf{q}) + \mathbf{g}(\mathbf{q}) = \tau \tag{.43}$$

and

$$\mathbf{B}(\mathbf{q}) = \begin{bmatrix} ((m_1+m_2)l_1^2 + m_2l_2^2 + 2m_2l_1l_2\cos\theta_2) & (m_2l_2^2 + m_2l_1l_2\cos\theta_2) \\ m_2l_2^2 + m_2l_1l_2\cos\theta_2 & m_2l_2^2 \end{bmatrix} \quad (.44)$$

$$\mathbf{C}(\dot{\mathbf{q}}) = \begin{bmatrix} -m_2l_1l_2\sin\theta_2(2\dot\theta_1\dot{theta}_2 + \dot\theta_2^2) \\ -m_2l_1l_2\sin\theta_2\dot\theta_1\dot{theta}_2 \end{bmatrix} \quad (.45)$$

$$\mathbf{g}(\mathbf{q}) = \begin{bmatrix} -(m_1+m_2)gl_1\sin\theta_1 - m_2gl_2\sin(\theta_1+\theta_2) \\ -m_2gl_2\sin(\theta_1+\theta_2) \end{bmatrix} \quad (.46)$$

where, m_1, m_2 are the mass of the links, l_1, l_2 are the link lengths and g is the gravity. Assume any suitable value for the parameters m_1, m_2, l_1 and l_2. Design an optimal controller using the SNAC algorithm as described in section 8.3.1 (Use the input matrix from the model).

3. Consider that the dynamics of the above manipulator are not known. Learn a TSK Fuzzy model which represents the above manipulator dynamics with arbitrary accuracy.

4. Consider a 2 DoF robotic manipulator given in Exercise 2. Now suppose the mathematical model od its dynamics is unknown. Design an optimal controller for the manipulator using the methodology presented in 8.3.2. Use the model in Exercise 2. for data generation.

5. In the above problem, learn the controller without applying the constraints (8.69) to (8.71) and check the performance of the controller.

Chapter 9

1. Design a motion encoding system which can generate velocity commands in the joint space to write the letter "B". Use kinesthetic demonstration and learn the model using DMP. Try out various function approximation techniques such as GMR, Fuzzy logic, RBFN to learn $f(\tau)$ and compare their performances.

2. Formulate a DMP model for a target which is non-static in the environment.

3. Suppose you have been provided 5 demonstrations for a pouring task. Which method of imitation learning would you prefer and why?

4. Replace the cost function in the SED formulation with the mean-square error as given in the following:

$$\underset{\theta}{\text{minimize}} \quad \frac{1}{\sum_{n=1}^{N} T^n} \sum_{n=1}^{N} \sum_{m=1}^{T^n} \|\dot{\mathbf{x}}^{m,n} - \hat{\dot{\mathbf{x}}}^{m,n}\| \quad (.47)$$

Exercises

with the required constraints and compare the performance with method described earlier. Consider any demonstration of your choice.

5. Consider a picking task demonstration and use C-FuzzStaMP to learn the task. Take different values of α and check the performance.

6. What will be the consequence if someone learns all the $\mathbf{A}^k$s in R-FuzzStaMP model as positive definite? Can you still achieve an asymptotically stable motion model? How would be the performance of the model?

7. Learn six tasks in a single motion model using GMR. Discuss how the robustness of the motion model can be improved.

8. Compare the performance of various methods presented in multitask learning. Consider the spatial error metric to evaluate the performance. Discuss how the performance varies among different regression techniques when the number of tasks increases.

Chapter 13

1. Consider a vector $x = [2.5 \quad 5.5 \quad 18]^T$ defined in a rotated frame, which has undergone a positive yaw rotation about the original robot frame, by an amount, $\psi = 2\pi/3$. Find the coordinates of the vector in the original robot coordinate frame.

2. Consider a vector $x = [3 \quad 8 \quad 2]^T$ defined in the rotated coordinate frame. Find the coordinates of the vector in the original frame, if the original frame is rotated by $\pi/8$ rad about the roll axis, $\pi/2$ rad about the pitch axis and $\pi/5$ rad about the yaw axis, and if the coordinates of the origin of the original frame with respect to the rotated frame be $\mathbf{P}^T = [15, 5, 45]^T$.

3. Consider a ground robot with the coordinates in ECEF frame given by $[3078.3, 4356.6, 3678]$ km. Find the corresponding latitude and longitude.

4. Design and implement a stable controller based on feedback linearization technique using the dynamic model of the robot given by (13.7) starting from $[0.5, 0.5]^T$ of the reference frame to track a lemniscates shape trajectory about the point $[1, 0]^T$, with a lobe length of 2 m, and an angular velocity of 0.3 rad/s.

5. Consider a two-wheel differential-driven mobile robot with the kinematic model given by (13.6). Design a stable backstepping controller to track a sinusoidal trajectory using the kinematic model of the robot. Also analyze the stability of the controller using Lyapunov stability theory.

6. Design and implement a stable controller for the two wheel differential-driven robot based on NTSMC and FNTSMC techniques to track a circular

trajectory of radius 3 m, and compare the performance of the controllers based on the speed of convergence.

7. Write a Matlab program for driving a nonholonomic mobile robot to chase a target moving in a sinusoidal trajectory. Assume the initial relative position of the target as $[0.5, 0.5]^T$ with respect to the robot. Use the kinematic model of the robot and the feedback linearization technique to design the controller.

8. Repeat the above simulation for a target following nonholonomic robot using sliding mode control technique based on the dynamic model.

Chapter 14

1. Consider a three-robot formation in leader-follower-based framework. Design a dynamic controller using Backstepping approach depicted in Chapter 13 for the trailing robots based on the dynamic model given by (13.7). The robots are assumed to make a triangular-shaped formation, in which the trailing robots are making a desired distance separation of 2 m with respect to the leader.

2. Consider a system of three robots. Design a kinematic controller for the system of robots using super twisting controller, and the robots are supposed to make a line formation of 1 m separation.

3. Consider a multi-agent system with three members. Design an APF-based path planner for the system to achieve the following objectives:

   a. The agents are supposed to make a triangular formation initially of formation size 2m.
   b. The formation should navigate to the target point at the location (15,15) in the global reference frame. Assume random initial positions for all the agents.
   c. The formation should switch to a line after 10s.
   d. Assume a point mass obstacle at location (5,10). The formation is supposed to retain the shape while avoiding the obstacle.

4. Consider a team of five robots with one leader and four followers. Design and implement the path planning and control scheme based on the proposed FAGNTSMC scheme for the trailing robots to satisfy the following objectives

   a. The followers are supposed to make a square shaped formation of 2 m separation between the consecutive members, and the formation is supposed to follow a sinusoidal trajectory.

Exercises   599

   b. Assume that one of the agents goes to complete failure on runtime, because of critical actuator fault. The rest of the agents are supposed to isolate the faulty member and reconfigure to a triangular formation. [hint: Make the velocity of the faulty agent to be zero, while the occurrence of the fault.]

5. Design and implement in Matlab an APF-based path planner and back stepping-based formation controller for a team of six robots in hexagonal formation to chase a target moving in a sinusoidal trajectory. Assume the target as a point mass and its initial relative position as $[5,5]^T$ with respect to the initial position of the formation centre.

6. Repeat the above simulation with fault tolerant APF and adaptive sliding mode technique, assuming that two of the agents are becoming faulty on run time, after twenty seconds. The agents are supposed to reconfigure to a square formation and continue chasing the target, after isolating the faulty member.

7. Suggest an APF-based path planning scheme for the flock of agents consisting of three groups with a formation size of four each, to follow a lemniscates shape trajectory with a lobe length of 2 m. Write the Matlab program to implement the algorithm.

8. Design an APF-based path planner for resilient flocking in a multi-agent system, and perform the simulation for the flock of agents with the formation sizes mentioned in the previous question. Improvise the algorithm, and repeat the simulation after including the obstacle avoidance and formation reconfiguration parts.

Chapter 15

1. Consider the system
$$\dot{x} = Ax + Bu, \qquad (.48)$$
where $A = \begin{bmatrix} 1 & -1 \\ 1 & 1 \end{bmatrix}$ and $B = \begin{bmatrix} 0 & 1 \end{bmatrix}^T$.

   (a) Design and implement feedback linearization based event triggered control scheme.
   (b) Analytically prove that the zeno phenomenon should not occur.
   (c) Design an smc-based event triggered control scheme for the system, and compare the results.

2. Consider a multi-agent system with the following dynamics

$$\dot{x}_i = Ax + Bu + d, \quad x_i \in \Re^n \quad (.49)$$

(a) Design and implement the sliding mode-based event triggered consensus protocol for a multi-agent system. Assume that the agents are communicating based on an undirected graph topology.

(b) Derive an analytic expression for the lower bound of the inter-execution time period.

3. Consider a multi-agent system with three members. Design and implement sliding mode-based event triggered control scheme for the system to achieve the following objectives:

   a. The agents are supposed to make a triangular formation of formation size 1 m.

   b. The formation should navigate to the target point at the location (50, 50) in the global frame of reference. Assume random initial positions for the agents.

   c. The formation should switch to a line after 10 s.

   d. Assume a point mass obstacle at location (5, 10) in the global frame. The system should regain the formation shape after avoiding the obstacle.

4. For a multi-agent system with the graph topology given by Figure 24, obtain the adjacency matrix, Laplacian and propose a finite time nonlinear consensus protocol.

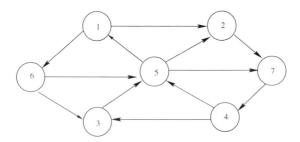

FIGURE 24: Graph.

5. Develop and implement a finite time event triggered consensus protocol based on super twisting sliding mode control technique depicted in chapter 13 for the consensus-based formation of a group of mobile agents following a directed graph communication topology. Analytically prove that the inter-execution time is lower bounded.

Exercises

6. Repeat the above simulation for a multi-agent system in leader-follower configuration consisting of four agents with one leader and three followers with the agent dynamics given by (15.12), and the relative state deviations of the followers w.r.t leader are as follows: $\boldsymbol{\delta}_1 = (1, -1)$, $\boldsymbol{\delta}_2 = (1, 1)$ and $\boldsymbol{\delta}_3 = (2, 0)$.

   Consider that the leader is free to navigate and the followers will reach into a consensus tracking the leader, and there is no feedback from the followers to the leader.

Chapter 16

1. Write a program to grab images from the camera, detect the object using color and SURF features, and perform background estimation using the frames. Assume that the object in the scene is moving. Repeat the simulation with challenging conditions such as illumination variation, scaling, occlusion, etc.

2. Design a visual servo controller for the mobile robot carrying a fixed camera to track a human in 2D plane based on sliding mode control technique, using the kinematic model of the robot and the pinhole model of the camera given in this chapter. The objective is to drive the mobile robot to keep the target human always in sight of the camera. This can be done by keeping the position of the human center on the heading direction and within a short distance from the mobile robot.

    a. Perform the simulation using human tracking algorithm based on SURF-based dynamic object model.

    b. Repeat the simulation using human tracking algorithm based on optical flow

3. Write a Matlab program to implement a fuzzy nonsingular fast terminal sliding mode controller for a mobile robot tracking a human based on SURF-based dynamic object model.

4. Develop an image Jacobian for the nonholonomic robot relating the feature motion with robot motion, and design image-based visual servo controller using sliding mode control technique for a human following mobile robot.

5. Assume that if the visual features being used move out from the field of view (FOV) of the camera due to motion of the robot or any other disturbances, propose a suitable estimation scheme, to restore the feature in the FOV, and repeat the simulation experiment with a image-based visual servo controller.

Bibliography

[1] W. Schultz, P. Dayan, and P. R. Montague, "A neural substrate of prediction and reward," *Science*, vol. 275, no. 5306, pp. 1593–1599, 1997.

[2] J. R. Hollerman and W. Schultz, "Dopamine neurons report an error in the temporal prediction of reward during learning," *Nature neuroscience*, vol. 1, no. 4, pp. 304–309, 1998.

[3] W. Schultz, "Neural coding of basic reward terms of animal learning theory, game theory, microeconomics and behavioural ecology," *Current opinion in neurobiology*, vol. 14, no. 2, pp. 139–147, 2004.

[4] M. Zhihong, A. P. Paplinski, and H. R. Wu, "A robust mimo terminal sliding mode control scheme for rigid robotic manipulators," *IEEE transactions on automatic control*, vol. 39, no. 12, pp. 2464–2469, 1994.

[5] L. M. Capisani and A. Ferrara, "Trajectory planning and second-order sliding mode motion/interaction control for robot manipulators in unknown environments," *IEEE Transactions on Industrial Electronics*, vol. 59, no. 8, pp. 3189–3198, Aug 2012.

[6] S. Schaal, "Is imitation learning the route to humanoid robots?" *Trends in cognitive sciences*, vol. 3, no. 6, pp. 233–242, 1999.

[7] W. Schultz, P. Dayan, and P. R. Montague, "A neural substrate of prediction and reward," *Science*, vol. 275, pp. 1593–1599, 1997.

[8] W. Schultz, "Neural coding of basic reward terms of animal learning theory, game theory, microeconomics and behavioural ecology," *Current Opinion in Neurobiology*, vol. 14, pp. 139–147, 2004.

[9] J. R. Hollerman and W. Schultz, "Dopamine neurons report an error in the temporal prediction of reward during learning," *Nature Neuroscience*, vol. 1, no. 4, pp. 304–309, August 1998.

[10] F. Chaumette and S. Hutchinson, "Visual servo control Part I : Basic approaches," *IEEE Robotics and Automation Magazine*, vol. 13, no. 4, pp. 82–90, December 2006.

[11] F. Chaumette and S. Hutchinson, "Visual servo control part II : Advanced approaches (tutorial)," *IEEE Robotics and Automation Magazine*, vol. 14, no. 1, pp. 109–118, March 2007.

[12] B. Siciliano, "Kinematic control of redundant robot manipulators : A tutorial," *Journal of Intelligent and Robotic systems*, vol. 4, no. 4, pp. 201–212, August 1990.

[13] D. E. DeMers and K. K. Kreutz-Delgado, "Solving the inverse kinematics problem for robots with excess degrees-of-freedom," http://citeseer.ist.psu.edu/375197.html.

[14] M. W. Spong, S. Hutchinson, and M. Vidyasagar, *Robot Modeling and Control*. John Wiley & sons Inc., 2005.

[15] D. E. Whitney, "Resolved motion rate control of manipulators and human prostheses," *IEEE Transactions on Man-Machine Systems*, vol. 10, no. 2, pp. 47–53, June 1969.

[16] J. Baillieul, "Kinematic programming alternatives for redundant manipulators," in *Robotics and Automation. Proceedings. 1985 IEEE International Conference on*, vol. 2, March 1985, pp. 722 – 728.

[17] A. Liegeois, "Automatic supervisory control of the configuration and behavior of multibody mechanisms," *IEEE Transactions of Systems, Man and Cybernetics*, vol. 7, no. 12, pp. 868–871, December 1977.

[18] T. F. Chan and R. V. Dubey, "A weighted least-norm based solution scheme for avoiding joint limits for redundant joint manipulators," *IEEE Transactions on Robotics and Automation*, vol. 11, no. 2, pp. 286–292, April 1995.

[19] T. Yoshikawa, "Manipulability and redundancy control of robotic mechanisms," in *Proceedings of IEEE International Conference on Robotics and Automation*, 1985, pp. 1004–1009.

[20] I. Gravagne and I. Walker, "On the structure of minimum effort solutions with application to kinematic redundancy resolution," *IEEE Transactions on Robotics and Automation*, vol. 16, no. 6, pp. 855 –863, December 2000.

[21] Y. Zhang, J. Wang, and Y. Xu, "A dual neural network for bi-criteria kinematic control redundant manipulators," *IEEE Transactions on Robotics and Automation*, vol. 18, no. 6, pp. 923–931, February 2002.

[22] L. Li, W. Gruver, Q. Zhang, and Z. Yang, "Kinematic control of redundant robots and the motion optimizability measure," *IEEE Transactions on Systems, Man, and Cybernetics, Part B: Cybernetics*, vol. 31, no. 1, pp. 155 –160, February 2001.

[23] D. N. Nenchev, "Redundancy resolution through local optimization: A review," *Journal of Robotic Systems*, December 1989.

[24] C. A. Klein and C. H. Huang, "Review of pseudoinverse control for use with kinematically redundant manipulators," *IEEE Transactions on Systems, Man and Cybernetics*, vol. SMC-13, pp. 245–250, 1983.

[25] C. Wampler, "Manipulator inverse kinematic solutions based on vector formulations and damped least-squares methods," *Systems, Man and Cybernetics, IEEE Transactions on*, vol. 16, no. 1, pp. 93–101, January 1986.

[26] C. Wampler and L. J. Leifer, "Application of damped least squares methods to resolved-rate and resolved-acceleration control of manipulators," *Journal of Dynamic Systems, Measurement, and Control*, vol. 110, no. 1, pp. 31–38, 1988.

[27] F. Caccavale, S. Chiaverini, and B. Siciliano, "Second-order kinematic control of robot manipulators with Jacobian damped least-squares inverse: theory and experiments," *IEEE/ASME Transactions on Mechatronics*, vol. 2, no. 3, pp. 188–194, September 1997.

[28] K. Glass, R. Colbaugh, D. Lim, and H. Seraji, "Real-time collision avoidance for redundant manipulators," *IEEE Transactions on Robotics and Automation*, vol. 11, no. 3, pp. 448–457, June 1995.

[29] S. Chiaverini, B. Siciliano, and O. Egeland, "Review of the damped least-squares inverse kinematics with experiments on an industrial robot manipulator," *IEEE Transactions on Control Systems Technology*, vol. 2, no. 2, pp. 123–134, June 1994.

[30] H. Seraji, "Configuration control of redundant manipulators: theory and implementation," *IEEE Transactions on Robotics and Automation*, vol. 5, no. 4, pp. 472–490, 1989.

[31] H. Seraji, "Task options for redundancy resolution using configuration control," in *Proceedings of 30th IEEE conference on Decision and Control*, Brighton, UK, December 1991, pp. 2793–2798.

[32] J. English and A. Maciejewski, "On the implementation of velocity control for kinematically redundant manipulators," *IEEE Transactions on Systems, Man and Cybernetics, Part A: Systems and Humans*, vol. 30, no. 3, pp. 233–237, May 2000.

[33] J. Wang, Q. Hu, and D. Jiang, "A Lagrangian network for kinematic control redundant robot manipulators," *IEEE Transactions on Neural networks*, vol. 10, no. 5, September 1999.

[34] Y. Xia and J. Wang, "A dual neural network for kinematic control redundant robot manipulators," *IEEE Transactions on Systems, Man and Cybernetics - Part B*, vol. 31, no. 1, February 2001.

[35] Y. Zhang, J. Wang, and Y. Xia, "A dual neural network for redundancy resolution of kinematically redundant manipulators subject to joint limits and joint velocity limits," *IEEE Transactions on Neural Networks*, vol. 14, no. 3, pp. 658 – 667, May 2003.

[36] Y. Zhang, S. S. Ge, and T. H. Lee, "A unified quadratic programming-based dynamical system approach to torque minimization of physically constrained redundant manipulators," *IEEE Transactions on Systems, Man, and Cybernetics - Part B: Cybernetics*, vol. 34, no. 5, pp. 2126–2132, October 2004.

[37] Y. Zhang and J. Wang, "Obstacle avoidance for kinematically redundant manipulators using a dual neural network," *IEEE Transactions on Systems, Man and Cybernetics - Part B:Cybernetics*, vol. 34, no. 1, pp. 752–759, February 2004.

[38] B. Cai and Y. Zhang, "Different-level redundancy-resolution and its equivalent relationship analysis for robot manipulators using gradient-descent and zhang et al. neural-dynamic methods," *IEEE Transactions on Industrial Electronics*, 2012.

[39] H. Ding and J. Wang, "Recurrent neural networks for minimum infinity-norm kinematic control redundant robot manipulators," *IEEE Transactions on Systems, Man and Cybernetics - Part A*, vol. 29, no. 3, May 1999.

[40] W. S. Tang and J. Wang, "A recurrent neural network for minimum infinity-norm kinematic control of redundant manipulators with an improved problem formulation and reduced architecture complexity," *IEEE Transactions on Systems, Man and Cybernetics - Part B : Cybernetics*, vol. 31, no. 1, pp. 98–105, February 2001.

[41] J. Hollerbach and K. Suh, "Redundancy resolution of manipulators through torque optimization," *IEEE Journal of Robotics and Automation*, vol. 3, no. 4, pp. 308 –316, August 1987.

[42] Y. Nakamura and H. Hanafusa, "Optimal redundancy control of robot manipulators," *International Journal of Robotic Research*, vol. 6, pp. 32– 42, 1987.

[43] K. Kazerounian and Z. Wang, "Global versus local optimization in redundancy resolution of robot manipulators," *International Journal of Robotic Research*, vol. 7, no. 3, pp. 227 – 246, 1988.

[44] K. Suh and J. Hollerbach, "Local versus global torque optimization of redundant manipulators," in *Robotics and Automation. Proceedings. 1987 IEEE International Conference on*, vol. 4, March 1987, pp. 619 – 624.

[45] M. Galicki, "Time-optimal controls of kinematically redundant manipulators with geometric constraints," *IEEE Transactions on Robotics and Automation*, vol. 16, no. 1, pp. 89–93, February 2000.

[46] D. Martin, J. Baillieul, and J. M. Hollerbach, "Resolution of kinematic redundancy using optimization," *IEEE Transactions on Robotics and Automation*, vol. 5, no. 4, pp. 529–533, 1989.

[47] S.-W. Kim, K.-B. Park, and J.-J. Lee, "Redundancy resolution of robot manipulators using optimal kinematic control," in *IEEE International Conference on Robotics and Automation*, vol. 1, 1994, pp. 683–688.

[48] Z. Ahmad and A. Guez, "On the solution to the inverse kinematic problem," in *Robotics and Automation, 1990. Proceedings., 1990 IEEE International Conference on*, vol. 3, May 1990, pp. 1692–1697.

[49] P. Martín and J. R. Millán, "Robot arm reaching through neural inversions and reinforcement learning," *Robotics and Autonomous Systems*, vol. 31, no. 4, pp. 227–246, 2000.

[50] R. V. Mayorga and P. Sanongboon, "An artificial neural network approach for inverse kinematics computation and singularities prevention of redundant manipulators," *Journal of Intelligent and Robotic Systems*, vol. 44, pp. 1–23, 2005.

[51] L. Behera and N. Kirubanandan, "A hybrid neural control scheme for visual-motor coordination," *IEEE Control Systems Magazine*, vol. 19, no. 4, pp. 34–41, August 1999.

[52] D. Kragic and H. I. Christensen, "Survey on visual servoing for manipulation," Computational Vision and Active Perception Laboratory, Tech. Rep., 2002.

[53] E. Malis and S. Benhimane, "A unified approach to visual tracking and servoing," *Robotics and Autonomous Systems*, vol. 52, no. 1, pp. 39–52, July 2005.

[54] A. Comport, E. Marchand, and F. Chaumette, "Statistically robust 2-D visual servoing," *IEEE Transactions on Robotics*, vol. 22, no. 2, pp. 415–420, April 2006.

[55] D. Kragic and H. I. Christensen, "Cue integration for visual servoing," *IEEE Transactions on Robotics and Automation*, vol. 17, no. 1, pp. 18–27, February 2001.

[56] W.-C. Chang, "Precise positioning of binocular eye-to-hand robotic manipulators," *Journal of Intelligent Robotics Systems*, vol. 49, pp. 219–236, 2007.

[57] C. C. Cheah, C. Liu, and J. J. E. Slotine, "Adaptive jacobian vision based control for robots with uncertain depth information," *Automatica*, vol. 46, no. 7, pp. 1228 – 1233, 2010.

[58] F. Conticelli and B. Allotta, "Discrete-time robot visual feedback in 3-d positioning tasks with depth adaptation," *IEEE Transactions on Mechatronics*, vol. 26, no. 4, pp. 684 –697, August 2010.

[59] E. Malis, Y. Mezouar, and P. Rives, "Robustness of image-based visual servoing with a calibrated camera in the presence of uncertainties in the three-dimensional structure," *IEEE Transactions on Robotics*, vol. 26, no. 1, pp. 112–120, February 2010.

[60] F. Chaumette, "Potential problems of stability and convergence in image based and position based visual servoing," in *D. Kreigman and G. Hager and S. Morse*, ser. Lecture notes in Control and Information Sciences, T. confluence of vision and control, Eds. New York: Springer-Verlag, 1998, vol. 237, pp. 66–78.

[61] J.-T. Lapreste, F. Jurie, M. Dhome, and F. Chaumette, "An efficient method to compute the inverse Jacobian matrix in visual servoing," in *Proceedings of IEEE International Conference on Robotics and Automation*, vol. 1. New Orleans, LA: IEEE, April 2004, pp. 727–732.

[62] F. Chaumette, P. Rives, and B. Espiau, "Positioning a robot with respect to an object, tracking it and estimating its velocity by visual servoing," in *Proceedings of the IEEE International Conference on Robotics and Automation (ICRA)*, vol. 3, 1991, pp. 2248–2253.

[63] B. Espiau, F. Chaumette, and P. Rives, "A new approach to visual servoing in robotics," *IEEE Transactions on Robotics and Automation*, vol. 8, no. 3, pp. 313–326, June 1992.

[64] S. Hutchinson, G. D. Hager, and P. I. Corke, "A tutorial on visual servo control," *IEEE Transactions on Robotics and Automation*, vol. 12, no. 5, pp. 651–670, October 1996.

[65] K. Hashimoto and T. Noritsugu, "Performance and sensitivity in visual servoing," in *Proceedings of the IEEE international Conference on Robotics and Automation (ICRA)*, vol. 2, 1998, pp. 2321–2326.

[66] M. Iwatsuki and N. Okiyama, "A new formulation of visual servoing based on cylindrical coordinate system," *IEEE Transactions on Robotics*, vol. 21, no. 2, pp. 266 – 273, April 2005.

[67] W. Wilson, C. Hulls, and G. Bell, "Relative end-effector control using cartesian position based servoing," *IEEE Transactions on Robotics and Automation*, vol. 12, no. 5, pp. 684–696, October 1996.

[68] F. Janabi-Sharifi and M. Marey, "A kalman-filter-based method for pose estimation in visual servoing," *IEEE Transactions on Robotics*, vol. 26, no. 5, pp. 939–947, October 2010.

[69] P. Wunsch, S. Winkler, and G. Hirzinger, "Real-time pose estimation of 3d objects from camera images using neural networks," in *Proceedings of IEEE International Conference on Robotics and Automation (ICRA)*, vol. 3, Albuquerque, New Mexico, 1997, pp. 3232–3237.

[70] T. Drummond and R. Cipolla, "Visual tracking and control using lie algebras," in *Proceedings of the Computer Society Conference on Computer Vision and Pattern Recognition (CVPR)*, vol. 2, Fort Collins, Colorado, 1999, pp. 652–657.

[71] P. Allen, A. Timcenko, B. Yoshimi, and P. Michelman, "Automated tracking and grasping of a moving object with a robotic hand-eye system," *IEEE Transactions on Robotics and Automation*, vol. 9, no. 2, pp. 152–165, April 1993.

[72] G. Taylor and L. Kleeman, "Hybrid position-based visual servoing with online calibration for a humanoid robot," in *Intelligent Robots and Systems, 2004. (IROS 2004). Proceedings. 2004 IEEE/RSJ International Conference on*, vol. 1, September 2004, pp. 686–691.

[73] W. Sepp, S. Fuchs, and G. Hirzinger, "Hierarchical featureless tracking for position-based 6-dof visual servoing," in *Intelligent Robots and Systems, 2006 IEEE/RSJ International Conference on*, oct. 2006, pp. 4310–4315.

[74] V. Lippiello, B. Siciliano, and L. Villani, "Position-Based Visual Servoing in industrial multirobot cells using a hybrid camera configuration," *IEEE Transactions on Robotics*, vol. 23, no. 1, pp. 73–86, February 2007.

[75] E. Malis, F. Chaumette, and S. Boudet, "2-1/2-d visual servoing," *IEEE Transactions on Robotics and Automation*, vol. 15, no. 2, pp. 238–250, April 1999.

[76] Y. Fang, A. Behal, W. E. Dixon, and D. M. Dawson, "Adaptive 2.5d visual servoing of kinematically redundant robot manipulators," in *Proceedings of the 41st Conference on Decision and Control*, Las Vegas, Nevada, USA, December 2002, pp. 2860–2865.

[77] P. I. Corke and S. Hutchinson, "A new partitioned approach to image based visual servo control," *IEEE Transactions on Robotics and Automation*, August 2001.

[78] N. Gans and S. Hutchinson, "Stable visual servoing through hybrid switched-system control," *IEEE Transactions on Robotics*, vol. 23, no. 3, pp. 530–540, June 2007.

[79] O. Tahri, Y. Mezouar, F. Chaumette, and P. Corke, "Decoupled Image-Based Visual Servoing for cameras obeying the unified projection model," *IEEE Transactions on Robotics*, vol. 26, no. 4, pp. 684–697, August 2010.

[80] J. Chen, D. Dawson, W. Dixon, and A. Behal, "Adaptive homography-based visual servo tracking for a fixed camera configuration with a camera-in-hand extension," *IEEE Transactions on Control Systems Technology*, vol. 13, no. 5, pp. 814–825, September 2005.

[81] F. Chaumette and E. Marchand, "A redundancy-based iterative approach for avoiding joint limits: Application to visual servoing," *IEEE Transactions on Robotics and Automation*, vol. 17, no. 5, pp. 719–730, October 2001.

[82] N. Mansard and F. Chaumette, "Visual servoing sequencing able to avoid obstacles," in *Robotics and Automation, 2005. ICRA 2005. Proceedings of the 2005 IEEE International Conference on*, April 2005, pp. 3143–3148.

[83] N. Mansard and F. Chaumette, "Directional redundancy for robot control," *IEEE Transactions on Automatic Control*, vol. 54, no. 6, pp. 1179–1192, June 2009.

[84] M. Waltz and K. Fu, "A heuristic approach to reinforcement learning control systems," *IEEE Transactions on Automatic Control*, vol. 10, no. 4, pp. 390–398, 1965.

[85] J. Mendel and R. McLaren, "8 reinforcement-learning control and pattern recognition systems," *Mathematics in Science and Engineering*, vol. 66, pp. 287–318, 1970.

[86] R. S. Sutton and A. G. Barto, *Reinforcement learning: An introduction*. MIT press Cambridge, 1998, vol. 1, no. 1.

[87] A. Segre and G. DeJong, "Explanation-based manipulator learning: Acquisition of planning ability through observation," in *Robotics and Automation. Proceedings. 1985 IEEE International Conference on*, vol. 2. IEEE, 1985, pp. 555–560.

[88] Y. Kuniyoshi, M. Inaba, and H. Inoue, "Teaching by showing: Generating robot programs by visual observation of human performance," in *Proc. of the 20th International Symp. on Industrial Robots*, 1989, pp. 119–126.

[89] P. Kormushev, S. Calinon, and D. G. Caldwell, "Reinforcement learning in robotics: Applications and real-world challenges," *Robotics*, vol. 2, no. 3, pp. 122–148, 2013.

[90] O. Brock and O. Khatib, "Elastic strips: A framework for integrated planning and execution," in *Experimental Robotics VI, ISER*, P. I. Corke and J. P. Trevelyan, Eds. Sydney, Australia: Springer-Verlag, Berlin, Heidelberg, Germany, 2000, Mar. 1999, pp. 329–338.

[91] J. S. Kelso, *Dynamic patterns: The self-organization of brain and behavior*. MIT press, 1997.

[92] S. Schaal, S. Kotosaka, and D. Sternad, "Nonlinear dynamical systems as movement primitives," in *Humanoids2000, First IEEE-RAS International Conference on Humanoid Robots*. cd-proceedings, 2000. [Online]. Available: http://www-slab.usc.edu/publications/S/schaal-ICHR2000.pdf

[93] A. Billard and G. Hayes, "Drama, a connectionist architecture for control and learning in autonomous robots," *Adaptive Behavior*, vol. 7, no. 1, pp. 35–63, 1999.

[94] A. I. Selverston, "Are central pattern generators understandable?" *Behavioral and Brain Sciences*, vol. 3, no. 4, pp. 535–540, 1980.

[95] A. Billard, S. Calinon, R. Dillmann, and S. Schaal, "Robot programming by demonstration," in *Handbook of Robotics*, B. Siciliano and O. Khatib, Eds. Secaucus, NJ, USA: Springer, 2008, pp. 1371–1394.

[96] A. J. Ijspeert, J. Nakanishi, and S. Schaal, "Movement imitation with nonlinear dynamical systems in humanoid robots," in *Robotics and Automation, 2002. Proceedings. ICRA'02. IEEE International Conference on*, vol. 2. IEEE, 2002, pp. 1398–1403.

[97] S. Calinon, F. D'halluin, E. L. Sauser, D. G. Caldwell, and A. G. Billard, "Learning and reproduction of gestures by imitation," *IEEE Robotics & Automation Magazine*, vol. 17, no. 2, pp. 44–54, 2010.

[98] D. Kulić, W. Takano, and Y. Nakamura, "Incremental learning, clustering and hierarchy formation of whole body motion patterns using adaptive hidden markov chains," *The International Journal of Robotics Research*, vol. 27, no. 7, pp. 761–784, 2008.

[99] S. Schaal, J. Peters, J. Nakanishi, and A. Ijspeert, "Learning movement primitives," *Robotics Research*, pp. 561–572, 2005.

[100] L. Righetti, J. Buchli, and A. J. Ijspeert, "Dynamic hebbian learning in adaptive frequency oscillators," *Physica D: Nonlinear Phenomena*, vol. 216, no. 2, pp. 269–281, 2006.

[101] S. Degallier, L. Righetti, S. Gay, and A. Ijspeert, "Toward simple control for complex, autonomous robotic applications: combining discrete and rhythmic motor primitives," *Autonomous Robots*, vol. 31, no. 2, pp. 155–181, 2011.

[102] J. Kober and J. Peters, "Movement templates for learning of hitting and batting," in *Learning Motor Skills*. Springer, 2014, pp. 69–82.

[103] D. Bullock and S. Grossberg, "The vite model: A neural command circuit for generating arm and articulator trajectories," *Dynamic patterns in complex systems*, pp. 305–326, 1988.

[104] P. Gaudiano and S. Grossberg, "Adaptive vector integration to endpoint: Self-organizing neural circuits for control of planned movement trajectories," *Human Movement Science*, vol. 11, no. 1, pp. 141–155, 1992.

[105] D. Bullock, R. M. Bongers, M. Lankhorst, and P. J. Beek, "A vector-integration-to-endpoint model for performance of viapoint movements," *Neural Networks*, vol. 12, no. 1, pp. 1–29, 1999.

[106] S. Grillner, "Locomotion in vertebrates: central mechanisms and reflex interaction," *Physiological reviews*, vol. 55, no. 2, pp. 247–304, 1975.

[107] M. H. Raibert, *Legged robots that balance*. MIT press, 1986.

[108] F. Delcomyn, "Neural basis of rhythmic behavior in animals," *Science*, vol. 210, no. 4469, pp. 492–498, 1980.

[109] E. Marder and D. Bucher, "Central pattern generators and the control of rhythmic movements," *Current biology*, vol. 11, no. 23, pp. R986–R996, 2001.

[110] M. Lukoševičius and H. Jaeger, "Reservoir computing approaches to recurrent neural network training," *Computer Science Review*, vol. 3, no. 3, pp. 127–149, 2009.

[111] B. A. Pearlmutter, "Gradient calculations for dynamic recurrent neural networks: A survey," *IEEE Transactions on Neural networks*, vol. 6, no. 5, pp. 1212–1228, 1995.

[112] L. Medsker and L. C. Jain, *Recurrent neural networks: design and applications*. CRC press, 1999.

[113] A. F. Atiya and A. G. Parlos, "New results on recurrent network training: unifying the algorithms and accelerating convergence," *IEEE transactions on neural networks*, vol. 11, no. 3, pp. 697–709, 2000.

[114] A. Ijspeert, J. Nakanishi, and S. Schaal, "Trajectory formation for imitation with nonlinear dynamical systems," in *IEEE International Conference on Intelligent Robots and Systems (IROS 2001)*, 2001, pp. 752–757. [Online]. Available: http://www-clmc.usc.edu/publications/I/ijspeert-IROS2001.pdf

[115] J.-J. E. Slotine and W. Li, *Applied nonlinear control.* Prentice hall Englewood Cliffs, NJ, 1991, vol. 199, no. 1.

[116] S. Boyd and L. Vandenberghe, *Convex optimization.* Cambridge university press, 2004.

[117] F. H. Clarke, *Optimization and nonsmooth analysis.* SIAM, 1990.

[118] M. S. Bazaraa, H. D. Sherali, and C. M. Shetty, *Nonlinear programming: theory and algorithms.* John Wiley & Sons, 2013.

[119] J. Nocedal and S. J. Wright, *Sequential quadratic programming.* Springer, 2006.

[120] J.-F. Bonnans, J. C. Gilbert, C. Lemaréchal, and C. A. Sagastizábal, *Numerical optimization: theoretical and practical aspects.* Springer Science & Business Media, 2006.

[121] D. E. Goldberg, *Genetic algorithms.* Pearson Education India, 2006.

[122] K. Deb, *Optimization for engineering design: Algorithms and examples.* PHI Learning Pvt. Ltd., 2012.

[123] R. Eberhart and J. Kennedy, "A new optimizer using particle swarm theory," in *Micro Machine and Human Science, 1995. MHS'95., Proceedings of the Sixth International Symposium on.* IEEE, 1995, pp. 39–43.

[124] P. J. Van Laarhoven and E. H. Aarts, "Simulated annealing," in *Simulated annealing: Theory and applications.* Springer, 1987, pp. 7–15.

[125] C. M. Bishop, *Pattern Recognition and Machine Learning.* Springer, 2006. [Online]. Available: http://research.microsoft.com/en-us/um/people/cmbishop/prml/

[126] G. J. McLachlan and K. E. Basford, *Mixture models: Inference and applications to clustering.* Marcel Dekker, 1988, vol. 84.

[127] G. McLachlan and D. Peel, *Finite mixture models.* John Wiley & Sons, 2004.

[128] A. P. Dempster, N. M. Laird, and D. B. Rubin, "Maximum likelihood from incomplete data via the em algorithm," *Journal of the royal statistical society. Series B (methodological)*, pp. 1–38, 1977.

[129] G. J. McLachlan and T. Krishnan, "The em algorithm and extensions. 1997," *Hoboken: Wiley and Sons Google Scholar.*

[130] "Schunk," http://www.schunk-modular-robotics.com/.

[131] "Amtec robotics," http://www.amtec-robotics.com/.

[132] F. i Digital Camera, "Unibrain inc." http://www.unibrain.com.

[133] R. Y. Tsai, "A versatile camera calibration technique for high-accuracy 3D machine vision metrology using off-the-shelf tv cameras and lenses," *IEEE Journal of Robotics and Automation*, vol. RA-3, no. 4, pp. 323–344, August 1987.

[134] R. Wilson, "Tsai camera calibration software," http://www.cs.cmu.edu/~rgw/TsaiCode.html.

[135] OpenCV, "Open source computer vision library," http://www.intel.com/technology/computing/opencv/.

[136] T. Kohonen, *Self-organizing Maps*. Springer, Heidelberg, 2001.

[137] J. Walter and H. Ritter, "Rapid learning with parametrized self-organizing maps," *Neurocomputing*, vol. 12, pp. 131–153, 1996.

[138] J. A. Walter, "PSOM network: Learning with few examples," in *Int. Conf. on Robotics and Automation*. Leuven, Belgium: IEEE, May 1998, pp. 2054–2059.

[139] J. Walter, C. Nolker, and H. Ritter, "The PSOM algorithm and applications," in *Proc. Int. ICSC Symposium on Neural Computation*, 2000, pp. 758–764.

[140] T. M. Martinetz, H. J. Ritter, and K. J. Schulten, "Three-dimensional neural net for learning visual motor coordination of a robot arm," *IEEE Transactions on Neural Networks*, vol. 1, no. 1, pp. 131–136, March 1990.

[141] J. A. Walter and K. J. Schulten, "Implementation of self-organizing neural networks for visual-motor control of an industrial robot," *IEEE Transactions on Neural Networks*, vol. 4, no. 1, pp. 86–95, January 1993.

[142] T. Martinetz, H. Ritter, and K. Schulten, "Learning of visuomotor-coordination of a robot arm with redundant degrees of freedom," in *Proc. of the Int. Conf. on Parallel Processing in Neural Systems and Computers (ICNC)*. Dusseldorf, Amsterdam: Elsevier, 1990, pp. 431–434.

[143] M. Han, N. Okada, and E. Kondo, "Coordination of an uncalibrated 3-d visuo-motor system based on multiple self-organizing maps," *JSME International Journal Series C*, vol. 49, no. 1, pp. 230–239, 2006.

[144] H. Zha, T. Onitsuka, and T. Nagata, "A self-organization learning algorithm for visuo-motor coordination in unstructured environment," *Artificial life and robotics*, vol. 1, no. 3, pp. 131–136, September 1997.

[145] X.-Z. Zheng and K. Ito, "Self-organized learning and its implementation of robot movements," in *IEEE Int. Conf. on SMC, 'Computational cybernetics and simulation'*, Orlando, FL, USA, 1997, pp. 281–286.

[146] F. Chaumette, "Image moments: A general and useful set of features for visual servoing," *IEEE Transactions on Robotics*, vol. 20, no. 4, pp. 713–723, August 2004.

[147] R. Sharma and S. Hutchinson, "Optimizing hand/eye configuration for visual-servo systems," in *Proc. of Int. Conf. on Robotics and Automation (ICRA)*. IEEE, May 1995, pp. 172–177.

[148] J. T. Feddema, C. S. G. Lee, and O. W. Mitchell, "Weighted selection of image features for resolved rate visual feedback control," *IEEE Trans. on Robotics and Automation*, vol. 7, no. 1, pp. 31–47, February 1991.

[149] G. Tevatia and S. Schaal, "Inverse kinematics of humanoid robots," in *Proc. of IEEE Int. Conf. on Robotics and Automation*, San Francisco, CA, April 2000, pp. 294–299.

[150] L. Behera and N. Kirubanandan, "A hybrid neural control scheme for visual-motor coordination," *IEEE Control System Magazine*, vol. 19, no. 4, pp. 34–41, 1999.

[151] V. R. Angulo and C. Torras, "Speeding up the learning of robot kinematics through function decomposition," *IEEE Transaction on Neural Networks*, vol. 16, no. 6, pp. 1504–1512, November 2005.

[152] S. H. M. Spong and M. Vidyasagar, *Robot Modeling and Control*. John Wiley & sons Inc., 2005.

[153] M. K. Hu, "Visual patter recognition by moment invariants," *IRE Trans. on Information Theory*, vol. 8, no. 2, pp. 179–187, February 1962.

[154] T. S. J. Flusser, "Affine moment invariants: a new tool for character recognition," Pattern Recognition, April 1994, available: http://www.sciencedirect.com/science/article/pii/0167865594900922.

[155] *Complex Objects Pose Estimation Based on Image Moment variants.* IEEE, April 2005.

[156] D. Cyganski and J. A. Orr, "Applications of tensor theory to object recognition and orientation determination," *IEEE Trans.on Pattern Analysis and Machine Intelligence*, vol. 7, no. 6, pp. 662–673, November 1985.

[157] C. H. Lo and H. S. Don, "3-d moment forms: their construction and application to object identification and positioning," *IEEE Trans.on Pattern Analysis and Machine Intelligence*, vol. 11, no. 10, pp. 1053–1064, October 1989.

[158] Z. Yang and F. Cohen, "Cross-weighted moments and affine invariants for image registration and matching," *IEEE Trans.on Pattern Analysis and Machine Intelligence*, vol. 21, no. 8, pp. 804–814, August 1999.

[159] J. Flusser and T. Suk, "Rotation moment invariants for recognition of symmetric objects," *IEEE Trans.on Image processing*, vol. 26, no. 5, pp. 3784–3796, December 2006.

[160] C. C. Chen, "Rotation moment invariants for recognition of symmetric objects," *IEEE Trans.on Image processing*, vol. 26, no. 5, pp. 683–686, December 1993.

[161] R. J. Prokop and A. P.Reeves, "A survey of moment-based techniques for unoccluded object representation and recognition," *CVGIP: Graph. Models Image Process*, vol. 54, no. 5, pp. 438–460, 2002.

[162] J.Stewart, *Calculus, 6th ed.* Thomson Learning Inc., 2008.

[163] *A first step toward visual servoing using image moments*, vol. 1. IEEE, 2002.

[164] F. Chaumette, "Image moments: A general and useful set of features for visual servoing," *IEEE Transactions on Robotics and Automation*, vol. 20, no. 4, pp. 713–723, August 2004.

[165] D. Ballard and C. Brown, *Computer Vision*. Prentice Hall, 1982.

[166] R. K. R. Jain and B. SChunk, *Machine Vision*. McGraw-Hill International Editions, 1995.

[167] F. Chaumette, "La relation vision-commande: theorie et application a des taches robotique," Ph.D. dissertation, IRISA, Ph.D. dissertation, University de Rennes 1, 1990.

[168] O. Tahri and F. Chaumette, "Image moments: generic descriptors for decoupled image-based visual servo," in *Proceedings. ICRA '04*, vol. 2. IEEE International Conference on Robotics and Automation, 2004, May 2004, pp. 1185–1190.

[169] "Open source computer vision library. [online]." Available: http://opencv.willowgarage.com.

[170] P. K. Patchaikani and L. Behera, "Visual servoing of redundant manipulator with Jacobian estimation using self-organizing map," *Robotics and Autonomous Systems*, vol. 58, no. 8, pp. 978–990, August 2010.

[171] K. Hosoda and M. Asada, "Versatile visual servoing without knowledge of true jacobian," in *Intelligent Robots and Systems '94. 'Advanced Robotic Systems and the Real World', IROS '94. Proceedings of the IEEE/RSJ/GI International Conference on*, vol. 1, September 1994, pp. 186–193.

[172] M. Asada, T. Tanaka, and K. Hosoda, "Visual tracking of unknown moving object by adaptive binocular visual servoing," in *Multisensor Fusion and Integration for Intelligent Systems, 1999. MFI '99. Proceedings. 1999 IEEE/SICE/RSJ International Conference on*, 1999, pp. 249 –254.

[173] A. Farahmand, A. Shademan, and M. Jagersand, "Global visual-motor estimation for uncalibrated visual servoing," in *Intelligent Robots and Systems, 2007. IROS 2007. IEEE/RSJ International Conference on*, vol. 2, November 2007, pp. 1969–1974.

[174] F. Nori, L. Natale, G. Sandini, and G. Metta, "Autonomous learning of 3d reaching in a humanoid robot," in *Intelligent Robots and Systems, 2007. IROS 2007. IEEE/RSJ International Conference on*, November 2007, pp. 1142–1147.

[175] L. Natale, F. Nori, G. Sandini, and G. Metta, "Learning precise 3d reaching in a humanoid robot," in *Development and Learning, 2007. ICDL 2007. IEEE 6th International Conference on*, July 2007, pp. 324 –329.

[176] N. Mansard, M. Lopes, J. Santos-Victor, and F. Chaumette, "Jacobian learning methods for tasks sequencing in visual servoing," in *Intelligent Robots and Systems, 2006 IEEE/RSJ International Conference on*, October 2006, pp. 4284–4290.

[177] T. M. Martinetz, H. J. Ritter, and K. J. Schulten, "Three-dimensional neural net for learning visual motor coordination of a robot arm," *IEEE Transactions on Neural Networks*, vol. 1, no. 1, pp. 131–136, March 1990.

[178] T. Martinetz, H. Ritter, and K. Schulten, "Learning of visuomotor-coordination of a robot arm with redundant degrees of freedom," in *Proceedings of the International Conference on Parallel Processing in Neural Systems and Computers*, 1990, pp. 431–434.

[179] M. Han, N. Okada, and E. Kondo, "Coordination of an uncalibrated 3-D visuo-motor system based on multiple self-organizing maps," *JSME International Journal Series C*, vol. 49, no. 1, pp. 230–239, 2006.

[180] G. Asuni, G. Teti, C. Laschi, E. Guglielmelli, and P. Dario, "A bio-inspired sensory-motor neural model for a neuro-robotic manipulation platform," in *Advanced Robotics, 2005. ICAR '05. Proceedings., 12th International Conference on*, July 2005, pp. 607–612.

[181] G. A. Barreto and A. F. R. Araujo, "Self-organizing feature maps for modeling and control of robotic manipulators," *Journal of Intelligent and Robotic Systems*, vol. 36, pp. 407–450, 2003.

[182] V. R. de Angulo and C. Torras, "Speeding up the learning of robot kinematics through function decomposition," *IEEE Transactions on Neural Networks*, vol. 16, no. 6, pp. 1504–1512, November 2005.

[183] S. Kumar, L. Behera, and T. McGinnity, "Kinematic control of a redundant manipulator using an inverse-forward adaptive scheme with a KSOM based hint generator," *Robotics and Autonomous Systems*, vol. 58, no. 5, pp. 622–633, May 2010.

[184] S. Kumar, P. K. Patchaikani, A. Dutta, and L. Behera, "Visual motor control of a 7DOF redundant manipulator using redundancy preserving learning network," *Robotica*, vol. 28, no. 6, pp. 795–810, 2010.

[185] T. Hesselroth, K. Sarkar, P. P. Smagt, and K. Schulten, "Neural network control of a pneumatic robot arm," *IEEE Transactions on Systems, Man and Cybernetics*, vol. 24, no. 1, pp. 28–38, January 1994.

[186] "Barrett Technology," http://web.barrett.com/support/WAM_Documentation/WAM_InertialSpecifications_AC-02.pdf, accessed: 2016-09-30.

[187] P. K. Patchaikani, L. Behera, and G. Prasad, "A single network adaptive critic based redundancy resolution for robot manipulators," *IEEE Transactions on Industrial Electronics*, vol. 58, pp. 1–13, 2012.

[188] J. C. Dunn, "A fuzzy relative of the isodata process and its use in detecting compact well-separated clusters," *Journal of Cybernetics*, vol. 3, no. 4, 1973.

[189] S. Zeki, "The representation of colours in the cerebral cortex," *Nature*, vol. 284, pp. 412–418, 1980.

[190] *The Self-Organizing Map*, vol. 78, no. 9. Proceedings of The IEEE, September 1990.

[191] L. Behera and I. Kar, *Intelligent Systems and Control: Principles and Applications*. Oxford University Press, 2009.

[192] T. Takagi and M. Sugeno, "Fuzzy identification of systems and its application to modeling and control," *IEEE Transactions on Systems Man and Cybernetics*, vol. 15, no. 1, pp. 116–132, January 1985.

[193] J.-S. Jang, C.-T.Sun, and E. Mizutani, *Neuro-Fuzzy and Soft Computing: A computational Approach to Learning and Machine Intelligence*. Pearson Prentice Hall, 2007.

[194] J. Espinosa, J. Vandewalle, and V. Wertz, *Fuzzy Logic, Identification and Predictive Control*. Springer-Verlag, 2005.

[195] J. Kluska, *Analytical Methods in Fuzzy Modelling and Control.* Springer-Verlag, 2009.

[196] J.-S. R. Jang, "Neuro-Fuzzy Modelling: Architecture, Analyses and Applications," Ph.D. dissertation, Department of Electrical and Computer Science University of California, 1984.

[197] S.-J. Kang, C.-H. Woo, H.-S. Hwang, and K. Woo, "Evolutionary design of fuzzy rule base for nonlinear system modeling and control," *Fuzzy Systems, IEEE Transactions on*, vol. 8, no. 1, pp. 37–45, February 2000.

[198] W. Pedrycz and M. Reformat, "Evolutionary fuzzy modeling," *Fuzzy Systems, IEEE Transactions on*, vol. 11, no. 5, pp. 652–665, October 2003.

[199] H. K. Khalil, *Nonlinear Systems, 2nd ed.*   Prentice Hall, 1996.

[200] S. Kumar, L. Behera, and T. McGinnity, "Kinematic control of a redundant manipulator using an inverse-forward adaptive scheme with a ksom based hint generator," *Robotics and Autonomous Systems*, vol. 58, no. 5, pp. 622–633, May 2010.

[201] P. Patchaikani, L. Behera, and G. Prasad, "A single network adaptive critic based redundancy resolution scheme for robot manipulators," *Robotics and Autonomous Systems*, vol. 58, no. 8, pp. 978–990, August 2009.

[202] C. C. Cheah, S. Hou, Y. Zhao, and J.-J. Slotine, "Adaptive vision and force tracking control for robots with constraint uncertainty," *Mechatronics, IEEE/ASME Transactions on*, vol. 15, no. 3, pp. 389–399, June 2010.

[203] H. Wang, Y.-H. Liu, W. Chen, and Z. Wang, "Adaptive Visual Servoing Using Point and Line Features With an Uncalibrated Eye-in-Hand Camera," *Robotics, IEEE Transactions on*, vol. 24, no. 4, pp. 843–857, August 2008.

[204] L. Hsu, R. R. Costa, and F. Lizarralde, "Lyapunov/passivity-based adaptive control of relative degree two mimo systems with an application to visual servoing," *IEEE Transactions on Automatic Control*, vol. 52, no. 2, pp. 364–371, February 2007.

[205] S. Islam and P. Liu, "Pd output feedback control design for industrial robotic manipulators," *Mechatronics, IEEE/ASME Transactions on*, vol. 16, no. 1, pp. 187–197, February 2011.

[206] J. Armstrong Piepmeier, G. McMurray, and H. Lipkin, "A dynamic jacobian estimation method for uncalibrated visual servoing," in *IEEE/ASME International Conference on Advanced Intelligent Mechatronics*, September 1999, pp. 944–949.

[207] C. Broyden, "A class of methods for solving nonlinear simultaneous equations," *Mathematics of Computation*, vol. PP, no. 19, pp. 577–593, 1965.

[208] M. Bonkovic, A. Hace, and K. Jezernik, "Population based uncalibrated visual servoing," *IEEE/ASME Transactions on Mechatronics*, vol. 13, no. 3, June 2008.

[209] Q. Fu, Z. Zhang, and J. Shi, "Uncalibrated visual servoing using more precise model," in *IEEE Conference on Robotics, Automation and Mechatronics*, September 2008, pp. 916–921.

[210] P. Patchaikani and L. Behera, "Visual servoing of redundant manipulator with jacobian matrix estimation using self-organizing map," *Robotics and Autonomous Systems*, vol. 58, no. 8, pp. 978–990, August 2010.

[211] P. Jiang, L. Bamforth, Z. Feng, J. Baruch, and Y. Chen, "Indirect iterative learning control for a discrete visual servo without a camera-robot model," *IEEE Transactions on Systems, Man, and Cybernetics, Part B: Cybernetics*, vol. 37, no. 4, pp. 863–876, August 2007.

[212] K. Ahn and H. Anh, "Inverse double narx fuzzy modeling for system identification," *Mechatronics, IEEE/ASME Transactions on*, vol. 15, no. 1, pp. 136–148, February 2010.

[213] P. Goncalves, L. Mendoca, J. Sousa, and J. Pinto, "Uncalibrated eye to hand visual servoing using inverse fuzzy models," *IEEE Transactions on Fuzzy Systems*, vol. 16, no. 2, pp. 341–353, April 2008.

[214] K. Tanaka, M. Tanaka, H. Ohtake, and H. Wang, "Shared nonlinear control in wireless-based remote stabilization: A theoretical approach," *Mechatronics, IEEE/ASME Transactions on*, vol. 3, no. 1, pp. 443–452, June 2012.

[215] R. Padhi, N. Unnikrishnan, X. Wang, and S. Balakrishnan, "A single network adaptive critic (SNAC) architecture for optimal control," *Neural Networks*, vol. 19, no. 10, pp. 1648–1660, 2006.

[216] R. Bellman, *Dynamic Programming*. NJ: Princeton university press, 1957.

[217] P. J. Werbos, "Approximate dynamic programming for real-time control and neural modeling," in *Handbook of Intelligent control*, D. A. White and D. A. Sofge, Eds. Multiscience Press, 1992.

[218] K. Doya, "Reinforcement learning in continuous time and space," *Neural computing*, vol. 12, no. 1, pp. 219–245, January 2000.

[219] L. C. B. III, "Reinforcement learning in continuous time: Advantage updating," in *International Joint conference on Neural Networks*. IEEE, June 1994.

[220] S. Kumar, R. Padhi, and L. Behera, "Direct adaptive control using single network adaptive critic," *IEEE International Conference on Systems of Systems Engineering*, 2007.

[221] P. K. Patchaikani, L. Behera, N. H. Siddique, and G. Prasad, "A T-S fuzzy based adaptive critic for continuous-time input affine nonlinear systems," in *Systems, Man and Cybernetics, 2009. SMC 2009. IEEE International Conference on*, October 2009, pp. 4329 –4334.

[222] G. N. Saridis and C. S. G. Lee, "An approximation theory of optimal control for trainable manipulators," *IEEE Transactions on Systems, Man and Cybernetics*, vol. SMC-9, no. 3, pp. 152–159, March 1979.

[223] K. G. Vamvoudakis and F. L. Lewis, "Online actor-critic algorithm to solve the continuous-time infinite horizon optimal control problem," *Automatica*, vol. 46, no. 5, pp. 878–888, 2010.

[224] T. Cheng, F. L. Lewis, and M. Abu-Khalaf, "Fixed-final-time-constrained optimal control of nonlinear systems using neural network HJB approach," *IEEE Transactions on Neural Networks*, vol. 18, no. 6, pp. 1725–1737, November 2007.

[225] A. Al-Tamimi, F. L. Lewis, and M. Abu-Khalaf, "Discrete-time nonlinear HJB solution using approximate dynamic programming: Convergence proof," *IEEE Transactions on Systems, Man and Cybernetics-Part B:Cybernetics*, vol. 38, no. 4, pp. 943–949, August 2008.

[226] H. Zghal, R. V. Dubey, and J. A. Euler, "Efficient gradient projection optimization for manipulators with multiple degrees of redundancy," in *Proceedings of IEEE International Conference on Robotics and Automation*, vol. 2, 1990, pp. 1006–1011.

[227] G. H. Golub and C. Reinsch, "Singular value decomposition and least square solutions," *Numerische Mathematik*, vol. 14, no. 5, pp. 403–420, 1970.

[228] P. Ioannou and J. Sun, *Robust Adaptive Control*. Upper Saddle River, NJ, USA: Prentice-Hall, 1996.

[229] S. H. Zak, *Systems and Control*. Oxford university press, 2002.

[230] S. Balakrishnan and V. Biega, "Adaptive-critic-based neural networks for aircraft optimal control," *Journal of Guidance, Control, and Dynamics*, vol. 19, no. 4, pp. 893–898, 1996.

[231] L. Behera and I. Kar, *Intelligent Systems and control principles and applications*. Oxford University Press, Inc., 2010.

[232] R. Padhi, S. Balakrishnan, and T. Randolph, "Adaptive-critic based optimal neuro control synthesis for distributed parameter systems," *Automatica*, vol. 37, no. 8, pp. 1223–1234, 2001, neural Network Feedback Control.

[233] R. Padhi, N. Unnikrishnan, X. Wang, and S. N. Balakrishnan, "A single network adaptic critic (snac) architecture for optimal control," *Neural Networks*, vol. 19, pp. 1648–1660, 2006.

[234] D. P. Bertsekas, *Dynamic Programming and Optimal Control*. Athena Scientific, Belmount, 1995.

[235] H. Modares, F. L. Lewis, and M. B. Naghibi-Sistani, "Adaptive optimal control of unknown constrained-input systems using policy iteration and neural networks," *IEEE Transactions on Neural Networks and Learning Systems*, vol. 24, no. 10, pp. 1513–1525, Oct 2013.

[236] A. F. Gómez-Skarmeta, M. Delgado, and M. Vila, "About the use of fuzzy clustering techniques for fuzzy model identification," *Fuzzy Sets and Systems*, vol. 106, no. 2, pp. 179–188, September 1999.

[237] A. Patnaik, S. Dutta, and L. Behera, "Data driven system identification using evolutionary algorithms," *Lecture Notes in Computer Science*, vol. 7665, pp. 568–576, 2012.

[238] K. Tanaka and H. O. Wang, *Fuzzy Control Syatems Design and Analysis*. New York: Wiley, 2001.

[239] B. J. Rhee and S. Won, "A new fuzzy lyapunov function approach for a takagi-sugeno fuzzy control system," *Fuzzy Sets and Systems*, vol. 157, pp. 1211–1228, 2006.

[240] S. Dutta and L. Behera, "Snac based near-optimal controller for robotic manipulator with unknown dynamics," in *Fuzzy Systems (FUZZ-IEEE), 2014 IEEE International Conference on*, July 2014, pp. 98–105.

[241] P. K. Patchaikani, "Learning based near-optimal redundancy resolution schemes for visually controlled robot manipulators," Ph.D. dissertation, Dept. Elect. Eng., Indian Institute of Technology Kanpur, India, Aug 2012.

[242] P. K. Patchaikani, L. Behera, and G. Prasad, "A single network adaptive critic-based redundancy resolution scheme for robot manipulators," *IEEE Transactions on Industrial Electronics*, vol. 59, no. 8, pp. 3241–3253, Aug 2012.

[243] J. J. Craig, *Introduction to robotics: Mechanics and Control.* Pearson, 2009.

[244] X. Li and C. C. Cheah, "Adaptive neural network control of robot based on a unified objective bound," *IEEE Transactions on Control Systems Technology*, vol. 22, no. 3, pp. 1032–1043, May 2014.

[245] S. Khansari-Zadeh and A. Billard, "Learning stable nonlinear dynamical systems with gaussian mixture models," *IEEE Transactions on Robotics*, vol. 27, no. 5, pp. 943–957, Oct 2011.

[246] S. Dutta, P. K. Patchaikani, and L. Behera, "Near-optimal controller for nonlinear continuous-time systems with unknown dynamics using policy iteration," *IEEE transactions on neural networks and learning systems*, vol. 27, no. 7, pp. 1537–1549, 2016.

[247] J. Lofberg, "Yalmip : a toolbox for modeling and optimization in matlab," in *2004 IEEE International Conference on Robotics and Automation (IEEE Cat. No.04CH37508)*, Sept 2004, pp. 284–289.

[248] A. Butz, "Higher order derivatives of Liapunov functions," *Automatic Control, IEEE Transactions on*, vol. 14, no. 1, pp. 111–112, Feb 1969.

[249] A. Ahmadi and P. Parrilo, "On higher order derivatives of Lyapunov functions," in *American Control Conference (ACC)*, June 2011, pp. 1313–1314.

[250] S. Vijayakumar, A. D'Souza, and S. Schaal, "Incremental online learning in high dimensions," *Neural Computation*, vol. 17, pp. 2602–2634, 2005.

[251] A. J. Smola and B. Schölkopf, "A tutorial on support vector regression," *Statistics and computing*, vol. 14, no. 3, pp. 199–222, 2004.

[252] C.-C. Chang and C.-J. Lin, "LIBSVM: A library for support vector machines," *ACM Transactions on Intelligent Systems and Technology*, vol. 2, pp. 27:1–27:27, 2011, software available at http://www.csie.ntu.edu.tw/~cjlin/libsvm.

[253] R. Diankov. (2013, Oct.) Ikfast: The robot kinematics compiler. [Online]. Available: http://openrave.org/docs/latest_stable/openravepy/ikfast/#ikfast-the-robot-kinematics-compiler

[254] R. Girshick, "Fast r-cnn," in *Proceedings of the IEEE international conference on computer vision*, 2015, pp. 1440–1448.

[255] S. Ren, K. He, R. Girshick, and J. Sun, "Faster r-cnn: Towards real-time object detection with region proposal networks," *IEEE transactions on pattern analysis and machine intelligence*, vol. 39, no. 6, pp. 1137–1149, 2017.

[256] W. Liu, D. Anguelov, D. Erhan, C. Szegedy, S. Reed, C.-Y. Fu, and A. C. Berg, "Ssd: Single shot multibox detector," in *European conference on computer vision*. Springer, 2016, pp. 21–37.

[257] H. Zhao, J. Shi, X. Qi, X. Wang, and J. Jia, "Pyramid scene parsing network," in *IEEE Conf. on Computer Vision and Pattern Recognition (CVPR)*, 2017, pp. 2881–2890.

[258] K. He, G. Gkioxari, P. Dollár, and R. Girshick, "Mask r-cnn," in *Computer Vision (ICCV), 2017 IEEE International Conference on*. IEEE, 2017, pp. 2980–2988.

[259] M. M. Adankon and M. Cheriet, "Support vector machine," in *Encyclopedia of biometrics*. Springer, 2009, pp. 1303–1308.

[260] W. S. McCulloch and W. Pitts, "A logical calculus of the ideas immanent in nervous activity," *The bulletin of mathematical biophysics*, vol. 5, no. 4, pp. 115–133, 1943.

[261] D. W. Ruck, S. K. Rogers, M. Kabrisky, M. E. Oxley, and B. W. Suter, "The multilayer perceptron as an approximation to a bayes optimal discriminant function," *IEEE Transactions on Neural Networks*, vol. 1, no. 4, pp. 296–298, 1990.

[262] C. Szegedy, S. Ioffe, V. Vanhoucke, and A. A. Alemi, "Inception-v4, inception-resnet and the impact of residual connections on learning." in *AAAI*, vol. 4, 2017, p. 12.

[263] C. Ledig, L. Theis, F. Huszár, J. Caballero, A. Cunningham, A. Acosta, A. P. Aitken, A. Tejani, J. Totz, Z. Wang, *et al.*, "Photo-realistic single image super-resolution using a generative adversarial network." in *CVPR*, vol. 2, no. 3, 2017, p. 4.

[264] Y. LeCun *et al.*, "Lenet-5, convolutional neural networks," *URL: http://yann. lecun. com/exdb/lenet*, p. 20, 2015.

[265] A. Krizhevsky, I. Sutskever, and G. E. Hinton, "Imagenet classification with deep convolutional neural networks," in *Advances in neural information processing systems*, 2012, pp. 1097–1105.

[266] M. D. Zeiler and R. Fergus, "Visualizing and understanding convolutional networks," in *European conference on computer vision*. Springer, 2014, pp. 818–833.

[267] C. Szegedy, W. Liu, Y. Jia, P. Sermanet, S. Reed, D. Anguelov, D. Erhan, V. Vanhoucke, and A. Rabinovich, "Going deeper with convolutions," in *Proceedings of the IEEE conference on computer vision and pattern recognition*, 2015, pp. 1–9.

[268] K. Simonyan and A. Zisserman, "Very deep convolutional networks for large-scale image recognition," *arXiv preprint arXiv:1409.1556*, 2014.

[269] K. He, X. Zhang, S. Ren, and J. Sun, "Deep residual learning for image recognition," in *Proceedings of the IEEE conference on computer vision and pattern recognition*, 2016, pp. 770–778.

[270] B. C. Russell, A. Torralba, K. P. Murphy, and W. T. Freeman, "Labelme: a database and web-based tool for image annotation," *International journal of computer vision*, vol. 77, no. 1, pp. 157–173, 2008.

[271] M. Everingham, S. A. Eslami, L. Van Gool, C. K. Williams, J. Winn, and A. Zisserman, "The pascal visual object classes challenge: A retrospective," *International journal of computer vision*, vol. 111, no. 1, pp. 98–136, 2015.

[272] H. Zhao, J. Shi, X. Qi, X. Wang, and J. Jia, "Pyramid scene parsing network," *arXiv preprint arXiv:1612.01105*, 2016.

[273] K. He, X. Zhang, S. Ren, and J. Sun, "Spatial pyramid pooling in deep convolutional networks for visual recognition," in *European Conference on Computer Vision*. Springer, 2014, pp. 346–361.

[274] S. Wu, S. Zhong, and Y. Liu, "Deep residual learning for image steganalysis," *Multimedia Tools and Applications*, pp. 1–17, 2017.

[275] F. Yu, V. Koltun, and T. Funkhouser, "Dilated residual networks," *arXiv preprint arXiv:1705.09914*, 2017.

[276] T.-Y. Lin, P. Dollár, R. Girshick, K. He, B. Hariharan, and S. Belongie, "Feature pyramid networks for object detection," *arXiv preprint arXiv:1612.03144*, 2016.

[277] Y. Xiang, T. Schmidt, V. Narayanan, and D. Fox, "Posecnn: A convolutional neural network for 6d object pose estimation in cluttered scenes," *arXiv preprint arXiv:1711.00199*, 2017.

[278] J. Mahler, F. T. Pokorny, B. Hou, M. Roderick, M. Laskey, M. Aubry, K. Kohlhoff, T. Kröger, J. Kuffner, and K. Goldberg, "Dex-net 1.0: A cloud-based network of 3d objects for robust grasp planning using a multi-armed bandit model with correlated rewards," in *Robotics and Automation (ICRA), 2016 IEEE International Conference on*. IEEE, 2016, pp. 1957–1964.

[279] J. Mahler, J. Liang, S. Niyaz, M. Laskey, R. Doan, X. Liu, J. A. Ojea, and K. Goldberg, "Dex-net 2.0: Deep learning to plan robust grasps with synthetic point clouds and analytic grasp metrics," *arXiv preprint arXiv:1703.09312*, 2017.

[280] L. Pinto and A. Gupta, "Supersizing self-supervision: Learning to grasp from 50k tries and 700 robot hours," in *Robotics and Automation (ICRA), 2016 IEEE International Conference on.* IEEE, 2016, pp. 3406–3413.

[281] S. Levine, P. Pastor, A. Krizhevsky, J. Ibarz, and D. Quillen, "Learning hand-eye coordination for robotic grasping with deep learning and large-scale data collection," *The International Journal of Robotics Research*, vol. 37, no. 4-5, pp. 421–436, 2018.

[282] A. Saxena, J. Driemeyer, and A. Y. Ng, "Robotic grasping of novel objects using vision," *Int. J. Rob. Res.*, vol. 27, no. 2, pp. 157–173, feb 2008. [Online]. Available: http://dx.doi.org/10.1177/0278364907087172

[283] I. Lenz, H. Lee, and A. Saxena, "Deep learning for detecting robotic grasps," *Int. J. Rob. Res.*, vol. 34, no. 4-5, pp. 705–724, apr 2015. [Online]. Available: http://dx.doi.org/10.1177/0278364914549607

[284] E. Johns, S. Leutenegger, and A. J. Davison, "Deep learning a grasp function for grasping under gripper pose uncertainty," in *Intelligent Robots and Systems (IROS), 2016 IEEE/RSJ International Conference on.* IEEE, 2016, pp. 4461–4468.

[285] A. ten Pas and R. Platt, "Localizing grasp affordances in 3-d points clouds using taubin quadric fitting," *CoRR*, vol. abs/1311.3192, 2013.

[286] A. Ten Pas and R. Platt, "Localizing handle-like grasp affordances in 3d point clouds," in *Experimental Robotics.* Springer, 2016, pp. 623–638.

[287] N. P. Koenig and A. Howard, "Design and use paradigms for gazebo, an open-source multi-robot simulator." in *IROS*, vol. 4. Citeseer, 2004, pp. 2149–2154.

[288] S. Chitta, I. Sucan, and S. Cousins, "Moveit![ros topics]," *IEEE Robotics & Automation Magazine*, vol. 19, no. 1, pp. 18–19, 2012.

[289] I. A. Sucan, M. Moll, and L. E. Kavraki, "The open motion planning library," *IEEE Robotics & Automation Magazine*, vol. 19, no. 4, pp. 72–82, 2012.

[290] S. Ren, K. He, R. Girshick, and J. Sun, "Faster r-cnn: Towards real-time object detection with region proposal networks," in *Advances in neural information processing systems*, 2015, pp. 91–99.

[291] D. Chaudhuri and A. Samal, "A simple method for fitting of bounding rectangle to closed regions," *Pattern recognition*, vol. 40, no. 7, pp. 1981–1989, 2007.

[292] A. t. Pas and R. Platt, "Using geometry to detect grasps in 3d point clouds," *arXiv preprint arXiv:1501.03100*, 2015.

[293] G. Vosselman, B. G. Gorte, G. Sithole, and T. Rabbani, "Recognising structure in laser scanner point clouds," *International archives of photogrammetry, remote sensing and spatial information sciences*, vol. 46, no. 8, pp. 33–38, 2004.

[294] T. Rabbani, F. Van Den Heuvel, and G. Vosselmann, "Segmentation of point clouds using smoothness constraint," *International archives of photogrammetry, remote sensing and spatial information sciences*, vol. 36, no. 5, pp. 248–253, 2006.

[295] Q. Chen, Q.-s. Sun, P. A. Heng, and D.-s. Xia, "A double-threshold image binarization method based on edge detector," *Pattern recognition*, vol. 41, no. 4, pp. 1254–1267, 2008.

[296] G. Jie and L. Ning, "An improved adaptive threshold canny edge detection algorithm," in *Computer Science and Electronics Engineering (ICCSEE), 2012 International Conference on*, vol. 1. IEEE, 2012, pp. 164–168.

[297] M. Gualtieri, A. ten Pas, K. Saenko, and R. Platt, "High precision grasp pose detection in dense clutter," *CoRR*, vol. abs/1603.01564, 2016. [Online]. Available: http://arxiv.org/abs/1603.01564

[298] S. Jain and B. Argall, "Grasp detection for assistive robotic manipulation," in *Robotics and Automation (ICRA), 2016 IEEE International Conference on*. IEEE, 2016, pp. 2015–2021.

[299] N. Somani, C. Cai, A. Perzylo, M. Rickert, and A. Knoll, *Object Recognition Using Constraints from Primitive Shape Matching*. Cham: Springer International Publishing, 2014, pp. 783–792. [Online]. Available: http://dx.doi.org/10.1007/978-3-319-14249-4_75

[300] M. Nieuwenhuisen, J. Stueckler, A. Berner, R. Klein, and S. Behnke, "Shape-primitive based object recognition and grasping," in *Robotics; Proceedings of ROBOTIK 2012; 7th German Conference on*, May 2012, pp. 1–5.

[301] G. Vezzani, U. Pattacini, and L. Natale, "A grasping approach based on superquadric models," in *Robotics and Automation (ICRA), 2017 IEEE International Conference on*. IEEE, 2017, pp. 1579–1586.

[302] Z. Zhang, "Microsoft kinect sensor and its effect," *IEEE multimedia*, vol. 19, no. 2, pp. 4–10, 2012.

[303] M. Draelos, Q. Qiu, A. Bronstein, and G. Sapiro, "Intel realsense= real low cost gaze," in *Image Processing (ICIP), 2015 IEEE International Conference on*. IEEE, 2015, pp. 2520–2524.

[304] I. Imaging, "Ensenso 3d cameras," https://en.ids-imaging.com/ensenso-stereo-3d-camera.html, 2018.

[305] A. Singh, J. Sha, K. S. Narayan, T. Achim, and P. Abbeel, "Bigbird: A large-scale 3d database of object instances," in *Robotics and Automation (ICRA), 2014 IEEE International Conference on*. IEEE, 2014, pp. 509–516.

[306] A. Aldoma, F. Tombari, L. Di Stefano, and M. Vincze, *A Global Hypotheses Verification Method for 3D Object Recognition*. Berlin, Heidelberg: Springer Berlin Heidelberg, 2012.

[307] S. G. F. Tombari, L. Di Stefano, "Rgb-d semantic segmentation dataset," http://vision.deis.unibo.it/fede/kinectDataset.html, 2011.

[308] "Willowgarage test set dataset," https://repo.acin.tuwien.ac.at/tmp/permanent/dataset_index.php.

[309] O. Kundu, "TCS grasping datasets for individual and multiple objects," https://sites.google.com/view/grasping-tcs-research/home, TATA Consultancy Services, 2018.

[310] Y. Luo, M. Zhou, and R. J. Caudill, "An integrated e-supply chain model for agile and environmentally conscious manufacturing," *IEEE/ASME Transactions On Mechatronics*, vol. 6, no. 4, pp. 377–386, 2001.

[311] Amazon.com, "The largest internet-based retailer in the world," http://www.amazon.com/.

[312] S. O'Connor, "Amazon unpacked," *Financial Times*, vol. 8, 2013.

[313] P. J. Reaidy, A. Gunasekaran, and A. Spalanzani, "Bottom-up approach based on internet of things for order fulfillment in a collaborative warehousing environment," *International Journal of Production Economics*, vol. 159, pp. 29–40, 2015.

[314] W. Ding, "Study of smart warehouse management system based on the iot," in *Intelligence Computation and Evolutionary Computation*. Springer, 2013, pp. 203–207.

[315] P. R. Wurman, R. D'Andrea, and M. Mountz, "Coordinating hundreds of cooperative, autonomous vehicles in warehouses," *AI magazine*, vol. 29, no. 1, p. 9, 2008.

[316] N. Mathew, S. L. Smith, and S. L. Waslander, "Planning paths for package delivery in heterogeneous multirobot teams," *IEEE Transactions on Automation Science and Engineering*, vol. 12, no. 4, pp. 1298–1308, 2015.

[317] V. Digani, L. Sabattini, C. Secchi, and C. Fantuzzi, "Ensemble coordination approach in multi-agv systems applied to industrial warehouses," *IEEE Transactions on Automation Science and Engineering*, vol. 12, no. 3, pp. 922–934, 2015.

[318] N. T. Truc and Y.-T. Kim, "Navigation method of the transportation robot using fuzzy line tracking and qr code recognition," *International Journal of Humanoid Robotics*, p. 1650027, 2016.

[319] H.-G. Xu, M. Yang, C.-X. Wang, and R.-Q. Yang, "Magnetic sensing system design for intelligent vehicle guidance," *IEEE/ASME Transactions on Mechatronics*, vol. 15, no. 4, pp. 652–656, 2010.

[320] A. Muis and K. Ohnishi, "Eye-to-hand approach on eye-in-hand configuration within real-time visual servoing," *IEEE/ASME transactions on Mechatronics*, vol. 10, no. 4, pp. 404–410, 2005.

[321] P. R. Wurman and J. M. Romano, "Amazon picking challenge 2015," *AI Magazine*, vol. 37, no. 2, pp. 97–99, 2016.

[322] C. Hernandez, M. Bharatheesha, W. Ko, H. Gaiser, J. Tan, K. van Deurzen, M. de Vries, B. Van Mil, J. van Egmond, R. Burger, *et al.*, "Team delft's robot winner of the amazon picking challenge 2016," in *Robot World Cup*. Springer, 2016, pp. 613–624.

[323] M. Quigley, K. Conley, B. Gerkey, J. Faust, T. Foote, J. Leibs, R. Wheeler, and A. Y. Ng, "Ros: an open-source robot operating system," in *ICRA workshop on open source software*, vol. 3. Kobe, Japan, 2009, p. 5.

[324] R. Tsai, "A versatile camera calibration technique for high-accuracy 3d machine vision metrology using off-the-shelf tv cameras and lenses," *IEEE Journal on Robotics and Automation*, vol. 3, no. 4, pp. 323–344, 1987.

[325] Z. Zhang, "A flexible new technique for camera calibration," *IEEE Transactions on pattern analysis and machine intelligence*, vol. 22, no. 11, pp. 1330–1334, 2000.

[326] M. R. Andersen, T. Jensen, P. Lisouski, A. K. Mortensen, M. K. Hansen, T. Gregersen, and P. Ahrendt, "Kinect depth sensor evaluation for computer vision applications," *Technical Report Electronics and Computer Engineering*, vol. 1, no. 6, 2015.

[327] C. L. Lawson and R. J. Hanson, *Solving least squares problems*. SIAM, 1995.

[328] K. S. Arun, T. S. Huang, and S. D. Blostein, "Least-squares fitting of two 3-d point sets," *IEEE Transactions on pattern analysis and machine intelligence*, no. 5, pp. 698–700, 1987.

[329] J. Matas, C. Galambos, and J. Kittler, "Robust detection of lines using the progressive probabilistic hough transform," *Comput. Vis. Image Underst.*, vol. 78, no. 1, pp. 119–137, Apr. 2000. [Online]. Available: http://dx.doi.org/10.1006/cviu.1999.0831

[330] G. Bradski et al., "The opencv library," *Doctor Dobbs Journal*, vol. 25, no. 11, pp. 120–126, 2000.

[331] A. K. Jain, N. K. Ratha, and S. Lakshmanan, "Object detection using gabor filters," *Pattern recognition*, vol. 30, no. 2, pp. 295–309, 1997.

[332] S. Belongie, J. Malik, and J. Puzicha, "Shape matching and object recognition using shape contexts," *IEEE transactions on pattern analysis and machine intelligence*, vol. 24, no. 4, pp. 509–522, 2002.

[333] D. G. Lowe, "Distinctive image features from scale-invariant keypoints," *International journal of computer vision*, vol. 60, no. 2, pp. 91–110, 2004.

[334] N. Dalal and B. Triggs, "Histograms of oriented gradients for human detection," in *Computer Vision and Pattern Recognition, 2005. CVPR 2005. IEEE Computer Society Conference on*, vol. 1. IEEE, 2005, pp. 886–893.

[335] R. Girshick, J. Donahue, T. Darrell, and J. Malik, "Region-based convolutional networks for accurate object detection and segmentation," *IEEE transactions on pattern analysis and machine intelligence*, vol. 38, no. 1, pp. 142–158, 2016.

[336] Y. Zhang, K. Sohn, R. Villegas, G. Pan, and H. Lee, "Improving object detection with deep convolutional networks via bayesian optimization and structured prediction," in *Proceedings of the IEEE Conference on Computer Vision and Pattern Recognition*, 2015, pp. 249–258.

[337] P. Felzenszwalb, D. McAllester, and D. Ramanan, "A discriminatively trained, multiscale, deformable part model," in *IEEE Conference on Computer Vision and Pattern Recognition, 2008. CVPR 2008.* IEEE, 2008, pp. 1–8.

[338] R. Girshick, J. Donahue, T. Darrell, and J. Malik, "Rich feature hierarchies for accurate object detection and semantic segmentation," in *Proceedings of the IEEE conference on computer vision and pattern recognition*, 2014, pp. 580–587.

[339] J. Redmon, S. Divvala, R. Girshick, and A. Farhadi, "You only look once: Unified, real-time object detection," in *Proceedings of the IEEE Conference on Computer Vision and Pattern Recognition*, 2016, pp. 779–788.

[340] B. Kehoe, A. Matsukawa, S. Candido, J. Kuffner, and K. Goldberg, "Cloud-based robot grasping with the google object recognition engine," in *Robotics and Automation (ICRA), 2013 IEEE International Conference on.* IEEE, 2013, pp. 4263–4270.

[341] D. Fischinger, M. Vincze, and Y. Jiang, "Learning grasps for unknown objects in cluttered scenes," in *Robotics and Automation (ICRA), 2013 IEEE International Conference on.* IEEE, 2013, pp. 609–616.

[342] A. Saxena, J. Driemeyer, and A. Y. Ng, "Robotic grasping of novel objects using vision," *The International Journal of Robotics Research*, vol. 27, no. 2, pp. 157–173, 2008.

[343] M. Gualtieri, A. ten Pas, K. Saenko, and R. Platt, "High precision grasp pose detection in dense clutter," in *Intelligent Robots and Systems (IROS), 2016 IEEE/RSJ International Conference on.* IEEE, 2016, pp. 598–605.

[344] J. Redmon and A. Angelova, "Real-time grasp detection using convolutional neural networks," in *Robotics and Automation (ICRA), 2015 IEEE International Conference on.* IEEE, 2015, pp. 1316–1322.

[345] I. Lenz, H. Lee, and A. Saxena, "Deep learning for detecting robotic grasps," *The International Journal of Robotics Research*, vol. 34, no. 4-5, pp. 705–724, 2015.

[346] G. P. Otto and T. K. Chau, "'region-growing'algorithm for matching of terrain images," *Image and vision computing*, vol. 7, no. 2, pp. 83–94, 1989.

[347] R. Schnabel, R. Wahl, and R. Klein, "Efficient ransac for point-cloud shape detection," in *Computer graphics forum*, vol. 26, no. 2. Wiley Online Library, 2007, pp. 214–226.

[348] N. J. Mitra and A. Nguyen, "Estimating surface normals in noisy point cloud data," in *Proceedings of the nineteenth annual symposium on Computational geometry.* ACM, 2003, pp. 322–328.

[349] P. Kovesi, "Shapelets correlated with surface normals produce surfaces," in *Computer Vision, 2005. ICCV 2005. Tenth IEEE International Conference on*, vol. 2. IEEE, 2005, pp. 994–1001.

[350] Z. Zivkovic, "Improved adaptive gaussian mixture model for background subtraction," in *Pattern Recognition, 2004. ICPR 2004. Proceedings of the 17th International Conference on*, vol. 2. IEEE, 2004, pp. 28–31.

[351] B. Siciliano, L. Sciavicco, L. Villani, and G. Oriolo, *Robotics: modelling, planning and control.* Springer Science & Business Media, 2010.

[352] D. E. Koditschek, "Robot planning and control via potential functions," *The robotics review*, p. 349, 1989.

[353] J. Barraquand, B. Langlois, and J. C. Latombe, "Numerical potential field techniques for robot path planning," *IEEE Transactions on Systems, Man, and Cybernetics*, vol. 22, no. 2, pp. 224–241, 1992.

[354] Y. K. Hwang and N. Ahuja, "A potential field approach to path planning," *IEEE Transactions on Robotics and Automation*, vol. 8, no. 1, pp. 23–32, 1992.

[355] N. Malone, A. Faust, B. Rohrer, R. Lumia, J. Wood, and L. Tapia, "Efficient motion-based task learning for a serial link manipulator," *Transaction on Control and Mechanical Systems*, vol. 3, no. 1, 2014.

[356] L. E. Kavraki, P. Svestka, J.-C. Latombe, and M. H. Overmars, "Probabilistic roadmaps for path planning in high-dimensional configuration spaces," *IEEE transactions on Robotics and Automation*, vol. 12, no. 4, pp. 566–580, 1996.

[357] F. Avnaim, J.-D. Boissonnat, and B. Faverjon, "A practical exact motion planning algorithm for polygonal objects amidst polygonal obstacles," in *Robotics and Automation, 1988. Proceedings., 1988 IEEE International Conference on*. IEEE, 1988, pp. 1656–1661.

[358] F. Lingelbach, "Path planning using probabilistic cell decomposition," in *Robotics and Automation, 2004. Proceedings. ICRA'04. 2004 IEEE International Conference on*, vol. 1. IEEE, 2004, pp. 467–472.

[359] S. M. LaValle and J. J. Kuffner Jr, "Rapidly-exploring random trees: Progress and prospects," 2000.

[360] J. Rosell, A. Pérez, A. Aliakbar, L. Palomo, N. García, *et al.*, "The kautham project: A teaching and research tool for robot motion planning," in *Proceedings of the 2014 IEEE Emerging Technology and Factory Automation (ETFA)*. IEEE, 2014, pp. 1–8.

[361] R. Diankov and J. Kuffner, "Openrave: A planning architecture for autonomous robotics," *Robotics Institute, Pittsburgh, PA, Tech. Rep. CMU-RI-TR-08-34*, vol. 79, 2008.

[362] D. Roberts, R. Wolff, O. Otto, and A. Steed, "Constructing a gazebo: supporting teamwork in a tightly coupled, distributed task in virtual reality," *Presence*, vol. 12, no. 6, pp. 644–657, 2003.

[363] J. Pan, S. Chitta, and D. Manocha, "Fcl: A general purpose library for collision and proximity queries," in *Robotics and Automation (ICRA), 2012 IEEE International Conference on*. IEEE, 2012, pp. 3859–3866.

[364] K. M. Wurm, A. Hornung, M. Bennewitz, C. Stachniss, and W. Burgard, "Octomap: A probabilistic, flexible, and compact 3d map representation for robotic systems," in *Proc. of the ICRA 2010 workshop on best practice in 3D perception and modeling for mobile manipulation*, vol. 2, 2010.

[365] C. Rennie, R. Shome, K. E. Bekris, and A. F. De Souza, "A dataset for improved rgbd-based object detection and pose estimation for warehouse

pick-and-place," *IEEE Robotics and Automation Letters*, vol. 1, no. 2, pp. 1179–1185, 2016.

[366] C. Eppner, S. Höfer, R. Jonschkowski, R. Martın-Martın, A. Sieverling, V. Wall, and O. Brock, "Lessons from the amazon picking challenge: Four aspects of building robotic systems," 2016.

[367] K.-T. Yu, N. Fazeli, N. Chavan-Dafle, O. Taylor, E. Donlon, G. D. Lankenau, and A. Rodriguez, "A summary of team mit's approach to the amazon picking challenge 2015," *arXiv preprint arXiv:1604.03639*, 2016.

[368] J. J. Craig, *Introduction to robotics: mechanics and control.* Pearson Prentice Hall Upper Saddle River, 2005, vol. 3.

[369] Universal Robots Support, "Axes for computing d-h parameter of an ur5 robot," https://www.universal-robots.com/how-tos-and-faqs/faq/ur-faq/actual-center-of-mass-for-robot-17264/.

[370] M. W. Spong and M. Vidyasagar, "Robot dynamics and control, 1989," *and*, vol. 247251, p. 141150, 1991.

[371] J. Wang, Y. Li, and X. Zhao, "Inverse kinematics and control of a 7-dof redundant manipulator based on the closed-loop algorithm," *International Journal of Advanced Robotic Systems*, vol. 7, no. 4, pp. 1–9, 2010.

[372] S. Jotawar, "An automated robotic pick and place system for a retail warehouse," https://www.youtube.com/watch?v=jQ4_poYAXZU, 2016.

[373] L. Behera, "IITK-TCS participation in amazon picking challenge 2016," https://sites.google.com/site/swagatkumar/home/apc_iitk_tcs, 2016.

[374] M. Soni and O. Kundu, "Demonstration of grasping algorithm," https://www.youtube.com/watch?v=lCxMGvCKe_g, 2016.

[375] N. Kejriwal and S. Jotawar, "Software codes developed by iitk-tcs team for amazon picking challenge 2016," https://github.com/amazon-picking-challenge/team_iitk_tcs, 2016.

[376] T. Xiao, J. Zhang, K. Yang, Y. Peng, and Z. Zhang, "Error-driven incremental learning in deep convolutional neural network for large-scale image classification," in *Proceedings of the 22nd ACM international conference on Multimedia.* ACM, 2014, pp. 177–186.

[377] H.-C. Shin, H. R. Roth, M. Gao, L. Lu, Z. Xu, I. Nogues, J. Yao, D. Mollura, and R. M. Summers, "Deep convolutional neural networks for computer aided detection: Cnn architectures, dataset characteristics and transfer learning," *IEEE transactions on medical imaging*, vol. 35, no. 5, pp. 1285–1298, 2016.

[378] J. Yosinski, J. Clune, Y. Bengio, and H. Lipson, "How transferable are features in deep neural networks?" in *Advances in neural information processing systems*, 2014, pp. 3320–3328.

[379] H. Herrero, J. L. Outón, U. Esnaola, D. Sallé, and K. L. de Ipiña, "State machine based architecture to increase flexibility of dual-arm robot programming," in *International Work-Conference on the Interplay Between Natural and Artificial Computation*. Springer, 2015, pp. 98–106.

[380] J. Bohren and S. Cousins, "The SMACH high-level executive [ROS news]," *IEEE Robotics & Automation Magazine*, vol. 17, no. 4, pp. 18–20, 2010.

[381] T. Field, "SMACH documentation," *Online at http://www.ros.org/wiki/smach/Documentation*, 2011.

[382] S. Calinon and A. Billard, "Active teaching in robot programming by demonstration," in *Robot and Human interactive Communication, 2007. RO-MAN 2007. The 16th IEEE International Symposium on*. IEEE, 2007, pp. 702–707.

[383] Y. Maeda, N. Ishido, H. Kikuchi, and T. Arai, "Teaching of grasp/graspless manipulation for industrial robots by human demonstration," in *Intelligent Robots and Systems, 2002. IEEE/RSJ International Conference on*, vol. 2. IEEE, 2002, pp. 1523–1528.

[384] T. P. Lillicrap, J. J. Hunt, A. Pritzel, N. Heess, T. Erez, Y. Tassa, D. Silver, and D. Wierstra, "Continuous control with deep reinforcement learning," *arXiv preprint arXiv:1509.02971*, 2015.

[385] S. Levine, P. Pastor, A. Krizhevsky, and D. Quillen, "Learning hand-eye coordination for robotic grasping with deep learning and large-scale data collection," *arXiv preprint arXiv:1603.02199*, 2016.

[386] F. Zhang, J. Leitner, M. Milford, B. Upcroft, and P. Corke, "Towards vision-based deep reinforcement learning for robotic motion control," *arXiv preprint arXiv:1511.03791*, 2015.

[387] G. Mohanarajah, D. Hunziker, R. D'Andrea, and M. Waibel, "Rapyuta: A cloud robotics platform," *IEEE Transactions on Automation Science and Engineering*, vol. 12, no. 2, pp. 481–493, 2015.

[388] B. Ranjbarsahraei, M. Roopaei, and S. Khosra, "Adaptive fuzzy formation control for a swarm of nonholonomic differentially driven vehicles," *Nonlinear Dynamics*, vol. 67, no. 4, pp. 2747–2757, Mar. 2012.

[389] S. X. Yanga, A. Zhu, G. Yuan, and M. Q.-H. Meng, "A bioinspired neurodynamics-based approach to tracking control of mobile robots," *IEEE Trans. Ind. Electron.*, vol. 59, no. 8, pp. 3211–3220, Aug. 2012.

[390] A. Loria, J. Dasdemir, and N. A. Jarquin, "Leader-follower formation and tracking control of mobile robots along straight paths," *IEEE Trans. Control Syst. Technol.*, vol. 24, no. 2, pp. 727–732, Mar. 2016.

[391] A. K. Das et al., "A vision-based formation control framework," *IEEE Trans. Robot. Autom.*, vol. 18, no. 5, pp. 813–825, Oct. 2002.

[392] G. Cook, *Mobile Robots: Navigation, Control and Remote Sensing*, 1st ed. Wiley-IEEE Press, 2011.

[393] J.-J. E. Slotine and W. Li, *Applied nonlinear control*, 1st ed. NJ: Prantice-Hall, Englewood Cliffs, 1991.

[394] V. Utkin, *Sliding Modes and Their Application in Variable Structure Systems*. Imported Publications, Incorporated, 1978.

[395] L. Wu, P. Shi, and X. Su, *Sliding Mode Control of Uncertain Parameter-Switching Hybrid Systems*, 1st ed. Wiley Publishing, 2014.

[396] S. T. Venkataraman and S. Gulati, "Control of nonlinear systems using terminal sliding modes," *ASME J. Dyn. Syst. Meas. Control*, vol. 115, no. 3, pp. 554–560, Sep. 1993.

[397] C. Edwards and S. Spurgeon, *Sliding mode control: theory and applications*. CRC Press, 1998.

[398] S. Yu, X. Yu, B. Shirinzadeh, and Z. Man, "Continuous finite time control for robotic manipulators with terminal sliding mode," *Automatica*, vol. 41, no. 11, pp. 1957–1964, Nov. 2005.

[399] L. Yang and J. Yang, "Nonsingular fast terminal sliding-mode control for nonlinear dynamical systems," *Int. J. Robust Nonlinear Control*, vol. 21, no. 16, pp. 1865–1879, Nov. 2011.

[400] S. S.-D. Xu, C.-C. Chen, and Z.-L. Wu, "Study of nonsingular fast terminal sliding-mode fault-tolerant control," *IEEE Trans. Ind. Electron.*, vol. 62, no. 6, pp. 3906–3913, Jun. 2015.

[401] H. Delavari, R. Ghaderi, A. Ranjbar, and S. Momani, "Fuzzy fractional order sliding mode controller for nonlinear systems," *Commun. Nonlinear Sci. Numer. Simul.*, vol. 15, no. 4, pp. 963–978, 2010.

[402] M. B. Delghavi, S. Shoja-Majidabad, and A. Yazdani, "Fractional-order sliding-mode control of islanded distributed energy resource systems," *IEEE Trans. Sustain. Energy*, vol. 7, no. 4, pp. 1482–1491, Oct. 2016.

[403] C. Izaguirre-Espinosa, A. J. Munoz-Vazquez, A. Sanchez-Orta, V. Parra-Vega, and P. Castillo, "Attitude control of quadrotors based on fractional sliding modes: theory and experiments," *IET Control Theory Appl.*, vol. 10, no. 7, pp. 825–832, 2016.

[404] C.-M. Chi and F. Gao, "Simulating fractional derivatives using matlab," *JSW*, vol. 8, no. 3, pp. 572–578, Mar. 2013.

[405] A. Levant, "Robust exact differentiation via sliding mode technique," *Automatica*, vol. 34, no. 3, pp. 379–384, Mar. 1998.

[406] A. Levant, "Higher-order sliding modes, differentiation and output-feedback control," *Int. J. Control*, vol. 76, no. 9/10, pp. 924–941, 2003.

[407] M. Defoort, T. Floquet, A. Kokosy, and W. Perruquetti, "A novel higher order sliding mode control scheme," *Syst. Control Lett.*, vol. 58, no. 2, pp. 102 – 108, Feb. 2009.

[408] G. Bartolini, A. Levant, A. Pisano, and E. Usai, "Adaptive second-order sliding mode control with uncertainty compensation," *Int. J. Control*, vol. 89, no. 9, pp. 1747–1758, Mar. 2016.

[409] V. Utkin, "Discussion aspects of high-order sliding mode control," *IEEE Trans. Autom. Control*, vol. 61, no. 3, pp. 829–833, Mar. 2016.

[410] J. Shao, G. Xie, and L. Wang, "Leader- following formation control of multiple mobile vehicles," *IET Control Theory Appl.*, vol. 1, no. 2, pp. 545–552, Mar. 2007.

[411] H. Xiao, Z. Li, and C. L. P. Chen, "Formation control of leader-follower mobile robots' systems using model predictive control based on neural-dynamic optimization," *IEEE Trans. Ind. Electron.*, vol. 63, no. 9, pp. 5752–5762, Sep. 2016.

[412] X. Lu, R. Lu, S. Chen, and J. Lu, "Finite-time distributed tracking control for multi-agent systems with a virtual leader," *IEEE Trans. Circuits Syst. I, Reg. Papers*, vol. 60, no. 2, pp. 352–362, Feb. 2013.

[413] H. Rezaee and F. Abdollahi, "A decentralized cooperative control scheme with obstacle avoidance for a team of mobile robots," *IEEE Trans. Ind. Electron.*, vol. 61, no. 1, pp. 1268–1282, Jan. 2014.

[414] J. Chen, M. Gauci, W. Li, A. Kolling, and R. Grob, "Occlusion-based cooperative transport with a swarm of miniature mobile robots," *IEEE Trans. Robot.*, vol. 31, no. 2, pp. 307–321, Apr. 2015.

[415] L. A. V. Reyes and H. G. Tanner, "Flocking, formation control, and path following for a group of mobile robots," *IEEE Trans. Control Syst. Technol.*, vol. 23, no. 4, pp. 1268–1282, Jul. 2015.

[416] K.-K. Oh, M.-C. Park, and H.-S. Ahn, "A survey of multi-agent formation control," *Automatica*, vol. 53, pp. 424–440, 2015.

[417] O. Khatib, "Real-time obstacle avoidance for manipulators and mobile robots," *Int. J. Rob. Res.*, vol. 5, no. 1, pp. 90–98, Apr. 1986.

[418] Y. Koren and J. Borenstein, "Potential field methods and their inherent limitations for mobile robot navigation," in *Proc. IEEE Conf. Robot. Autom.*, vol. 2, Apr 1991, pp. 1398–1404.

[419] O. Khatib, "Real-time obstacle avoidance for manipulators and mobile robots," in *Proc. IEEE Conf. Robot. Autom. (ICRA)*, vol. 2, Mar. 1985, pp. 500–505.

[420] V. Gazi, "Swarm aggregations using artificial potentials and sliding-mode control," *IEEE Trans. Robot.*, vol. 21, pp. 1208–1214, Dec. 2005.

[421] V. Gazi and R. Ordonez, "Target tracking using artificial potentials and sliding mode control," *Int. J. Control*, vol. 80, pp. 1626–1635, Oct. 2007.

[422] R. R. Nair and L. Behera, "Swarm aggregation using artificial potential field and fuzzy sliding mode control with adaptive tuning technique," in *Proc. Amer. Control Conf. (ACC)*, Jun. 2012, pp. 6184–6189.

[423] V. Gazi, "Swarm aggregations using artificial potentials and sliding-mode control," *IEEE Trans. Robot.*, vol. 21, no. 6, pp. 1208–1214, Dec. 2005.

[424] J.-O. Kim and P. K. Khosla, "Real-time obstacle avoidance using harmonic potential functions," *IEEE Trans. Robot. Autom.*, vol. 8, no. 3, pp. 338–349, Jun. 1992.

[425] L. McFetridge and M. Y. Ibrahim, "New technique of mobile robot navigation using a hybrid adaptive fuzzy potential field approach," *Comput. Ind. Eng.*, vol. 35, no. 3, pp. 471–474, 1998.

[426] J. H. Chuang and N. Ahuja, "Path planning using the newtonian potential," in *Proc. IEEE Conf. on Robot. Autom.*, Apr. 1991, pp. 558–563.

[427] M. C. Lee and M. G. Park, "Artificial potential field based path planning for mobile robots using a virtual obstacle concept," in *Proc. IEEE/ASME Conf. Adv. Intell. Mechatronics (AIM 2003)*, vol. 2, Jul. 2003, pp. 735–740.

[428] V. Gazi and K. M. Passino, "A class of attractions/repulsion functions for stable swarm aggregations," *Int. J. Control*, vol. 77, no. 18, pp. 1567–1579, 2004.

[429] D. Izzo and L. Pettazzi, "Autonomous and distributed motion planning for satellite swarm," *J. Guid. Control Dynam.*, vol. 30, no. 2, pp. 449–459, Mar.-Apr. 2007.

[430] S. S. Ge and Y. J. Cui, "New potential functions for mobile robot path planning," *IEEE Tran. Robot. Autom.*, vol. 16, no. 5, pp. 615–620, Oct. 2000.

[431] Y. Yang, C. Hua, and X. Guan, "Finite time control design for bilateral teleoperation system with position synchronization error constrained," *IEEE Trans. Cybern.*, vol. 46, no. 3, pp. 609–619, Mar. 2016.

[432] S. S.-D. Xu, C.-C. Chen, and Z.-L. Wu, "Study of nonsingular fast terminal sliding-mode fault-tolerant control," *IEEE Trans. Ind. Electron.*, vol. 62, no. 6, pp. 3906–3913, Jun. 2015.

[433] R. Zhang, L. Dong, and C. Sun, "Adaptive nonsingular terminal sliding mode control design for near space hypersonic vehicles," *IEEE/CAA J. Autom. Sin.*, vol. 1, no. 2, pp. 155–161, Apr. 2014.

[434] V. Gazi, "Swarm aggregations using artificial potentials and sliding-mode control," *IEEE Trans. Robot.*, vol. 21, no. 6, pp. 1208–1214, Dec. 2005.

[435] R. R. Nair, H. Karki, A. Shukla, L. Behera, and M. Jamshidi, "Fault-tolerant formation control of nonholonomic robots using fast adaptive gain nonsingular terminal sliding mode control," *IEEE Systems Journal*, vol. 13, no. 1, pp. 1006–1017, Mar 2019.

[436] R. R. Nair, L. Behera, V. Kumar, and M. Jamshidi, "Multi-satellite formation control for remote sensing applications using artificial potential field and adaptive fuzzy sliding mode control," *IEEE Syst. J.*, vol. 9, no. 2, pp. 508–518, Jun. 2015.

[437] R. J. Wai, "Fuzzy sliding-mode control using adaptive tuning technique," *IEEE Trans. Ind. Electron.*, vol. 54, no. 1, pp. 586–594, Feb. 2007.

[438] M. Guo, D. V. Dimarogonas, and K. H. Johansson, "Distributed real-time fault detection and isolation for cooperative multi-agent systems," in *Proc. Amer. Control Conf. (ACC)*, 2012, pp. 5270–5275.

[439] "Fault tolerant multi-robot formation control," https://youtu.be/4-S8ztofrQQ.

[440] A. K. Behera and B. Bandyopadhyay, "Event-triggered sliding mode control for a class of nonlinear systems," *Int. J. Control*, vol. 0, no. 0, pp. 1–16, Jul. 2016.

[441] D. V. Dimarogonas, E. Frazzoli, and K. H. Johansson, "A distributed event-triggered control for multi-agent systems," *IEEE Trans. Autom. Control*, vol. 57, no. 5, pp. 1291–1296, May. 2012.

[442] Y. Fan, L. Liu, G. Feng, and Y. Wang, "Self-triggered consensus for multi-agent systems with zeno-free triggers," *IEEE Trans. Autom. Control*, vol. 60, no. 10, pp. 2779–2784, Oct. 2015.

[443] T.-H. Cheng, Z. Kan, J. M. Shea, and W. E. Dixon, "Decentralized event-triggered control for leader-follower consensus," in *Proc. 53rd IEEE Conf. Decision Control*, Dec. 2014, pp. 1244–1249.

[444] W. Hu, L. Liu, , and G. Feng, "Consensus of linear multi-agent systems by distributed event-triggered strategy," *IEEE Trans. Cybern.*, vol. 46, no. 1, pp. 148–157, Jan. 2016.

[445] Y. Zhu, X. Guan, X. Luo, and S. Li, "Finite-time consensus of multi-agent system via nonlinear event-triggered control strategy," *IET Control Theory Appl.*, vol. 9, no. 17, pp. 2548–2552, Nov. 2015.

[446] Q. Liu, Z. Wang, X. He, and D. H. Zhou, "Event-based h_∞ consensus control of multi-agent systems with relative output feedback: The finite-horizon case," *IEEE Trans. Autom. Control*, vol. 60, no. 9, pp. 2553–2558, Sep. 2015.

[447] S. Yu and X. Long, "Finite-time consensus for second-order multi-agent systems with disturbances by integral sliding mode," *Automatica*, vol. 54, pp. 158–165, Apr. 2015.

[448] S. Rao and D. Ghose, "Sliding mode control-based autopilots for leaderless consensus of unmanned aerial vehicles," *IEEE Trans. Control Syst. Technol.*, vol. 22, no. 5, pp. 1964–1972, Sep. 2015.

[449] A. K. Behera and B. Bandyopadhyay, "Self-triggering-based sliding-mode control for linear systems," *IET Control Theory Appl.*, vol. 9, no. 17, pp. 2541–2547, Nov. 2015.

[450] R. R. Nair, L. Behera, and S. Kumar, "Event-triggered finite-time integral sliding mode controller for consensus-based formation of multirobot systems with disturbances," *IEEE Transactions on Control Systems Technology*, vol. 27, no. 1, pp. 39–47, Jan 2019.

[451] S. Yu, X. Yu, B. Shirinzadeh, and Z. Man, "Continuous finite time control for robotic manipulators with terminal sliding mode," *Automatica*, vol. 41, no. 11, pp. 1957–1964, Nov. 2005.

[452] W. Gao and J. C. H. Rao, "Variable structure control of nonlinear systems: A new approach," *IEEE Trans. Ind. Electron.*, vol. 40, no. 1, pp. 45–55, Feb. 1993.

[453] S. Li, H. Du, and X. Lin, "Finite-time consensus algorithm for multi-agent systems with double-integrator dynamics," *Automatica*, vol. 47, no. 8, pp. 1706–1712, Mar. 2011.

[454] X. Lu, R. Lua, S. Chen, and J. Lu, "Finite-time distributed tracking control for multi-agent systems with a virtual leader," *IEEE Trans. Circuits Syst. I, Reg. Papers*, vol. 60, no. 2, pp. 352–362, Feb. 2013.

[455] R. R. Nair and L. Behera, "Robust adaptive gain nonsingular fast terminal sliding mode control for spacecraft formation flying," in *Proc. 54th IEEE Conf. Decision Control*, 2015, pp. 5314–5319.

[456] R. D. Robinett, D. G. Wilson, G. R. Eisler, and J. E. Hurtado, *Applied Dynamic Programming for Optimization of Dynamical Systems*. SIAM, 2005.

[457] J. Van de Weijer, T. Gevers, and A. D. Bagdanov, "Boosting color saliency in image feature detection," *IEEE Transactions on Pattern Analysis and Machine Intelligence (PAMI)*, vol. 28, no. 1, pp. 150–156, 2006.

[458] R. Lukac and k. N. Plataniotis, *Color Image Processing: Methods and Applications*. CRC Press, 2006, ch. Color Feature Detection: An Overview.

[459] J. Shotton, J. Winn, C. Rother, and A. Criminisi, "Textonboost for image understanding: multi-class object recognition and segmentation by jointly modeling texture, layout, and context," *International Journal on Computer Vision*, vol. 81, no. 1, pp. 2–23, 2008.

[460] J. Winn, A. Criminisi, and T. Minka, "Object categorization by learned universal visual dictionary," in *IEEE International Conference on Computer vision (ICCV)*, vol. 2, 2005, pp. 1800–1807.

[461] B. S. Manjunath and W.-Y. Ma, "Texture features for browsing and retrieval of image data," *IEEE Transactions on Pattern Analysis and Machine Intelligence (PAMI)*, vol. 18, no. 8, pp. 837–842, 1996.

[462] T. Ojala, M. Pietikainen, and T. Maenpaa, "Multiresolution grayscale and rotation invariant texture classification with local binary patterns," *IEEE Transactions on Pattern Analysis and Machine Intelligence (PAMI)*, vol. 24, no. 7, pp. 972–987, 2002.

[463] B. K. Horn and B. G. Schunk, "Determining optical flow," *Artificial Intelligence*, vol. 17, pp. 185–203, 1981.

[464] B. Lucas and T. Kanade, "An iterative image registration technique with an application to stereo vision," *IJCAI*, vol. 81, pp. 674–679, 1981.

[465] P. Sabzmeydani and G. Mori, "Detecting pedestrians by learning shapelet features," in *IEEE Conference on Computer Vision and Pattern Recognition (CVPR)*. IEEE, 2007, pp. 1–8.

[466] D. G. Lowe, "Distinctive image features from scale-invariant keypoints," *Internation Journal of Computer Vision*, vol. 60, no. 2, pp. 91–110, January 2004.

[467] H. Bay, A. Ess, T. Tuytelaars, and L. V. Gool, "Speeded-up robust features (SURF)," *Computer Vision and Image Understanding, Elsevier*, vol. 110, pp. 346–359, December 2008.

[468] N. Dalal and B. Triggs, "Histograms of oriented gradients for human detection," in *CVPR*, 2005, pp. 886–893.

[469] C. Feichtenhofer and A. Pinz, "Spatio-temporal good features to track," in *IEEE International Conference on Computer Vision Workshops*. IEEE, 2013, pp. 246–253.

[470] L. Juan and O. Gwun, "A comparison of SIFT, PCA-SIFT and SURF," *International Journal of Image Processing (IJIP)*, vol. 3, no. 4, pp. 143–152, 2009.

[471] Z. Bing, Y. Wang, J. Hou, H. Lu, and H. Chen, "Research of tracking robot based on SURF features," in *International Conference on Natural Computation (ICNC)*. Yantai, Shandong: IEEE, 2010, pp. 3523–3527.

[472] S. Gu, Y. Zheng, and C. Tomasi, "Efficient visual object tracking with online nearest neighbor classifier," in *Asian Conference on Computer Vision (ACCV)*. Springer, 2010, pp. 271–282.

[473] D.-N. Ta, W.-C. Chen, N. Gelfand, and K. Pulli, "SURFTrac: Efficient tracking and continuous object recognition using local feature descriptors," in *IEEE Conference on Computer Vision and Pattern Recognition (CVPR)*. Miami,FL: IEEE, 2009, pp. 2937–2944.

[474] W. He, T. Yamashita, L. Hongtao, and S. Lao, "SURF tracking," in *International Conference on Computer Vision*. Kyoto: IEEE, 2009, pp. 1586–1592.

[475] M. Gupta, S. Garg, S. Kumar, and L. Behera, "An on-line visual human tracking algorithm using SURF-based dynamic object model," in *International Conference on Image Processing (ICIP)*. IEEE, 2013, pp. 3875–3879.

[476] Y. Liu and H. Zhang, "Indexing visual features: Real-time loop closure detection using a tree structure," in *IEEE International Conference on Robotics and Automation (ICRA)*. IEEE, May 2012, pp. 3613–3618.

[477] L. Baird and A. W. Moore, "Gradient descent for general reinforcement learning," *Advances in neural information processing systems*, pp. 968–974, 1999.

[478] J. Zhang, J. Fang, and J. Lu, "Mean-shift algorithm integrating with SURF for tracking," in *Natural Computation*. Shanghai: IEEE, 2011, pp. 960–963.

[479] D. Comaniciu, V. Ramesh, and P. Meer, "Kernel-based object tracking," *IEEE Transactions on Pattern Analysis and Machine Intelligence*, vol. 25, no. 4, pp. 1–14, April 2003.

[480] A. Ahmadi, M. R. Daliri, A. Nodehi, and A. Qorbani, "Objects recognition using the histogram based on descriptors of SIFT and SURF," *Journal of Basic and Applied Scientific Research*, vol. 2, no. 9, pp. 8612–8616, 2012.

[481] B. Thomee, E. M. Bakker, and M. S. Lew, "TOP-SURF: a visual words toolkit," in *Proc. of International Conference on Multimedia*. New York: ACM, 2010, pp. 1473–1476.

[482] T. Nicosevici and R. Garcia, "Automatic visual bag-of-words for online robot navigation and mapping," *IEEE Transactions on Robotics*, vol. 28, no. 4, pp. 886–898, August 2012.

[483] S. Garg and S. Kumar, "Mean-shift based object tracking algorithm using SURF features," in *International Conference on Signal Processing, Robotics and Automation*. WSEAS, 2013, pp. 187–194.

[484] R. Adams and L. Bischof, "Seeded region growing," *IEEE Transactions on Pattern Analysis and Machine Intelligence*, vol. 16, no. 6, pp. 641–647, 1994.

[485] I. Leichter, M. Lindenbaum, and E. Rivlin, "Tracking by affine kernel transformations using color and boundary cues," *IEEE Transactions on Pattern Analysis and Machine Intelligence*, vol. 31, no. 1, pp. 164–171, 2009.

[486] D. Zhou and D. Hu, "A robust object tracking algorithm based on surf," in *Wireless Communications & Signal Processing (WCSP), 2013 International Conference on*. IEEE, 2013, pp. 1–5.

[487] M. Muja and D. G. Lowe, "Fast approximate nearest neighbors with automatic algorithm configuration." *VISAPP (1)*, vol. 2, 2009.

[488] S. Baker and I. Matthews, "Lucas-kanade 20 years on: A unifying framework," *International Journal of Computer Vision*, pp. 221–255, 2004.

[489] Z. Kalal, J. Matas, and K. Mikolajczyk, "Online learning of robust object detectors during unstable tracking," in *IEEE ICCV Workshops*. Kyoto: IEEE, 2009, pp. 1417–1424.

[490] I. Matthew and S. Baker, "Active appearance models revisited," *International Journal of Computer Vision*, vol. 60, no. 2, pp. 135–164, November 2004.

Bibliography

[491] J. Li, Y. Wang, and Y. Wang, "Visual tracking and learning using speeded up robust features," *Pattern Recognition Letters*, vol. 33, no. 16, pp. 2094–2101, December 2012.

[492] D. A. Ross, J. Lim, and R.-S. Lin, "Incremental learning for robust visual tracking," *International Journal of Computer Vision*, vol. 77, no. 1-3, pp. 125–141, May 2008.

[493] W. Kloihofer and M. Kampel, "Interest point based tracking," in *International Conference on Pattern Recognition (ICPR)*. ACM, 2010, pp. 3549–3552.

[494] K. F. Sim and K. Sundaraj, "Human motion tracking of athlete using optical flow and artificial markers," in *International Conference on Intelligent and Advanced Systems (ICIAS)*. Kualampur, Malaysia: IEEE, 2010, pp. 1–4.

[495] C.-C. Lien, S.-J. Lin, C.-Y. Ma, and Y.-W. Lin, "SURF-badge-based target tracking," in *World Academy of Science, Engineering and Technology*, vol. 77, 2013, pp. 877–883.

[496] Wiki, "Pinhole camera model," http://en.wikipedia.org/wiki/Pinhole_camera_model.

[497] Y. Kanayama, Y. Kimura, F. Miyazaki, and T. Noguchi, "A stable tracking control method for an autonomous mobile robot," in *Robotics and Automation, 1990. Proceedings., 1990 IEEE International Conference on*. IEEE, 1990, pp. 384–389.

[498] M. W. Spong, S. Hutchinson, and M. Vidyasagar, *Robot modeling and control*. Wiley New York, 2006, vol. 3.

[499] J.-H. Jean and F.-L. Lian, "Robust visual servo control of a mobile robot for object tracking using shape parameters," *IEEE Transactions on Control Systems Technology*, vol. 20, no. 6, pp. 1461–1472, Nov 2012.

[500] A. Mekonnen, C. Briand, F. Lerasle, and A. Herbulot, "Fast hog based person detection devoted to a mobile robot with a spherical camera," in *Intelligent Robots and Systems (IROS), 2013 IEEE/RSJ International Conference on*. IEEE, 2013, pp. 631–637.

[501] A. Ess, B. Leibe, and L. V. Gool, "Depth and appearance for mobile scene analysis," in *International Conference on Computer Vision*, October 2007, pp. 1–8.

[502] M. Gupta, "Demonstration video of surf-based human tracking algorithm," https://www.youtube.com/watch?v=wUxhAQeWXGg&feature=youtu.be, July 2014.

[503] M. Gupta, "Surf-based human tracking from a mobile robot platform," https://www.youtube.com/watch?v=J2I2E38cB-g, Sept. 2015.

[504] Q. Wang, F. Chen, W. Xu, and M.-H. Yang, "An experimental comparison of online object-tracking algorithms," in *SPIE Optical Engineering+ Applications*. International Society for Optics and Photonics, 2011, pp. 81 381A–81 381A.

[505] A. Mobilerobots, "Mobile robots," http://www.mobilerobots.com/Mobile_Robots.asp.

Index

3D point cloud, 425, 433, 445
7 DOF Manipulator, 6, 54, 139

Activation function, 387
Adaptive critic, 289
  Update, 304
Adaptive distributed T-S fuzzy PD controller, 208
AlexNet, 400
Artificial Neural Network, 387
Artificial potential function, 511–513
Autonomous picking and stowing, 454
Axis Assignment, 428

Backpropagation, 390
Backstepping, 495
Barrett Hand, 195
Barrett Wam Manipulator, 45, 46
  DH Parameters, 49
  Dynamic model, 47, 49
Behavior-based approach, 508

C-FuzzStaMP, 328
Camera model, 38
  Calibration, 38
  Computation of image feature velocity, 39
  Transformation from Cartesian space to vision space, 36
Caputo Approximation, 503
Cartesian space, 18
Chattering, 549
Collision avoidance potential, 514, 516
Consensus, 541
Consensus protocol, 544

Continuous-time optimal control problem, 241
  Adaptive critic, 242
  CT-SNAC, 242
Convex optimization problem, 23
Convolutional Neural Networks, 395, 399, 428, 453
CT-SNAC, 242
  Critic network, 246
    Weight update law, 243
  Value function approximation with T-S fuzzy model-based critic network, 246

Damped Least Square Method, 475
Data augmentation, 427
Data augmentation technique, 405
Deep learning techniques, 395
Deep Neural Networks, 386
Differential-drive mobile robots, 485
Discrete-time optimal control problem, 230
  Adaptive critic, 231
  Adaptive critic
    Dual network training, 232
    DT-SNAC, 175, 234, 261
    Single network adaptive critic, 234
DT-SNAC, 234
  Costate vector modeling with T-S fuzzy model-based critic network
    Nearest neighbor heuristic, 236
  Critic network, 236
Dynamic controller, 4
Dynamic Movement Primitives, 320

645

Earth-Centered Earth-Fixed
co-ordinate system, 491
Edge Points, 435
Energy dissipation rate (EDR), 354
Event triggered consensus, 539, 543
Event triggered control, 537
Event triggered sliding mode control, 544
Expectation maximization (EM), 26
Experimental setup, 31

Fast Adaptive Fuzzy NTSMC, 524
Fast Adaptive Gain NTSMC, 519
Fast Nonsingular terminal sliding mode Control, 501
Fast reaching law, 520
Faster RCNN , 481
Fault tolerant Formation control, 527
Feedback linearization, 493
Finite time consensus, 546
Fitness function, 345
FuzzStaMP, 327
Fuzzy Lyapunov function, 329
Fuzzy sliding mode control, 524
Fuzzy-based fast reaching law, 525

Gaussian mixture model, 25
Gaussian Mixture Regression, 324
motion modeling, 364
Genetic algorithm, 24
Crossover, 25
Fitness, 25
Mutation, 25
Selection, 25
Goal Seeking Potential, 516
Grasp Decide Index (GDI), 430
Grasp pose detection, 434
Grasp region, 424
Grasping, 201, 445
Final Pose selection, 431
Gripping and suction system, 453, 457
Grunwald-Letnikov Approximation, 502

Hamilton-Jacobi-Bellman equation, 287
Higher order sliding mode control, 503
Human Following Mobile Robot, 555
Human Following Robot, 589
Human Tracking Algorithm
Optical flow, 576
Human tracking algorithm, 558
out-of-plane rotations, 567
Pose change detection, 567

Image moment interaction matrix, 132, 134
Image moment velocity, 126, 134
Image moments, 122, 132
Imitation Learning, 319
Imitation learning, 16
Inverse kinematics, 474

Kalman Filter based motion predictor, 578
KD-tree classifier, 572
Kinematic control, 4, 6
Pseudo-inverse based, 7
Kinematic Jacobian, 7
Kinesthetic teaching, 18
Kohonen Self Organizing Map, 56–58, 60, 199
KSOM based kinematic control, 148
Approximation of inverse Jacobian
Empirical verification, 152
Approximation of inverse Jacobian, 151
Experimental results, 160
Network architecture, 149
Redundancy resolution, 159
Visual servoing, 156
Stability analysis, 157
Weight update, 149
KSOM-SC Architecture
Adaptive Sub-clustering, 94
Modified Neighborhood, 99
Reaching points, 103

Index 647

Real-time experiment, 108
Redundancy Resolution, 98
Tracking a line, 105
Tracking Ellipse, 107

Laplacian, 541, 550
Leader-follower-based approach, 507, 519, 542
Learning, 17
Learning-Based Inverse Kinematic Control, 66, 69
Learning-based visual automation, 5
Learning-based visual control, 40
schematic, 40
Linear quadratic regulator, 285
Locally weighted projection regression, 367
Lyapunov theory, 21, 297

Manipulator
Kinematic control, 121
Manipulators, 31
kinematic model, 43, 45
Measurement error, 550
MLP, 389
Mobile robot, 485
Dynamic Model, 486
Kinematic Model, 486
Model-based grasping, 425
Model-Based Visual Servoing, 113
Multi agent system, 512, 541
Multi-Class Segmentation, 417
Multi-layer perceptron, 389
Multi-robot Formation, 507, 530
Multi-robot system, 549
Multiple task-equilibrium, 375
Multitasking, 375

Neural networks, 387
Newton-Euler Algorithm, 48
NN, 387
Nonholonomic mobile robots, 485
Nonholonomic robot, 549
Null space optimization, 473

Object detection, 465
Object recognition, 462, 478, 563
Object Verification, 463
Obstacle avoidance potential, 528
Occlusion, 574
Occlusion detection, 564
Optical Flow, 557
Optical flow tracker, 580
Optimization, 22

Pinhole camera model, 582
Pinhole camera projection, 128
Policy iteration scheme, 291, 292
PowerCube manipulator, 31, 152, 160, 195, 276
Kinematic configuration, 32
Kinematic constraints, 33

R-CNN, 428
R-FuzzStaMP, 335
Rack detection, 460
RANSAC, 565
Rapidly-exploring Random Trees, 427
Reaching condition, 497
Reaching law, 498
Redundancy resolution
Extended Jacobian method, 8
Redundancy Preserving Networks, 55
Redundancy resolution, 6
Global optimization, 9
Neural Network, 10
Null space control, 8
Optimization based method, 9
Redundancy Resolution Criteria
Lazy Arm motion, 98
Minimum angle norm, 98
Minimum Condition Number, 98
Region growing algorithm, 437
RELU, 397
ResNet50, 428
Resource optimal control, 537
Riemann-Liouville Approximation, 503
Robot Attitude-Pitch, 488

Robot Attitude-Roll, 487
Robot Attitude-Yaw, 489

SAKE EZGripper, 427
Self triggered control, 537
Sigmoid function, 387, 388
Single Network Adaptive Critic, 229, 283
Single Shot Detection, 416
Sliding Mode Control, 496
    Nonsingular Fast Terminal SMC, 501
    Nonsingular Terminal SMC, 500
    Fractional Order SMC, 502
    Higher order SMC, 503
    Terminal SMC, 499
SNAC based redundancy resolution, 175, 261
    from Cartesian space
        Discrete-time input affine system representation of forward kinematics, 257
        Modeling the primary and additional tasks as an integral cost function, 259
    from Cartesian space, 261
        Computational complexity, 264
        T-S fuzzy model-based critic, 262
        Schematic, 261
    from vision space, 174, 175
        discrete-time input affine system representation of forward kinematics, 174
        Control challenges, 177
        KSOM based critic, 185
        Schematic, 176
        T-S fuzzy model-based critic, 179
Stability, 21
Stochastic Gradient Descent (SGD), 404
Suction, 478
Suction-based end-effector, 469

Support vector regression, 373
SURF
    Dynamic object model, 559
    Error Recovery Module, 572
    Human tracking algorithm, 562
    Object model description, 560
    Template pool, 561
    Template Update Module, 571

T-S Fuzzy Model, 205–207, 295
T-S fuzzy PD controller, 208, 211
    Offline learning algorithm, 209
    Online adaptation algorithm, 212
Target window prediction, 564
Teleoperation, 19
Template initialization, 569
Tensorflow, 392
Time triggered control, 537
Triggering function, 545
Tsai's algorithm, 38
Two-finger parallel-jaw gripper, 433

Unicycle model, 486
UR10 Robot-DH Parameters, 43

VGG networks, 402
Virtual-leader-based approach, 508
Vision-based control, 3
Vision-Based Grasping, 423
Vision-based object detection, 404
Vision space, 116, 120, 121
Visual servoing
    Eye-to-hand, 11
Visual control of redundant manipulator, 13, 172, 174
Visual Motor Control
    Schematic, 90
Visual Perception, 385
Visual servo controller, 581
Visual servoing, 11, 122, 128, 138, 205
    3-D Servoing, 12
    2-1/2-D visual servoing, 13
    Eye-in-hand, 11

Index

Image based (IBVS), 12
Position based (PBVS), 12
Visual servoing controller, 123, 139
Visual tracking, 555
features, 556

Warehouse Automation
Grasping, 465

Warehouse automation, 3, 412, 453
Autonomous picking and stowing, 456
Motion planning, 466
Suction, 470
Weighted adjacency matrix, 541

Zeno behavior, 539
ZFNet, 401